Ordinary Differential Equations

A Dynamical Point of View

Stephen Wiggins

University of Bristol, UK

World Scientific

NEW JERSEY • LONDON • SINGAPORE • BEIJING • SHANGHAI • HONG KONG • TAIPEI • CHENNAI • TOKYO

Published by

World Scientific Publishing Co. Pte. Ltd.
5 Toh Tuck Link, Singapore 596224
USA office: 27 Warren Street, Suite 401-402, Hackensack, NJ 07601
UK office: 57 Shelton Street, Covent Garden, London WC2H 9HE

Library of Congress Control Number: 2023035278

British Library Cataloguing-in-Publication Data
A catalogue record for this book is available from the British Library.

ORDINARY DIFFERENTIAL EQUATIONS
A Dynamical Point of View

ISBN 978-981-128-154-9 (hardcover)
ISBN 978-981-128-268-3 (paperback)
ISBN 978-981-128-155-6 (ebook for institutions)
ISBN 978-981-128-156-3 (ebook for individuals)

For any available supplementary material, please visit
https://www.worldscientific.com/worldscibooks/10.1142/13548#t=suppl

Preface

This book consists of ten weeks of material given as a course on ordinary differential equations (ODEs) for second year mathematics majors at the University of Bristol. It is the first course devoted solely to differential equations that these students will take. An obvious question is "why does there need to be another textbook on ODEs"? From one point of view the answer is certainly that it is not needed. The classic textbooks of Coddington and Levinson, Hale, and Hartman[1] provide a thorough exposition of the topic and are essential references for mathematicians, scientists and engineers who encounter and must understand ODEs in the course of their research. However, these books are not ideal for use as a textbook for a student's first exposure to ODEs beyond the basic calculus course (more on that shortly). Their depth and mathematical thoroughness often leave students that are relatively new to the topic feeling overwhelmed and grasping for the essential ideas within the topics that are covered. Of course, (probably) no one would consider using these texts for a second year course in ODEs. That's not really an issue, and there is a large market for ODE texts for second year mathematics students (and new texts continue to appear each year). I spent some time examining some of these texts (many which sell for well over a hundred dollars) and concluded that none of them really would "work" for the course that I wanted to deliver. So, I decided to write my own notes, which has turned into this small book. I have taught

[1]E. A. Coddington and N. Levinson. *Theory of Ordinary Differential Equations*. Krieger, 1984; J. K. Hale. *Ordinary Differential Equations*. Dover, 2009; and P. Hartman. *Ordinary Differential Equations*. Society for Industrial and Applied Mathematics, 2002.

this course for three years now. There are typically about 160 students in the class, in their second year, and I have been somewhat surprised, and pleased, by how the course has been received by the students. So now I will explain a bit about my rationale, requirements, and goals for the course.

In the UK students come to University to study mathematics with a good background in calculus and linear algebra. Many have "seen" some basic ODEs already. In their first year, students have a year long course in calculus where they encounter the typical first order ODEs, second order linear constant coefficient ODEs, and two-dimensional first order linear matrix ODEs. This material tends to form a substantial part of the traditional second year course in ODEs and since I can consider the material as "already seen, at least, once", it allows me to develop the course in a way that makes contact with more contemporary concepts in ODEs and to touch on a variety of research issues. This is very good for our program since many students will do substantial projects that approach research level and require varying amounts of knowledge of ODEs.

This book consists of 10 chapters, and the course is 12 weeks long. Each chapter is covered in a week, and in the remaining two weeks I summarize the entire course, answer lots of questions, and prepare the students for the exam. I do not cover the material in the appendices in the lectures. Some of it is basic material that the students have already seen that I include for completeness and other topics are "tasters" for more advanced material that students will encounter in later courses or in their project work. Students are very curious about the notion of "chaos", and I have included some material in an appendix on that concept. The focus in that appendix is only to connect it with ideas that have been developed in this course related to ODEs and to prepare them for more advanced courses in dynamical systems and ergodic theory that are available in their third and fourth years.

There is a significant transition from first to second year mathematics at Bristol. For example, the first year course in calculus teaches a large number of techniques for performing analytical computations, e.g. the usual set of tools for computing derivatives and integrals of functions of one, and more variables. Armed with a large set of computational skills, the second year makes the transition to "thinking about mathematics" and "creating mathematics". The course in ODEs is ideal for making this transition. It is a course in ordinary differential "equations", and equations are what mathematicians learn how to solve. It follows then that students take the course with the expectation of learning how to solve ODEs. Therefore it is a bit disconcerting when I tell them that it is likely that almost all of the ODEs that they encounter throughout their career as a mathematician will *not* have analytical solutions. Moreover, even if they do have analytical solutions the complexity of the analytical solutions, even for "simple" ODEs, is not likely to yield much insight into the nature of the behavior of the solutions of ODEs. This last statement provides the entry into the nature of the course, which is based on the "vision of Poincaré" — rather than seeking to find specific solutions of ODEs, we seek to understand how *all* possible solutions are related in their behavior in the geometrical setting of phase space. In other words, this course has been designed to be a beginning course in ODEs from the dynamical systems point of view.

I am grateful to all of the students who have taken this course over the past few years. Teaching this particular course has been a very rewarding experience for me and I very much enjoyed discussing this material with them during office hours.

While this book originated during my time in the city of Bristol, United Kingdom, I have completed it in the past few months during my stay at the United States Naval Academy in Annapolis, Maryland. During this time I have used the book in teaching the course SM222, Differential Equations with Matrices. This is a second year course at the Naval Academy and the students have all had three courses in calculus, and most have had a one semester course in matrix theory. This was an adequate background and I was very happy with the student engagement with the course. Discussions with midshipmen Luke Artz, Brown Bedard, Thomas (Cohen) Bruner, Alexander (Alex) Coelho, Natali Dilts, Bianka Harris, Haley Harris, Francis (Frank) Hollahan, Sarabeth Joyner, Emilyrose Kosanovich, George Main, Rocco Patel, and Aurora Sandoval contributed to the refinement of explanations in many sections as well as the inclusion of quite a few new exercises. I am grateful for their input. Finally, I am very grateful for the support of the William R. Davis '68 Chair in the Department of Mathematics at the United States Naval Academy.

Contents

List of Figures

List of Tables

Chapter 1

Getting Started: The Language of ODEs

THIS IS A COURSE ABOUT ORDINARY DIFFERENTIAL EQUATIONS (ODES). So we begin by defining what we mean by this term.[1]

Definition 1 (Ordinary differential equation). An ordinary differential equation (ODE) is an equation for a function of one variable that involves ("ordinary") derivatives of the function (and, possibly, known functions of the same variable).

We give several examples below:

1. $\frac{d^2x}{dt^2} + \omega^2 x = 0$,
2. $\frac{d^2x}{dt^2} - \alpha x \frac{dx}{dt} - x + x^3 = \sin \omega t$,
3. $\frac{d^2x}{dt^2} - \mu(1 - x^2)\frac{dx}{dt} + x = 0$,
4. $\frac{d^3f}{d\eta^3} + f\frac{d^2f}{d\eta^2} + \beta\left(1 - \left(\frac{d^2f}{d\eta^2}\right)^2\right) = 0$,
5. $\frac{d^4y}{dx^4} + x^2\frac{d^2y}{dx^2} + x^5 = 0$.

[1]The material in these lectures can be found in most textbooks on ODEs. It consists mainly of terminology and definitions. Generally such "defintion and terminology" introductions can be tedious and a bit boring. We probably have not entirely avoided that trap, but the material does introduce much of the essential language and some concepts that permeate the rest of the course. If you skip it, you will need to come back to it.

ODEs can be succinctly written by adopting a more compact notation for the derivatives. We rewrite the examples above with this shorthand notation.

1. $\ddot{x} + \omega^2 x = 0$,
2. $\ddot{x} - \alpha x \dot{x} - x + x^3 = \sin \omega t$,
3. $\ddot{x} - \mu(1 - x^2)\dot{x} + x = 0$,
4. $f''' + ff'' + \beta(1 - (f'')^2) = 0$,
5. $y'''' + x^2 y'' + x^5 = 0$.

Now that we have defined the notion of an ODE, we will need to develop some additional concepts in order to more deeply describe the structure of ODEs. The notions of "structure" are important since we will see that they play a key role in how we understand the nature of the behavior of solutions of ODEs.

Definition 2 (Dependent variable). The value of the function, e.g. for Example 1, $x(t)$.

Definition 3 (Independent variable). The argument of the function, e.g. for Example 1, t.

We summarize a list of the dependent and independent variables in the five examples of ODEs given above.

Table 1.1. Identifying the independent and dependent variables for several examples.

Example	Dependent variable	Independent variable
1	x	t
2	x	t
3	x	t
4	f	η
5	y	x

The notion of "order" is an important characteristic of ODEs.

Definition 4 (Order of an ODE). The number associated with the largest derivative of the dependent variable in the ODE.

We give the order of each of the ODEs in the five examples above.

Distinguishing between the independent and dependent variables enables us to define the notion of autonomous and nonautonomous ODEs.

Table 1.2. Identifying the order of the ODE for several examples.

Example	Order
1	second order
2	second order
3	second order
4	third order
5	fourth order

Definition 5 (Autonomous, Nonautonomous). An ODE is said to be *autonomous* if none of the coefficients (i.e. functions) multiplying the dependent variable, or any of its derivatives, depend explicitly on the independent variable, and also if no terms not depending on the dependent variable or any of its derivatives depend explicitly on the independent variable. Otherwise, it is said to be *nonautonomous.*

Or, more succinctly, an ODE is autonomous if the independent variable does not explicitly appear in the equation. Otherwise, it is nonautonomous.

We apply this definition to the five examples above, and summarize the results in the table below:

Table 1.3. Identifying autonomous and nonautonomous ODEs for several examples.

Example	
1	autonomous
2	nonautonomous
3	autonomous
4	autonomous
5	nonautonomous

All scalar ODEs, i.e. the value of the dependent variable is a scalar, can be written as first order equations where the new dependent variable is a vector having the same dimension as the order of the ODE. This is done by constructing a vector whose components consist of the dependent variable and all of its derivatives are *below* the highest order. This vector is the new dependent variable. We illustrate this for the five examples above.

1.

$$\begin{aligned}\dot{x} &= v,\\ \dot{v} &= -\omega^2 x, \quad (x, v) \in \mathbb{R} \times \mathbb{R}.\end{aligned}$$

2.

$$\begin{aligned}\dot{x} &= v,\\ \dot{v} &= \alpha x v + x - x^3 + \sin\omega t, \quad (x, v) \in \mathbb{R} \times \mathbb{R}.\end{aligned}$$

3.

$$\begin{aligned}\dot{x} &= v,\\ \dot{v} &= \mu(1 - x^2)v - x, \quad (x, v) \in \mathbb{R} \times \mathbb{R}.\end{aligned}$$

4.

$$\begin{aligned}f' &= v,\\ f'' &= u,\\ f''' &= -ff'' - \beta(1 - (f'')^2)\end{aligned}$$

or

$$\begin{aligned}f' &= v,\\ v' &= f'' = u,\\ u' &= f''' = -fu - \beta(1 - u^2)\end{aligned}$$

or

$$\begin{pmatrix} f' \\ v' \\ u' \end{pmatrix} = \begin{pmatrix} v \\ u \\ -fu - \beta(1 - u^2) \end{pmatrix}, \quad (f, v, u) \in \mathbb{R} \times \mathbb{R} \times \mathbb{R}.$$

5.

$$\begin{aligned}y' &= w,\\ y'' &= v,\\ y''' &= u,\\ y'''' &= -x^2y'' - x^5\end{aligned}$$

or

$$\begin{aligned}y' &= w,\\ w' &= y'' = v,\\ v' &= y''' = u,\\ u' &= y'''' = -x^2v - x^5\end{aligned}$$

or

$$\begin{pmatrix} y' \\ w' \\ v' \\ u' \end{pmatrix} = \begin{pmatrix} w \\ v \\ u \\ -x^2 v - x^5 \end{pmatrix}, \quad (y, w, v, u) \in \mathbb{R} \times \mathbb{R} \times \mathbb{R} \times \mathbb{R}$$

Therefore without loss of generality, the general form of the ODE that we will study can be expressed as a *first order vector ODE*:

$$\dot{x} = f(x), \quad x(t_0) \equiv x_0, \quad x \in \mathbb{R}^n, \quad \text{autonomous}, \tag{1.1}$$

$$\dot{x} = f(x, t), \quad x(t_0) \equiv x_0 \quad x \in \mathbb{R}^n, \quad \text{nonautonomous}, \tag{1.2}$$

where $x(t_0) \equiv x_0$ is referred to as the *initial condition.*

This first order vector form of ODEs allows us to discuss many properties of ODEs in a way that is independent of the order of the ODE. It also lends itself to a natural geometrical description of the solutions of ODEs that we will see shortly.

A key characteristic of ODEs is whether or not they are linear or nonlinear.

Definition 6 (Linear and Nonlinear ODEs). An ODE is said to be *linear* if it is a linear function of the dependent variable. If it is not linear, it is said to be *nonlinear.*

Note that the independent variable *does not* play a role in whether or not the ODE is linear or nonlinear.

Table 1.4. Identifying linear and nonlinear ODEs for several examples.

Example	
1	linear
2	nonlinear
3	nonlinear
4	nonlinear
5	linear

When written as a first order vector equation the (vector) space of dependent variables is referred to as the phase space of the ODE. The ODE then has the geometric interpretation as a vector field on phase space. The structure of phase space, e.g. its dimension and geometry, can have a significant influence on the nature of solutions of ODEs. We will encounter

ODEs defined on different types of phase space, and of different dimensions. Some examples are given in the following lists.

1-dimension

1. $\mathbb{R}$ — the real line,
2. $I \subset \mathbb{R}$ — an interval on the real line,
3. S^1 — the circle.

"Solving" One-dimensional Autonomous ODEs. Formally (we will explain what that means shortly) an expression for the solution of a one-dimensional autonomous ODE can be obtained by integration. We explain how this is done, and what it means. Let $\mathcal{P}$ denote one of the one-dimensional phase spaces described above. We consider the autonomous vector field defined on $\mathcal{P}$ as follows:

$$\dot{x} = \frac{dx}{dt} = f(x), \quad x(t_0) = x_0, \quad x \in \mathcal{P}. \tag{1.3}$$

This is an example of a one-dimensional separable ODE which can be written as follows:

$$\int_{x(t_0)}^{x(t)} \frac{dx'}{f(x')} = \int_{t_0}^{t} dt' = t - t_0. \tag{1.4}$$

If we can compute the integral on the left-hand side of (1.4), then it may be possible to solve for $x(t)$. However, we know that not all functions $\frac{1}{f(x)}$ can be integrated. This is what we mean by we can "formally" solve for the solution for this example. We may not be able to represent the solution in a form that is useful.

The higher dimensional phase spaces that we will consider will be constructed as Cartesian products of these three basic one-dimensional phase spaces.

2-dimensions

1. $\mathbb{R}^2 = \mathbb{R} \times \mathbb{R}$ — the plane,
2. $\mathbb{T}^2 = \mathbb{S} \times \mathbb{S}$ — the two torus,
3. $C = I \times \mathbb{S}$ — the (finite) cylinder,
4. $C = \mathbb{R} \times \mathbb{S}$ — the (infinite) cylinder.

In many applications of ODEs the independent variable has the interpretation of time, which is why the variable t is often used to denote the independent variable. Dynamics is the study of how systems change in time. When written as a first order system ODEs are often referred to as dynamical systems, and ODEs are said to generate *vector fields on phase space*. For this reason the phrases *ODE* and *vector field* tend to be used synonomously. Moreover, this geometrical view leads to an alternate, but synonomous, terminology for solutions of ODEs. In particular, solutions of ODEs may be referred to as *trajectories* or *orbits*.

Several natural questions arise when analyzing an ODE. "Does the ODE have a solution?" "Are solutions unique?" (And what does "unique" mean?) The standard way of treating this in an ODE course is to "prove a big theorem" about existence and uniqueness. Rather, than do that (you can find the proof in hundreds of books, as well as in many sites on the internet), we will consider some examples that illustrate the main issues concerning what these questions mean, and afterwards we will describe sufficent conditions for an ODE to have a unique solution (and then consider what "uniqueness" means).

First, do ODEs have solutions? Not necessarily, as the following example shows.

Example 1 (An example of an ODE that has no solutions).

Consider the following ODE defined on $\mathbb{R}$:

$$\dot{x}^2 + x^2 + t^2 = -1, \quad x \in \mathbb{R}.$$

This ODE has no solutions since the left-hand side is non-negative and the right-hand side is strictly negative.

Then you can ask the question — "if the ODE has solutions, are they unique?" Again, the answer is "not necessarily", as the following example shows.

Example 2 (An example illustrating the meaning of uniqueness).

$$\dot{x} = ax, \quad x \in \mathbb{R}, \tag{1.5}$$

where a is an arbitrary constant. The solution is given by

$$x(t) = ce^{at}. \tag{1.6}$$

So we see that there are an infinite number of solutions, depending upon the choice of the constant c. So what could uniqueness of solutions mean? If we evaluate the solution (1.6) at $t = 0$ we see that

$$x(0) = c. \tag{1.7}$$

Substituting this into the solution (1.6), the solution has the form:

$$x(t) = x(0)e^{at}. \tag{1.8}$$

From the form of (1.8) we can see exactly what "uniquess of solutions" means. For a given initial condition, there is *exactly one* solution of the ODE satisfying that initial condition.

Example 3. An example of an ODE with non-unique solutions.

Consider the following ODE defined on $\mathbb{R}$:

$$\dot{x} = 3x^{\frac{2}{3}}, \quad x(0) = 0, \quad x \in \mathbb{R}. \tag{1.9}$$

It is easy to see that a solution satisfying $x(0) = 0$ is $x = 0$. However, one can verify directly by substituting into the equation that the following is also a solution satisfying $x(0) = 0$:

$$x(t) = \begin{cases} 0, & t \leq a \\ (t-a)^3, & t > a \end{cases} \tag{1.10}$$

for any $a > 0$. Hence, in this example, there are an infinite number of solutions satisfying *the same initial condition*. This example illustrates precisely what we mean by uniqueness. Given an initial condition, only one ("uniqueness") solution satisfies the initial condition at the chosen initial time.

There is another question that comes up. If we have a unique solution does it exist *for all time*? Not necessarily, as the following example shows.

Example 4. An example of an ODE with unique solutions that exists only for a finite time.

Consider the following ODE on $\mathbb{R}$:

$$\dot{x} = x^2, \quad x(0) = x_0, \quad x \in \mathbb{R}. \tag{1.11}$$

We can easily integrate this equation (it is separable) to obtain the following solution satisfying the initial condition:

$$x(t) = \frac{x_0}{1 - x_0 t} \tag{1.12}$$

The solution becomes infinite, or "does not exist" or "blows up" at $\frac{t}{x_0}$. This is what "does not exist" means. So the solution only exists for a finite time, and this "time of existence" depends on the initial condition.

These three examples contain the essence of the "existence issues" for ODEs that will concern us. They are the "standard examples" that can be found in many textbooks.[2,3]

Now we will state the standard "existence and uniqueness" theorem for ODEs. The statement is an example of the power and flexibility of expressing a general ODE as a first order vector equation. The statement is valid for any (finite) dimension.

We consider the general vector field on $\mathbb{R}^n$

$$\dot{x} = f(x,t), \quad x(t_0) = x_0, \quad x \in \mathbb{R}^n. \tag{1.13}$$

It is important to be aware that for the general result we are going to state it does not matter whether or not the ODE is autonomous or nonautonomous.

We define the domain of the vector field. Let $U \subset \mathbb{R}^n$ be an open set and let $I \subset \mathbb{R}$ be an interval. Then we express that the n-dimensional vector field is defined on this domain as follows:

$$\begin{aligned} f\,:\, U \times I &\to \mathbb{R}^n, \\ (x,t) &\to f(x,t) \end{aligned} \tag{1.14}$$

We need a definition to describe the "regularity" of the vector field.

Definition 7 (C^r function). We say that $f(x,t)$ is C^r on $U \times I \subset \mathbb{R}^n \times \mathbb{R}$ if it is r times differentiable and each derivative is a continuous function (on the same domain). If $r = 0$, $f(x,t)$ is just said to be continuous.

---

[2] J. K. Hale. *Ordinary Differential Equations.* Dover, 2009; P. Hartman. *Ordinary Differential Equations.* Society for industrial and Applied Mathematics, 2002; E. A. Coddington and N. Levinson. *Theory of Ordinary Differential Equations.* Krieger, 1984.

[3] The "Existence and Uniqueness Theorem" is, traditionally, a standard part of ODE courses beyond the elementary level. This theorem, whose proof can be found in numerous texts (including those mentioned in the Preface), will not be given in this course. There are several reasons for this. One is that it requires considerable time to construct a detailed and careful proof, and I do not feel that this is the best use of time in a 12 week course. The other reason (not unrelated to the first) is that I do not feel that the understanding of the detailed "Existence and Uniqueness" proof is particularly important at this stage of the students education. The subject has grown so much in the last forty years, especially with the merging of the "dynamical systems point of view" with the subject of ODEs, that there is just not enough time to devote to all the topics that students "should know". However, it is important to know what it means for an ODE to have a solution, what uniqueness of solutions means, and general conditions for when an ODE has unique solutions.

Now we can state *sufficient conditions* for (1.13) to have a unique solution. We suppose that $f(x,t)$ is C^r, $r \geq 1$. We choose any point $(x_0, t_0) \in U \times I$. Then there exists a unique solution of (1.13) satisfying this initial condition. We denote this solution by $x(t, t_0, x_0)$, and reflect in the notation that it satisfies the initial condition by $x(t_0, t_0, x_0) = x_0$. This unique solution exists for a time interval centered at the initial time t_0, denoted by $(t_0 - \epsilon,\, t_0 + \epsilon)$, for some $\epsilon > 0$. Moreover, this solution, $x(t, t_0, x_0)$, is a C^r function of t, t_0, x_0. Note that from Example 4 ϵ may depend on x_0. This also explains how a solution "fails to exist" — it becomes unbounded ("blow up") in a finite time.

Finally, we remark that existence and uniqueness of ODEs is the mathematical manifestation of determinism. If the initial condition is specified (with 100% accuracy), then the past and the future are uniquely determined. The key phrase here is "100% accuracy". Numbers cannot be specified with 100% accuracy. There will always be some imprecision in the specification of the initial condition. Chaotic dynamical systems are deterministic dynamical systems having the property that imprecisions in the initial conditions may be magnified by the dynamical evolution, leading to seemingly random behavior (even though the system is completely deterministic).

Problem Set 1

1. For each of the ODEs below, write it as a first order system, state the dependent and independent variables, state any parameters in the ODE (i.e. unspecified constants) and state whether it is linear or nonlinear, and autonomous or nonautonomous,

   (a)
   $$\ddot{\theta} + \delta\dot{\theta} + \sin\theta = F\cos\omega t, \quad \theta \in S^1.$$

   (b)
   $$\ddot{\theta} + \delta\dot{\theta} + \theta = F\cos\omega t, \quad \theta \in S^1.$$

   (c)
   $$\frac{d^3y}{dx^3} + x^2 y\frac{dy}{dx} + y = 0, \quad x \in \mathbb{R}^1.$$

   (d)
   $$\ddot{x} + \delta\dot{x} + x - x^3 = \theta,$$
   $$\ddot{\theta} + \sin\theta = 0, \quad (x,\theta) \in \mathbb{R}^1 \times S^1.$$

(e)

$$
\begin{aligned}
\ddot{\theta} + \delta\dot{\theta} + \sin\theta &= x, \\
\dot{x} - x + x^3 &= 0, \quad (\theta, x) \in S^1 \times \mathbb{R}^1
\end{aligned}
$$

2. Consider the vector field:

$$\dot{x} = 3x^{\frac{2}{3}}, \quad x(0) \neq 0, \quad x \in \mathbb{R}.$$

Does this vector field have unique solutions?[4]

3. Consider the vector field:

$$\dot{x} = -x + x^2, \quad x(0) = x_0, \quad x \in \mathbb{R}.$$

Determine the time interval of existence of all solutions as a function of the initial condition, x_0.[5]

4. Consider the vector field:

$$\dot{x} = a(t)x + b(t), \quad x \in \mathbb{R}.$$

Determine sufficient conditions on the coefficients $a(t)$ and $b(t)$ for which the solutions will exist for all time. Do the results depend on the initial condition?[6]

5. Consider the following two ODEs:

$$\frac{dy}{dx} = -\frac{1}{x}y + x, \tag{1.15}$$

and

$$
\begin{aligned}
\dot{x} &= x, \\
\dot{y} &= -y + x^2.
\end{aligned} \tag{1.16}
$$

---

[4]Note the similarity of this exercise to Example 3. The point of this exercise is to think about how issues of existence and uniqueness depend on the initial condition.

[5]Here is the point of this exercise: $\dot{x} = -x$ has solutions that exist for all time (for any initial condition), $\dot{x} = x^2$ has solutions that "blow up in finite time". What happens when you "put these two together"? In order to answer this you will need to solve for the solution, $x(t, x_0)$.

[6]The "best" existence and uniqueness theorems are when you can analytically solve for the "exact" solution of an ODE for arbitrary initial conditions. This is the type of ODE where that can be done. It is a first order, linear inhomogeneous ODE that can be solved using an "integrating factor" (see Appendix B for help with this). However, you will need to argue that the integrals obtained from this procedure "make sense".

and

(a) For each ODE state the dependent variable, the independent variable and whether it is linear or nonlinear, autonomous or nonautonomous.
(b) Find the explicit solutions of each ODE for a general initial condition. (Hint: using the techniques in Appendix B may be helpful.)
(c) Describe in what sense these two ODEs are "the same".

6. Recall the definition of second order linear ODE given in this chapter (see also Appendix C):

$$m\frac{d^2s}{dt^2} = (a_0 + a_1(t))s + (b_0 + b_1(t))\dot{s} + c_0 + c_1(t), \tag{1.17}$$

where a_0, b_0, c_0 are constants, and $a_1(t)$, $b_1(t)$, $c_1(t)$ are functions of t. First, consider the situation where $c_0 = c_1(t) = 0$, i.e.

$$m\frac{d^2s}{dt^2} = (a_0 + a_1(t))s + (b_0 + b_1(t))\dot{s}. \tag{1.18}$$

In this case the linear ODE is said to be homogeneous.

(a) Suppose $s_1(t)$ is a solution of (1.18), and let k_1 denote a constant (real number). Prove that $k_1 s_1(t)$ is also a solution of (1.18).
(b) Suppose $s_1(t)$ and $s_2(t)$ are solutions of (1.18), and let k_1 and k_2 denote constants (real numbers). Prove that $k_1 s_1(t) + k_2 s_2(t)$ is also a solution of (1.18). This is the *superposition principle* for linear homogeneous ODE's.
(c) Do these two results hold for (1.17)?

7. Are the following second order ODEs linear or nonlinear?

(a) $m\ddot{s} = -s + \cos t$,
(b) $m\ddot{s} = -s^2 + \cos t$,
(c) $m\ddot{s} = -s \cos t$,
(d) $m\ddot{s} = -t^2 s$,
(e) $m\ddot{s} = -s + s^2$,

8. Consider the following ODE:

$$\dot{x} = ax, \quad x(0) = x_0,$$

where x, $a \in \mathbb{R}$.

(a) Show how to compute the solution of this ODE.
(b) In chapter we have considered "uniqueness of solutions of ODEs", as well as the fact that ODEs have "many" solutions. Explain these two (seemingly) contradictory statements in the context of this example.

9. Consider the following ODE:

$$\dot{x} = x^2, \quad x(0) = x_0, \quad x \in \mathbb{R}$$

with solution given by:

$$x(t) = \frac{x_0}{1 - x_0\, t}.$$

As we showed in this chapter, this is an example of an ODE where solutions may exist only for a finite time interval. But are there any solutions that exist "for all time", i.e. for an infinite time interval?

10. Solve the following ODE:

$$m\ddot{s} = g, \quad s(0) = s_0,\ \dot{s}(0) = 0,$$

where g is a constant.

11. Solve the following ODE:

$$m\ddot{s} = \sin t, \quad s(0) = s_0, \quad \dot{s}(0) = 0.$$

12. Consider the following ODE:

$$m\ddot{s} = s - s^2.$$

Find the function of s and $\dot{s}$ that a solution of this ODE must satisfy (hint: Appendix C will be helpful).

13. Consider Newton's equations:

$$m\frac{d^2 s}{dt^2} = F(s).$$

Define a new time variable, τ, which is related to the "old" time t by:

$$t = \sqrt{m}\tau.$$

Use the chain rule to show that with respect to the new time the ODE becomes:

$$\frac{d^2 s}{d\tau^2} = F(s),$$

i.e. the constant disappears. This is referred to as *rescaling time.*

14. Consider the following nonlinear ODE:

$$\ddot{s} = s - s^2.$$

Suppose $s_1(t)$ and $s_2(t)$ are solutions, and let k_1 and k_2 denote constants.[7]

(a) Is $k_1 s_1(t)$ a solution?
(b) Is $k_1 s_1(t) + k_2 s_2(t)$ a solution?

[7]This exercise shows that nonlinear ODEs do not (generally) obey a superposition principle.

Chapter 2

Special Structure and Solutions of ODEs

A CONSISTENT THEME THROUGHOUT ALL OF ODES IS THAT "SPECIAL STRUCTURE" OF THE *equations* CAN REVEAL INSIGHT INTO THE NATURE OF THE *solutions*. Here we look at a very basic and important property of autonomous equations:

> *"Time shifts of solutions of autonomous ODEs are also solutions of the ODE (but with a different initial condition)".*

Now we will show how to see this.

THROUGHOUT THIS COURSE WE WILL ASSUME THAT EXISTENCE AND UNIQUENESS OF SOLUTIONS HOLD ON A DOMAIN AND TIME INTERVAL SUFFICIENT FOR OUR ARGUMENTS AND CALCULATIONS.

We start by establishing the setting. We consider an autonomous vector field defined on $\mathbb{R}^n$:

$$\dot{x} = f(x), \quad x(0) = x_0, \quad x \in \mathbb{R}^n, \tag{2.1}$$

with solution denoted by:

$$x(t, 0, x_0), \quad x(0, 0, x_0) = x_0.$$

Here we are taking the initial time to be $t_0 = 0$. We will see, shortly, that for autonomous equations this can be done without loss of generality.

Now we choose $s \in \mathbb{R}$ ($s \neq 0$, which is to be regarded as a fixed constant). We must show the following:

$$\dot{x}(t+s) = f(x(t+s)) \tag{2.2}$$

This is what we mean by the phrase time shifts of solutions are solutions. This relation follows immediately from the chain rule calculation:

$$\frac{d}{dt} = \frac{d}{d(t+s)} \frac{d(t+s)}{dt} = \frac{d}{d(t+s)}. \tag{2.3}$$

Finally, we need to determine the initial condition for the time shifted solution. For the original solution we have:

$$x(t, 0, x_0), \quad x(0, 0, x_0) = x_0, \tag{2.4}$$

and for the time shifted solution we have:

$$x(t+s, 0, x_0), \quad x(s, 0, x_0). \tag{2.5}$$

It is for this reason that, without loss of generality, for autonomous vector fields we can take the initial time to be $t_0 = 0$. This allows us to simplify the arguments in the notation for solutions of autonomous vector fields, i.e. $x(t, 0, x_0) \equiv x(t, x_0)$ with $x(0, 0, x_0) = x(0, x_0) = x_0$.

Example 5 (An example illustrating the time-shift property of autonomous vector fields). Consider the following one-dimensional autonomous vector field:

$$\dot{x} = \lambda x, \quad x(0) = x_0, \quad x \in \mathbb{R}, \ \lambda \in \mathbb{R}. \tag{2.6}$$

The solution is given by:

$$x(t, 0, x_0) = x(t, x_0) = e^{\lambda t} x_0. \tag{2.7}$$

The time shifted solution is given by:

$$x(t+s, x_0) = e^{\lambda(t+s)} x_0. \tag{2.8}$$

We see that it is a solution of the ODE with the following calculations:

$$\frac{d}{dt} x(t+s, x_0) = \lambda e^{\lambda(t+s)} x_0 = \lambda x(t+s, x_0), \tag{2.9}$$

with initial condition:

$$x(s, x_0) = e^{\lambda s} x_0. \tag{2.10}$$

In summary, we see that the solutions of autonomous vector fields satisfy the following three properties:

1. $x(0, x_0) = x_0$
2. $x(t, x_0)$ is C^r in x_0
3. $x(t + s, x_0) = x(t, x(s, x_0))$.

Property one just reflects the notation we have adopted. Property 2 is a statement of the properties arising from existence and uniqueness of solutions. Property 3 uses two characteristics of solutions. One is the "time shift" property for autonomous vector fields that we have proven. The other is "uniquess of solutions" since the left-hand side and the right-hand side of Property 3 satisfy the same initial condition at $t = 0$.

These three properties are the defining properties of a flow, i.e. a one-parameter group of transformations of the phase space. In other words, we view the solutions as defining a *map* of points in phase space. The group property arises from property 3, i.e. the time-shift property. In order to emphasize this "map of phase space" property we introduce a general notation for the flow as follows:

$$x(t, x_0) \equiv \phi_t(\cdot),$$

where the "$\cdot$" in the argument of $\phi_t(\cdot)$ reflects the fact that the flow is a function on the phase space. With this notation the three properties of a flow are written as follows:

1. $\phi_0(\cdot)$ is the identity map
2. $\phi_t(\cdot)$ is C^r for each t
3. $\phi_{t+s}(\cdot) = \phi_t \circ \phi_s(\cdot)$.

We often use the phrase "the flow generated by the (autonomous) vector field". Autonomous is in parentheses as it is understood that when we are considering flows then we are considering the solutions of autonomous vector fields. This is because nonautonomous vector fields do not necessarily satisfy the time-shift property, as we now show with an example.

Example 6 (An example of a nonautonomous vector field not having the time-shift property). Consider the following one-dimensional vector field on $\mathbb{R}$:

$$\dot{x} = \lambda t x, \quad x(0) = x_0, \quad x \in \mathbb{R}, \ \lambda \in \mathbb{R}.$$

This vector field is separable and the solution is easily found to be:

$$x(t, 0, x_0) = x_0 e^{\frac{\lambda}{2} t^2}.$$

The time shifted "solution" is given by:

$$x(t+s, 0, x_0) = x_0 e^{\frac{\lambda}{2}(t+s)^2}.$$

We show that this does not satisfy the vector field with the following calculation:

$$\begin{aligned} \frac{d}{dt} x(t+s, 0, x_0) &= x_0 e^{\frac{\lambda}{2}(t+s)^2} \lambda(t+s). \\ &\neq \lambda t x(t+s, 0, x_0). \end{aligned}$$

Perhaps a more simple example illustrating that nonautonomous vector fields do not satisfy the time-shift property is the following.

Example 7. Consider the following one-dimensional nonautonomous vector field:

$$\dot{x} = e^t, \quad x \in \mathbb{R}.$$

The solution is given by:

$$x(t) = e^t.$$

It is easy to verify that the time-shifted function:

$$x(t+s) = e^{t+s},$$

does not satisfy the equation.

In the study of ODEs certain types of solutions have achieved a level of prominence largely based on their significance in applications. They are

- equilibrium solutions,
- periodic solutions,
- heteroclinic solutions,
- homoclinic solutions.

We define each of these.

Definition 8 (Equilibrium). A *point* in phase space $x = \bar{x} = \mathbb{R}^n$ that is a solution of the ODE, i.e.

$$f(\bar{x}) = 0, \quad f(\bar{x}, t) = 0,$$

is called an equilibrium point. These may also be referred to as *fixed points.*

For example, $x = 0$ is an equilibrium point for the following autonomous and nonautonomous one-dimensional vector fields, respectively,

$$\dot{x} = x, \quad x \in \mathbb{R},$$
$$\dot{x} = tx, \quad x \in \mathbb{R}.$$

A periodic solution is simply a solution that is periodic in time. Its definition is the same for both autonomous and nonautonomous vector fields.

Definition 9 (Periodic solutions). A solution $x(t, t_0, x_0)$ is periodic if there exists a $T > 0$ such that

$$x(t, t_0, x_0) = x(t + T, t_0, x_0).$$

Homoclinic and heteroclinic solutions are important in a variety of applications. Their definition is not so simple as the definitions of equilibrium and periodic solutions since they can be defined and generalized to many different settings. We will only consider these special solutions for autonomous vector fields, and solutions homoclinic or heteroclinic to equilibrium solutions.

Definition 10 (Homoclinic and Heteroclinic Solutions). Suppose $\bar{x}_1$ and $\bar{x}_2$ are equilibrium points of an autonomous vector field, i.e.

$$f(\bar{x}_1) = 0, \quad f(\bar{x}_2) = 0.$$

A trajectory $x(t, t_0, x_0)$ is said to be heteroclinic to $\bar{x}_1$ and $\bar{x}_2$ if

$$\begin{aligned} \lim_{t \to \infty} x(t, t_0, x_0) &= \bar{x}_2, \\ \lim_{t \to -\infty} x(t, t_0, x_0) &= \bar{x}_1. \end{aligned} \tag{2.11}$$

If $\bar{x}_1 = \bar{x}_2$ the trajectory is said to be homoclinic to $\bar{x}_1 = \bar{x}_2$.

Example 8.[1] Here we give an example illustrating equilibrium points and heteroclinic orbits. Consider the following one-dimensional autonomous vector field on $\mathbb{R}$:

$$\dot{x} = x - x^3 = x(1 - x^2), \quad x \in \mathbb{R}. \tag{2.12}$$

This vector field has three equilibrium points at $x = 0, \pm 1$.

In Fig. 2.1 we show the graph of the vector field (2.12) in panel (a) and the phase line dynamics in panel (b).

The solid black dots in panel (b) correspond to the equilibrium points and these, in turn, correspond to the zeros of the vector field shown in panel (a). Between its zeros, the vector field has a fixed sign (i.e. positive or negative), corresponding to $\dot{x}$ being either increasing or decreasing. This is indicated by the direction of the arrows in panel (b).

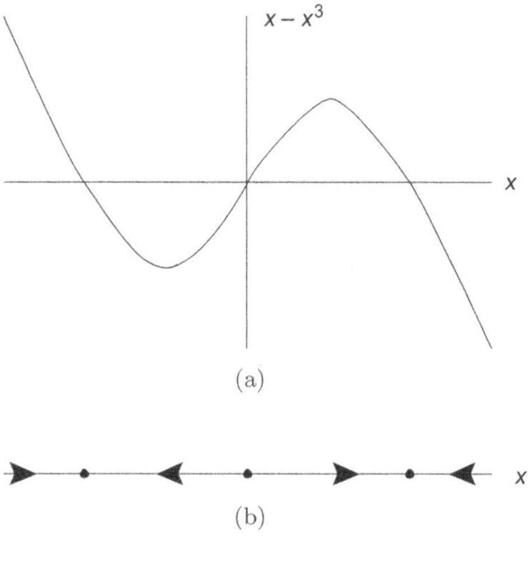

Fig. 2.1. (a) Graph of the vector field. (b) The phase space.

[1]There is a question that we will return to throughout this course. What does it mean to "solve" an ODE? We would argue that a more "practical" question might be, "what does it mean to understand the nature of all possible solutions of an ODE?". But don't you need to be able to answer the first question before you can answer the second? We would argue that Fig. 2.1 gives a complete "qualitative" understanding of (2.12) in a manner that is much simpler than one could directly obtain from its solutions. In fact, it would be an instructive exercise to first solve (2.12) and from the solutions sketch as in Fig. 2.1. This may seem a bit confusing, but it is even more instructive to think about, and understand, what it means.

Our discussion about trajectories, as well as this example, brings us to a point where it is natural to introduce the important notion of an invariant set. While this is a general idea that applies to both autonomous and nonautonomous systems, in this course we will only discuss this notion in the context of autonomous systems. Accordingly, let $\phi_t(\cdot)$ denote the flow generated by an autonomous vector field.

Definition 11 (Invariant Set). A set $M \subset \mathbb{R}^n$ is said to be *invariant* if

$$x \in M \Rightarrow \phi_t(x) \in M \quad \forall t.$$

In other words, a set is invariant (with respect to a flow) if you start in the set, and remain in the set, forever.

If you think about it, it should be clear that invariant sets are sets of trajectories. Any single trajectory is an invariant set. The entire phase space is an invariant set. The most interesting cases are those "in between". Also, it should be clear that the union of any two invariant sets is also an invariant set (just apply the definition of invariant set to the union of two, or more, invariant sets).

There are certain situations where we will be interested in sets that are invariant only for positive time-positive invariant sets.

Definition 12 (Positive Invariant Set). A set $M \subset \mathbb{R}^n$ is said to be *positive invariant* if

$$x \in M \Rightarrow \phi_t(x) \in M \quad \forall t > 0.$$

There is a similar notion of negative invariant sets, but the generalization of this from the definition of positive invariant sets should be obvious, so we will not write out the details.

Concerning Example 8, the three equilibrium points are invariant sets, as well as the closed intervals $[-1, 0]$ and $[0, 1]$. Are there other invariant sets?

Problem Set 2

1. Consider the following ODE:

$$\dot{x} = ax, \quad x(0) = x_0,$$

where x, $a \in \mathbb{R}$.

(a) Show how to how compute the solution of this ODE.
(b) Show that if you shift the time of the solution of this ODE, i.e. let $t \to t + s$, for some fixed s, then the result is still a solution of the ODE.
(c) How does the initial condition of the ODE, i.e. the value of the solution at $t = 0$, compare to the initial condition of the solution of the ODE without the time shift?

2. Consider an autonomous vector field on the plane having an equilibrium point with a homoclinic orbit connecting the equilibrium point, as illustrated in Fig. 2.2. We assume that existence and uniqueness of solutions hold. Can a trajectory starting at any point on the homoclinic orbit reach the equilibrium point in a finite time? (You must justify your answer.)[2]

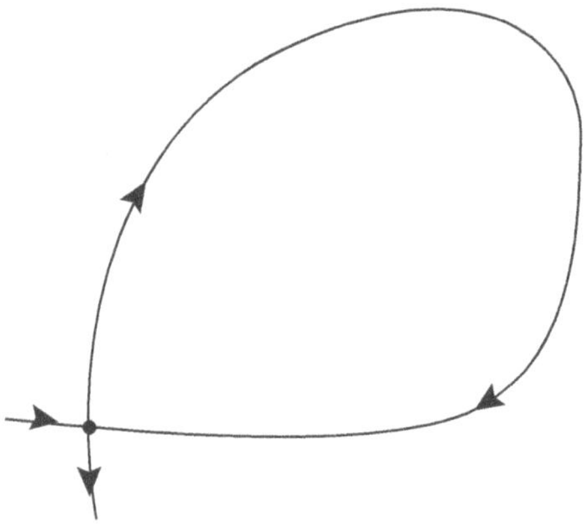

Fig. 2.2. Homoclinic orbit connecting an equilibrium point.

3. Can an autonomous vector field on $\mathbb{R}$ that has no equilibrium points have periodic orbits? We assume that existence and uniqueness of solutions hold. (You must justify your answer.)[3]

[2]The main points to take into account for this problem are the fact that two trajectories cannot cross (in a finite time), and that an equilibrium point is a trajectory.

[3]The main points to take into account in this problem is that the phase space is $\mathbb{R}$ and using this with the implication that trajectories of autonomous ODEs "cannot cross".

4. Can a nonautonomous vector field on $\mathbb{R}$ that has no equilibrium points have periodic orbits? We assume that existence and uniqueness of solutions hold. (You must justify your answer.)[4]
5. Can an autonomous vector field on the circle that has no equilibrium points have periodic orbits? We assume that existence and uniqueness of solutions hold. (You must justify your answer.)[5]
6. Consider the following autonomous vector field on the plane:

$$\dot{x} = -\omega y,$$
$$\dot{y} = \omega x, \quad (x, y) \in \mathbb{R}^2,$$

where $\omega > 0$.

- Show that the flow generated by this vector field is given by[6]:

$$\begin{pmatrix} x(t) \\ y(t) \end{pmatrix} = \begin{pmatrix} \cos \omega t & -\sin \omega t \\ \sin \omega t & \cos \omega t \end{pmatrix} \begin{pmatrix} x_0 \\ y_0 \end{pmatrix}.$$

- Show that the flow obeys the time shift property.
- Give the initial condition for the time shifted flow.

7. Consider the following autonomous vector field on the plane:

$$\dot{x} = \lambda y,$$
$$\dot{y} = \lambda x, \quad (x, y) \in \mathbb{R}^2,$$

where $\lambda > 0$.

- Show that the flow generated by this vector field is given by:

$$\begin{pmatrix} x(t) \\ y(t) \end{pmatrix} = \begin{pmatrix} \cosh \lambda t & \sinh \lambda t \\ \sinh \lambda t & \cosh \lambda t \end{pmatrix} \begin{pmatrix} x_0 \\ y_0 \end{pmatrix}.$$

- Show that the flow obeys the time shift property.
- Give the initial condition for the time shifted flow.

---

[4]It is probably easiest to answer this problem by constructing a specific example.
[5]The main point to take into account here is that the phase space is "periodic".
[6]Recall that the flow is obtained from the solution of the ODE for an arbitrary initial condition.

8. Show that the time shift property for autonomous vector fields implies that trajectories cannot "cross each other", i.e. intersect, in phase space.
9. Show that the union of two invariant sets is an invariant set.
10. Show that the intersection of two invariant sets is an invariant set.
11. Show that the complement of a positive invariant set is a negative invariant set.

Chapter 3

Behavior Near Trajectories and Invariant Sets: Stability

Consider the general nonautonomous vector field in n dimensions:

$$\dot{x} = f(x, t), \quad x \in \mathbb{R}^n, \tag{3.1}$$

and let $\bar{x}(t, t_0, x_0)$ be a solution of this vector field.

MANY QUESTIONS IN ODES CONCERN UNDERSTANDING THE BEHAVIOR OF NEIGHBORING SOLUTIONS NEAR A GIVEN, CHOSEN SOLUTION. We will develop the general framework for considering such questions by transforming (3.1) to a form that allows us to explicitly consider these issues.

We consider the following (time dependent) transformation of variables:

$$x = y + \bar{x}(t, t_0, x_0). \tag{3.2}$$

We wish to express (3.1) in terms of the y variables. It is important to understand what this will mean in terms of (3.2). For y small it means that x is near the solution of interest, $\bar{x}(t, t_0, x_0)$. In other words, expressing the vector field in terms of y will provide us with an explicit form of the vector field for studying the behavior near $\bar{x}(t, t_0, x_0)$. Towards this end, we begin by transforming (3.1) using (3.2) as follows:

$$\dot{x} = \dot{y} + \dot{\bar{x}} = f(x, t) = f(y + \bar{x}, t), \tag{3.3}$$

or,

$$\begin{aligned} \dot{y} &= f(y + \bar{x}, t) - \dot{\bar{x}}, \\ &= f(y + \bar{x}, t) - f(\bar{x}, t) \equiv g(y, t), \quad g(0, t) = 0. \end{aligned} \tag{3.4}$$

Hence, we have shown that solutions of (3.1) near $\bar{x}(t, t_0, x_0)$ are equivalent to solutions of (3.4) near $y = 0$.

The first question we want to ask related to the behavior near $\bar{x}(t, t_0, x_0)$ is whether or not this solution is *stable*? However, first we need to mathematically define what is meant by this term "stable". Now we should know that, without loss of generality, we can discuss this question in terms of the zero solution of (3.4).

We begin by defining the notion of "Lyapunov stability" (or just "stability").

Definition 13 (Lyapunov Stability). $y = 0$ is said to be Lyapunov stable at t_0 if given $\epsilon > 0$ there exists a $\delta = \delta(t_0, \epsilon)$ such that

$$|y(t_0)| < \delta \Rightarrow |y(t)| < \epsilon, \quad \forall t > t_0 \tag{3.5}$$

If a solution is not Lyapunov stable, then it is said to be unstable.

Definition 14 (Unstable). If $y = 0$ is not Lyapunov stable, then it is said to be unstable.

Then we have the notion of *asymptotic stability.*

Definition 15 (Asymptotic stability). $y = 0$ is said to be asymptotically stable at t_0 if:

1. it is Lyapunov stable at t_0,
2. there exists $\delta = \delta(t_0) > 0$ such that:

$$|y(t_0)| < \delta \Rightarrow \lim_{t \to \infty} |y(t)| = 0. \tag{3.6}$$

We have several comments about these definitions.

- Roughly speaking, a Lyapunov stable solution means that if you start close to that solution, you stay close — forever. Asymptotic stability not only means that you start close and stay close forever, but that you actually get "closer and closer" to the solution.
- Stability is an infinite time concept.
- If the ODE is autonomous, then the quantity $\delta = \delta(t_0, \epsilon)$ can be chosen to be independent of t_0.
- The *definitions* of stability do not tell us how to prove that a solution is stable (or unstable). We will learn two techniques for analyzing this question — linearization and Lyapunov's (second) method.

- Why is Lyapunov stability included in the definition of asymptotic stability? Because it is possible to construct examples where nearby solutions do get closer and closer to the given solution as $t \to \infty$, but in the process there are intermediate intervals of time where nearby solutions make "large excursions" away from the given solution.

"Stability" is a notion that applies to a "neighborhood" of a trajectory.[1] At this point we want to formalize various notions related to distance and neighborhoods in phase space. For simplicity in expressing these ideas we will take as our phase space $\mathbb{R}^n$. Points in this phase space are denoted $x \in \mathbb{R}^n$, $x \equiv (x_1, \dots, x_n)$. The norm, or length, of x, denoted $|x|$ is defined as:

$$|x| = \sqrt{x_1^2 + x_2^2 + \cdots + x_n^2} = \sqrt{\sum_{i=1}^{n} x_i^2}.$$

The distance between two points in $x, y \in \mathbb{R}^n$ is defined as:

$$\begin{aligned} d(x,y) \equiv |x-y| &= \sqrt{(x_1-y_1)^2 + \cdots + (x_n-y_n)^2}, \\ &= \sqrt{\sum_{i=1}^{n}(x_i - y_i)^2}. \end{aligned} \tag{3.7}$$

Distance between points in $\mathbb{R}^n$ should be somewhat familiar, but now we introduce a new concept, the distance between a point and a set. Consider a set M, $M \subset \mathbb{R}^n$, let $p \in \mathbb{R}^n$. Then the distance from p to M is defined as follows:

$$\text{dist}(p, M) \equiv \inf_{x \in M} |p - x|. \tag{3.8}$$

We remark that it follows from the definition that if $p \in M$, then $\text{dist}(p, M) = 0$.

We have previously defined the notion of an invariant set. Roughly speaking, invariant sets are comprised of trajectories. We now have the background to discuss the notion of stability of invariant sets. Recall, that

[1]The notion that stability of a trajectory is a property of solutions in a *neighborhood* of a trajectory often causes confusion. To avoid confusion it is important to be clear about the notion of a "neighborhood of a trajectory", and then to realize that for solutions that are Lyapunov (or asymptotically) stable *all* solutions in the neighborhood have the same behavior as $t \to \infty$.

the notion of invariant set was only developed for autonomous vector fields. So we consider an autonomous vector field:

$$\dot{x} = f(x), \quad x \in \mathbb{R}^n, \tag{3.9}$$

and denote the flow generated by this vector field by $\phi_t(\cdot)$. Let M be a closed invariant set (in many applications we may also require M to be bounded) and let $U \supset M$ denoted a neighborhood of M.

The definition of Lyapunov stability of an invariant set is as follows:

Definition 16 (Lyapunov Stability of M). M is said to be Lyapunov stable if for any neighborhood $U \supset M$, $x \in U \Rightarrow \phi_t(x) \in U$, $\forall t > 0$.

Similarly, we have the following definition of asymptotic stability of an invariant set.

Definition 17 (Asymptotic Stability of M). M is said to be asymptotically stable if

1. it is Lyapunov stable,
2. there exists a neighborhood $U \supset M$ such that $\forall x \in U$, $\text{dist}(\phi_t(x), M) \to 0$ as $t \to \infty$.

In the dynamical systems approach to ordinary differential equations some alternative terminology is typically used.

Definition 18 (Attracting Set). If M is asymptotically stable it is said to be an attracting set.

The significance of attracting sets is that they are the "observable" regions in phase space since they are regions to which trajectories evolve in time. The set of points that evolve towards a specific attracting set is referred to as the basin of attraction for that invariant set.

Definition 19 (Basin of Attraction). Let $\mathcal{B} \subset \mathbb{R}^n$ denote the set of all points, $x \in B \subset \mathbb{R}^n$ such that

$$\text{dist}(\phi_t(x), M) \to 0 \quad \text{as } t \to \infty.$$

Then $\mathcal{B}$ is called the basin of attraction of M.

We now consider an example that allows us to explicitly explore these ideas.[2]

Example 9 Consider the following autonomous vector field on the plane:

$$\dot{x} = -x,$$
$$\dot{y} = y^2(1-y^2) \equiv f(y), \quad (x,y) \in \mathbb{R}^2. \tag{3.10}$$

First, it is useful to note that the x and y components of (3.10) are independent. Consequently, this may seem like a trivial example. However, we will see that such examples provide a great deal of insight, especially since they allow for simple computations of many of the mathematical ideas.

In Fig. 3.1 we illustrate the flow of the x and y components of (3.10) separately.

The two-dimensional vector field (3.10) has equilibrium points at:

$$(x,y) = (0,0), \quad (0,1), \quad (0,-1).$$

In this example it is easy to identify three *invariant horizontal lines* (examples of invariant sets). Since $y = 0$ implies that $\dot{y} = 0$, this implies that the x-axis is invariant. Since $y = 1$ implies that $\dot{y} = 0$, this implies that the line $y = 1$ is invariant. Since $y = -1$ implies that $\dot{y} = 0$, which implies that the line $y = -1$ is invariant. This is illustrated in Fig. 3.2.[3] Below we provide some additional invariant sets for (3.10). It is instructive to understand

[2]Initially, this type of problem (two independent, one-dimensional autonomous vector fields) might seem trivial and like a completely academic problem. However, we believe that there is quite a lot of insight that can be gained from such problems (that has been the case for the author). Generally, it is useful to think about breaking a problem up into smaller, understandable, pieces and then putting the pieces back together. Problems like this provide a controlled way of doing this. But also, these problems allow for exact computation by hand of concepts that do not lend themselves to such computations in the types of ODEs arising in typical applications. This gives some level of confidence that you understand the concept. Also, such examples could serve as useful benchmarks for numerical computations, since checking numerical methods against equations where you have an analytical solution to the equation can be very helpful.

[3]Make sure you understand why these constraints on the coordinates imply the existence of invariant lines.

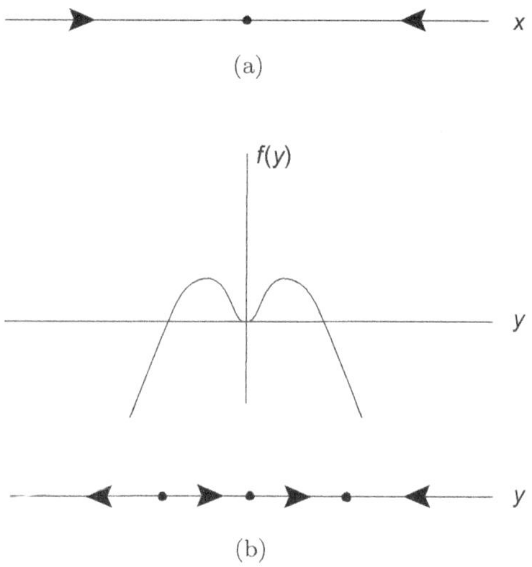

Fig. 3.1. (a) The phase "line" of the x component of (3.10). (b) The graph of $f(y)$ (top) and the phase "line" of the y component of (3.10) directly below.

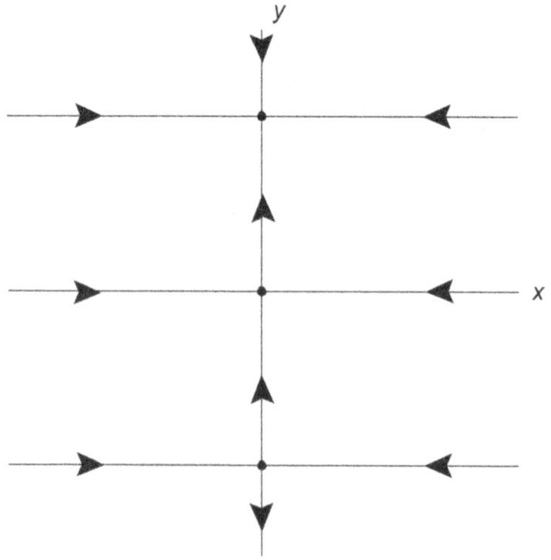

Fig. 3.2. Phase plane of (3.10). The black dots indicate equilibrium points.

why they are invariant, and whether or not there are other invariant sets.

Additional invariant sets for (3.10).

$\{(x, y) | -\infty < x < 0, -\infty < y < -1\}$,

$\{(x, y) | 0 < x < \infty, -\infty < y < -1\}$,

$\{(x, y) | -\infty < x < 0, -1 < y < 0\}$,

$\{(x, y) | 0 < x < \infty, -1 < y < 0\}$,

$\{(x, y) | -\infty < x < 0, 0 < y < 1\}$,

$\{(x, y) | 0 < x < \infty, 0 < y < 1\}$,

$\{(x, y) | -\infty < x < 0, 1 < y < \infty\}$,

$\{(x, y) | 0 < x < \infty, 1 < y < \infty\}$,

Problem Set 3

1. Consider the following autonomous vector field on $\mathbb{R}$:

$$\dot{x} = x - x^3, \quad x \in \mathbb{R}. \tag{3.11}$$

   - Compute all equilibria and determine their stability, i.e. are they Lyapunov stable, asymptotically stable, or unstable?
   - Compute the flow generated by (3.11) and verify the stability results for the equilibria directly from the flow.

2. Consider the following three autonomous ODEs on $\mathbb{R}$:

$$\dot{x} = -x, \quad \dot{x} = -x^2, \quad \dot{x} = -x^3.$$

   $x = 0$ is an equilibrium point for each vector field. For each vector field explain if $x = 0$ is Lyapunov stable, unstable, or asymptotically stable.
3. Consider the following autonomous vector field on $\mathbb{R}$:

$$\dot{x} = 2.$$

   Determine all of the equilibria for this vector field and discuss their stability.
4. Consider the following autonomous vector field on $\mathbb{R}$:

$$\dot{x} = 0.$$

   Is the closed unit interval, i.e. $[0, 1]$, and invariant set for this vector field? If so, discuss its stability. Does this vector field have other invariant sets?

5. Consider the following autonomous ODE on $\mathbb{R}$,

$$\dot{x} = \frac{1}{x}.$$

Sketch the direction of the vector field on $\mathbb{R}$ and determine all equilibria and characterize their stability.

6. Consider an autonomous vector field on $\mathbb{R}^n$[4]:

$$\dot{x} = f(x), \quad x \in \mathbb{R}^n. \tag{3.12}$$

Suppose $M \subset \mathbb{R}^n$ is a bounded, invariant set for (3.12). Let $\phi_t(\cdot)$ denote the flow generated by (3.12). Suppose $p \in \mathbb{R}^n$, $p \notin M$. Is it possible for

$$\phi_t(p) \in M,$$

for some finite t?

7. Consider the following vector field on the plane:

$$\dot{x} = x - x^3,$$
$$\dot{y} = -y, \quad (x, y) \in \mathbb{R}^2. \tag{3.13}$$

   (a) Determine zero-dimensional, one-dimensional, and two-dimensional invariant sets.
   (b) Determine the attracting sets and their basins of attraction.
   (c) Describe the heteroclinic orbits and compute analytical expressions for the heteroclinic orbits.
   (d) Does the vector field have periodic orbits?[5]
   (e) Sketch the phase portrait.[6]

8. Consider the following three one-dimensional autonomous vector fields:

   (a) $\dot{\theta} = \sin\theta$,
   (b) $\dot{\theta} = 1 + \sin\theta$,
   (c) $\dot{\theta} = 2 + \sin\theta$,

---

[4]This problem is "essentially the same" as Problem 1 from Problem Set 2.

[5]Keep in mind here that a trajectory is periodic if it is periodic, with the same period, *in each component.*

[6]There is a point to consider early on in this course. What exactly does "sketch the phase portrait mean"? It means sketching trajectories through different initial conditions in the phase space in such a way that a "complete picture" of all *qualitatively distinct* trajectories is obtained. The phrase *qualitatively distinct* is the key here since you can only sketch trajectories for a finite number of initial conditions (unless you sketch an invariant set or manifold) and only a finite length of the trajectory, unless the trajectory is a fixed point, periodic, homoclinic, or heteroclinic.

where θ is an angular variable. We consider these vector fields defined on

$$-\pi \leq \theta \leq \pi, \quad \mod 2\pi,$$

and mod 2π means that angles that differ by integer multiples of 2π are considered as the same value.

(a) Calculate the equilibria of these vector fields and sketch the phase portraits.
(b) Do any of these vector fields have periodic orbits?

9. Consider the following ODE:

$$\dot{x} = \sqrt{x}.$$

where x is a real number.

(a) What is the phase space of this ODE?
(b) What are the equilibrium points of this ODE?
(c) Sketch the equilibrium points and indicate their stability in the phase space of this ODE.

10. Consider the following two-dimensional linear autonomous ODE on $\mathbb{R}^2$,

$$\dot{x} = y,$$
$$\dot{y} = -x.$$

Consider the transformation from cartesian to polar coordinates defined by $x = r\cos\theta$, $y = r\sin\theta$. Express this ODE in polar coordinates in the form of $\dot{r}$ and $\dot{\theta}$ and show that the trajectories of the ODE in polar coordinates are circles. From this argue that the origin is Lyapunov stable.

Chapter 4

Behavior Near Trajectories: Linearization

NOW WE ARE GOING TO DISCUSS A METHOD FOR ANALYZING STABILITY THAT UTILIZES LINEARIZATION ABOUT THE OBJECT WHOSE STABILITY IS OF INTEREST. For now, the "objects of interest" are specific solutions of a vector field. The structure of the solutions of linear, constant coefficient systems is covered in many ODE textbooks. My favorite is the book by Hirsch *et al.*[1] It covers all of the linear algebra needed for analyzing linear ODEs that you probably did not cover in your linear algebra course. The book by Arnold[2] is also very good, but the presentation is more compact, with fewer examples.

We begin by considering a general nonautonomous vector field:

$$\dot{x} = f(x, t), \quad x \in \mathbb{R}^n, \tag{4.1}$$

and we suppose that

$$\bar{x}(t, t_0, x_0), \tag{4.2}$$

is the solution of (4.1) for which we wish to determine its stability properties. As when we introduced the definitions of stability, we proceed by localizing the vector field about the solution of interest. We do this by

[1] M. W. Hirsch, S. Smale, and R. L. Devaney. *Differential Equations, Dynamical Systems, and an Introduction to Chaos*. Academic Press, 2012.

[2] V. I. Arnold. *Ordinary Differential Equations*. M.I.T. Press, Cambridge, 1973. ISBN 0262010372.

introducing the change of coordinates

$$x = y + \bar{x}$$

for (4.1) as follows:

$$\dot{x} = \dot{y} + \dot{\bar{x}} = f(y + \bar{x}, t),$$

or

$$\begin{aligned} \dot{y} &= f(y + \bar{x}, t) - \dot{\bar{x}}, \\ &= f(y + \bar{x}, t) - f(\bar{x}, t), \end{aligned} \tag{4.3}$$

where we omit the arguments of $\bar{x}(t, t_0, x_0)$ for the sake of a less cumbersome notation. Next we Taylor expand $f(y + \bar{x}, t)$ in y about the solution $\bar{x}$, but we will only require the leading order terms explicitly[3]:

$$f(y + \bar{x}, t) = f(\bar{x}, t) + Df(\bar{x}, t)y + \mathcal{O}(|y|^2), \tag{4.4}$$

where Df denotes the derivative (i.e. Jacobian matrix) of the vector valued function f and $\mathcal{O}(|y|^2)$ denotes higher order terms in the Taylor expansion that we will not need in explicit form. Substituting this into (4.4) gives:

$$\begin{aligned} \dot{y} &= f(y + \bar{x}, t) - f(\bar{x}, t), \\ &= f(\bar{x}, t) + Df(\bar{x}, t)y + \mathcal{O}(|y|^2) - f(\bar{x}, t), \\ &= Df(\bar{x}, t)y + \mathcal{O}(|y|^2). \end{aligned} \tag{4.5}$$

Keep in mind that we are interested in the behavior of solutions near $\bar{x}(t, t_0, x_0)$, i.e. for y small. Therefore, in that situation it seems reasonable that neglecting the $\mathcal{O}(|y|^2)$ in (4.5) would be an approximation that would provide us with the particular information that we seek. For example, would it provide sufficient information for us to determine stability? In particular,

$$\dot{y} = Df(\bar{x}, t)y, \tag{4.6}$$

is referred to as the linearization of the vector field $\dot{x} = f(x, t)$ about the solution $\bar{x}(t, t_0, x_0)$.

[3]For the necessary background that you will need on Taylor expansions see Appendix A.

Before we answer the question as to whether or not (4.1) provides an adequate approximation to solutions of (4.5) for y "small", we will first study linear vector fields on their own.

Linear vector fields can also be classified as nonautonomous or autonomous. Nonautonomous linear vector fields are obtained by linearizing a nonautonomous vector field about a solution (and retaining only the linear terms). They have the general form:

$$\dot{y} = A(t)y, \quad y(0) = y_0, \tag{4.7}$$

where

$$A(t) \equiv Df(\bar{x}(t, t_0, x_0), t) \tag{4.8}$$

is a $n \times n$ matrix. They can also be obtained by linearizing an autonomous vector field about a time-dependent solution.

An autonomous linear vector field is obtained by linearizing an autonomous vector field about an equilibrium point. More precisely, let $\dot{x} = f(x)$ denote an autonomous vector field and let $x = x_0$ denote an equilibrium point, i.e. $f(x_0) = 0$. The linearized autonomous vector field about this equilibrium point has the form:

$$\dot{y} = Df(x_0)y, \quad y(0) = y_0, \tag{4.9}$$

or

$$\dot{y} = Ay, \quad y(0) = y_0, \tag{4.10}$$

where $A \equiv Df(x_0)$ is a $n \times n$ matrix of real numbers. This is significant because (4.10) can be solved using techniques of linear algebra, but (4.7), generally, cannot be solved in this manner. Hence, we will now describe the general solution of (4.10).

The general solution of (4.10) is given by:

$$y(t) = e^{At}y_0. \tag{4.11}$$

In order to verify that this is the solution, we merely need to substitute into the right-hand side and the left-hand side of (4.10) and show that equality holds. However, first we need to explain what e^{At} is, i.e. the exponential of the $n \times n$ matrix A (by examining (4.11) it should be clear that if (4.11) is to make sense mathematically, then e^{At} must be a $n \times n$ matrix).

Just like the exponential of a scalar, the exponential of a matrix is defined through the exponential series as follows:

$$\begin{aligned} e^{At} &\equiv \mathbb{I} + At + \frac{1}{2!}A^2t^2 + \cdots + \frac{1}{n!}A^nt^n + \cdots, \\ &= \sum_{i=0}^{\infty} \frac{1}{i!}A^it^i, \end{aligned} \tag{4.12}$$

where $\mathbb{I}$ denotes the $n \times n$ identity matrix. But we still must answer the question, "does this exponential series involving products of matrices make mathematical sense"? Certainly we can compute products of matrices and multiply them by scalars. But we have to give meaning to an infinite sum of such mathematical objects. We do this by defining the norm of a matrix and then considering the convergence of the series in norm. When this is done the "convergence problem" is exactly the same as that of the exponential of a scalar. Therefore the exponential series for a matrix converges absolutely for all t, and therefore it can be differentiated with respect to t term-by-term, and the resulting series of derivatives also converges absolutely.

Next we need to argue that (4.11) is a solution of (4.10). If we differentiate the series (4.12) term by term, we obtain that:

$$\frac{d}{dt}e^{At} = Ae^{At} = e^{At}A, \tag{4.13}$$

where we have used the fact that the matrices A and e^{At} commute (this is easy to deduce from the fact that A commutes with any power of A).[4] It then follows from this calculation that:

$$\dot{y} = \frac{d}{dt}e^{At}y_0 = Ae^{At}y_0 = Ay. \tag{4.14}$$

Therefore the general problem of solving (4.10) is equivalent to computing e^{At}, and we now turn our attention to this task.

First, suppose that A is a diagonal matrix, say

$$A = \begin{pmatrix} \lambda_1 & 0 & \cdots & 0 \\ 0 & \lambda_2 & \cdots & 0 \\ 0 & 0 & \ddots & 0 \\ 0 & 0 & \cdots & \lambda_n \end{pmatrix}. \tag{4.15}$$

[4] See Appendix B for another derivation of the solution of (4.10).

Then it is easy to see by substituting A into the exponential series (4.12) that:

$$e^{At} = \begin{pmatrix} e^{\lambda_1 t} & 0 & \cdots & 0 \\ 0 & e^{\lambda_2 t} & & 0 \\ 0 & 0 & \ddots & 0 \\ 0 & 0 & \cdots & e^{\lambda_n t} \end{pmatrix}. \tag{4.16}$$

Therefore our strategy will be to transform coordinates so that in the new coordinates A becomes diagonal (or as "close as possible" to diagonal, which we will explain shortly). Then e^{At} will be easily computable in these coordinates. Once that is accomplished, then we use the inverse of the transformation to transform the solution back to the original coordinate system.

Now we make these ideas precise. We let

$$y = Tu, \quad u \in \mathbb{R}^n, \ y \in \mathbb{R}^n, \tag{4.17}$$

where T is a $n \times n$ matrix whose precise properties will be developed in the following.

This is a typical approach in ODEs. We propose a general coordinate transformation of the ODE, and then we construct it in a way that gives the properties of the ODE that we desire. Substituting (4.17) into (4.10) gives:

$$\dot{y} = T\dot{u} = Ay = ATu, \tag{4.18}$$

T will be constructed in a way that makes it invertible, so that we have:

$$\dot{u} = T^{-1}ATu, \quad u(0) = T^{-1}y(0). \tag{4.19}$$

To simplify the notation we let:

$$\Lambda = T^{-1}AT, \tag{4.20}$$

or

$$A = T\Lambda T^{-1}. \tag{4.21}$$

Substituting (4.21) into the series for the matrix exponential (4.12) gives:

$$\begin{aligned} e^{At} &= e^{T\Lambda T^{-1} t}, \\ &= \mathbb{1} + T\Lambda T^{-1}t + \frac{1}{2!}\left(T\Lambda T^{-1}\right)^2 t^2 + \cdots + \frac{1}{n!}\left(T\Lambda T^{-1}\right)^n t^n + \cdots. \end{aligned} \tag{4.22}$$

Now note that for any positive integer n we have:

$$\left(T\Lambda T^{-1}\right)^n = \underbrace{\left(T\Lambda T^{-1}\right)\left(T\Lambda T^{-1}\right)\cdots\left(T\Lambda T^{-1}\right)\left(T\Lambda T^{-1}\right)}_{\text{n factors}},$$
$$= T\lambda^n T^{-1}. \tag{4.23}$$

Substituting this into (4.22) gives:

$$e^{At} = \sum_{n=0}^{\infty} \frac{1}{n!}\left(T\Lambda T^{-1}\right)^n t^n,$$
$$= T\left(\sum_{n=0}^{\infty} \frac{1}{n!}\Lambda^n t^n\right) T^{-1},$$
$$= Te^{\Lambda t}T^{-1}, \tag{4.24}$$

or

$$e^{At} = Te^{\Lambda t}T^{-1}. \tag{4.25}$$

Now we arrive at our main result. If T is constructed so that

$$\Lambda = T^{-1}AT, \tag{4.26}$$

is diagonal, then it follows from (4.16) and (4.25) that e^{At} can always be computed. So the ODE problem of solving (4.10) becomes a problem in linear algebra. But can a general $n \times n$ matrix A always be diagonalized? If you have had a course in linear algebra, you know that the answer to this question is "no". There is a theory of the (real) that will apply here. However, that would take us into too great a diversion for this course. Instead, we will consider the three standard cases for 2×2 matrices. That will suffice for introducing the main ideas without getting bogged down in linear algebra. Nevertheless, it cannot be avoided entirely. You will need to be able to compute eigenvalues and eigenvectors of 2×2 matrices, and understand their meaning.

The three cases of 2×2 matrices that we will consider are characterized by their eigenvalues:

- two real eigenvalues, diagonalizable A,
- two identical eigenvalues, nondiagonalizable A,
- a complex conjugate pair of eigenvalues.

In the table below we summarize the form that these matrices can be transformed to (referred to as that of A) and the resulting exponential of this canonical form.

eigenvalues of A	canonical form, Λ	e^{Λ}
λ, μ real, diagonalizable	$\begin{pmatrix} \lambda & 0 \\ 0 & \mu \end{pmatrix}$	$\begin{pmatrix} e^{\lambda} & 0 \\ 0 & e^{\mu} \end{pmatrix}$
$\lambda = \mu$ real, nondiagonalizable	$\begin{pmatrix} \lambda & 1 \\ 0 & \lambda \end{pmatrix}$	$\left(\mathbb{I} + \begin{pmatrix} 0 & 1 \\ 0 & 0 \end{pmatrix}\right) \begin{pmatrix} e^{\lambda} & 0 \\ 0 & e^{\lambda} \end{pmatrix}$
complex conjugate pair, $\alpha \pm i\beta$	$\begin{pmatrix} \alpha & -\beta \\ \beta & \alpha \end{pmatrix}$	$e^{\alpha} \begin{pmatrix} \cos\beta & -\sin\beta \\ \sin\beta & \cos\beta \end{pmatrix}$

Once the transformation to Λ has been carried out, we will use these results to deduce e^{Λ}.

Problem Set 4

1. Suppose Λ is a $n \times n$ matrix and T is a $n \times n$ invertible matrix. Use mathematical induction to show that:

$$\left(T^{-1}\Lambda T\right)^k = T^{-1}\Lambda^k T,$$

for all natural numbers k, i.e., $k = 1, 2, 3, \ldots$.

2. Suppose A is a $n \times n$ matrix. Use the exponential series to give an argument that:

$$\frac{d}{dt}e^{At} = Ae^{At}.$$

(You are allowed to use $e^{A(t+h)} = e^{At}e^{Ah}$ without proof, as well as the fact that A and e^{At} commute, without proof.)

3. Consider the following linear autonomous vector field:

$$\dot{x} = Ax, \quad x(0) = x_0, \quad x \in \mathbb{R}^n,$$

where A is a $n \times n$ matrix of real numbers.

- Show that the solutions of this vector field exist for all time.
- Show that the solutions are infinitely differentiable with respect to the initial condition, x_0.

See Footnote 5.

4. Consider the following linear autonomous vector field on the plane:

$$\begin{pmatrix} \dot{x}_1 \\ \dot{x}_2 \end{pmatrix} = \begin{pmatrix} 0 & 1 \\ 0 & 0 \end{pmatrix} \begin{pmatrix} x_1 \\ x_2 \end{pmatrix}, \quad (x_1(0), x_2(0)) = (x_{10}, x_{20}). \tag{4.27}$$

   (a) Describe the invariant sets.
   (b) Sketch the phase portrait.
   (c) Is the origin stable or unstable? Why?

5. Consider the following linear autonomous vector field on the plane:

$$\begin{pmatrix} \dot{x}_1 \\ \dot{x}_2 \end{pmatrix} = \begin{pmatrix} 0 & 0 \\ 0 & 0 \end{pmatrix} \begin{pmatrix} x_1 \\ x_2 \end{pmatrix}, \quad (x_1(0), x_2(0)) = (x_{10}, x_{20}). \tag{4.28}$$

   (a) Describe the invariant sets.
   (b) Sketch the phase portrait.
   (c) Is the origin stable or unstable? Why?

6. Compute e^A, where

$$A = \begin{pmatrix} \lambda & 1 \\ 0 & \lambda \end{pmatrix}.$$

   Hint. Write

$$A = \underbrace{\begin{pmatrix} \lambda & 0 \\ 0 & \lambda \end{pmatrix}}_{\equiv S} + \underbrace{\begin{pmatrix} 0 & 1 \\ 0 & 0 \end{pmatrix}}_{\equiv N}.$$

   Then

$$A \equiv S + N, \quad \text{and} \quad NS = SN.$$

[5]The next two problems often give students difficulties. There are no hard calculations involved. Just a bit of thinking. The solutions for each ODE can be obtained easily. Once these are obtained you just need to think about what they mean in terms of the concepts involved in the questions that you are asked, e.g. Lyapunov stability means that if you "start close, you stay close — forever".

Use the binomial expansion fo compute $(S+N)^n$, $n \geq 1$,

$$(S+N)^n = \sum_{k=0}^{n} \binom{n}{k} S^k N^{n-k},$$

where

$$\binom{n}{k} \equiv \frac{n!}{k!(n-k)!}$$

and substitute the results into the exponential series.[6]

7. Let

$$A = \begin{pmatrix} \alpha & -\beta \\ \beta & \alpha \end{pmatrix}$$

Show that

$$e^A = e^\alpha \begin{pmatrix} \cos\beta & -\sin\beta \\ \sin\beta & \cos\beta \end{pmatrix}$$

Hint: The eigenvalues of A are $\alpha \pm i\beta$. Compute the *complex* eigenvectors for each eigenvalue. These should be

$$\begin{pmatrix} 1 \\ -i \end{pmatrix}$$

and

$$\begin{pmatrix} 1 \\ i \end{pmatrix},$$

respectively. Form the transformation matrix T with these complex eigenvectors as the columns. Compute T^{-1}.

Then

$$A = T\Lambda T^{-1},$$

where

$$\Lambda = \begin{pmatrix} \alpha + i\beta & 0 \\ 0 & \alpha - i\beta \end{pmatrix}.$$

---

[6] N is an example of a nilpotent matrix. A square matrix is said to be nilpotent if N^k is the zero matrix (i.e. a matrix all of whose entries are zero) for some positive integer k. The smallest such k is called the index of N or the degree of N.

Now since Λ is diagonal we have

$$e^{\Lambda} = \begin{pmatrix} e^{(\alpha+i\beta)} & 0 \\ 0 & e^{(\alpha-i\beta)} \end{pmatrix} = e^{\alpha} \begin{pmatrix} e^{i\beta} & 0 \\ 0 & e^{-i\beta} \end{pmatrix}.$$

Now use Euler's formula:

$$e^{i\beta} = \cos\beta + i\sin\beta, \quad e^{-i\beta} = \cos\beta - i\sin\beta,$$

and substitute this into the expression for e^{Λ}. Finally, you can work out

$$e^{A} = Te^{\Lambda}T^{-1}.$$

8. The equation for a damped and driven harmonic oscillator is given by:

$$m\ddot{x} + \gamma\dot{x} + kx = f(t).$$

where m, γ, k are positive and $f(t)$ is a bounded continuous function.

We want to first simplify the constants a bit. We divide through by m:

$$\ddot{x} + \frac{\gamma}{m}\dot{x} + \frac{k}{m}x = \frac{f(t)}{m}.$$

We then redefine certain of the constants:

$$\omega_0 = \sqrt{\frac{k}{m}}, \quad 2\zeta\omega_0 = \frac{\gamma}{m}, \quad g(t) = \frac{f(t)}{m},$$

after which the equation becomes:

$$\ddot{x} + 2\zeta\omega_0\dot{x} + \omega_0^2 x = g(t),$$

and this is the equation that this problem will be dealing with.

(a) Show that this equation can be written in the form:

$$\dot{u} = Au + h(t),$$

where $u \in \mathbb{R}^2$ and A is a 2×2 matrix.

What is the general form of the solutions of ODEs of this type? (Recall Appendix B.)

(b) Compute the eigenvalues of A.

Chapter 5

Behavior Near Equilibria: Linearization

Now we will consider several examples for solving, and understanding, the nature of the solutions, of

$$\dot{x} = Ax, \quad x \in \mathbb{R}^2. \tag{5.1}$$

For all of the examples, the method for solving the system is the same.

Step 1. Compute the eigenvalues of A.
Step 2. Compute the eigenvectors of A.
Step 3. Use the eigenvectors of A to form the transformation matrix T.
Step 4. Compute $\Lambda = T^{-1}AT$.
Step 5. Compute $e^{At} = Te^{\Lambda t}T^{-1}$.

Once we have computed e^{At} we have the solution of (5.1) through any initial condition x_0 since $x(t)$, $x(0) = x_0$, is given by $x(t) = e^{At}x_0$.[1]

Example 10. We consider the following linear, autonomous ODE:

$$\begin{pmatrix} \dot{x}_1 \\ \dot{x}_2 \end{pmatrix} = \begin{pmatrix} 2 & 1 \\ 1 & 2 \end{pmatrix} \begin{pmatrix} x_1 \\ x_2 \end{pmatrix}, \tag{5.2}$$

where

$$A \equiv \begin{pmatrix} 2 & 1 \\ 1 & 2 \end{pmatrix}. \tag{5.3}$$

[1]Most of the linear algebra techniques necessary for this material is covered in Appendix A.

Step 1. Compute the eigenvalues of A.

The eigenvalues of A, denote by λ, are given by the solutions of the characteristic polynomial:

$$\det\begin{pmatrix} 2-\lambda & 1 \\ 1 & 2-\lambda \end{pmatrix} = (2-\lambda)^2 - 1 = 0,$$
$$= \lambda^2 - 4\lambda + 3 = 0, \tag{5.4}$$

or

$$\lambda_{1,2} = 2 \pm \frac{1}{2}\sqrt{16-12} = 3,\ 1.$$

Step 2. Compute the eigenvectors of A.

For each eigenvalue, we compute the corresponding eigenvector. The eigenvector correponding to the eigenvalue 3 is found by solving:

$$\begin{pmatrix} 2 & 1 \\ 1 & 2 \end{pmatrix}\begin{pmatrix} x_1 \\ x_2 \end{pmatrix} = 3\begin{pmatrix} x_1 \\ x_2 \end{pmatrix}, \tag{5.5}$$

or,

$$2x_1 + x_2 = 3x_1, \tag{5.6}$$
$$x_1 + 2x_2 = 3x_2. \tag{5.7}$$

Both of these equations yield the same equation since the two equations are dependent:

$$x_2 = x_1. \tag{5.8}$$

Therefore we take as the eigenvector corresponding to the eigenvalue 3:

$$\begin{pmatrix} 1 \\ 1 \end{pmatrix}. \tag{5.9}$$

Next we compute the eigenvector corresponding to the eigenvalue 1. This is given by a solution to the following equations:

$$\begin{pmatrix} 2 & 1 \\ 1 & 2 \end{pmatrix}\begin{pmatrix} x_1 \\ x_2 \end{pmatrix} = \begin{pmatrix} x_1 \\ x_2 \end{pmatrix}, \tag{5.10}$$

or

$$2x_1 + x_2 = x_1, \tag{5.11}$$

$$x_1 + 2x_2 = x_2 \tag{5.12}$$

Both of these equations yield the same equation:

$$x_2 = -x_1. \tag{5.13}$$

Therefore we take as the eigenvector corresponding to the eigenvalue 1:

$$\begin{pmatrix} 1 \\ -1 \end{pmatrix}. \tag{5.14}$$

Step 3. Use the eigenvectors of *A* to form the transformation matrix *T*.

For the columns of T we take the eigenvectors corresponding the the eigenvalues 1 and 3:

$$T = \begin{pmatrix} 1 & 1 \\ -1 & 1 \end{pmatrix}, \tag{5.15}$$

with the inverse given by:

$$T^{-1} = \frac{1}{2}\begin{pmatrix} 1 & -1 \\ 1 & 1 \end{pmatrix}. \tag{5.16}$$

Step 4. Compute $\Lambda = T^{-1}AT$.

We have:

$$\begin{aligned} T^{-1}AT &= \frac{1}{2}\begin{pmatrix} 1 & -1 \\ 1 & 1 \end{pmatrix}\begin{pmatrix} 2 & 1 \\ 1 & 2 \end{pmatrix}\begin{pmatrix} 1 & 1 \\ -1 & 1 \end{pmatrix}, \\ &= \frac{1}{2}\begin{pmatrix} 1 & -1 \\ 1 & 1 \end{pmatrix}\begin{pmatrix} 1 & 3 \\ -1 & 3 \end{pmatrix}, \\ &= \begin{pmatrix} 1 & 0 \\ 0 & 3 \end{pmatrix} \equiv \Lambda. \end{aligned} \tag{5.17}$$

Therefore, in the $u_1 - u_2$ coordinates (5.2) becomes:

$$\begin{pmatrix} \dot{u}_1 \\ \dot{u}_2 \end{pmatrix} = \begin{pmatrix} 1 & 0 \\ 0 & 3 \end{pmatrix}\begin{pmatrix} u_1 \\ u_2 \end{pmatrix}. \tag{5.18}$$

In the $u_1 - u_2$ coordinates it is easy to see that the origin is an unstable equilibrium point.

Step 5. Compute $e^{At} = Te^{\Lambda t}T^{-1}$.

We have:

$$\begin{aligned} e^{At} &= \frac{1}{2}\begin{pmatrix} 1 & 1 \\ -1 & 1 \end{pmatrix}\begin{pmatrix} e^t & 0 \\ 0 & e^{3t} \end{pmatrix}\begin{pmatrix} 1 & -1 \\ 1 & 1 \end{pmatrix}, \\ &= \frac{1}{2}\begin{pmatrix} 1 & 1 \\ -1 & 1 \end{pmatrix}\begin{pmatrix} e^t & -e^t \\ e^{3t} & e^{3t} \end{pmatrix}, \\ &= \frac{1}{2}\begin{pmatrix} e^t + e^{3t} & -e^t + e^{3t} \\ -e^t + e^{3t} & e^t + e^{3t} \end{pmatrix}. \end{aligned} \tag{5.19}$$

We see that the origin is also *unstable* in the original $x_1 - x_2$ coordinates. It is referred to as a source, and this is characterized by the fact that all of the eigenvalues of A have positive real part. The phase portrait is illustrated in Fig. 5.1.

We remark this it is possible to infer the behavior of e^{At} as $t \to \infty$ from the behavior of $e^{\Lambda t}$ as $t \to \infty$ since T does not depend on t.

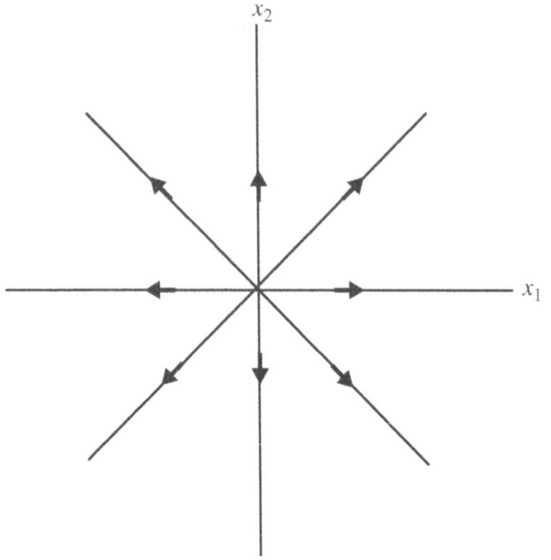

Fig. 5.1. Phase plane of (5.19). The origin is unstable — a source.

Example 11. We consider the following linear, autonomous ODE:

$$\begin{pmatrix} \dot{x}_1 \\ \dot{x}_2 \end{pmatrix} = \begin{pmatrix} -1 & -1 \\ 9 & -1 \end{pmatrix} \begin{pmatrix} x_1 \\ x_2 \end{pmatrix}, \tag{5.20}$$

where

$$A \equiv \begin{pmatrix} -1 & -1 \\ 9 & -1 \end{pmatrix}. \tag{5.21}$$

Step 1. Compute the eigenvalues of A.

The eigenvalues of A, denote by λ, are given by the solutions of the characteristic polynomial:

$$\begin{aligned} \det \begin{pmatrix} -1-\lambda & -1 \\ 9 & -1-\lambda \end{pmatrix} &= (-1-\lambda)^2 + 9 = 0, \\ &= \lambda^2 + 2\lambda + 10 = 0. \end{aligned} \tag{5.22}$$

or,

$$\lambda_{1,\,2} = -1 \pm \frac{1}{2}\sqrt{4-40} = -1 \pm 3i.$$

The eigenvectors are complex, so we know it is not diagonalizable over the real numbers. What this means is that we cannot find real eigenvectors so that it can be transformed to a form where there are real numbers on the diagonal, and zeros in the off diagonal entries. The best we can do is to transform it to a form where the real parts of the eigenvalue are on the diagonal, and the imaginary parts are on the off diagonal locations, but the off diagonal elements differ by a minus sign.

Step 2. Compute the eigenvectors of A.

The eigenvector of A corresponding to the eigenvector $-1-3i$ is the solution of the following equations:

$$\begin{pmatrix} -1 & -1 \\ 9 & -1 \end{pmatrix} \begin{pmatrix} x_1 \\ x_2 \end{pmatrix} = (-1-3i) \begin{pmatrix} x_1 \\ x_2 \end{pmatrix} \tag{5.23}$$

or,

$$-x_1 - x_2 = -x_1 - 3ix_1, \tag{5.24}$$

$$9x_1 - x_2 = -x_2 - 3ix_2. \tag{5.25}$$

A solution to these equations is given by:

$$\begin{pmatrix} 1 \\ 3i \end{pmatrix} = \begin{pmatrix} 1 \\ 0 \end{pmatrix} + i \begin{pmatrix} 0 \\ 3 \end{pmatrix}.$$

Step 3. Use the eigenvectors of A to form the transformation matrix T.

For the first column of T we take the real part of the eigenvector corresponding to the eigenvalue $-1-3i$, and for the second column we take the complex part of the eigenvector:

$$T = \begin{pmatrix} 1 & 0 \\ 0 & 3 \end{pmatrix}, \tag{5.26}$$

with inverse

$$T^{-1} = \begin{pmatrix} 1 & 0 \\ 0 & \frac{1}{3} \end{pmatrix}. \tag{5.27}$$

Step 4. Compute $\Lambda = T^{-1}AT$.

We have:

$$\begin{aligned} T^{-1}AT &= \begin{pmatrix} 1 & 0 \\ 0 & \frac{1}{3} \end{pmatrix} \begin{pmatrix} -1 & -1 \\ 9 & -1 \end{pmatrix} \begin{pmatrix} 1 & 0 \\ 0 & 3 \end{pmatrix}, \\ &= \begin{pmatrix} 1 & 0 \\ 0 & \frac{1}{3} \end{pmatrix} \begin{pmatrix} -1 & -3 \\ 9 & -3 \end{pmatrix}, \\ &= \begin{pmatrix} -1 & -3 \\ 3 & -1 \end{pmatrix} \equiv \Lambda. \end{aligned} \tag{5.28}$$

With Λ in this form, we know from the previous chapter that:

$$e^{\Lambda t} = e^{-t} \begin{pmatrix} \cos 3t & -\sin 3t \\ \sin 3t & \cos 3t \end{pmatrix}. \tag{5.29}$$

Then we have:

$$e^{At} = Te^{\Lambda t}T^{-1}.$$

From this expression we can conclude that $e^{At} \to 0$ as $t \to \infty$. Hence the origin is asymptotically stable. It is referred to as a sink and it is characterized by the real parts of the eigenvalues of A being negative. The phase plane is sketched in Fig. 5.2.

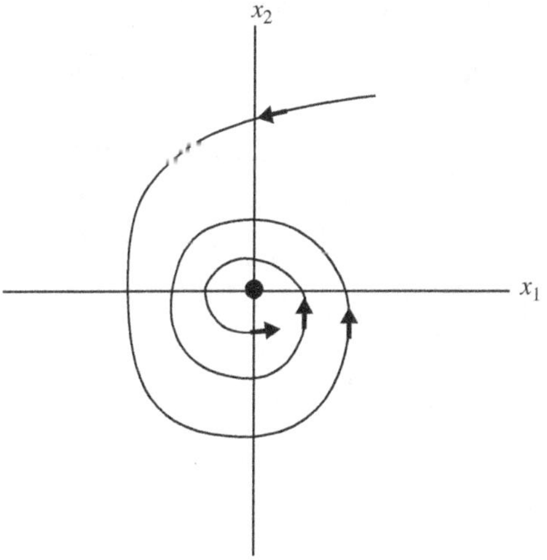

Fig. 5.2. Phase plane of (5.20). The origin is a sink.

Example 12. We consider the following linear autonomous ODE:

$$\begin{pmatrix} \dot{x}_1 \\ \dot{x}_2 \end{pmatrix} = \begin{pmatrix} -1 & 1 \\ 1 & 1 \end{pmatrix} \begin{pmatrix} x_1 \\ x_2 \end{pmatrix}, \tag{5.30}$$

where

$$A = \begin{pmatrix} -1 & 1 \\ 1 & 1 \end{pmatrix}. \tag{5.31}$$

Step 1. Compute the eigenvalues of *A*.

The eigenvalues are given by the solution of the characteristic equation:

$$\begin{aligned} \det \begin{pmatrix} -1-\lambda & 1 \\ 1 & 1-\lambda \end{pmatrix} &= (-1-\lambda)(1-\lambda) - 1 = 0, \\ &= \lambda^2 - 2 = 0, \end{aligned} \tag{5.32}$$

which are:

$$\lambda_{1,\,2} = \pm\sqrt{2}$$

Step 2. Compute the eigenvectors of A.

The eigenvector corresponding to the eigenvalue $\sqrt{2}$ is given by the solution of the following equations:

$$\begin{pmatrix} -1 & 1 \\ 1 & 1 \end{pmatrix} \begin{pmatrix} x_1 \\ x_2 \end{pmatrix} = \sqrt{2} \begin{pmatrix} x_1 \\ x_2 \end{pmatrix} \tag{5.33}$$

or

$$-x_1 + x_2 = \sqrt{2}x_1, \tag{5.34}$$

$$x_1 + x_2 = \sqrt{2}x_2. \tag{5.35}$$

A solution is given by:

$$x_2 = (1 + \sqrt{2})x_1,$$

corresponding to the eigenvector:

$$\begin{pmatrix} 1 \\ 1 + \sqrt{2} \end{pmatrix}.$$

The eigenvector corresponding to the eigenvalue $-\sqrt{2}$ is given by the solution to the following equations:

$$\begin{pmatrix} -1 & 1 \\ 1 & 1 \end{pmatrix} \begin{pmatrix} x_1 \\ x_2 \end{pmatrix} = -\sqrt{2} \begin{pmatrix} x_1 \\ x_2 \end{pmatrix}, \tag{5.36}$$

or

$$-x_1 + x_2 = -\sqrt{2}x_1, \tag{5.37}$$

$$x_1 + x_2 = -\sqrt{2}x_2. \tag{5.38}$$

A solution is given by:

$$x_2 = (1 - \sqrt{2})x_1,$$

corresponding to the eigenvector:

$$\begin{pmatrix} 1 \\ 1 - \sqrt{2} \end{pmatrix}.$$

Step 3. Use the eigenvectors of A to form the transformation matrix T.

For the columns of T we take the eigenvectors corresponding to the eigenvalues $\sqrt{2}$ and $-\sqrt{2}$.

$$T = \begin{pmatrix} 1 & 1 \\ 1+\sqrt{2} & 1-\sqrt{2} \end{pmatrix}, \tag{5.39}$$

with T^{-1} given by:

$$T^{-1} = -\frac{1}{2\sqrt{2}} \begin{pmatrix} 1-\sqrt{2} & -1 \\ -1-\sqrt{2} & 1 \end{pmatrix}. \tag{5.40}$$

Step 4. Compute $\Lambda = T^{-1}AT$.

We have:

$$\begin{aligned} T^{-1}AT &= -\frac{1}{2\sqrt{2}} \begin{pmatrix} 1-\sqrt{2} & -1 \\ -1-\sqrt{2} & 1 \end{pmatrix} \begin{pmatrix} -1 & 1 \\ 1 & 1 \end{pmatrix} \begin{pmatrix} 1 & 1 \\ 1+\sqrt{2} & 1-\sqrt{2} \end{pmatrix}, \\ &= -\frac{1}{2\sqrt{2}} \begin{pmatrix} 1-\sqrt{2} & -1 \\ -1-\sqrt{2} & 1 \end{pmatrix} \begin{pmatrix} \sqrt{2} & -\sqrt{2} \\ 2+\sqrt{2} & 2-\sqrt{2} \end{pmatrix}, \\ &= -\frac{1}{2\sqrt{2}} \begin{pmatrix} -4 & 0 \\ 0 & 4 \end{pmatrix} = \begin{pmatrix} \sqrt{2} & 0 \\ 0 & -\sqrt{2} \end{pmatrix} \equiv \Lambda. \end{aligned} \tag{5.41}$$

Therefore in the $u_1 - u_2$ coordinates (5.30) takes the form:

$$\begin{pmatrix} \dot{u}_1 \\ \dot{u}_2 \end{pmatrix} = \begin{pmatrix} \sqrt{2} & 0 \\ 0 & -\sqrt{2} \end{pmatrix} \begin{pmatrix} u_1 \\ u_2 \end{pmatrix}. \tag{5.42}$$

The phase portrait of (5.42) is shown in Fig. 5.3.

It is easy to see that the origin is unstable for (5.42). In Fig. 5.3 we see that the origin has the structure of a saddle point, and we want to explore this idea further.

In the $u_1 - u_2$ coordinates the span of the eigenvector corresponding to the eigenvalue $\sqrt{2}$ is given by $u_2 = 0$, i.e. the u_1-axis. The span of the eigenvector corresponding to the eigenvalue $-\sqrt{2}$ is given by $u_1 = 0$, i.e. the u_2-axis. Moreover, we can see from the form of (5.42) that these coordinate axes are invariant. The u_1-axis is referred to as the unstable subspace, denoted by E^u, and the u_2-axis is referred to as the stable subspace, denoted by E^s. In other words, the unstable subspace is the span

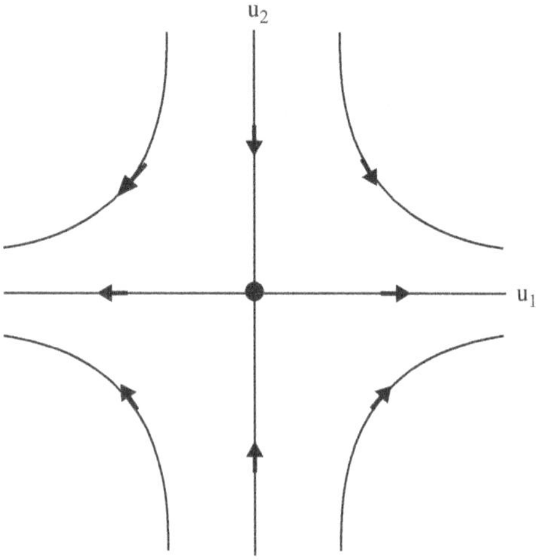

Fig. 5.3. Phase portrait of (5.42).

of the eigenvector corresponding to the eigenvalue with positive real part and the stable subspace is the span of the eigenvector corresponding to the eigenvalue having negative real part. The stable and unstable subspaces are invariant subspaces with respect to the flow generated by (5.42).

The stable and unstable subspaces correspond to the coordinate axes in the coordinate system given by the eigenvectors. Next we want to understand how they would appear in the original x_1–x_2 coordinates. This is accomplished by transforming them to the original coordinates using the transformation matrix (5.39).

We first transform the unstable subspace from the u_1–u_2 coordinates to the x_1–x_2 coordinates. In the u_1–u_2 coordinates points on the unstable subspace have coordinates $(u_1, 0)$. Acting on these points with T gives:

$$T\begin{pmatrix} u_1 \\ 0 \end{pmatrix} = \begin{pmatrix} 1 & 1 \\ 1+\sqrt{2} & 1-\sqrt{2} \end{pmatrix}\begin{pmatrix} u_1 \\ 0 \end{pmatrix} = \begin{pmatrix} x_1 \\ x_2 \end{pmatrix}, \tag{5.43}$$

which gives the following relation between points on the unstable subspace in the u_1–u_2 coordinates to points in the x_1–x_2 coordinates:

$$u_1 = x_1 \tag{5.44}$$

$$(1+\sqrt{2})u_1 = x_2, \tag{5.45}$$

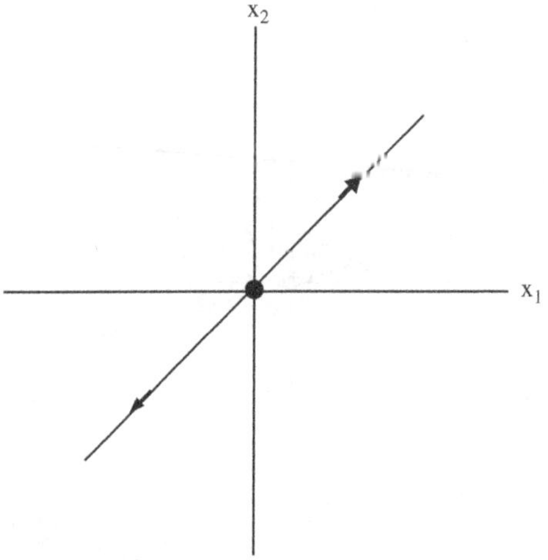

Fig. 5.4. The unstable subspace in the original coordinates.

or

$$(1 + \sqrt{2})x_1 = x_2. \tag{5.46}$$

This is the equation for the unstable subspace in the x_1–x_2 coordinates, which we illustrate in Fig. 5.4.

Next we transform the stable subspace from the u_1–u_2 coordinates to the x_1–x_2 coordinates. In the u_1–u_2 coordinates points on the stable subspace have coordinates $(0, u_2)$. Acting on these points with T gives:

$$T\begin{pmatrix} 0 \\ u_2 \end{pmatrix} = \begin{pmatrix} 1 & 1 \\ 1+\sqrt{2} & 1-\sqrt{2} \end{pmatrix}\begin{pmatrix} 0 \\ u_2 \end{pmatrix} = \begin{pmatrix} x_1 \\ x_2 \end{pmatrix}, \tag{5.47}$$

which gives the following relation between points on the stable subspace in the u_1–u_2 coordinates to points in the x_1–x_2 coordinates:

$$u_2 = x_1 \tag{5.48}$$

$$(1 - \sqrt{2})u_2 = x_2 \tag{5.49}$$

or

$$(1 - \sqrt{2})x_1 = x_2. \tag{5.50}$$

This is the equation for the stable subspace in the x_1–x_2 coordinates, which we illustrate in Fig. 5.5.

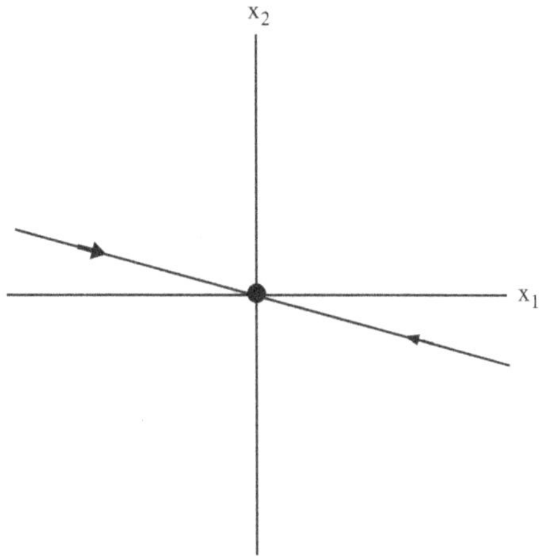

Fig. 5.5. The stable subspace in the original coordinates.

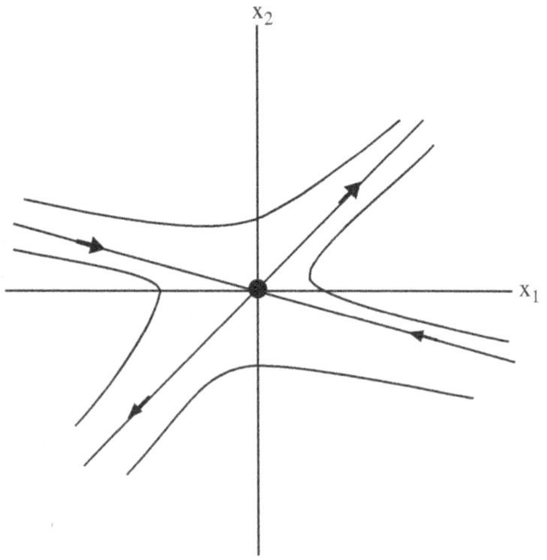

Fig. 5.6. The stable and unstable subspaces in the original coordinates.

In Fig. 5.6 we illustrate both the stable and the unstable subspaces in the original coordinates.

Now we want to discuss some general results from these three examples. For all three examples, the real parts of the eigenvalues of A were nonzero, and stability of the origin was determined by the sign of the real parts of the eigenvalues, e.g. for Example 10, the origin was unstable (the real parts of the eigenvalues of A were positive), for Example 11, the origin was stable (the real parts of the eigenvalues of A were negative), and for Example 12, the origin was unstable (A had one positive eigenvalue and one negative eigenvalue). This is generally true for all linear, autonomous vector fields. We state this more formally.

Consider a linear, autonomous vector field on $\mathbb{R}^n$:

$$\dot{y} = Ay, \quad y(0) = y_0, \quad y \in \mathbb{R}^n. \tag{5.51}$$

Then if A has no eigenvalues having zero real parts, the stability of the origin is determined by the real parts of the eigenvalues of A. If all of the real parts of the eigenvalues are strictly less than zero, then the origin is asymptotically stable. If at least one of the eigenvalues of A has real part strictly larger than zero, then the origin is unstable.

There is a term applied to this terminology that permeates all of dynamical systems theory.

Definition 20 (Hyperbolic Equilibrium Point). The origin of (5.51) is said to be hyperbolic if none of the real parts of the eigenvalues of A have zero real parts.

It follows that hyperbolic equilibria of linear, autonomous vector fields on $\mathbb{R}^n$ can be either sinks, sources, or saddles. The key point is that the eigenvalues of A *all* have nonzero real parts.

If we restrict ourselves to two dimensions, it is possible to make a (short) list of all of the distinct canonical forms for A. These are given by the following six 2×2 matrices.

The first is a diagonal matrix with real, nonzero eigenvalues λ, $\mu \neq 0$, i.e. the origin is a hyperbolic fixed point:

$$\begin{pmatrix} \lambda & 0 \\ 0 & \mu \end{pmatrix}. \tag{5.52}$$

In this case the orgin can be a sink if both eigenvalues are negative, a source if both eigenvalues are positive, and a saddle if the eigenvalues have opposite sign.

The next situation corresponds to complex eigenvalues, with the real part, α, and imaginary part, β, both being nonzero. In this case, the equilibrium point is hyperbolic, and a sink for $\alpha < 0$, and a source for $\alpha > 0$.

The sign of β does not influence stability:

$$\begin{pmatrix} \alpha & \beta \\ -\beta & \alpha \end{pmatrix}. \tag{5.53}$$

Next we consider the case when the eigenvalues are real, identical, and nonzero, but the matrix is nondiagonalizable, i.e. two eigenvectors cannot be found. In this case, the origin is hyperbolic for $\lambda \neq 0$, and is a sink for $\lambda < 0$ and a source for $\lambda > 0$:

$$\begin{pmatrix} \lambda & 1 \\ 0 & \lambda \end{pmatrix}. \tag{5.54}$$

Next, we consider some cases corresponding to the origin being non-hyperbolic that would have been possible to include in the discussion of earlier cases, but it is more instructive to explicitly point out these cases separately.

We first consider the case where A is diagonalizable with one nonzero real eigenvalue and one zero eigenvalue:

$$\begin{pmatrix} \lambda & 0 \\ 0 & 0 \end{pmatrix}. \tag{5.55}$$

We consider the case where the two eigenvalues are purely imaginary, $\pm i\sqrt{\beta}$. In this case the origin is referred to as a center.

$$\begin{pmatrix} 0 & \beta \\ -\beta & 0 \end{pmatrix}. \tag{5.56}$$

For completeness, we consider the case where both eigenvalues are zero and A is diagonal.

$$\begin{pmatrix} 0 & 0 \\ 0 & 0 \end{pmatrix}. \tag{5.57}$$

Finally, we want to expand on the discussion related to the geometrical aspects of Example 12. Recall that for that example the span of the eigenvector corresponding to the eigenvalue with negative real part was an invariant subspace, referred to as the stable subspace. Trajectories with initial conditions in the stable subspace decayed to zero at an exponential rate as $t \to +\infty$. The stable invariant subspace was denoted by E^s. Similarly, the span of the eigenvector corresponding to the eigenvalue with positive real part was an invariant subspace, referred to as the unstable subspace.

Trajectories with initial conditions in the unstable subspace decayed to zero at an exponential rate as $t \to -\infty$. The unstable invariant subspace was denoted by E^u.

We can easily see that (5.52) has this behavior when λ and μ have opposite signs. If λ and μ are both negative, then the span of the eigenvectors corresponding to these two eigenvalues is $\mathbb{R}^2$, and the entire phase space is the stable subspace. Similarly, if λ and μ are both positive, then the span of the eigenvectors corresponding to these two eigenvalues is $\mathbb{R}^2$, and the entire phase space is the unstable subspace.

The case (5.53) is similar. For that case there is not a pair of real eigenvectors corresponding to each of the complex eigenvalues. The vectors that transform the original matrix to this canonical form are referred to as *generalized eigenvectors*. If $\alpha < 0$ the span of the generalized eigenvectors is $\mathbb{R}^2$, and the entire phase space is the stable subspace. Similarly, if $\alpha > 0$ the span of the generalized eigenvectors is $\mathbb{R}^2$, and the entire phase space is the unstable subspace. The situation is similar for (5.54). For $\lambda < 0$ the entire phase space is the stable subspace, for $\lambda > 0$ the entire phase space is the unstable subspace.

The case (5.55) is different. The span of the eigenvector corresponding to λ is the stable subspace for $\lambda < 0$, and the unstable subspace for $\lambda > 0$. The space of the eigenvector corresponding to the zero eigenvalue is referred to as the center subspace.

For the case (5.56) there are no two real eigenvectors leading to the resulting canonical form. Rather, there are two generalized eigenvectors associated with this pair of complex eigenvalues having zero real part. The span of these two eigenvectors is a two-dimensional center subspace corresponding to $\mathbb{R}^2$. An equilibrium point with purely imaginary eigenvalues is referred to as a center.

Finally, the case (5.57) is included for completeness. It is the zero vector field where $\mathbb{R}^2$ is the center subspace.

We can characterize stability of the origin in terms of the stable, unstable, and center subspaces. The origin is asymptotically stable if $E^u = \emptyset$ and $E^c = \emptyset$. The origin is unstable if $E^u \neq \emptyset$.

Problem Set 5

1. Suppose A is a $n \times n$ matrix of real numbers. Show that if λ is an eigenvalue of A with eigenvector e, then $\bar{\lambda}$ is an eigenvalue of A with eigenvector $\bar{e}$.

2. Consider the matrices:

$$A_1 = \begin{pmatrix} 0 & -\omega \\ \omega & 0 \end{pmatrix}, \quad A_2 = \begin{pmatrix} 0 & \omega \\ -\omega & 0 \end{pmatrix}, \quad \omega > 0. \tag{5.58}$$

Sketch the trajectories of the associated linear autonomous ordinary differential equations:

$$\begin{pmatrix} \dot{x}_1 \\ \dot{x}_2 \end{pmatrix} = A_i \begin{pmatrix} x_1 \\ x_2 \end{pmatrix}, \quad i = 1, 2. \tag{5.59}$$

3. Consider the matrix

$$A = \begin{pmatrix} -1 & -1 \\ 9 & -1 \end{pmatrix} \tag{5.60}$$

(a) Show that the eigenvalues and eigenvectors are given by:

$$\begin{aligned} -1 - 3i : \quad \begin{pmatrix} 1 \\ 3i \end{pmatrix} &= \begin{pmatrix} 1 \\ 0 \end{pmatrix} + i \begin{pmatrix} 0 \\ 3 \end{pmatrix} \\ -1 + 3i : \quad \begin{pmatrix} 1 \\ -3i \end{pmatrix} &= \begin{pmatrix} 1 \\ 0 \end{pmatrix} - i \begin{pmatrix} 0 \\ 3 \end{pmatrix}. \end{aligned} \tag{5.61}$$

(b) Consider the four matrices:

$$\begin{aligned} T_1 &= \begin{pmatrix} 1 & 0 \\ 0 & -3 \end{pmatrix}, \quad T_2 = \begin{pmatrix} 1 & 0 \\ 0 & 3 \end{pmatrix}, \\ T_3 &= \begin{pmatrix} 0 & 1 \\ -3 & 0 \end{pmatrix}, \quad T_4 = \begin{pmatrix} 0 & 1 \\ 3 & 0 \end{pmatrix}. \end{aligned} \tag{5.62}$$

Compute $\Lambda_i = T_i^{-1} A T_i$, $i = 1, \ldots, 4$.

(c) Discuss the form of T in terms of the eigenvectors of A.[2]

4. Consider the following two-dimensional linear autonomous vector field:

$$\begin{pmatrix} \dot{x}_1 \\ \dot{x}_2 \end{pmatrix} = \begin{pmatrix} -2 & 1 \\ -5 & 2 \end{pmatrix} \begin{pmatrix} x_1 \\ x_2 \end{pmatrix}, \quad (x_1(0), x_2(0)) = (x_{10}, x_{20}). \tag{5.63}$$

Show that the origin is Lyapunov stable. Compute and sketch the trajectories.

[2]The point ot this problem is to show that when you have complex eigenvalues (in the 2×2 case) there is a great deal of freedom in how you compute the real canonical form.

5. Consider the following two-dimensional linear autonomous vector field:

$$\begin{pmatrix} \dot{x}_1 \\ \dot{x}_2 \end{pmatrix} = \begin{pmatrix} 1 & 2 \\ 2 & 1 \end{pmatrix} \begin{pmatrix} x_1 \\ x_2 \end{pmatrix}, \quad (x_1(0), x_2(0)) = (x_{10}, x_{20}). \tag{5.64}$$

Show that the origin is a saddle. Compute the stable and unstable subspaces of the origin in the original coordinates, i.e. the $x_1 - x_2$ coordinates. Sketch the trajectories in the phase plane.

6. Consider the following two-dimensional autonomous ODE on $\mathbb{R}^2$,

$$\begin{pmatrix} \dot{x}_1 \\ \dot{x}_2 \end{pmatrix} = \begin{pmatrix} 1 & 0 \\ 0 & -1 \end{pmatrix} \begin{pmatrix} x_1 \\ x_2 \end{pmatrix},$$

with initial condition

$$(x_1(0), x_2(0)) = (x_{10}, x_{20}).$$

(a) Show that the x_1-axis, i.e. $x_2 = 0$, is invariant, and sketch the direction of flow on this line.
(b) Show that the x_2-axis, i.e. $x_1 = 0$, is invariant, and sketch the direction of flow on this line.
(c) Compute e^{At}, where

$$A = \begin{pmatrix} 1 & 0 \\ 0 & -1 \end{pmatrix}.$$

(d) Use e^{At} to express the solution of this ODE.
(e) Is $(x_1, x_2) = (0, 0)$ an equilibrium point? If it is an equilbrium point, what is the nature of its stability?

7. Consider the following two-dimensional linear autonomous ODE on $\mathbb{R}^2$,

$$\dot{x} = y,$$
$$\dot{y} = x.$$

(a) Show that the trajectories lie on the curves:

$$\frac{1}{2}\left(y^2 - x^2\right) = C, \tag{5.65}$$

where C is a constant.
(b) Show that

$$x - y = 0, \quad \text{and} \quad x + y = 0.$$

are invariant lines. How do these invariant lines relate to (5.65) and the constant C?

(c) Sketch these invariant lines in the phase plane and indicate the direction of flow along these lines. Sketch some trajectories between these invariant lines, and indicate the direction of flow along the lines that you sketch.
(d) How do these invariant lines relate to the stable and unstable subspaces of the origin?

8. Consider the following two-dimensional linear autonomous ODE on $\mathbb{R}^2$,

$$\dot{x} = y,$$
$$\dot{y} = -x.$$

(a) Show that the trajectories lie on the curves:

$$\frac{1}{2}\left(y^2 + x^2\right) = C, \tag{5.66}$$

where C is a constant.

9. Consider the following two-dimensional autonomous ODE on $\mathbb{R}^2$,

$$\begin{pmatrix} \dot{x}_1 \\ \dot{x}_2 \end{pmatrix} = \begin{pmatrix} -1 & -1 \\ 3 & 3 \end{pmatrix} \begin{pmatrix} x_1 \\ x_2 \end{pmatrix} \tag{5.67}$$

and we define

$$A = \begin{pmatrix} -1 & -1 \\ 3 & 3 \end{pmatrix}.$$

(a) Compute e^{At}.
(b) Give equations for the invariant subspaces of (5.67) in the x_1–x_2 coordinates and describe their stability, i.e. are they stable, unstable, or center subspaces?

10. Consider the following two-dimensional autonomous ODE on $\mathbb{R}^2$,

$$\begin{pmatrix} \dot{x}_1 \\ \dot{x}_2 \end{pmatrix} = \begin{pmatrix} 1 & 1 \\ 1 & 1 \end{pmatrix} \begin{pmatrix} x_1 \\ x_2 \end{pmatrix} \tag{5.68}$$

and we define

$$A = \begin{pmatrix} 1 & 1 \\ 1 & 1 \end{pmatrix}.$$

(a) What do you know about the eigenvalues and eigenvectors of A *without doing any calculations*?
(b) Compute e^{At}.

(c) Give equations for the invariant subspaces of (5.68) in the x_1–x_2 coordinates and describe their stability, i.e. are they stable, unstable, or center subspaces?

11. Consider the following two dimensional autonomous ODE on $\mathbb{R}^2$,

$$\begin{pmatrix} \dot{x}_1 \\ \dot{x}_2 \end{pmatrix} = \begin{pmatrix} -2 & -2 \\ 6 & 6 \end{pmatrix} \begin{pmatrix} x_1 \\ x_2 \end{pmatrix} \tag{5.69}$$

and we define

$$A = \begin{pmatrix} -2 & -2 \\ 6 & 6 \end{pmatrix}.$$

(a) Compute e^{At}.

(b) Give equations for the invariant subspaces of (5.69) in the x_1–x_2 coordinates and describe their stability, i.e. are they stable, unstable, or center subspaces?

12. Consider the following two-dimensional autonomous ODE on $\mathbb{R}^2$,

$$\begin{pmatrix} \dot{x}_1 \\ \dot{x}_2 \end{pmatrix} = \begin{pmatrix} -1 & 3 \\ -1 & 3 \end{pmatrix} \begin{pmatrix} x_1 \\ x_2 \end{pmatrix} \tag{5.70}$$

and we define

$$A = \begin{pmatrix} -1 & 3 \\ -1 & 3 \end{pmatrix}.$$

(a) Compute e^{At}.

(b) Give equations for the invariant subspaces of (5.70) in the x_1–x_2 coordinates and describe their stability, i.e. are they stable, unstable, or center subspaces?

Chapter 6

Stable and Unstable Manifolds of Hyperbolic Equilibria

For hyperbolic equilibria of autonomous vector fields, the linearization captures the local behavior near the equilibria for the nonlinear vector field. We describe the results justifying this statement in the context of two-dimensional autonomous systems.[1]

We consider a C^r, $r \geq 1$ two-dimensional autonomous vector field of the following form:

$$
\begin{aligned}
\dot{x} &= f(x, y), \\
\dot{y} &= g(x, y), \quad (x, y) \in \mathbb{R}^2.
\end{aligned} \tag{6.1}
$$

Let $\phi_t(\cdot)$ denote the flow generated by (6.1). Suppose (x_0, y_0) is a hyperbolic equilibrium point of this vector field, i.e. the two eigenvalues of the Jacobian matrix:

$$
\begin{pmatrix}
\frac{\partial f}{\partial x}(x_0, y_0) & \frac{\partial f}{\partial y}(x_0, y_0) \\
\frac{\partial g}{\partial x}(x_0, y_0) & \frac{\partial g}{\partial y}(x_0, y_0)
\end{pmatrix},
$$

[1]An extremely complete and thorough exposition of "hyperbolic theory" is given in A. Katok and B. Hasselblatt. *Introduction to the Modern Theory of Dynamical Systems*, Volume 54. Cambridge University Press, 1997.

have nonzero real parts. There are three cases to consider:

- (x_0, y_0) is a source for the linearized vector field,
- (x_0, y_0) is a sink for the linearized vector field,
- (x_0, y_0) is a saddle for the linearized vector field.

We consider each case individually.[2]

(x_0, y_0) is a source.

In this case (x_0, y_0) is a source for (6.1). More precisely, there exists a neighborhood $\mathcal{U}$ of (x_0, y_0) such that for any $p \in \mathcal{U}$, $\phi_t(p)$ leaves $\mathcal{U}$ as t increases.

(x_0, y_0) is a sink.

In this case (x_0, y_0) is a sink for (6.1). More precisely, there exists a neighborhood $\mathcal{S}$ of (x_0, y_0) such that for any $p \in \mathcal{S}$, $\phi_t(p)$ approaches (x_0, y_0) at an exponential rate as t increases. In this case (x_0, y_0) is an example of an attracting set and its basin of attraction is given by:

$$\mathcal{B} \equiv \bigcup_{t \leq 0} \phi_t\,(\mathcal{S}).$$

(x_0, y_0) is a saddle.

For the case of hyperbolic saddle points, the saddle point structure is still retained near the equilibrium point for nonlinear systems. We now explain precisely what this means. In order to do this we will need to examine (6.1) more closely. In particular, we will need to transform (6.1) to a coordinate system that "localizes" the behavior near the equilibrium point and specifically displays the structure of the linear part. We have already done this in the previous two chapters in examining the behavior near specific solutions, so we will not repeat those details.

Transforming locally near (x_0, y_0) in this manner, we can express (6.1) in the following form:

$$\begin{pmatrix} \dot{\xi} \\ \dot{\eta} \end{pmatrix} = \begin{pmatrix} -\alpha & 0 \\ 0 & \beta \end{pmatrix} \begin{pmatrix} \xi \\ \eta \end{pmatrix} + \begin{pmatrix} u(\xi,\eta) \\ v(\xi,\eta) \end{pmatrix}, \quad \alpha,\, \beta > 0, \quad (\xi,\eta) \in \mathbb{R}^2, \quad (6.2)$$

[2]Details of the stability results for the hyperbolic source and sink can be found in M. W. Hirsch, S. Smale, and R. L. Devaney. *Differential Equations, Dynamical Systems, and an Introduction to Chaos*. Academic Press, 2012.

where the Jacobian at the origin,

$$\begin{pmatrix} -\alpha & 0 \\ 0 & \beta \end{pmatrix} \tag{6.3}$$

reflects the hyperbolic nature of the equilibrium point. The linearization of (6.1) about the origin is given by:

$$\begin{pmatrix} \dot{\xi} \\ \dot{\eta} \end{pmatrix} = \begin{pmatrix} -\alpha & 0 \\ 0 & \beta \end{pmatrix} \begin{pmatrix} \xi \\ \eta \end{pmatrix}. \tag{6.4}$$

It is easy to see for the linearized system that

$$E^s = \{(\xi, \eta) \mid \eta = 0\}, \tag{6.5}$$

is the invariant stable subspace and

$$E^u = \{(\xi, \eta) \mid \xi = 0\}, \tag{6.6}$$

is the invariant unstable subspace.

We now state how this saddle point structure is inherited by the nonlinear system by stating the results of the *stable and unstable manifold theorem for hyperbolic equilibria* for two-dimensional autonomous vector fields.[3]

First, we consider two intervals of the coordinate axes containing the origin as follows:

$$I_\xi \equiv \{-\epsilon < \xi < \epsilon\}, \tag{6.7}$$

and

$$I_\eta \equiv \{-\epsilon < \eta < \epsilon\}, \tag{6.8}$$

for some small $\epsilon > 0$. A neighborhood of the origin is constructed by taking the cartesian product of these two intervals:

$$B_\epsilon \equiv \left\{(\xi, \eta) \in \mathbb{R}^2 \mid (\xi, \eta) \in I_\xi \times I_\eta\right\}, \tag{6.9}$$

[3]The stable and unstable manifold theorem for hyperbolic equilibria of autonomous vector fields is a fundamental result. The proof requires some preliminary work to develop the appropriate mathematical setting in order for the proof to proceed. Of course, before one proves a result one must have a through understanding of the result that one hopes to prove. For this reason we focus on explicit examples here. See, for example E. A. Coddington and N. Levinson. *Theory of Ordinary Differential Equations.* Krieger, 1984; P. Hartman. *Ordinary Differential Equations.* Society for industrial and Applied Mathematics, 2002; J. K. Hale. *Ordinary Differential Equations.* Dover, 2009; and C. Chicone. *Ordinary Differential Equations with Applications.* Springer, 2000.

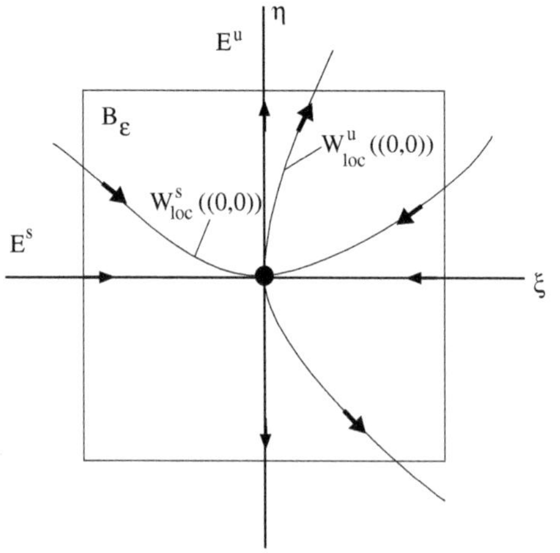

Fig. 6.1. The neighborhood of the origin, B_ϵ, showing the local stable and unstable manifolds.

and it is illustrated in Fig. 6.1. The stable and unstable manifold theorem for hyperbolic equilibrium points of autonomous vector fields states the following.

There exists a C^r curve, given by the graph of a function of the ξ variables:

$$\eta = S(\xi), \quad \xi \in I_\xi, \tag{6.10}$$

This curve has three important properties.

It passes through the origin, i.e. $S(0) = 0$.
It is tangent to E^s at the origin, i.e. $\frac{dS}{d\xi}(0) = 0$.
It is locally invariant in the sense that any trajectory starting on the curve approaches the origin at an exponential rate as $t \to \infty$, and it leaves B_ϵ as $t \to -\infty$.

Moreover, the curve satisfying these three properties is *unique*. For these reasons, this curve is referred to as the local stable manifold of the origin, and it is denoted by:

$$W^s_{\text{loc}}((0,0)) = \{(\xi,\eta) \in B_\epsilon \,|\, \eta = S(\xi)\}. \tag{6.11}$$

Similarly, there exists another C^r curve, given by the graph of a function of the η variables:

$$\xi = U(\eta), \quad \eta \in I_\eta, \tag{6.12}$$

This curve has three important properties.

It passes through the origin, i.e. $U(0) = 0$.
It is tangent to E^u at the origin, i.e. $\frac{dU}{d\eta}(0) = 0$.
It is locally invariant in the sense that any trajectory starting on the curve approaches the origin at an exponential rate as $t \to -\infty$, and it leaves B_ϵ as $t \to \infty$.

For these reasons, this curve is referred to as the local unstable manifold of the origin, and it is denoted by:

$$W^u_{\text{loc}}((0,0)) = \{(\xi, \eta) \in B_\epsilon \,|\, \xi = U(\eta)\}. \tag{6.13}$$

The curve satisfying these three properties is *unique*.

These local stable and unstable manifolds are the "seeds" for the global stable and unstable manifolds that are defined as follows:

$$W^s((0,0)) \equiv \bigcup_{t \le 0} \phi_t(W^s_{\text{loc}}((0,0))), \tag{6.14}$$

and

$$W^u((0,0)) \equiv \bigcup_{t \ge 0} \phi_t(W^u_{\text{loc}}((0,0))). \tag{6.15}$$

Now we will consider a series of examples showing how these ideas are used.

Example 13. We consider the following autonomous, nonlinear vector field on the plane:

$$\dot{x} = x,$$
$$\dot{y} = -y + x^2, \quad (x, y) \in \mathbb{R}^2. \tag{6.16}$$

This vector field has an equilibrium point at the origin, $(x, y) = (0, 0)$. The Jacobian of the vector field evaluated at the origin is given by:

$$\begin{pmatrix} 1 & 0 \\ 0 & -1 \end{pmatrix}. \tag{6.17}$$

From this calculation we can conclude that the origin is a hyperbolic saddle point. Moreover, the x-axis is the unstable subspace for the linearized vector field and the y-axis is the stable subspace for the linearized vector field.

Next we consider the nonlinear vector field (6.16). By inspection, we see that the y-axis (i.e. $x = 0$) is the global stable manifold for the origin. We next consider the unstable manifold. Dividing the second equation by the first equation in (6.16) gives:

$$\frac{\dot{y}}{\dot{x}} = \frac{dy}{dx} = -\frac{y}{x} + x. \tag{6.18}$$

This is a linear nonautonomous equation.[4] A solution of this equation passing through the origin is given by:

$$y = \frac{x^2}{3}. \tag{6.19}$$

It is also tangent to the unstable subspace at the origin. It is the global unstable manifold.[5]

We examine this statement further. It is easy to compute the flow generated by (6.16). The x component can be solved and substituted into the y component to yield a first order linear nonautonomous equation. Hence, the flow generated by (6.16) is given by[6]:

$$\begin{aligned} x(t, x_0) &= x_0 e^t, \\ y(t, y_0) &= \left(y_0 - \frac{x_0^2}{3}\right) e^{-t} + \frac{x_0^2}{3} e^{2t}. \end{aligned} \tag{6.20}$$

The global unstable manifold of the origin is the set of initial conditions having the property that the trajectories through these initial conditions approach the origin at an exponential rate as $t \to -\infty$. On examining the two components of (6.20), we see that the x component approaches zero as $t \to -\infty$ for any choice of x_0. However, the y component will only approach

[4]This linear first order equation can be solved in the usual way with an integrating factor. See Appendix B for details of this procedure.

[5]You should verify that this curve is invariant. In terms of the vector field, invariance means that the vector field is tangent to the curve. Why?

[6]Again, the equation is solved in the usual way by using an integrating factor, see Appendix B for details.

zero as $t \to -\infty$ if y_0 and x_0 are chosen such that

$$y_0 = \frac{x_0^2}{3}. \tag{6.21}$$

Hence (6.21) is the global unstable manifold of the origin.[7]

Example 14. Consider the following nonlinear autonomous vector field on the plane:

$$\begin{aligned} \dot{x} &= x - x^3, \\ \dot{y} &= -y, \qquad (x, y) \in \mathbb{R}^2. \end{aligned} \tag{6.22}$$

Note that the x and y components evolve independently.

The equilibrium points and the Jacobians associated with their linearizations are given as follows:

$$(x, y) = (0, 0); \quad \begin{pmatrix} 1 & 0 \\ 0 & -1 \end{pmatrix}; \quad \text{saddle} \tag{6.23}$$

$$(x, y) = (\pm 1, 0); \quad \begin{pmatrix} -2 & 0 \\ 0 & -1 \end{pmatrix}; \quad \text{sinks} \tag{6.24}$$

We now compute the global stable and unstable manifolds of these equilibria. We begin with the saddle point at the origin.

$$(0,0): \quad \begin{aligned} W^s\left((0,0)\right) &= \{(x,y) | x = 0\} \\ W^u\left((0,0)\right) &= \{(x,y) | -1 < x < 1,\ y = 0\} \end{aligned} \tag{6.25}$$

For the sinks, the global stable manifold is synonomous with the basin of attraction for the sink.

$$(1,0): \quad W^s\left((1,0)\right) = \{(x,y) | x > 0\} \tag{6.26}$$

$$(-1,0): \quad W^s\left((-1,0)\right) = \{(x,y) | x < 0\} \tag{6.27}$$

[7]Equation (6.19) characterizes the unstable manifold as an invariant curve passing through the origin and tangent to the unstable subspace at the origin. Equation (6.21) characterizes the unstable manifold in terms of the asymptotic behavior of trajectories (as $t \to -\infty$) whose initial conditions satisfy a particular constraint, and that constraint is that they are on the unstable manifold of the origin.

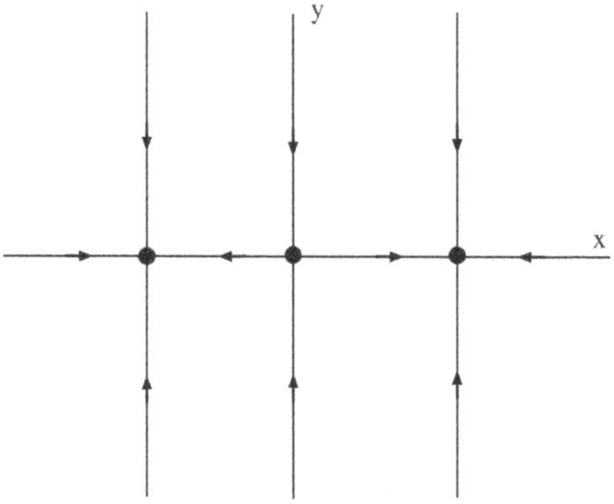

Fig. 6.2. Invariant manifold structure of (6.22). The black dots indicate equilibrium points.

Example 15. In this example we consider the following nonlinear autonomous vector field on the plane:

$$\begin{aligned} \dot{x} &= -x, \\ \dot{y} &= y^2(1-y^2), \qquad (x,y) \in \mathbb{R}^2. \end{aligned} \tag{6.28}$$

Note that the x and y components evolve independently.

The equilibrium points and the Jacobians associated with their linearizations are given as follows:

$$(x,y) = (0,0),\ (0,\pm 1) \tag{6.29}$$

$$(x,y) = (0,0); \quad \begin{pmatrix} -1 & 0 \\ 0 & 0 \end{pmatrix}; \quad \text{not hyperbolic} \tag{6.30}$$

$$(x,y) = (0,1); \quad \begin{pmatrix} -1 & 0 \\ 0 & -2 \end{pmatrix}; \quad \text{sink} \tag{6.31}$$

$$(x,y) = (0,-1); \quad \begin{pmatrix} -1 & 0 \\ 0 & 2 \end{pmatrix}; \quad \text{saddle} \tag{6.32}$$

We now compute the global invariant manifold structure for each of the equilibria, beginning with $(0,0)$.

$$(0,0); \quad \begin{aligned} W^s\left((0,0)\right) &= \{(x,y) | y = 0\} \\ W^c\left((0,0)\right) &= \{(x,y) | x = 0,\ -1 < y < 1\} \end{aligned} \tag{6.33}$$

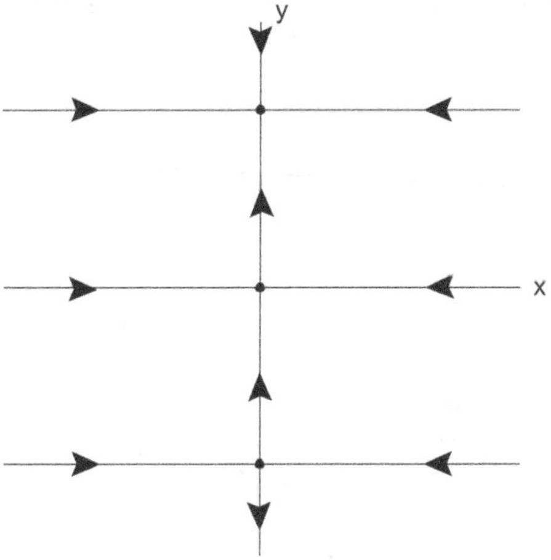

Fig. 6.3. Invariant manifold structure of (6.28). The black dots indicate equilibrium points.

The x-axis is clearly the global stable manifold for this equilibrium point. The segment on the y-axis between -1 and 1 is invariant, but it does not correspond to a hyperbolic direction. It is referred to as the *center manifold of the origin*, and we will learn much more about invariant manifolds associated with nonhyperbolic directions later.

The equilibrium point $(0, 1)$ is a sink. Its global stable manifold (basin of attraction) is given by:

$$(0,1); \quad W^s\left((0,1)\right) = \left\{(x,y)|y>0\right\}. \tag{6.34}$$

The equilibrium point $(0, -1)$ is a saddle point with global stable and unstable manifolds given by:

$$(0,-1); \quad \begin{aligned} &W^s\left((0,-1)\right) = \left\{(x,y)|y=-1\right\} \\ &W^u\left((0,-1)\right) = \left\{(x,y)|x=0,\ -\infty<y<0\right\} \end{aligned} \tag{6.35}$$

Example 16. In this example, we consider the following nonlinear autonomous vector field on the plane:

$$\begin{aligned} &\dot{x} = y, \\ &\dot{y} = x - x^3 - \delta y, \quad (x,y) \in \mathbb{R}^2, \quad \delta \geq 0, \end{aligned} \tag{6.36}$$

where $\delta > 0$ is to be viewed as a parameter. The equilibrium points are given by:

$$(x, y) = (0, 0),\ (\pm 1, 0). \tag{6.37}$$

We want to classify the linearized stability of the equilibria. The Jacobian of the vector field is given by:

$$A = \begin{pmatrix} 0 & 1 \\ 1 - 3x^2 & -\delta \end{pmatrix}, \tag{6.38}$$

and the eigenvalues of the Jacobian are:

$$\lambda_{\pm} = -\frac{\delta}{2} \pm \frac{1}{2}\sqrt{\delta^2 + 4 - 12x^2}. \tag{6.39}$$

We evaluate this expression for the eigenvalues at each of the equilibria to determine their linearized stability.

$$(0, 0);\quad \lambda_{\pm} = -\tfrac{\delta}{2} \pm \tfrac{1}{2}\sqrt{\delta^2 + 4}. \tag{6.40}$$

Note that

$$\delta^2 + 4 > \delta^2,$$

therefore the eigenvalues are always real and of opposite sign. This implies that $(0, 0)$ is a saddle.

$$(\pm 1, 0)\quad \lambda_{\pm} = -\tfrac{\delta}{2} \pm \tfrac{1}{2}\sqrt{\delta^2 - 8}. \tag{6.41}$$

First, note that

$$\delta^2 - 8 < \delta^2.$$

This implies that these two fixed points are always sinks. However, there are two subcases.

$\delta^2 - 8 < 0$: The eigenvalues have a nonzero imaginary part.
$\delta^2 - 8 \geq 0$: The eigenvalues are purely real.

In Fig. 6.4 we sketch the local invariant manifold structure for these two cases.

In Fig. 6.5 we sketch the global invariant manifold structure for the two cases. In the coming lectures we will learn how we can justify this figure. However, note the role that the *stable* manifold of the saddle plays in defining the basins of attractions of the two sinks.

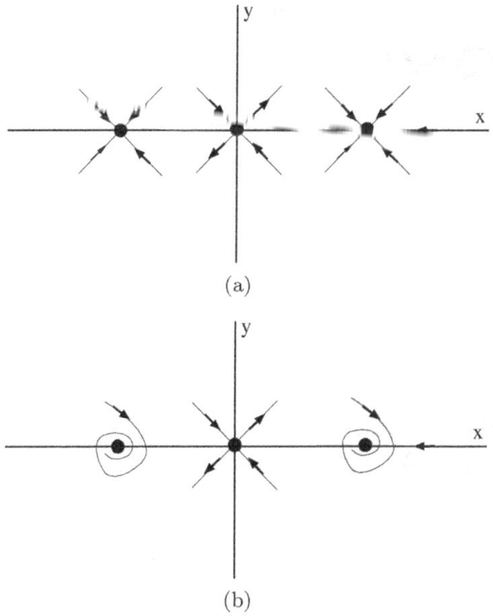

Fig. 6.4. Local invariant manifold structure of (6.36). The black dots indicate equilibrium points. (a) $\delta^2 - 8 \geq 0$, (b) $\delta^2 - 8 < 0$.

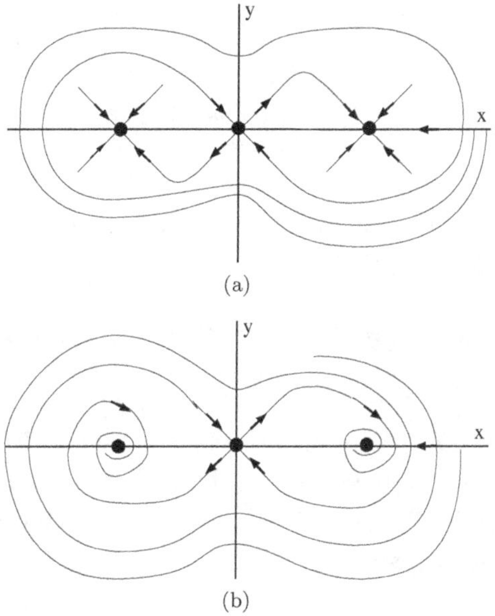

Fig. 6.5. A sketch of the global invariant manifold structure of (6.36). The black dots indicate equilibrium points. (a) $\delta^2 - 8 \geq 0$, (b) $\delta^2 - 8 < 0$.

Problem Set 6

1. Consider the C^r, $r \geq 1$, autonomous vector field on $\mathbb{R}^2$:

$$\dot{x} = f(x),$$

with flow

$$\phi_t(\cdot),$$

and let $x = \bar{x}$ denote a hyperbolic saddle type equilibrium point for this vector field. We denote the local stable and unstable manifolds of this equilibrium point by:

$$W^s_{\text{loc}}(\bar{x}), \quad W^u_{\text{loc}}(\bar{x}),$$

respectively. The global stable and unstable manifolds of $\bar{x}$ are defined by:

$$W^s(\bar{x}) \equiv \bigcup_{t \leq 0} \phi_t\left(W^s_{\text{loc}}(\bar{x})\right),$$

$$W^u(\bar{x}) \equiv \bigcup_{t \geq 0} \phi_t\left(W^u_{\text{loc}}(\bar{x})\right).$$

(a) Show that $W^s(\bar{x})$ and $W^u(\bar{x})$ are invariant sets.
(b) Suppose that $p \in W^s(\bar{x})$, show that $\phi_t(p) \to \bar{x}$ at an exponential rate as $t \to \infty$.
(c) Suppose that $p \in W^u(\bar{x})$, show that $\phi_t(p) \to \bar{x}$ at an exponential rate as $t \to -\infty$.

2. Consider the C^r, $r \geq 1$, autonomous vector field on $\mathbb{R}^2$ having a hyperbolic saddle point. Can its stable and unstable manifolds intersect at an *isolated* point (which is *not* a fixed point of the vector field) as shown in Fig. 6.6?

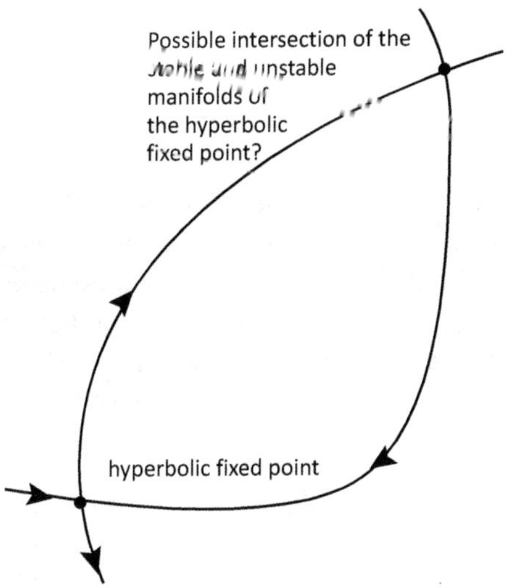

Fig. 6.6. Possible intersection of the stable and unstable manifolds of a hyperbolic fixed point?

3. Consider the following autonomous vector field on the plane:

$$\dot{x} = \alpha x,$$
$$\dot{y} = \beta y + \gamma x^{n+1}, \tag{6.42}$$

where $\alpha < 0$, $\beta > 0$, γ is a real number, and n is a positive integer.

(a) Show that the origin is a hyperbolic saddle point.
(b) Compute and sketch the stable and unstable subspaces of the origin.
(c) Show that the stable and unstable subspaces are invariant under the linearized dynamics.
(d) Show that the flow generated by this vector field is given by:

$$x(t, x_0) = x_0 e^{\alpha t},$$
$$y(t, x_0, y_0) = e^{\beta t}\left(y_0 - \frac{\gamma x_0^{n+1}}{\alpha(n+1) - \beta}\right) + \left(\frac{\gamma x_0^{n+1}}{\alpha(n+1) - \beta}\right) e^{\alpha(n+1)t}$$

(e) Compute the global stable and unstable manifolds of the origin from the flow.

(f) Show that the global stable and unstable manifolds that you have computed are invariant.

(g) Sketch the global stable and unstable manifolds and discuss how they depend on γ and n.

4. Suppose $\dot{x} = f(x), \quad x \in \mathbb{R}^n$ is a C^r vector field having a hyperbolic fixed point, $x = x_0$, with a homoclinic orbit. Describe the homoclinic orbit in terms of the stable and unstable manifolds of x_0.
5. Suppose $\dot{x} = f(x), \quad x \in \mathbb{R}^n$ is a C^r vector field having hyperbolic fixed points, $x = x_0$ and x_1, with a heteroclinic orbit connecting x_0 and x_1. Describe the heteroclinic orbit in terms of the stable and unstable manifolds of x_0 and x_1.

Chapter 7

Lyapunov's Method and the LaSalle Invariance Principle

We will next learn a method for determining stability of equilibria which may be applied when stability information obtained from the linearization of the ODE is not sufficient for determining stability information for the nonlinear ODE. The book by LaSalle[1] is an excellent supplement to this lecture. This is Lyapunov's method (or Lyapunov's second method, or the method of Lyapunov functions).[2] We begin by describing the framework for the method in the setting that we will use.

We consider a general C^r, $r \geq 1$ autonomous ODE

$$\dot{x} = f(x), \quad x \in \mathbb{R}^n, \tag{7.1}$$

having an equilibrium point at $x = \bar{x}$, i.e.

$$f(\bar{x}) = 0. \tag{7.2}$$

For a scalar valued function defined on $\mathbb{R}^n$

$$\begin{aligned} V : \mathbb{R}^n &\to \mathbb{R}, \\ x &\mapsto V(x), \end{aligned} \tag{7.3}$$

[1] J. P. LaSalle. *The Stability of Dynamical Systems*, Volume 25. SIAM, 1976.

[2] The original work of Lyapunov is reprinted as below and an excellent perspective of Lyapunov's work is given in Parks as well A. M. Lyapunov. *General Problem of the Stability of Motion.* Control Theory and Applications Series. Taylor & Francis, 1992; and P. C. Parks. A. M. Lyapunov's stability theory — 100 years on. *IMA Journal of Mathematical Control and Information*, 9(4):275–303, 1992.

we define the derivative of (7.3) along trajectories of (7.1) by:

$$\frac{d}{dt}V(x) = \dot{V}(x) = \nabla V(x) \cdot \dot{x}, \\ = \nabla V(x) \cdot f(x). \tag{7.4}$$

We can now state Lyapunov's theorem on the stability of the equilibrium point $x = \bar{x}$.

Theorem 1. *Consider the following C^r $(r \geq 1)$ autonomous vector field on $\mathbb{R}^n$:*

$$\dot{x} = f(x), \quad x \in \mathbb{R}^n. \tag{7.5}$$

Let $x = \bar{x}$ be an equilibrium point of (7.5) and let $V : U \to \mathbb{R}$ be a C^1 function defined in some neighborhood U of $\bar{x}$ such that:

1. $V(\bar{x}) = 0$ *and* $V(x) > 0$ *if* $x \neq \bar{x}$.
2. $\dot{V}(x) \leq 0$ *in* $U - \{\bar{x}\}$

Then $\bar{x}$ is Lyapunov stable. Moreover, if

$$\dot{V}(x) < 0 \quad in\ U - \{\bar{x}\}$$

then $\bar{x}$ is asymptotically stable.

The function $V(x)$ is referred to as a Lyapunov function.

We now consider an example.

Example 17.

$$\dot{x} = y, \\ \dot{y} = -x - \epsilon x^2 y, \quad (x, y) \in \mathbb{R}^2, \tag{7.6}$$

where ϵ is a parameter. It is clear that $(x, y) = (0, 0)$ is an equilibrium point of (7.6) and we want to determine the nature of its stability.

We begin by linearizing (7.6) about this equilibrium point. The matrix associated with this linearization is given by:

$$A = \begin{pmatrix} 0 & 1 \\ -1 & 0 \end{pmatrix}, \tag{7.7}$$

and its eigenvalues are $\pm i$. Hence, the origin is not hyperbolic and therefore the information provided by the linearization of (7.6) about $(x, y) = (0, 0)$ does not provide information about stability of $(x, y) = (0, 0)$ for the nonlinear system (7.6).

Therefore we will attempt to apply Lyapunov's method to determine the stability of the origin.

We take as a Lyapunov function:

$$V(x, y) = \frac{1}{2}(x^2 + y^2). \tag{7.8}$$

Note that $V(0,0) = 0$ and $V(x, y) > 0$ in any neighborhood of the origin. Moreover, we have:

$$\begin{aligned} \dot{V}(x, y) &= \frac{\partial V}{\partial x}\dot{x} + \frac{\partial V}{\partial y}\dot{y}, \\ &= xy + y(-x - \epsilon x^2 y), \\ &= -\epsilon x^2 y^2, \\ &\leq 0, \quad \text{for } \epsilon \geq 0. \end{aligned} \tag{7.9}$$

Hence, it follows from Theorem 1 that the origin is Lyapunov stable.[3]

Next we will introduce the LaSalle invariance principle.[4] Rather than focus on the particular question of stability of an equilibrium solution as in Lyapunov's method, the LaSalle invariance principle gives conditions that describe the behavior as $t \to \infty$ of all solutions of an autonomous ODE.

We begin with an autonomous ODE defined on $\mathbb{R}^n$:

$$\dot{x} = f(x), \quad x \in \mathbb{R}^n \tag{7.10}$$

where $f(x)$ is C^r, $(r \geq 1)$. Let $\phi_t(\cdot)$ denote the flow generated by (7.10) and let $\mathcal{M} \subset \mathbb{R}^n$ denote a positive invariant set that is compact (i.e. closed and bounded in this setting). Suppose we have a scalar valued function

$$V : \mathbb{R}^n \to \mathbb{R}, \tag{7.11}$$

such that

$$\dot{V}(x) \leq 0 \quad \text{in } \mathcal{M} \tag{7.12}$$

(Note the "less than or equal to" in this inequality.)

[3]The obvious question that comes up at this point is, *how do I find the Lyapunov function?* We discuss this issue in Appendix C.

[4]The original paper is a technical report that can be found on the internet. See also J. P. LaSalle. An invariance principle in the theory of stability. Technical Report 66-1, Brown University, 1966; J. P. LaSalle. *The Stability of Dynamical Systems*, Volume 25. SIAM, 1976; and I. Barkana. Defending the beauty of the invariance principle. *International Journal of Control*, 87(1):186–206, 2014.

Let

$$E = \left\{ x \in \mathcal{M} \mid \dot{V}(x) = 0 \right\}, \tag{7.13}$$

and

$$M = \left\{ \begin{array}{l} \text{the union of all trajectories that start} \\ \text{in } E \text{ and remain in } E \text{ for all } t \geq 0 \end{array} \right\} \tag{7.14}$$

Now we can state LaSalle's invariance principle.

Theorem 2. *For all* $x \in \mathcal{M}$, $\phi_t(x) \to M$ *as* $t \to \infty$.

We will now consider an example.

Example 18. Consider the following vector field on $\mathbb{R}^2$:

$$\begin{aligned} \dot{x} &= y, \\ \dot{y} &= x - x^3 - \delta y, \quad (x, y) \in \mathbb{R}^2, \quad \delta > 0. \end{aligned} \tag{7.15}$$

This vector field has three equilibrium points — a saddle point at $(x, y) = (0, 0)$ and two sinks at $(x, y) = (\pm, 1, 0)$.

Consider the function

$$V(x, y) = \frac{y^2}{2} - \frac{x^2}{2} + \frac{x^4}{4}, \tag{7.16}$$

and its level sets:

$$V(x, y) = C.$$

We compute the derivative of V along trajectories of (7.15):

$$\begin{aligned} \dot{V}(x, y) &= \frac{\partial V}{\partial x}\dot{x} + \frac{\partial V}{\partial y}\dot{y}, \\ &= (-x + x^3)y + y(x - x^3 - \delta y). \\ &= -\delta y^2, \end{aligned} \tag{7.17}$$

from which it follows that

$$\dot{V}(x, y) \leq 0 \quad \text{on } V(x, y) = C.$$

Therefore, for C sufficiently large, the corresponding level set of V bounds a compact positive invariant set, $\mathcal{M}$, containing the three equilibrium points of (7.15).

Next we determine the nature of the set

$$E = \{(x, y) \in \mathcal{M} \mid \dot{V}(x, y) = 0\}. \tag{7.18}$$

Using (7.17) we see that:

$$E = \{(x, y) \in \mathcal{M} \mid y = 0 \cap \mathcal{M}\}. \tag{7.19}$$

The only points in E that remain in E for all time are:

$$M = \{(\pm 1, 0), \quad (0, 0)\}. \tag{7.20}$$

Therefore it follows from Theorem 2 that given any initial condition in $\mathcal{M}$, the trajectory starting at that initial condition approaches one of the three equilibrium points as $t \to \infty$.

Autonomous Vector Fields on the Plane; Bendixson's Criterion and the Index Theorem

Now we will consider some useful results that apply to vector fields on the plane.

First we will consider a simple, and easy to apply, criterion that rules out the existence of periodic orbits for autonomous vector fields *on the plane* (e.g. it is not valid for vector fields on the two torus).

We consider a C^r, $r \geq 1$ vector field on the plane of the following form:

$$\begin{aligned} \dot{x} &= f(x, y), \\ \dot{y} &= g(x, y), \quad (x, y) \in \mathbb{R}^2. \end{aligned} \tag{7.21}$$

The following criterion due to Bendixson provides a simple, computable condition that rules out the existence of periodic orbits in certain regions of $\mathbb{R}^2$.

Theorem 3 (Bendixson's Criterion). *If on a simply connected region $D \subset \mathbb{R}^2$ the expression*

$$\frac{\partial f}{\partial x}(x, y) + \frac{\partial g}{\partial y}(x, y). \tag{7.22}$$

is not identically zero and does not change sign then (7.21) *has no periodic orbits lying entirely in D.*

Example 19. A proof of this result can be found in Footnote 5. We consider the following nonlinear autonomous vector field on the plane:

$$\begin{aligned}\dot{x} &= y \equiv f(x,y),\\ \dot{y} &= x - x^3 - \delta y \equiv g(x,y), \quad (x,y) \in \mathbb{R}^2, \quad \delta > 0.\end{aligned} \tag{7.23}$$

Computing (7.22) gives:

$$\frac{\partial f}{\partial x} + \frac{\partial g}{\partial y} = -\delta. \tag{7.24}$$

Therefore, this vector field has no periodic orbits for $\delta \neq 0$.

Example 20. We consider the following linear autonomous vector field on the plane:

$$\begin{aligned}\dot{x} &= ax + by \equiv f(x,y),\\ \dot{y} &= cx + dy \equiv g(x,y), \quad (x,y) \in \mathbb{R}^2, \quad a,\, b,\, c,\, d \in \mathbb{R}\end{aligned} \tag{7.25}$$

Computing (7.22) gives:

$$\frac{\partial f}{\partial x} + \frac{\partial g}{\partial y} = a + d. \tag{7.26}$$

Therefore, for $a + d \neq 0$, this vector field has no periodic orbits.

Next we will consider the index theorem. The index theorem is a very useful tool for determining whether or not a phase portrait that you have computed is "complete" or "correct" in the sense that it provides a constraint on the stability types of the equilibria with respect to the number of equilibria as well as the number and stability types of equilibria that can be contained within a closed trajectory. Proofs of the index theorem are not so common in ordinary differential equations textbooks as they rely on concepts that are more fully developed in differential topology[6] or algebraic topology.[7] This further highlights the breadth of mathematical ideas that are used in the study of ODEs.

---

[5]S. Wiggins. *Introduction to Applied Nonlinear Dynamical Systems and Chaos*, Volume 2. Springer Science & Business Media, 2003.

[6]V. Guillemin and A. Pollack. *Differential Topology*, Volume 370. American Mathematical Soc., 2010.

[7]W. Fulton. *Algebraic Topology: A First Course*, Volume 153. Springer Science & Business Media, 2013.

Theorem 4. *Inside any periodic orbit there must be* at least *one fixed point. If there is only one, then it must be a sink, source, or center. If all the fixed points inside the periodic orbit are hyperbolic, then there must be an odd number,* $2n+1$, *of which* n *are saddles, and* $n+1$ *are either sinks or sources.*

Example 21. We consider the following nonlinear autonomous vector field on the plane:

$$\begin{aligned} \dot{x} &= y \equiv f(x,y), \\ \dot{y} &= x - x^3 - \delta y + x^2 y \equiv g(x,y), \quad (x,y) \in \mathbb{R}^2, \end{aligned} \tag{7.27}$$

where $\delta > 0$. The equilibrium points are given by:

$$(x,y) = (0,0),\ (\pm 1, 0).$$

The Jacobian of the vector field, denoted by A, is given by:

$$A = \begin{pmatrix} 0 & 1 \\ 1 - 3x^2 + 2xy & -\delta + x^2 \end{pmatrix} \tag{7.28}$$

Using the general expression for the eigenvalues for a 2×2 matrix A:

$$\lambda_{1,2} = \frac{tr\,A}{2} \pm \frac{1}{2}\sqrt{(tr\,A)^2 - 4\,det\,A},$$

we obtain the following expression for the eigenvalues of the Jacobian:

$$\lambda_{1,2} = \frac{-\delta + x^2}{2} \pm \frac{1}{2}\sqrt{(-\delta + x^2)^2 + 4(1 - 3x^2 + 2xy)} \tag{7.29}$$

If we substitute the locations of the equilibria into this expression we obtain the following values for the eigenvalues of the Jacobian of the vector field evaluated at the respective equilibria:

$$(0,0) \quad \lambda_{1,2} = -\frac{\delta}{2} \pm \frac{1}{2}\sqrt{\delta^2 + 4} \tag{7.30}$$

$$(\pm 1, 0) \quad \lambda_{1,2} = \frac{-\delta + 1}{2} \pm \frac{1}{2}\sqrt{(-\delta + 1)^2 - 8} \tag{7.31}$$

From these expressions, we conclude that $(0,0)$ is a saddle for all values of δ and $(\pm 1, 0)$ are

$$\begin{aligned} &\text{sinks for} && \delta > 1 \\ &\text{centers for} && \delta = 1 \\ &\text{sources for} && 0 < \delta < 1 \end{aligned} \tag{7.32}$$

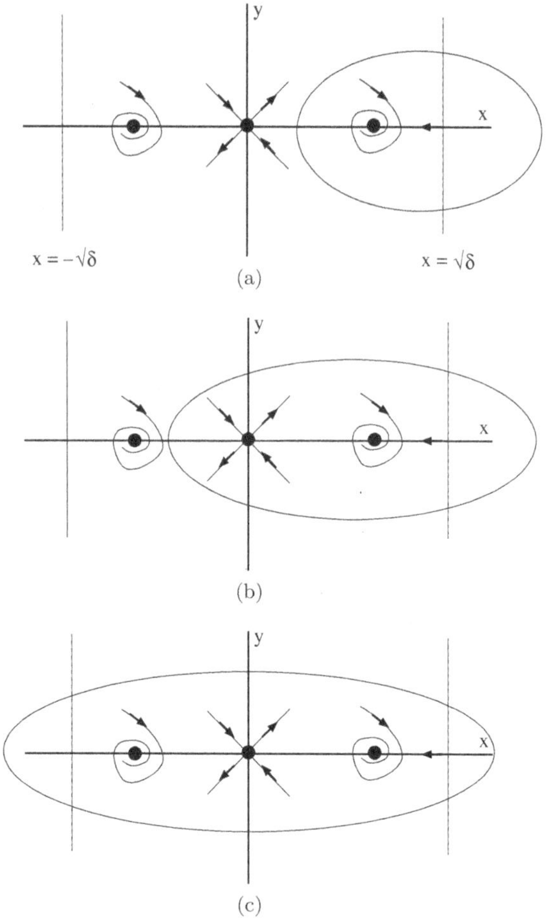

Fig. 7.1. The case $\delta > 1$. Possibilities for periodic orbits that satisfy Bendixson's criterion. However, (b) is not possible because it violates the index theorem.

Now we will use Bendixson's criterion and the index theorem to determine regions in the phase plane where periodic orbits *may* exist. For this example, (7.22) is given by:

$$-\delta + x^2. \tag{7.33}$$

Hence the two vertical lines $x = -\sqrt{\delta}$ and $x = \sqrt{\delta}$ divide the phase plane into three regions where periodic orbits *cannot* exist entirely in one of these regions (or else Bendixson's criterion would be violated). There are two

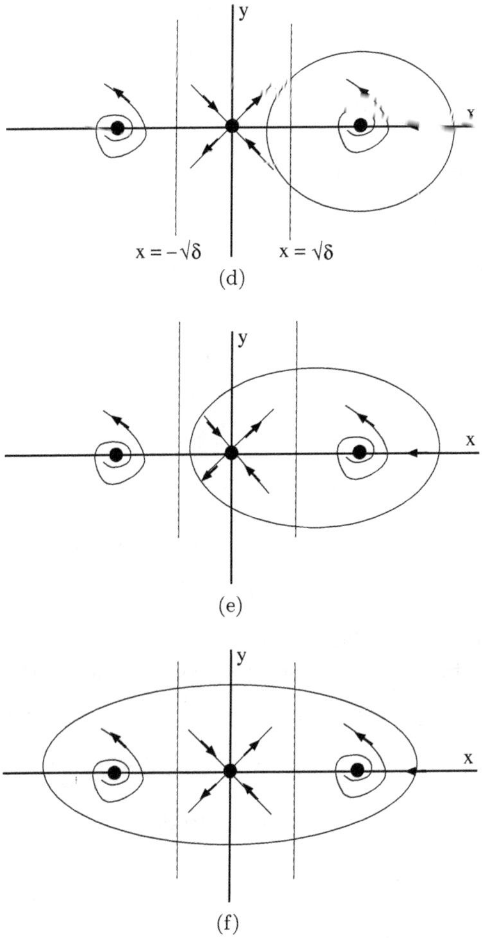

Fig. 7.2. The case $0 < \delta < 1$. Possibilities for periodic orbits that satisfy Bendixson's criterion. However, (e) is not possible because it violates the index theorem.

cases to be considered for the location of these vertical lines with respect to the equilibria: $\delta > 1$ and $0 < \delta < 1$.

In Fig. 7.1 we show three possibilities (they do *not* exhaust all possible cases) for the existence of periodic orbits that would satisfy Bendixson's criterion in the case $\delta > 1$. However, (b) is not possible because it violates the index theorem.

In Fig. 7.2 we show three possibilities (they do *not* exhaust all possible cases) for the existence of periodic orbits that would satisfy Bendixson's criterion in the case $0 < \delta < 1$. However, (e) is not possible because it violates the index theorem.

Problem Set 7

1. Consider the following autonomous vector field on the plane:

$$\dot{x} = y,$$
$$\dot{y} = x - x^3 - \delta y, \quad \delta \geq 0, \quad (x, y) \in \mathbb{R}^2.$$

   Use Lyapunov's method to show that the equilibria $(x, y) = (\pm 1, 0)$ are Lyapunov stable for $\delta = 0$ and asymptotically stable for $\delta > 0$.

2. Consider the following autonomous vector field on the plane:

$$\dot{x} = y,$$
$$\dot{y} = -x - \varepsilon x^2 y, \quad \varepsilon > 0,\ (x, y) \in \mathbb{R}^2.$$

   Use the LaSalle invariance principle to show that

$$(x, y) = (0, 0),$$

   is asymptotically stable.

3. Consider the following autonomous vector field on the plane:

$$\dot{x} = y,$$
$$\dot{y} = x - x^3 - \alpha x^2 y, \quad \alpha > 0, \quad (x, y) \in \mathbb{R}^2,$$

   use the LaSalle invariance principle to describe the fate of all trajectories as $t \to \infty$.

4. Consider the following autonomous vector field on the plane:

$$\dot{x} = y,$$
$$\dot{y} = x - x^3 + \alpha x y, \quad (x, y) \in \mathbb{R}^2,$$

   where α is a real parameter. Determine the equilibria and discuss their linearized stability as a function of α.

5. Consider the following autonomous vector field on the plane:

$$\dot{x} = ax + by,$$
$$\dot{y} = cx + dy, \quad (x, y) \in \mathbb{R}^2, \tag{7.34}$$

   where a, b, c, $d \in \mathbb{R}$. In the questions below you are asked to give conditions on the constants a, b, c, and d so that particular dynamical phenomena are satisfied. You do *not* have to give all possible conditions on the constants in order for the dynamical condition to be satisfied. One condition will be sufficient, but you must justify your answer.

(a) Give conditions on a, b, c, d for which the vector field has no periodic orbits.
(b) Give conditions on a, b, c, d for which all of the orbits are periodic.
(c) Using

$$V(x,y) = \frac{1}{2}(x^2 + y^2)$$

as a Lyapunov function, give conditions on a, b, c, d for which $(x,y) = (0,0)$ is asymptotically stable.
(d) Give conditions on a, b, c, d for which $x = 0$ is the stable manifold of $(x,y) = (0,0)$ and $y = 0$ is the unstable manifold of $(x,y) = (0,0)$.

6. Consider the following autonomous vector field on the plane:

$$\begin{aligned} \dot{x} &= y, \\ \dot{y} &= -x - \frac{x^2 y}{2}, \quad (x,y) \in \mathbb{R}^2. \end{aligned} \tag{7.35}$$

(a) Determine the linearized stability of $(x,y) = (0,0)$.
(b) Describe the invariant manifold structure for the linearization of (7.35) about $(x,y) = (0,0)$.
(c) Using $V(x,y) = \frac{1}{2}(x^2 + y^2)$ as a Lyapunov function, what can you conclude about the stability of the origin? Does this agree with the linearized stability result obtained above? Why or why not?
(d) Using the LaSalle invariance principle, determine the fate of a trajectory starting at an arbitrary initial condition as $t \to \infty$? What does this result allow you to conclude about stability of $(x,y) = (0,0)$?

7. Consider the following ODE:

$$\begin{aligned} \dot{x} &= 0, \\ \dot{y} &= -y - x^2 y, \quad (x,y) \in \mathbb{R}^2. \end{aligned}$$

(a) Compute the flow generated by this vector field and show that all trajectories approach the x-axis as $t \to \infty$.
(b) Using the vector field and the trajectories show that $(0,0)$ is a Lyapunov stable equilibrium point.

(c) Using the function $V(x, y) = \frac{1}{2}\left(x^2 + y^2\right)$ as a Lyapunov function, use Lyapunov's theorem to show that $(x, y) = (0, 0)$ is Lyapunov stable.

(d) Show that the LaSalle Invariance principle implies that all solutions approach the x-axis as $t \to \infty$.

8. Consider the following two-dimensional autonomous ODE on $\mathbb{R}^2$,

$$\dot{x} = -x,$$
$$\dot{y} = -x^2 y, \quad (x, y) \in \mathbb{R}^2.$$

Using $V(x, y) = \frac{1}{2}\left(x^2 + y^2\right)$ as a Lyapunov function, show that the origin is a stable equilibrium point.

9. Consider the following ODE:

$$\dot{x} = y + xz.$$
$$\dot{y} = x - x^3 - \delta y - xz,$$
$$\dot{z} = -z + xyz, \tag{7.36}$$

where $\delta > 0$ is a parameter. Show that all trajectories with initial conditions $(x_0, y_0, 0)$ approach one of the points $(1, 0, 0)$, $(-1, 0, 0)$, $(0, 0, 0)$. (Hint: Apply the LaSalle Invariance principle and use the function $V(x, y) = \frac{y^2}{2} - \frac{x^2}{2} + \frac{x^4}{4}$).

10. Consider the ODE:

$$\dot{x} = -x, \quad x \in \mathbb{R}.$$

Illustrate five ways to show that the equilibrium point $x = 0$ is asymptotically stable. (Hint: Behavior of the explicit analytical form of the trajectories, the sign of the vector field, linearization, Lyapunov's method, the LaSalle invariance principle.)

Chapter 8

Bifurcation of Equilibria, I

WE WILL NOW STUDY THE TOPIC OF BIFURCATION OF EQUILIBRIA OF AUTONOMOUS VECTOR FIELDS, or "what happens as an equilibrium point loses hyperbolicity as a parameter is varied?" We will study this question through a series of examples, and then consider what the examples teach us about the "general situation" (and what this might be).

Example 22 (The Saddle-Node Bifurcation). Consider the following nonlinear, autonomous vector field on $\mathbb{R}^2$:

$$\begin{aligned} \dot{x} &= \mu - x^2, \\ \dot{y} &= -y, \quad (x, y) \in \mathbb{R}^2 \end{aligned} \tag{8.1}$$

where μ is a (real) parameter. The equilibrium points of (8.1) are given by:

$$(x, y) = (\sqrt{\mu}, 0), \, (-\sqrt{\mu}, 0). \tag{8.2}$$

It is easy to see that there are no equilibrium points for $\mu < 0$, one equilibrium point for $\mu = 0$, and two equilibrium points for $\mu > 0$.

The Jacobian of the vector field evaluated at each equilibrium point is given by:

$$(\sqrt{\mu}, 0): \quad \begin{pmatrix} -2\sqrt{\mu} & 0 \\ 0 & -1 \end{pmatrix}, \tag{8.3}$$

from which it follows that the equilibria are hyperbolic and asymptotically stable for $\mu > 0$, and nonhyperbolic for $\mu = 0$.

$$(-\sqrt{\mu}, 0): \begin{pmatrix} 2\sqrt{\mu} & 0 \\ 0 & -1 \end{pmatrix} \tag{8.4}$$

from which it follows that the equilibria are hyperbolic saddle points for $\mu > 0$, and nonhyperbolic for $\mu = 0$. We emphasize again that there are no equilibrium points for $\mu < 0$.

As a result of the "structure" of (8.1) we can easily represent the behavior of the equilibria as a function of μ in a bifurcation diagram. That is, since the x and y components of (8.1) are "decoupled", and the change in the number and stability of equilibria us completely captured by the x coordinates, we can plot the x component of the vector field as a function of μ, as we show in Fig. 8.1.

In Fig. 8.2 we illustrate the bifurcation of equilibria for (8.1) in the x–y plane.

This type of bifurcation is referred to as a saddle-node bifurcation (occasionally it may also be referred to as a fold bifurcation or tangent bifurcation, but these terms are used less frequently).

The key characteristic of the saddle-node bifurcation is the following. As a parameter (μ) is varied, the number of equilibria change from zero to two, and the change occurs at a parameter value corresponding to the two equilibria coalescing into one *nonhyperbolic equilibrium.*

μ is called the bifurcation parameter and $\mu = 0$ is called the bifurcation point.

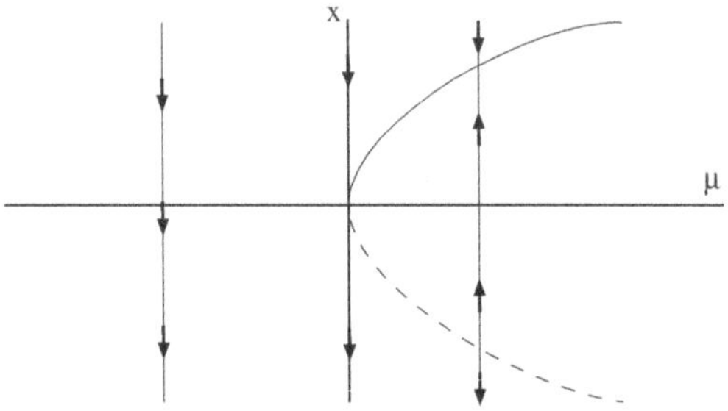

Fig. 8.1. Bifurcation diagram for (8.1) in the μ–x plane. The curve of equilibria is given by $\mu = x^2$. The dashed line denotes the part of the curve corresponding to unstable equilibria, and the solid line denotes the part of the curve corresponding to stable equilibria.

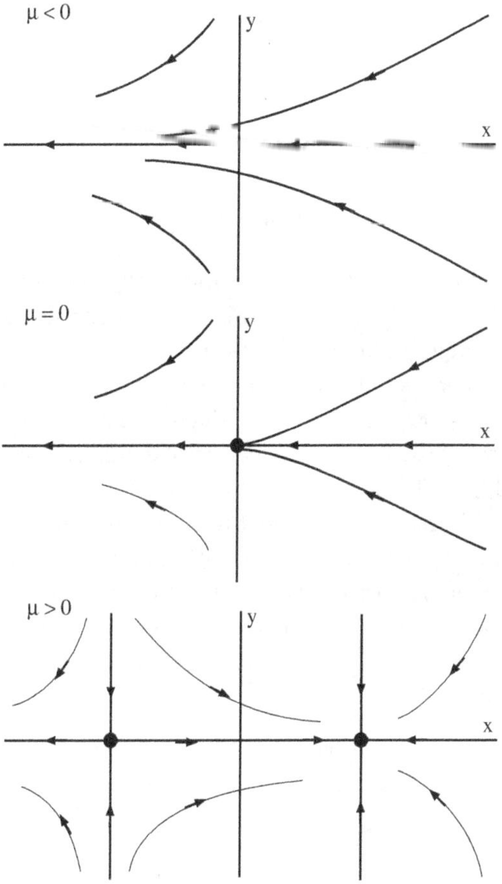

Fig. 8.2. Bifurcation diagram for (8.1) in the x–y plane for $\mu < 0$, $\mu = 0$, and $\mu > 0$. Compare with Fig. 8.1.

Example 23 (The Transcritical Bifurcation). Consider the following nonlinear, autonomous vector field on $\mathbb{R}^2$:

$$\dot{x} = \mu x - x^2,$$
$$\dot{y} = -y, \quad (x, y) \in \mathbb{R}^2, \tag{8.5}$$

where μ is a (real) parameter. The equilibrium points of (8.5) are given by:

$$(x, y) = (0, 0),\ (\mu, 0). \tag{8.6}$$

The Jacobian of the vector field evaluated at each equilibrium point is given by:

$$(0,0): \quad \begin{pmatrix} \mu & 0 \\ 0 & -1 \end{pmatrix} \tag{8.7}$$

$$(\mu,0): \quad \begin{pmatrix} -\mu & 0 \\ 0 & -1 \end{pmatrix} \tag{8.8}$$

from which it follows that $(0,0)$ is asymptotically stable for $\mu < 0$, and a hyperbolic saddle for $\mu > 0$, and $(\mu, 0)$ is a hyperbolic saddle for $\mu < 0$ and asymptotically stable for $\mu > 0$. These two lines of fixed points cross at $\mu = 0$, at which there is only one, nonhyperbolic fixed point.

In Fig. 8.3 we show the bifurcation diagram for (8.5) in the μ–x plane.

In Fig. 8.4 we illustrate the bifurcation of equilibria for (8.5) in the x–y plane for $\mu < 0$, $\mu = 0$, and $\mu > 0$.

This type of bifurcation is referred to as a transcritical bifurcation.

The key characteristic of the transcritical bifurcation is the following. As a parameter (μ) is varied, the number of equilibria change from two to one, and back to two, and the change in number of equilibria occurs at a parameter value corresponding to the two equilibria coalescing into one *nonhyperbolic equilibrium*.

Example 24 (The (Supercritical) Pitchfork Bifurcation). Consider the following nonlinear, autonomous vector field on $\mathbb{R}^2$:

$$\begin{aligned} \dot{x} &= \mu x - x^3, \\ \dot{y} &= -y, \quad (x,y) \in \mathbb{R}^2, \end{aligned} \tag{8.9}$$

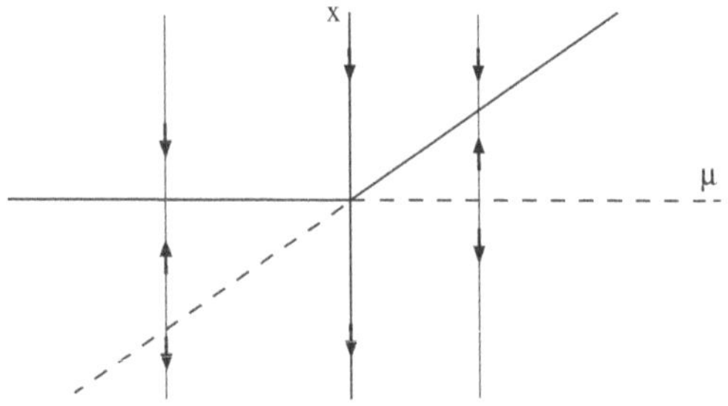

Fig. 8.3. Bifurcation diagram for (8.5) in the μ–x plane. The curves of equilibria are given by $\mu = x$ and $x = 0$. The dashed line denotes unstable equilibria, and the solid line denotes stable equilibria.

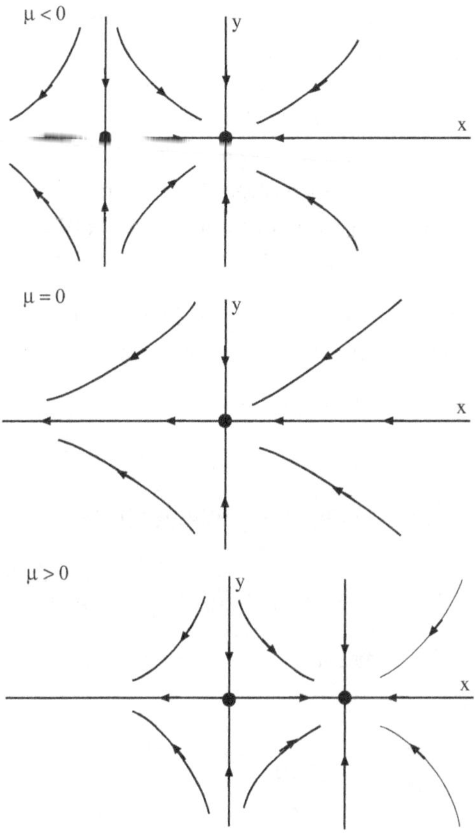

Fig. 8.4. Bifurcation diagram for (8.1) in the x–y plane for $\mu < 0$, $\mu = 0$, and $\mu > 0$. Compare with Fig. 8.3.

where μ is a (real) parameter. The equilibrium points of (8.9) are given by:

$$(x, y) = (0, 0),\ (\sqrt{\mu}, 0),\ (-\sqrt{\mu}, 0) \tag{8.10}$$

The Jacobian of the vector field evaluated at each equilibrium point is given by:

$$(0, 0): \quad \begin{pmatrix} \mu & 0 \\ 0 & -1 \end{pmatrix} \tag{8.11}$$

$$(\pm\sqrt{\mu}, 0): \quad \begin{pmatrix} -2\mu & 0 \\ 0 & -1 \end{pmatrix} \tag{8.12}$$

from which it follows that $(0,0)$ is asymptotically stable for $\mu < 0$, and a hyperbolic saddle for $\mu > 0$, and $(\pm\sqrt{\mu}, 0)$ are asymptotically stable for $\mu > 0$, and do not exist for $\mu < 0$. These two curves of fixed points pass through zero at $\mu = 0$, at which there is only one, nonhyperbolic fixed point.

In Fig. 8.5 we show the bifurcation diagram for (8.9) in the μ–x plane.

In Fig. 8.6, we illustrate the bifurcation of equilibria for (8.9) in the x–y plane for $\mu < 0$, $\mu = 0$, and $\mu > 0$.

Example 25 (The (Subcritical) Pitchfork Bifurcation). Consider the following nonlinear, autonomous vector field on $\mathbb{R}^2$:

$$\begin{aligned} \dot{x} &= \mu x + x^3, \\ \dot{y} &= -y, \quad (x, y) \in \mathbb{R}^2, \end{aligned} \tag{8.13}$$

where μ is a (real) parameter. The equilibrium points of (8.9) are given by:

$$(x, y) = (0, 0), \ (\sqrt{-\mu}, 0), \ (-\sqrt{-\mu}, 0). \tag{8.14}$$

The Jacobian of the vector field evaluated at each equilibrium point is given by:

$$(0,0): \quad \begin{pmatrix} \mu & 0 \\ 0 & -1 \end{pmatrix} \tag{8.15}$$

$$(\pm\sqrt{-\mu}, 0): \quad \begin{pmatrix} -2\mu & 0 \\ 0 & -1 \end{pmatrix} \tag{8.16}$$

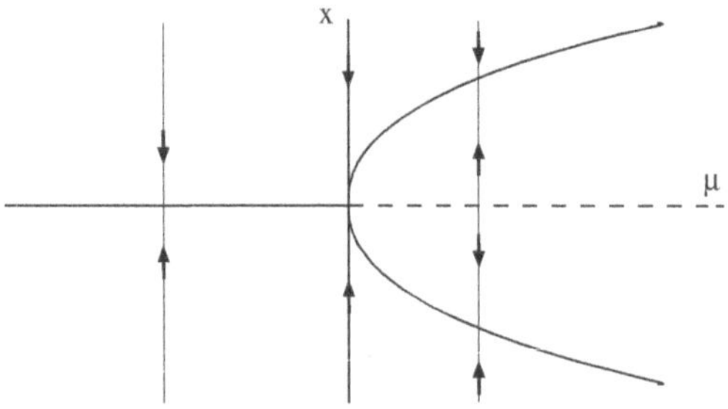

Fig. 8.5. Bifurcation diagram for (8.9) in the μ–x plane. The curves of equilibria are given by $\mu = x^2$, and $x = 0$. The dashed line denotes unstable equilibria, and the solid line denotes stable equilibria.

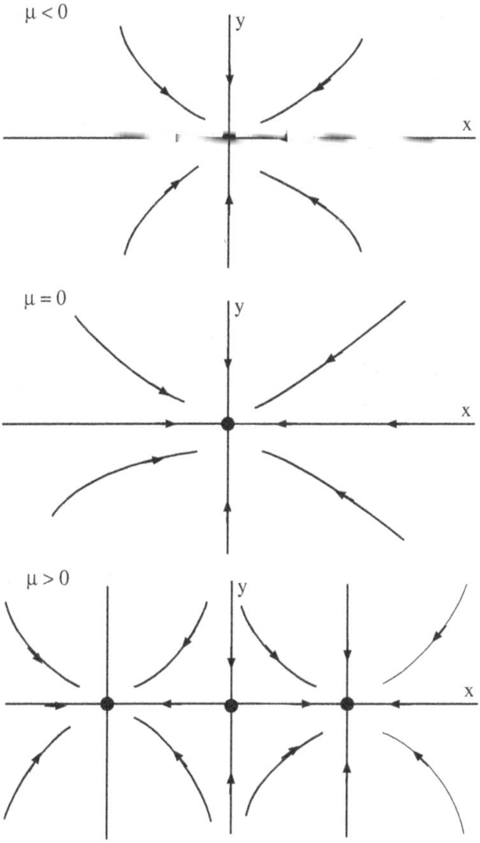

Fig. 8.6. Bifurcation diagram for (8.9) in the x–y plane for $\mu < 0$, $\mu = 0$, and $\mu > 0$. Compare with Fig. 8.5.

from which it follows that $(0, 0)$ is asymptotically stable for $\mu < 0$, and a hyperbolic saddle for $\mu > 0$, and $(\pm\sqrt{-\mu}, 0)$ are hyperbolic saddles for $\mu < 0$, and do not exist for $\mu > 0$. These two curves of fixed points pass through zero at $\mu = 0$, at which there is only one, nonhyperbolic fixed point.

In Fig. 8.7, we show the bifurcation diagram for (8.13) in the μ–x plane.

In Fig. 8.8, we illustrate the bifurcation of equilibria for (8.9) in the x–y plane for $\mu < 0$, $\mu = 0$, and $\mu > 0$.

We note that the phrase supercritical pitchfork bifurcation is also referred to as a soft loss of stability and the phrase subcritical pitchfork

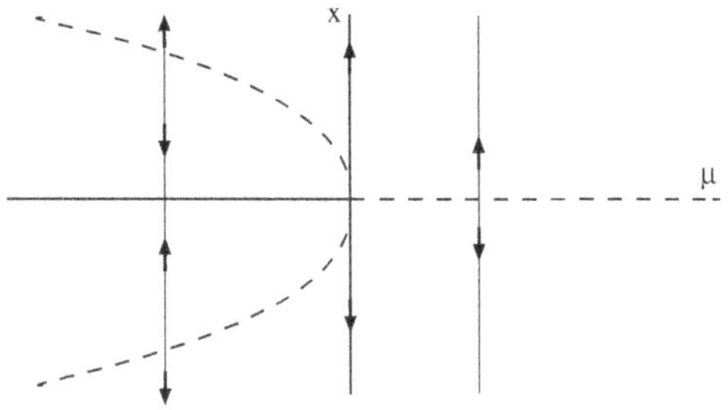

Fig. 8.7. Bifurcation diagram for (8.13) in the μ–x plane. The curves of equilibria are given by $\mu = -x^2$, and $x = 0$. The dashed curves denote unstable equilibria, and the solid line denotes stable equilibria.

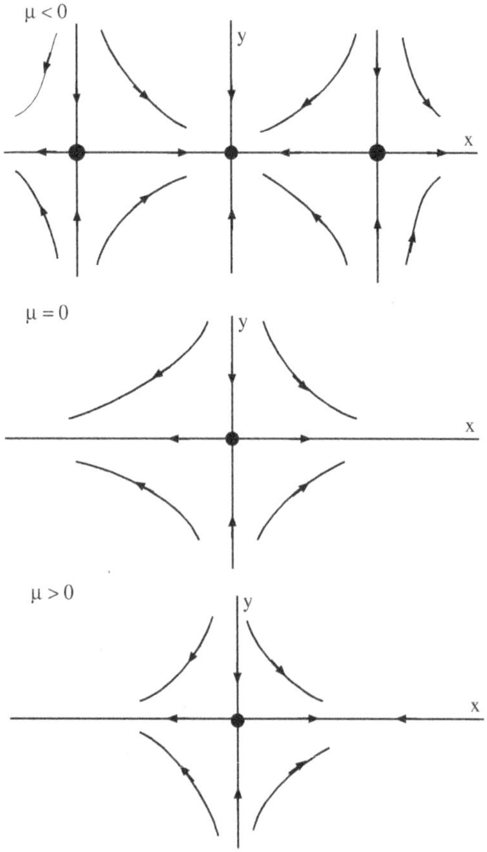

Fig. 8.8. Bifurcation diagram for (8.13) in the x–y plane for $\mu < 0$, $\mu = 0$, and $\mu > 0$. Compare with Fig. 8.7.

bifurcation is referred to as a hard loss of stability. What this means is the following. In the supercritical pitchfork bifurcation as μ goes from negative to positive the equilibrium point loses stability, but as μ increases past zero the trajectories near the origin are bounded in how far away from the origin they can move. In the subcritical pitchfork bifurcation, the origin loses stability as μ increases from negative to positive, but trajectories near the unstable equilibrium can become unbounded.

It is natural to ask the question,

"WHAT IS COMMON ABOUT THESE THREE EXAMPLES OF BIFURCATIONS OF FIXED POINTS OF ONE-DIMENSIONAL AUTONOMOUS VECTOR FIELDS?"

We note the following:

- A necessary (but *not* sufficient) for bifurcation of a fixed point is nonhyperbolicity of the fixed point.
- The "nature" of the bifurcation (e.g. numbers and stability of fixed points that are created or destroyed) is determined by the form of the nonlinearity.

But we could go further and ask what is in common about these examples that could lead to a definition of the bifurcation of a fixed point for autonomous vector fields? From the common features we give the following definition.

Definition 21 (Bifurcation of a fixed point of a one-dimensional autonomous vector field). We consider a one-dimensional autonomous vector field depending on a parameter, μ. We assume that at a certain parameter value it has a fixed point that is not hyperbolic. We say that a bifurcation occurs at that "nonhyperbolic parameter value" if for μ in a neighborhood of that parameter value the number of fixed points and their stability changes.[1]

Finally, we finish the discussion of bifurcations of a fixed point of one-dimensional autonomous vector fields with an example showing that a nonhyperbolic fixed point may not bifurcate as a parameter is varied, i.e. nonhyperbolicity is a necessary, but not sufficient, condition for bifurcation.

---

[1]Consider this definition in the context of our examples (and pay particular attention to the transcritical bifurcation).

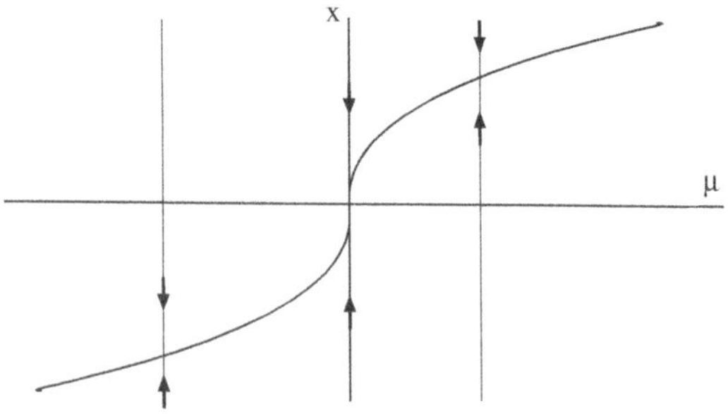

Fig. 8.9. Bifurcation diagram for (8.17).

Example 26. We consider the one-dimensional autonomous vector field:

$$\dot{x} = \mu - x^3, \quad x \in \mathbb{R}, \tag{8.17}$$

where μ is a parameter. This vector field has a nonhyperbolic fixed point at $x = 0$ for $\mu = 0$. The curve of fixed points in the μ–x plane is given by $\mu = x^3$, and the Jacobian of the vector field is $-3x^2$, which is strictly negative at all fixed points, except for the nonhyperbolic fixed point at the origin.

In Fig. 8.9 we plot the fixed points as a function of μ.

We see that there is no change in the number or stability of the fixed points for $\mu > 0$ and $\mu < 0$. Hence, no bifurcation.

Problem Set 8

1. Consider the following autonomous vector fields on the plane depending on a scalar parameter μ. Verify that each vector field has a fixed point at $(x, y) = (0, 0)$ for $\mu = 0$. Determine the linearized stability of this fixed point. Determine the nature (i.e. stability and number) of the fixed points for μ in a neighborhood of zero. (In other words, carry out a bifurcation analysis.) Sketch the flow in a neighborhood of each fixed point for values of μ corresponding to changes in stability and/or numbers of fixed points.

   (a)

$$\dot{x} = \mu + 10x^2,$$
$$\dot{y} = x - 5y.$$

(b)

$$
\begin{aligned}
\dot{x} &= \mu x + 10x^2, \\
\dot{y} &= x - 2y.
\end{aligned}
$$

(c)

$$
\begin{aligned}
\dot{x} &= \mu x + x^5, \\
\dot{y} &= -y.
\end{aligned}
$$

2. Consider the following two-dimensional autonomous ODE:

$$
\begin{aligned}
\dot{x} &= \mu x - x^2 - x^2 y, \\
\dot{y} &= x^2 y + y, \quad (x, y) \in \mathbb{R}^2.
\end{aligned}
$$

where μ is a parameter.

(a) Find all equilibria and show that they lie on the x-axis.
(b) Show that the x-axis is an invariant manifold?
(c) Show that a transcritical bifurcation occurs as a function of μ.
(d) Draw an appropriate bifurcation diagram.

3. Consider the ODE:

$$
\begin{aligned}
\dot{x} &= \mu + xy - x^2, \\
\dot{y} &= xy - y^3 - \nu y, \quad (x, y) \in \mathbb{R}^2,
\end{aligned}
\tag{8.18}
$$

where μ and ν are parameters.

(a) Compute a one-dimensional invariant manifold for (8.18).
(b) Show that a saddle-node bifurcation occurs in this problem. What is the bifurcation parameter? Sketch the bifurcation diagram and indicate the stability of the equilibria.

4. Consider the following two-dimensional autonomous ODE:

$$
\begin{aligned}
\dot{x} &= -x^3 + xy^2 + \mu x, \\
\dot{y} &= xy + \mu y, \quad (x, y) \in \mathbb{R}^2,
\end{aligned}
$$

where μ is a parameter.

(a) Find all equilibria.
(b) Show that the x-axis and the y-axis are invariant manifolds
(c) Show that a pitchfork bifurcation occurs as a function of μ.
(d) Draw an appropriate bifurcation diagram.

Chapter 9

Bifurcation of Equilibria, II

We have examined fixed points of one-dimensional autonomous vector fields where the matrix associated with the linearization of the vector field about the fixed point has a zero eigenvalue.

IT IS NATURAL TO ASK "ARE THERE MORE COMPLICATED BIFURCATIONS, AND WHAT MAKES THEM COMPLICATED?". If one thinks about the examples considered so far, there are two possibilities that could complicate the situation. One is that there could be more than one eigenvalue of the linearization about the fixed point with zero real part (which would necessarily require a consideration of higher dimensional vector fields), and the other would be more complicated nonlinearity (or a combination of these two). Understanding how dimensionality and nonlinearity contribute to the "complexity" of a bifurcation (and what that might mean) is a very interesting topic, but beyond the scope of this course. This is generally a topic explored in graduate level courses on dynamical systems theory that emphasize bifurcation theory. Here we are mainly just introducing the basic ideas and issues with examples that one might encounter in applications. Towards that end we will consider an example of a bifurcation that is very important in applications — the Poincaré–Andronov–Hopf bifurcation (or just Hopf bifurcation as it is more commonly referred to). This is a bifurcation of a fixed point of an autonomous vector field where the fixed point is nonhyperbolic as a result of the Jacobian having a pair of purely imaginary eigenvalues, $\pm i\,\omega$, $\omega \neq 0$. Therefore this type of bifurcation requires (at least two dimensions), and it is not characterized by a change in the number of fixed points, but by the creation of time dependent periodic solutions. We will analyze this situation by considering a specific example. References solely devoted to the Hopf bifurcation are the books of Marsden and McCracken[1] and Hassard, Kazarinoff, and Wan.[2]

We consider the following nonlinear autonomous vector field on the plane:

$$\begin{aligned} \dot{x} &= \mu x - \omega y + (ax - by)(x^2 + y^2), \\ \dot{y} &= \omega x + \mu y + (bx + ay)(x^2 + y^2), \quad (x, y) \in \mathbb{R}^2, \end{aligned} \tag{9.1}$$

where we consider a, b, ω as fixed constants and μ as a variable parameter. The origin, $(x, y) = (0, 0)$ is a fixed point, and we want to consider its stability. The matrix associated with the linearization about the origin is given by:

$$\begin{pmatrix} \mu & -\omega \\ \omega & \mu \end{pmatrix}, \tag{9.2}$$

and its eigenvalues are given by:

$$\lambda_{1,2} = \mu \pm i\omega. \tag{9.3}$$

Hence, as a function of μ the origin has the following stability properties:

$$\begin{cases} \mu < 0 & \text{sink} \\ \mu = 0 & \text{center} \\ \mu > 0 & \text{source} \end{cases} \tag{9.4}$$

The origin is not hyperbolic at $\mu = 0$, and there is a change in stability as μ changes sign. We want to analyze the behavior near the origin, both in phase space and in parameter space, in more detail.

Towards this end we transform (9.1) to polar coordinates using the standard relationship between cartesian and polar coordinates:

$$x = r\cos\theta, \quad y = r\sin\theta \tag{9.5}$$

---

[1] J. E. Marsden and M. McCracken. *The Hopf Bifurcation and Its Applications.* Applied Mathematical Sciences. Springer-Verlag, 1976.

[2] B. D. Hassard, N. D. Kazarinoff, and Y.-H. Wan. *Theory and Applications of Hopf Bifurcation*, Volume 41. CUP Archive, 1981.

Differentiating these two expressions with respect to t, and substituting into (9.1) gives:

$$\dot{x} = \dot{r}\cos\theta - r\dot{\theta}\sin\theta = \mu r\cos\theta - \omega r\sin\theta + (ar\cos\theta - br\sin\theta)r^2, \tag{9.6}$$

$$\dot{y} = \dot{r}\sin\theta + r\dot{\theta}\cos\theta = \omega r\cos\theta + \mu r\sin\theta + (br\cos\theta + ar\sin\theta)r^2, \tag{9.7}$$

from which we obtain the following equations for $\dot{r}$ and $\dot{\theta}$:

$$\dot{r} = \mu r + ar^3, \tag{9.8}$$

$$r\dot{\theta} = \omega r + br^3, \tag{9.9}$$

where (9.8) is obtained by multiplying (9.6) by $\cos\theta$ and (9.7) by $\sin\theta$ and adding the two results, and (9.9) is obtained by multiplying (9.6) by $-\sin\theta$ and (9.7) by $\cos\theta$ and adding the two results. Dividing both equations by r gives the two equations that we will analyze:

$$\dot{r} = \mu r + ar^3, \tag{9.10}$$

$$\dot{\theta} = \omega + br^2. \tag{9.11}$$

Note that (9.10) has the form of the pitchfork bifurcation that we studied earlier. However, it is very important to realize that we are dealing with the equations in polar coordinates and to understand what they reveal to us about the dynamics in the original cartesian coordinates. To begin with, we must keep in mind that $r \geq 0$.

Note that (9.10) is independent of θ, i.e. it is a one-dimensional, autonomous ODE which we rewrite below:

$$\dot{r} = \mu r + ar^3 = r(\mu + ar^2). \tag{9.12}$$

The fixed points of this equation are:

$$r = 0, \quad r = \sqrt{-\frac{\mu}{a}} \equiv r^+. \tag{9.13}$$

(Keep in mind that $r \geq 0$.) Substituting r^+ into (9.10) and (9.11) gives:

$$\dot{r^+} = \mu r^+ + a(r)^{+3} = 0,$$

$$\dot{\theta} = \omega + b\left(-\frac{\mu}{a}\right) \tag{9.14}$$

The θ component can easily be solved (using $r^+ = \sqrt{-\frac{\mu}{a}}$), after which we obtain:

$$\theta(t) = \left(\omega - \frac{\mu b}{a}\right) t + \theta(0). \tag{9.15}$$

Therefore r does not change in time at $r = r^+$ and θ evolves linearly in time. But θ is an angular coordinate. This implies that $r = r^+$ corresponds to a periodic orbit.[3]

Using this information, we analyze the behavior of (9.12) by constructing the bifurcation diagram. There are two cases to consider: $a > 0$ and $a < 0$.

In Fig. 9.1 we sketch the zeros of (9.12) as a function of μ for $a > 0$.

We see that a periodic orbit bifurcates from the nonhyperbolic fixed point at $\mu = 0$. The periodic orbit is unstable and exists for $\mu < 0$. In Fig. 9.2 we illustrate the dynamics in the $x - y$ phase plane.

In Fig. 9.3 we sketch the zeros of (9.12) as a function of μ for $a < 0$.

We see that a periodic orbit bifurcates from the nonhyperbolic fixed point at $\mu = 0$. The periodic orbit is stable in this case and exists for $\mu > 0$. In Fig. 9.4 we illustrate the dynamics in the $x - y$ phase plane.

In this example, we have seen that a nonhyperbolic fixed point of a two-dimensional autonomous vector field, where the nonhyperbolicity arises from the fact that the linearization at the fixed point has a pair of pure imaginary eigenvalues, $\pm i\omega$, can lead to the creation of periodic orbits as a parameter is varied. This is an example of what is generally called the

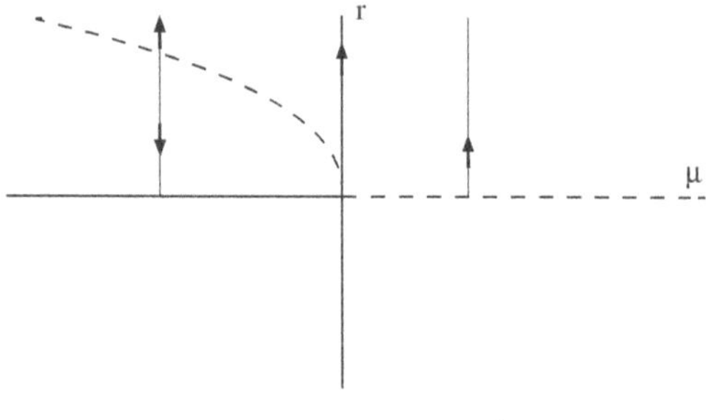

Fig. 9.1. The zeros of (9.12) as a function of μ for $a > 0$.

[3]At this point it is very useful to "pause" and think about the reasoning that led to the conclusion that $r = r^+$ is a periodic orbit.

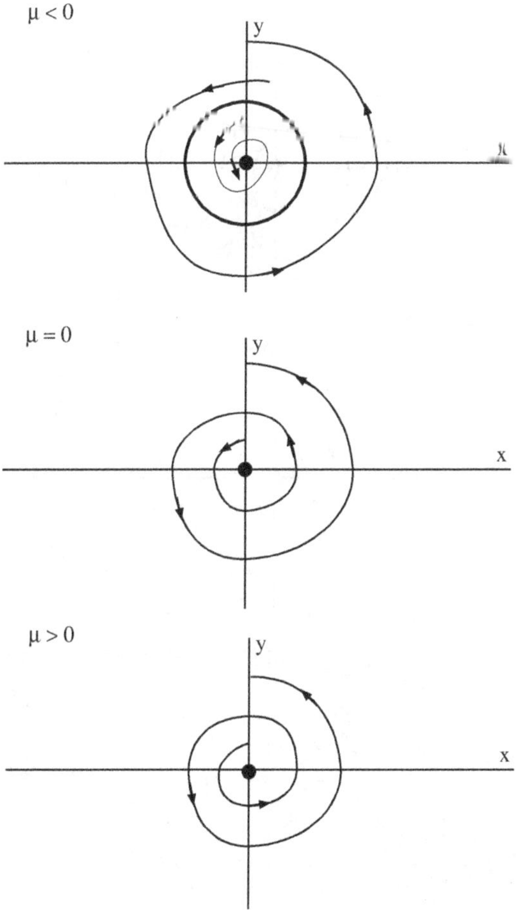

Fig. 9.2. Phase plane as a function of μ in the $x-y$ plane for $a>0$.

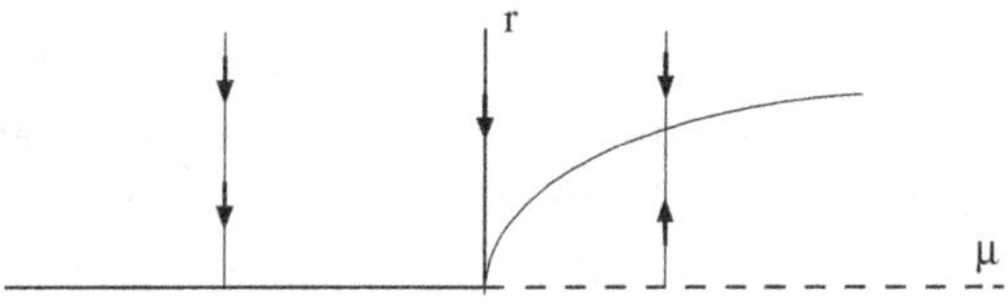

Fig. 9.3. The zeros of (9.12) as a function of μ for $a<0$.

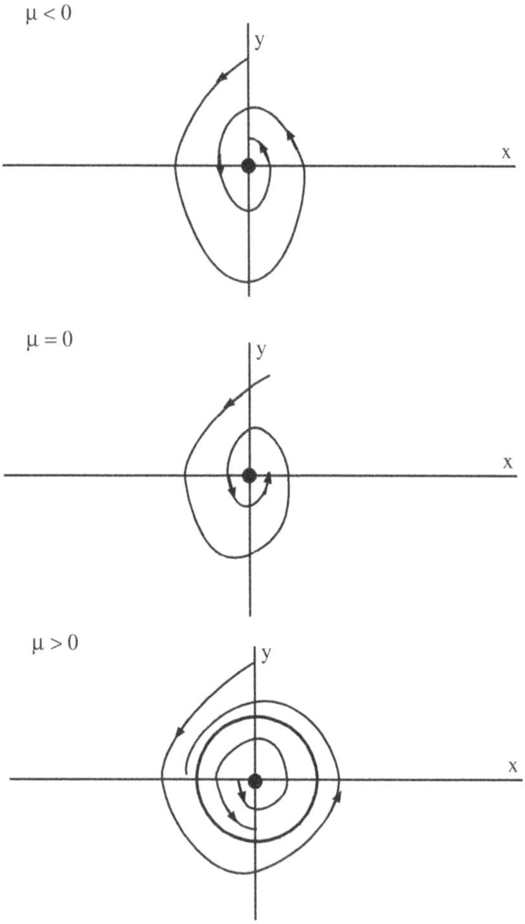

Fig. 9.4. Phase plane as a function of μ in the $x - y$ plane for $a < 0$.

Hopf bifurcation. It is the first example we have seen of the bifurcation of an equilibrium solution resulting in time-dependent solutions.

At this point it is useful to summarize the nature of the conditions resulting in a Hopf bifurcation for a two-dimensional, autonomous vector field. To begin with, we need a fixed point where the Jacobian associated with the linearization about the fixed point has a pair of pure imaginary eigenvalues. This is a necessary condition for the Hopf bifurcation. Now just as for bifurcations of fixed points of one-dimensional autonomous vector fields (e.g. the saddle-node, transcritical, and pitchfork bifurcations that we have studied in the previous chapter) the nature of the bifurcation for parameter values in a neighborhood of the bifurcation parameter is determined by the form of the nonlinearity of the vector field.[4] We developed the idea of the Hopf bifurcation in the context of (9.1), and in that example, the stability of the bifurcating periodic orbit was given by the sign of the coefficient a (stable for $a < 0$, unstable for $a > 0$). In the example the stability coefficient was evident from the simple structure of the nonlinearity. In more complicated examples, i.e. more complicated nonlinear terms, the determination of the stability coefficient is more algebraically intensive. Explicit expressions for the stability coefficient are given in many dynamical systems texts. For example, it is given in Guckenheimer and Holmes[5] and Wiggins.[6] The complete details of the calculation of the stability coefficient are carried out in the reference in Footnote 7. Problem 2 at the end of this chapter explores the nature of the Hopf bifurcation, e.g. the number and stability of bifurcating periodic orbits, for different forms of nonlinearity.

Next, we return to the examples of bifurcations of fixed points in one-dimensional vector fields and give two examples of one-dimensional vector fields where more than one of the bifurcations we discussed earlier can occur.

Example 27. Consider the following one-dimensional autonomous vector field depending on a parameter μ:

$$\dot{x} = \mu - \frac{x^2}{2} + \frac{x^3}{3}, \quad x \in \mathbb{R}. \tag{9.16}$$

The fixed points of this vector field are given by:

$$\mu = \frac{x^2}{2} - \frac{x^3}{3}, \tag{9.17}$$

and are plotted in Fig. 9.5.

[4] It would be very insightful to think about this statement in the context of the saddle-node, transcritical, and pitchfork bifurcations that we studied in the previous chapter.

[5] J. Guckenheimer and P. J. Holmes. *Nonlinear Oscillations, Dynamical Systems, and Bifurcations of Vector Fields*, Volume 42. Springer Science & Business Media, 2013.

[6] S. Wiggins. *Introduction to Applied Nonlinear Dynamical Systems and Chaos*, Volume 2. Springer Science & Business Media, 2003.

[7] B. D. Hassard, N. D. Kazarinoff, and Y.-H. Wan. *Theory and Applications of Hopf Bifurcation*, Volume 41. CUP Archive, 1981.

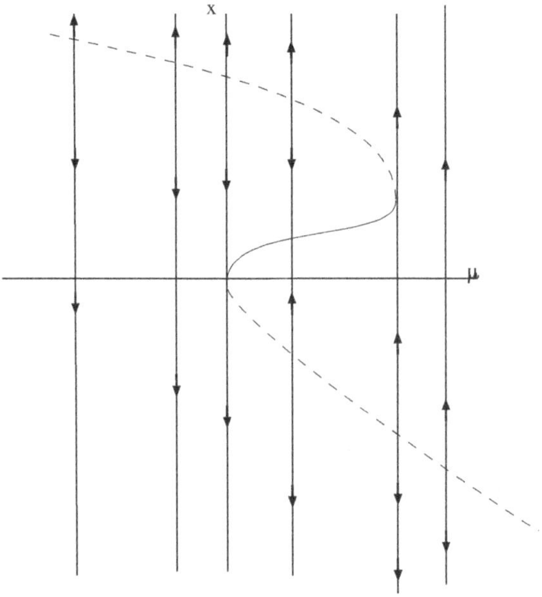

Fig. 9.5. Fixed points of (9.16) plotted in the $\mu - x$ plane.

The curve plotted is where the vector field is zero. Hence, it is positive to the right of the curve and negative to the left of the curve. From this we conclude the stability of the fixed points as shown in the figure.

There are two saddle node bifurcations that occur where $\frac{d\mu}{dx}(x) = 0$. These are located at

$$(x, \mu) = (0, 0), \left(1, \frac{1}{6}\right). \tag{9.18}$$

Example 28. Consider the following one-dimensional autonomous vector field depending on a parameter μ:

$$\begin{aligned} \dot{x} &= \mu x - \frac{x^3}{2} + \frac{x^4}{3}, \\ &= x\left(\mu - \frac{x^2}{2} + \frac{x^3}{3}\right) \end{aligned} \tag{9.19}$$

The fixed points of this vector field are given by:

$$\mu = \frac{x^2}{2} - \frac{x^3}{3}, \tag{9.20}$$

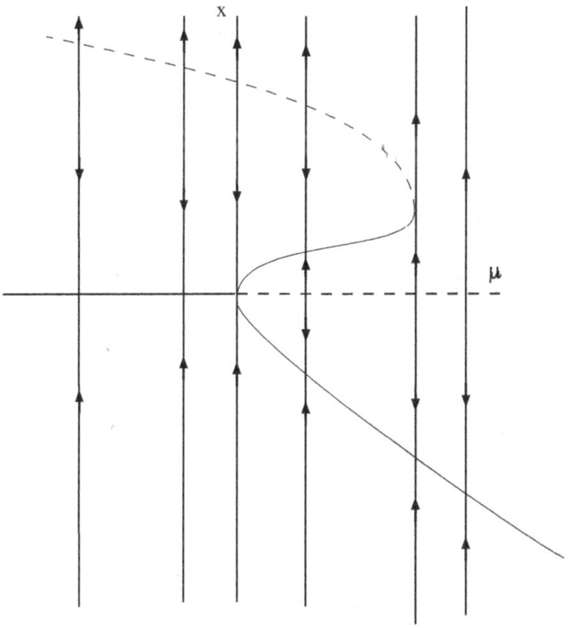

Fig. 9.6. Fixed points of (9.19) plotted in the $\mu - x$ plane.

and

$$x = 0, \tag{9.21}$$

and are plotted in the $\mu - x$ plane in Fig. 9.6.

In this example we see that there is a pitchfork bifurcation and a saddle-node bifurcation.

Problem Set 9

1. Consider the following autonomous vector field on the plane:

$$\dot{x} = \mu x - 3y - x(x^2 + y^2)^3,$$
$$\dot{y} = 3x + \mu y - y(x^2 + y^2)^3,$$

where μ is a parameter. Analyze possible bifurcations at $(x, y) = (0, 0)$ for μ in a neighborhood of zero. (Hint: Use polar coordinates.)

2. These exercises are from the book by Marsden and McCracken.[8] Consider the following vector fields expressed in polar coordinates, i.e. $(r, \theta) \in \mathbb{R}^+ \times S^1$, depending on a parameter μ. Analyze the stability of the origin and the stability of all bifurcating periodic orbits as a function of μ.

   (a)
$$\begin{aligned}\dot{r} &= -r(r-\mu)^2,\\ \dot{\theta} &= 1.\end{aligned}$$

   (b)
$$\begin{aligned}\dot{r} &= r(\mu - r^2)(2\mu - r^2)^2,\\ \dot{\theta} &= 1.\end{aligned}$$

   (c)
$$\begin{aligned}\dot{r} &= r(r+\mu)(r-\mu),\\ \dot{\theta} &= 1.\end{aligned}$$

   (d)
$$\begin{aligned}\dot{r} &= \mu r(r^2 - \mu),\\ \dot{\theta} &= 1.\end{aligned}$$

   (e)
$$\begin{aligned}\dot{r} &= -\mu^2 r(r+\mu)^2(r-\mu)^2,\\ \dot{\theta} &= 1.\end{aligned}$$

3. Consider the following vector field:

$$\dot{x} = \mu x - \frac{x^3}{2} + \frac{x^5}{4}, \quad x \in \mathbb{R},$$

   where μ is a parameter. Classify all bifurcations of equilibria and, in the process of doing this, determine all equilibria and their stability type.

[8] J. E. Marsden and M. McCracken. *The Hopf Bifurcation and Its Applications*. Applied Mathematical Sciences. Springer-Verlag, 1976.

4. Consider the following ODE:

$$\begin{aligned}\dot{x} &= \mu x - \omega y + (ax - by)\sqrt{x^2+y^2},\\ \dot{y} &= \omega x + \mu y + (bx + ay)\sqrt{x^2+y^2}, \quad (x,y) \in \mathbb{R}^2.\end{aligned} \tag{9.22}$$

   (a) Using the polar coordinate transformation $x = r\cos\theta$, $y = r\sin\theta$, transform (9.22) to polar coordinates.
   (b) Construct and describe bifurcation diagrams in the $r-\mu$ plane indicating the equilibria and periodic orbits for the cases $a < 0$, $a > 0$.

5. Consider the following ODE:

$$\begin{aligned}\dot{x} &= \mu x - \omega y + \left(-\frac{1}{2}x - y\right)(x^2+y^2) + \left(\frac{1}{4}x + y\right)(x^2+y^2)^2,\\ \dot{y} &= \omega x + \mu y + \left(x - \frac{1}{2}y\right)(x^2+y^2) + \left(-x + \frac{1}{4}y\right)(x^2+y^2)^2,\end{aligned} \tag{9.23}$$

   where $(x,y) \in \mathbb{R}^2$.

   (a) Using the polar coordinate transformation $x = r\cos\theta$, $y = r\sin\theta$, transform (9.23) to polar coordinates.
   (b) Construct and describe bifurcation diagrams in the $r-\mu$ plane indicating the equilibria and periodic orbits, and their stability.

Chapter 10

Center Manifold Theory

THIS CHAPTER IS ABOUT CENTER MANIFOLDS, DIMENSIONAL REDUCTION, AND STABILITY OF FIXED POINTS OF AUTONOMOUS VECTOR FIELDS.[1] We begin with a motivational example.

Example 29. Consider the following linear, autonomous vector field on $\mathbb{R}^c \times \mathbb{R}^s$:

$$\dot{x} = Ax,$$
$$\dot{y} = By, \quad (x, y) \in \mathbb{R}^c \times \mathbb{R}^s, \tag{10.1}$$

where A is a $c \times c$ matrix of real numbers having eigenvalues with zero real part and B is a $s \times s$ matrix of real numbers having eigenvalues with negative real part. Suppose we are interested in stability of the nonhyperbolic fixed point $(x, y) = (0, 0)$. Then that question is determined by the nature of stability of $x = 0$ in the lower dimensional vector field:

$$\dot{x} = Ax, \quad x \in \mathbb{R}^c. \tag{10.2}$$

[1]Expositions of center manifold theory are mostly found in advanced dynamical systems textbooks. It is likely true that all such expositions have their roots in the monographs by Henry and Carr. The monograph of Henry is a bit obscure, but it is a seminal work in the field. The monograph by Carr is a real jewel. All of the theorems in this chapter are taken from Carr, where the proofs can also be found.

D. Henry. *Geometric Theory of Semilinear Parabolic Equations.* Lecture Notes in Mathematics. Springer Berlin Heidelberg, 1993; and J. Carr. *Applications of Centre Manifold Theory,* Volume 35. Springer Science & Business Media, 2012.

This follows from the nature of the eigenvalues of B, and the properties that x and y are decoupled in (10.1) and that it is linear. More precisely, the solution of (10.1) is given by:

$$\begin{pmatrix} x(t, x_0) \\ y(t, y_0) \end{pmatrix} = \begin{pmatrix} e^{At} x_0 \\ e^{Bt} y_0 \end{pmatrix}. \tag{10.3}$$

From the assumption of the real parts of the eigenvalues of B having negative real parts, it follows that:

$$\lim_{t\to\infty} e^{Bt} y_0 = 0.$$

In fact, 0 is approached at an exponential rate in time. Therefore it follows that stability, or asymptotic stability, or instability of $x = 0$ for (10.2) implies stability, or asymptotic stability, or instability of $(x, y) = (0, 0)$ for (10.1).

It is natural to ask if such a dimensional reduction procedure holds for nonlinear systems. This might seem unlikely since, in general, nonlinear systems are coupled and the superposition principle of linear systems does not hold. However, we will see that this is not the case.

INVARIANT MANIFOLDS LEAD TO A FORM OF DECOUPLING THAT RESULTS IN A DIMENSIONAL REDUCTION PROCEDURE that gives, essentially, the same result as is obtained for this motivational linear example.[2] This is the topic of center manifold theory that we now develop.

We begin by describing the set-up. It is important to realize that when applying these results to a vector field, it must be in the following form.

$$\begin{aligned} \dot{x} &= Ax + f(x, y), \\ \dot{y} &= By + g(x, y), \quad (x, y) \in \mathbb{R}^c \times \mathbb{R}^s, \end{aligned} \tag{10.4}$$

[2]In fact, this is one of the main uses of invariant manifolds. They can play an essential role in developing dimensional reduction schemes. The "manifold" part is important because it is desirable for the reduced dimensional system to have properties where the usual techniques of calculus can be applied.

where the matrices A and B have the following properties:

$A-$ $c \times c$ matrix of real numbers
having eigenvalues with zero real parts,

$B-$ $s \times s$ matrix of real numbers
having eigenvalues with negative real parts,

and f and g are nonlinear functions. That is, they are of order two or higher in x and y, as expressed in the following properties:

$$\begin{aligned} f(0,0) &= 0, \qquad Df(0,0) = 0, \\ g(0,0) &= 0, \qquad Dg(0,0) = 0, \end{aligned} \tag{10.5}$$

and they are C^r, r as large as required (we will explain what this means when we explicitly use this property later on).

With this set-up $(x, y) = (0, 0)$ is a fixed point for (10.4) and we are interested in its stability properties.

The linearization of (10.4) about the fixed point is given by:

$$\begin{aligned} \dot{x} &= Ax, \\ \dot{y} &= By, \quad (x, y) \in \mathbb{R}^c \times \mathbb{R}^s. \end{aligned} \tag{10.6}$$

The fixed point is nonhyperbolic. It has a c-dimensional invariant center subspace and a s-dimensional invariant stable subspace given by:

$$E^c = \{(x, y) \in \mathbb{R}^c \times \mathbb{R}^s \,|\, y = 0\}, \tag{10.7}$$

$$E^s = \{(x, y) \in \mathbb{R}^c \times \mathbb{R}^s \,|\, x = 0\}, \tag{10.8}$$

respectively.

For the nonlinear system (10.4) there is a s-dimensional, C^r local invariant stable manifold passing through the origin and tangent to E^s at the origin. Moreover, trajectories in the local stable manifold inherit their behavior from trajectories in E^s under the linearized dynamics in the sense that they approach the origin at an exponential rate in time.

Similarly, there is a c-dimensional C^r local invariant center manifold that passes through the origin and is tangent to E^c at the origin. Hence,

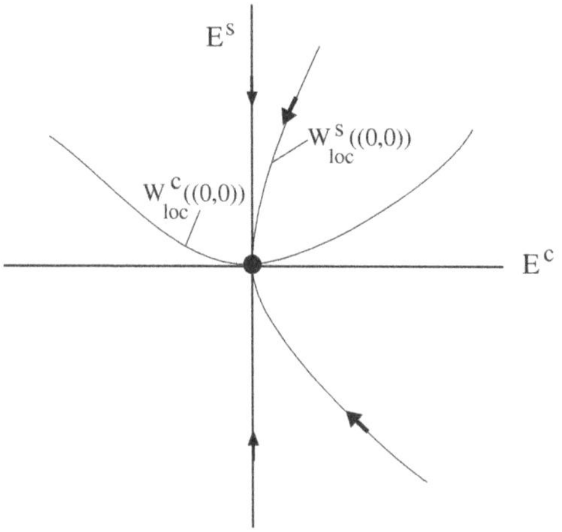

Fig. 10.1. The geometry of the stable and center manifolds near the origin.

the center manifold has the form:

$$W^c_{\text{loc}}((0,0)) = \{(x,y) \in \mathbb{R}^c \times \mathbb{R}^s \,|\, y = h(x),\, h(0) = 0,\, Dh(0) = 0\}, \tag{10.9}$$

which is valid in a neighborhood of the origin, i.e. for $|x|$ sufficiently small.[3]

We illustrate the geometry in Fig. 10.1.

The application of the center manifold theory for analyzing the behavior of trajectories near the origin is based on three theorems:

- existence of the center manifold and the vector field restricted to the center manifold,
- stability of the origin restricted to the center manifold and its relation to the stability of the origin in the full dimensional phase space,
- obtaining an approximation to the center manifold.

Theorem 5 (Existence and Restricted Dynamics). *There exists a C^r center manifold of $(x,y) = (0,0)$ for (10.4). The dynamics of (10.4)*

[3] At this point it would be insightful to consider (10.9) and make sure you understand how it embodies the geometrical properties of the local center manifold of the origin that we have described.

restricted to the center manifold is given by:

$$\dot{u} = Au + f(u, h(u)), \quad u \in \mathbb{R}^c, \tag{10.10}$$

for $|u|$ *sufficiently small.*

A natural question that arises from the statement of this theorem is "why did we use the variable 'u' when it would seem that 'x' would be the more natural variable to use in this situation"? Understanding the answer to this question will provide some insight and understanding to the nature of solutions near the origin and the geometry of the invariant manifolds near the origin. The answer is that "x and y are already used as variables for describing the coordinate axes in (10.4)". We do not want to confuse a point in the center manifold with a point on the coordinate axis. A point on the center manifold is denoted by $(x, h(x))$. The coordinate u denotes a parametric representation of points along the center manifold. Moreover, we will want to compare trajectories of (10.10) with trajectories in (10.4). This will be confusing if x is used to denote a point on the center manifold. However, when computing the center manifold and when considering (i.e. (10.10)) it is traditional to use the coordinate 'x', i.e. the coordinate describing the points in the center *subspace*. This does not cause ambiguities since we can name a coordinate anything we want. However, it would cause ambiguities when comparing trajectores in (10.4) with trajectories in (10.10), as we do in the next theorem.[4]

Theorem 6. *(i) Suppose that the zero solution of* (10.10) *is stable (asymptotically stable) (unstable), then the zero solution of* (10.4) *is also stable (asymptotically stable) (unstable). (ii) Suppose that the zero solution of* (10.10) *is stable. Then if* $(x(t), y(t))$ *is a solution of* (10.4) *with* $(x(0), y(0))$ *sufficiently small, then there is a solution* $u(t)$ *of* (10.10) *such that as* $t \to \infty$

$$\begin{aligned} x(t) &= u(t) + \mathcal{O}(e^{-\gamma t}), \\ y(t) &= h(u(t)) + \mathcal{O}(e^{-\gamma t}), \end{aligned} \tag{10.11}$$

where $\gamma > 0$ *is a constant.*

[4]If this long paragraph is confusing it would be fruitful to spend some time considering each point. There is some useful insight to be gained.

Part (i) of this theorem says that stability properties of the origin in the center manifold imply *the same* stability properties of the origin in the full dimensional equations. Part (ii) gives much more precise results for the case that the origin is stable. It says that trajectories starting at initial conditions sufficiently close to the origin asymptotically approach a trajectory in the center manifold.

Now we would like to compute the center manifold so that we can use these theorems in specific examples. In general, it is not possible to compute the center manifold. However, it is possible to approximate it to "sufficiently high accuracy" so that we can verify the stability results of Theorem 6 can be confirmed. We will show how this can be done. The idea is to derive an equation that the center manifold must satisfy, and then develop an approximate solution to that equation.

We develop this approach step-by-step.

The center manifold is realized as the graph of a function,

$$y = h(x), \quad x \in \mathbb{R}^c, \ y \in \mathbb{R}^s, \tag{10.12}$$

i.e. any point (x_c, y_c), sufficiently close to the origin, that is in the center manifold satisfies $y_c = h(x_c)$. In addition, the center manfold passes through the origin ($h(0) = 0$) and is tangent to the center subspace at the origin ($Dh(0) = 0$).

Invariance of the center manifold implies that the graph of the function $h(x)$ must also be invariant with respect to the dynamics generated by (10.4). Differentiating (10.12) with respect to time shows that $(\dot{x}, \dot{y})$ at any point on the center manifold satisfies:

$$\dot{y} = Dh(x)\dot{x}. \tag{10.13}$$

THIS IS JUST THE ANALYTICAL MANIFESTATION OF THE FACT THAT INVARIANCE OF A SURFACE WITH RESPECT TO A VECTOR FIELD IMPLIES THAT THE VECTOR FIELD MUST BE TANGENT TO THE SURFACE.[5]

We will now use these properties to derive an equation that must be satisfied by the local center manifold.

The starting point is to recall that any point on the local center manifold obeys the dynamics generated by (10.4). Substituting $y = h(x)$

[5] It would be very insightful to think about this statement in the context of the examples from earlier chapters that involved determining invariant sets and invariant manifolds.

into (10.1) gives:

$$\dot{x} = Ax + f(x, h(x)), \tag{10.14}$$

$$\dot{y} = Bh(x) + g(x, h(x)), \quad (x, y) \in \mathbb{R}^c \times \mathbb{R}^s. \tag{10.15}$$

Substituting (10.14) and (10.15) into the invariance condition $\dot{y} = Dh(x)\dot{x}$ gives:

$$Bh(x) + g(x, h(x)) = Dh(x)\,(Ax + f(x, h(x))), \tag{10.16}$$

or

$$Dh(x)\,(Ax + f(x, h(x))) - Bh(x) - g(x, h(x)) \equiv \mathcal{N}(h(x) = 0. \tag{10.17}$$

This is an equation for $h(x)$. By construction, the solution implies invariance of the graph of $h(x)$, and we seek a solution satisfying the additional conditions $h(0) = 0$ and $Dh(0) = 0$. The basic result on approximation of the center manifold is given by the following theorem.

Theorem 7 (Approximation). *Let $\phi : \mathbb{R}^c \to \mathbb{R}^s$ be a C^1 mapping with*

$$\phi(0) = 0, \quad D\phi(0) = 0,$$

such that

$$\mathcal{N}(\phi(x)) = \mathcal{O}(|x|^q) \quad as \quad x \to 0,$$

for some $q > 1$. Then

$$|h(x) - \phi(x)| = \mathcal{O}(|x|^q) \quad as \quad x \to 0.$$

The theorem states that if we can find an approximate solution of (10.17) to a specified degree of accuracy, then that approximate solution is actually an approximation to the local center manifold, to the same degree of accuracy.[6]

[6]This answers the question of why we said that the original vector field, (10.4), was C^r, "r as large as required". r will need to be as large as needed in order to obtain a sufficiently accurate approximation to the local center manifold. "Sufficiently accurate" is determined by our ability to deduce stability properties of the zero solution of the vector field restricted to the center manifold.

We now consider some examples showing how these results are applied.

Example 30. We consider the following autonomous vector field on the plane:

$$\begin{aligned} \dot{x} &= x^2 y - x^5, \\ \dot{y} &= -y + x^2, \quad (x, y) \in \mathbb{R}^2. \end{aligned} \tag{10.18}$$

or, in matrix form:

$$\begin{pmatrix} \dot{x} \\ \dot{y} \end{pmatrix} = \begin{pmatrix} 0 & 0 \\ 0 & -1 \end{pmatrix} \begin{pmatrix} x \\ y \end{pmatrix} + \begin{pmatrix} x^2 y - x^5 \\ x^2 \end{pmatrix}. \tag{10.19}$$

We are interested in determining the nature of the stability of $(x, y) = (0, 0)$. The Jacobian associated with the linearization about this fixed point is:

$$\begin{pmatrix} 0 & 0 \\ 0 & -1 \end{pmatrix},$$

which is nonhyperbolic, and therefore the linearization does not suffice to determine stability.

The vector field is in the form of (10.4)

$$\begin{aligned} \dot{x} &= Ax + f(x, y), \\ \dot{y} &= By + g(x, y), \quad (x, y) \in \mathbb{R} \times \mathbb{R}, \end{aligned} \tag{10.20}$$

where

$$A = 0, \ B = -1, \ f(x, y) = x^2 y - x^5, \ g(x, y) = x^2. \tag{10.21}$$

We assume a center manifold of the form:

$$y = h(x) = ax^2 + bx^3 + \mathcal{O}(x^4), \tag{10.22}$$

which satisfies $h(0) = 0$ ("passes through the origin") and $Dh(0) = 0$ (tangent to E^c at the origin). A center manifold of this type will require the vector field to be at least C^3 (hence, the meaning of the phrase C^r, r as large as necessary).

Substituting this expression into the equation for the center manifold (10.17) (using (10.21)) gives:

$$\begin{aligned}&(2ax + 3bx^2 + \mathcal{O}(x^3))(ax^4 + bx^5 + \mathcal{O}(x^6) - x^5)\\&\quad + ax^2 + bx^3 + \mathcal{O}(x^4) - x^2 = 0.\end{aligned} \tag{10.23}$$

In order for this equation to be satisfied the coefficients on each power of x must be zero. Through third order this gives:

$$\begin{aligned}&x^2 : a - 1 = 0 \Rightarrow a = 1,\\&x^3 : b = 0.\end{aligned} \tag{10.24}$$

Substituting these values into (10.22) gives the following expression for the center manifold through third order:

$$y = x^2 + \mathcal{O}(x^4). \tag{10.25}$$

Therefore the vector field restricted to the center manifold is given by:

$$\dot{x} = x^4 + \mathcal{O}(x^5). \tag{10.26}$$

Hence, for x sufficiently small, $\dot{x}$ is positive for $x \neq 0$, and therefore the origin is unstable. We illustrate the flow near the origin in Fig. 10.2.

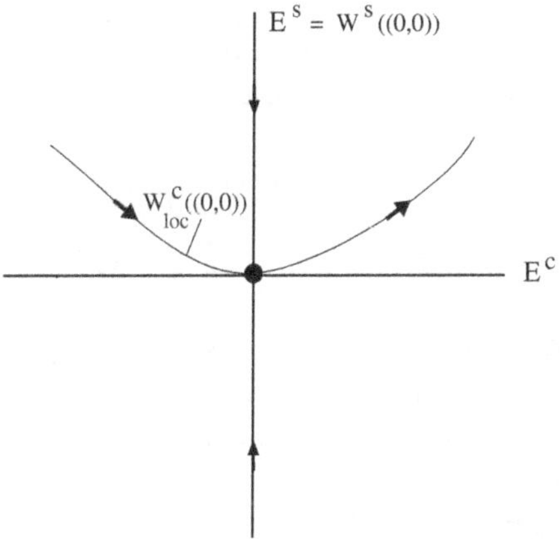

Fig. 10.2. The flow near the origin for (10.19).

Example 31. We consider the following autonomous vector field on the plane:

$$\begin{aligned}\dot{x} &= xy,\\ \dot{y} &= -y + x^3, \quad (x, y) \in \mathbb{R}^2,\end{aligned} \tag{10.27}$$

or, in matrix form:

$$\begin{pmatrix}\dot{x}\\ \dot{y}\end{pmatrix} = \begin{pmatrix}0 & 0\\ 0 & -1\end{pmatrix}\begin{pmatrix}x\\ y\end{pmatrix} + \begin{pmatrix}xy\\ x^3\end{pmatrix} \tag{10.28}$$

We are interested in determining the nature of the stability of $(x, y) = (0, 0)$. The Jacobian associated with the linearization about this fixed point is:

$$\begin{pmatrix}0 & 0\\ 0 & -1\end{pmatrix}.$$

which is nonhyperbolic, and therefore the linearization does not suffice to determine stability.

The vector field is in the form of (10.1)

$$\begin{aligned}\dot{x} &= Ax + f(x, y),\\ \dot{y} &= By + g(x, y), \quad (x, y) \in \mathbb{R} \times \mathbb{R},\end{aligned} \tag{10.29}$$

where

$$A = 0,\ B = -1,\ f(x, y) = xy,\ g(x, y) = x^3. \tag{10.30}$$

We assume a center manifold of the form:

$$y = h(x) = ax^2 + bx^3 + \mathcal{O}(x^4) \tag{10.31}$$

which satisfies $h(0) = 0$ ("passes through the origin") and $Dh(0) = 0$ (tangent to E^c at the origin). A center manifold of this type will require the vector field to be at least C^3 (hence, the meaning of the phrase C^r, r as large as necessary).

Substituting this expression into the equation for the center manifold (10.17) (using (10.30)) gives:

$$(2ax + 3bx^2 + \mathcal{O}(x^3))(ax^3 + bx^4 + \mathcal{O}(x^5)) + ax^2 + bx^3 + \mathcal{O}(x^4) - x^3 = 0. \tag{10.32}$$

In order for this equation to be satisfied the coefficients on each power of x must be zero. Through third order this gives:

$$\begin{aligned} x^2 &: a = 0, \\ x^3 &: b - 1 = 0 \Rightarrow b = 1. \end{aligned} \tag{10.33}$$

Substituting these values into (10.31) gives the following expression for the center manifold through third order:

$$y = x^3 + \mathcal{O}(x^4). \tag{10.34}$$

Therefore the vector field restricted to the center manifold is given by:

$$\dot{x} = x^4 + \mathcal{O}(x^5). \tag{10.35}$$

Since $\dot{x}$ is positive for x sufficiently small, the origin is unstable. We illustrate the flow near the origin in Fig. 10.3.

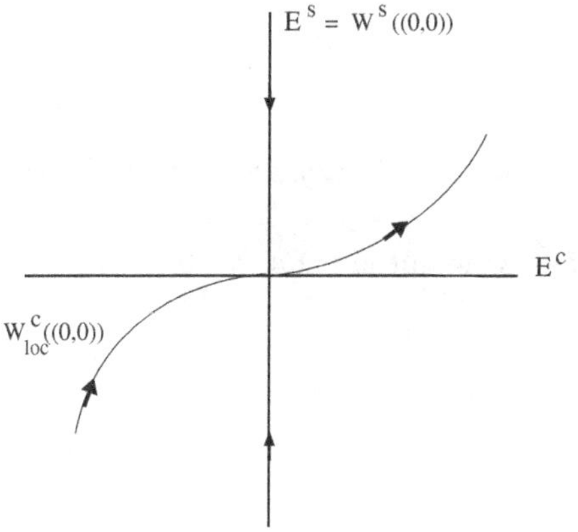

Fig. 10.3. The flow near the origin for (10.27).

Problem Set 10

1. Consider the following autonomous vector field on the plane:
$$\dot{x} = x^2 y - x^3,$$
$$\dot{y} = -y + x^3, \quad (x, y) \in \mathbb{R}^2.$$
Determine the stability of $(x, y) = (0, 0)$ using center manifold theory.[7]
2. Consider the following autonomous vector field on the plane:
$$\dot{x} = x^2,$$
$$\dot{y} = -y + x^2, \quad (x, y) \in \mathbb{R}^2.$$
Determine the stability of $(x, y) = (0, 0)$ using center manifold theory. Does the fact that solutions of $\dot{x} = x^2$ "blow up in finite time" influence your answer (why or why not)?
3. Consider the following autonomous vector field on the plane:
$$\dot{x} = -x + y^2,$$
$$\dot{y} = -2x^2 + 2xy^2, \quad (x, y) \in \mathbb{R}^2.$$
Show that $y = x^2$ is an invariant manifold. Show that there is a trajectory connecting $(0, 0)$ to $(1, 1)$, i.e. a heteroclinic trajectory.
4. Consider the following autonomous vector field on $\mathbb{R}^3$:
$$\dot{x} = y,$$
$$\dot{y} = -x - x^2 y,$$
$$\dot{z} = -z + xz^2, \quad (x, y, z) \in \mathbb{R}^3.$$
Determine the stability of $(x, y, z) = (0, 0, 0)$ using center manifold theory.[8]
5. Consider the following autonomous vector field on $\mathbb{R}^3$:
$$\dot{x} = y,$$
$$\dot{y} = -x - x^2 y + zxy,$$
$$\dot{z} = -z + xz^2, \quad (x, y, z) \in \mathbb{R}^3.$$

[7] When the equation restricted to the center manifold is one-dimensional, then stability can be deduced from the sign of the one-dimensional vector field near the origin.

[8] In this problem the vector field restricted to the center manifold is two-dimensional. The only techniques we learned for determining stability of a nonhypebolic fixed points for vector fields with more than one dimension was Lyapunov's method and the LaSalle invariance principle.

Determine the stability of $(x, y, z) = (0, 0, 0)$ using center manifold theory.

6. Consider the following autonomous vector field on $\mathbb{R}^3$:

$$\begin{aligned}\dot{x} &= y,\\ \dot{y} &= -x + zy^2,\\ \dot{z} &= -z + xz^2, \quad (x, y, z) \in \mathbb{R}^3.\end{aligned}$$

Determine the stability of $(x, y, z) = (0, 0, 0)$ using center manifold theory.

7. Consider the following ODE:

$$\begin{aligned}\dot{x} &= -x + y^2,\\ \dot{y} &= xy^2 - y^5.\end{aligned}$$

$(0, 0)$ is an equilibrium point.

(a) Compute the Jacobian of the ODE at the origin.
(b) Show that the linearized ODE has a one-dimensional stable subspace and a one-dimensional center subspace, and sketch them.
(c) Use center manifold theory to determine the nature of the stability of $(0, 0)$.

8. Consider the ODE:

$$\begin{aligned}\dot{x} &= xy,\\ \dot{y} &= -y - x^2 - 2y^3, \quad (x, y) \in \mathbb{R}^2.\end{aligned}$$

(a) Show that $(x, y) = (0, 0)$ is the only equilibrium point and it is not hyperbolic.
(b) Show that $x = 0$ is an invariant manifold.
(c) Using $V(x, y) = \frac{1}{2}\left(x^2 + y^2\right)$ what does Lyapunov's method allow you to conclude about stability of $(x, y) = (0, 0)$?
(d) Now apply the Lasalle invariant principle using $V(x, y) = \frac{1}{2}\left(x^2 + y^2\right)$. What does it allow you to conclude about stability of $(x, y) = (0, 0)$?
(e) Apply center manifold theory to analyze stability of $(x, y) = (0, 0)$. How does your conclusion compare with what you can conclude from the application of Lyapunov's method and the LaSalle invariance principle?

9. Consider the ODE:

$$\dot{x} = xy,$$
$$\dot{y} = -y + x^2 - +2y^2, \quad (x, y) \in \mathbb{R}^2.$$

(a) Show that $(x, y) = (0, 0)$ a nonhyperbolic equilibrium point.
(b) Use center manifold theory to analyze the stability of $(x, y) = (0, 0)$.
(c) Show that $x = 0$ is an invariant manifold.
(d) Find all the remaining equilibria and classify their stability.
(e) Sketch the phase portrait.

10. Consider the ODE:

$$\dot{x} = -\sin x - y^2,$$
$$\dot{y} = xy, \quad (x, y) \in \mathbb{R}^2.$$

(a) Show that $(x, y) = (0, 0)$ a nonhyperbolic equilibrium point.
(b) Use center manifold theory to analyze stability of $(x, y) = (0, 0)$.
(c) Show that $y = 0$ is an invariant manifold.
(d) Sketch the phase portrait near $(x, y) = (0, 0)$.

A

Jacobians, Inverses of Matrices, and Eigenvalues

In this appendix we collect together some results on Jacobians and inverses and eigenvalues of 2×2 matrices that are used repeatedly in the material.

First, we consider the Taylor expansion of a vector valued function of two variables, denoted as follows:

$$H(x, y) = \begin{pmatrix} f(x, y) \\ g(x, y) \end{pmatrix}, \quad (x, y) \in \mathbb{R}^2. \tag{A.1}$$

More precisely, we will need to Taylor expand such functions through second order:

$$H(x_0 + h, y_0 + k) = H(x_0, y_0) + DH(x_0, y_0) \begin{pmatrix} h \\ k \end{pmatrix} + \mathcal{O}(2). \tag{A.2}$$

The Taylor expansion of a scalar valued function of one variable should be familiar to most students at this level. Possibly there is less familiarity with the Taylor expansion of a vector valued function of a vector variable. However, to compute this, we just Taylor expand each component of the function (which is a scalar valued function of a vector variable) in each variable, holding the other variable fixed for the expansion in that particular variable, and then we gather the results for each component into matrix form.

Carrying this procedure out for the $f(x,y)$ component of (A.1) gives:

$$\begin{aligned} f(x_0+h, y_0+k) &= f(x_0, y_0+k) + \frac{\partial f}{\partial x}(x_0, y_0+k)h + \mathcal{O}(h^2), \\ &= f(x_0, y_0) + \frac{\partial f}{\partial y}(x_0, y_0)k + \mathcal{O}(k^2) + \frac{\partial f}{\partial x}(x_o, y_0)h \\ &\quad + \mathcal{O}(hk) + \mathcal{O}(h^2). \end{aligned} \tag{A.3}$$

The same procedure can be applied to $g(x,y)$. Recombining the terms back into the vector expresson for (A.1) gives:

$$\begin{aligned} H(x_0+h, y_0+k) &= \begin{pmatrix} f(x_0, y_0) \\ g(x_0, y_0) \end{pmatrix} \\ &\quad + \begin{pmatrix} \frac{\partial f}{\partial x}(x_0, y_0) & \frac{\partial f}{\partial y}(x_0, y_0) \\ \frac{\partial g}{\partial x}(x_0, y_0) & \frac{\partial g}{\partial y}(x_0, y_0) \end{pmatrix} \begin{pmatrix} h \\ k \end{pmatrix} + \mathcal{O}(2). \end{aligned} \tag{A.4}$$

Hence, the Jacobian of (A.1) at (x_0, y_0) is:

$$\begin{pmatrix} \frac{\partial f}{\partial x}(x_0, y_0) & \frac{\partial f}{\partial y}(x_0, y_0) \\ \frac{\partial g}{\partial x}(x_0, y_0) & \frac{\partial g}{\partial y}(x_0, y_0) \end{pmatrix}, \tag{A.5}$$

which is a 2×2 matrix of real numbers.

We will need to compute the inverse of such matrices, as well as its eigenvalues.

We denote a general 2×2 matrix of real numbers:

$$A = \begin{pmatrix} a & b \\ c & d \end{pmatrix}, \quad a,\, b,\, c,\, d \in \mathbb{R}. \tag{A.6}$$

It is easy to verify that the inverse of A is given by:

$$A^{-1} = \frac{1}{ad-bc} \begin{pmatrix} d & -b \\ -c & a \end{pmatrix}. \tag{A.7}$$

Let $\mathbb{I}$ denote the 2×2 identity matrix. Then the eigenvalues of A are the solutions of the characteristic equation:

$$\det\,(A - \lambda \mathbb{I}) = 0. \tag{A.8}$$

where "det" is notation for the determinant of the matrix. This is a quadratic equation in λ which has two solutions:

$$\lambda_{1,2} = \frac{\mathrm{tr}\, A}{2} \pm \frac{1}{2}\sqrt{(\mathrm{tr}\, A)^2 - 4 \det A}, \tag{A.9}$$

where we have used the notation:

$$\mathrm{tr}\, A \equiv \mathrm{trace}\, A = a + d, \quad \det A \equiv \mathrm{determinant}\, A = ad - bc.$$

B

Integration of Some Basic Linear ODEs

In this appendix we collect together a few common ideas related to solving, explicitly, linear inhomogeneous differential equations. Our discussion is organized around a series of examples.

Example 32. Consider the one-dimensional, autonomous linear vector field:

$$\dot{x} = ax, \quad x,\, a \in \mathbb{R}. \tag{B.1}$$

We often solve problems in mathematics by transforming them into simpler problems that we already know how to solve. Towards this end, we introduce the following (time-dependent) transformation of variables:

$$x = ue^{at}. \tag{B.2}$$

Differentiating this expression with respect to t, and using (B.1), gives the following ODE for u:

$$\dot{u} = 0, \tag{B.3}$$

which is trivial to integrate, and gives:

$$u(t) = u(0), \tag{B.4}$$

and it is easy to see from (36) that:

$$u(0) = x(0). \tag{B.5}$$

Using (36), as well as (B.4) and (B.5), it follows that:

$$x(t)e^{-at} = u(t) = u(0) = x(0), \tag{B.6}$$

or

$$x(t) = x(0)e^{at}. \tag{B.7}$$

Example 33. Consider the following linear inhomogeneous nonautonomous ODE (due to the presence of the term $b(t)$):

$$\dot{x} = ax + b(t), \quad a, x \in \mathbb{R}, \tag{B.8}$$

where $b(t)$ is a scalar valued function of t, whose precise properties we will consider a bit later. We will use exactly the same strategy and change of coordinates as in the previous example:

$$x = ue^{at}. \tag{B.9}$$

Differentiating this expression with respect to t, and using (B.8), gives:

$$\dot{u} = e^{-at}b(t). \tag{B.10}$$

(Compare with (B.3).) Integrating (B.10) gives:

$$u(t) = u(0) + \int_0^t e^{-at'}b(t')dt'. \tag{B.11}$$

Now using (B.9) (with the consequence $u(0) = x(0)$) with (B.11) gives:

$$x(t) = x(0)e^{at} + e^{at}\int_0^t e^{-at'}b(t')dt'. \tag{B.12}$$

Finally, we return to the necessary properties of $b(t)$ in order for this unique solution of (B.8) to "make sense". Upon inspection of (B.12) it is clear that all that is required is for the integrals involving $b(t)$ to be well-defined. Continuity is a sufficient condition.

Example 34. Consider the one-dimensional, nonautonomous linear vector field:

$$\dot{x} = a(t)x, \quad x \in \mathbb{R}, \tag{B.13}$$

where $a(t)$ is a scalar valued function of t whose precise properties will be considered later. The similarity between (B.1) and (B.13) should be evident. We introduce the following (time-dependent) transformation of variables (compare with (36)):

$$x = ue^{\int_0^t a(t')dt'}. \tag{B.14}$$

Differentiating this expression with respect to t, and substituting (B.13) into the result gives:

$$\begin{aligned} \dot{x} &= \dot{u}e^{\int_0^t a(t')dt'} + ua(t)e^{\int_0^t a(t')dt'}, \\ &= \dot{u}e^{\int_0^t a(t')dt'} + a(t)x, \end{aligned} \tag{B.15}$$

which reduces to:

$$\dot{u} = 0. \tag{B.16}$$

Integrating this expression, and using (B.14), gives:

$$u(t) = u(0) = x(0) = x(t)e^{-\int_0^t a(t')dt'}, \tag{B.17}$$

or

$$x(t) = x(0)e^{\int_0^t a(t')dt'}. \tag{B.18}$$

As in the previous example, all that is required for the solution to be well-defined is for the integrals involving $a(t)$ to exist. Continuity is a sufficient condition.

Example 35. Consider the one-dimensional inhomogeneous nonautonomous linear vector field:

$$\dot{x} = a(t)x + b(t), \quad x \in \mathbb{R}, \tag{B.19}$$

where $a(t)$, $b(t)$ are scalar valued functions whose required properties will be considered at the end of this example. We make the same transformation as (B.14):

$$x = ue^{\int_0^t a(t')dt'}, \tag{B.20}$$

from which we obtain:

$$\dot{u} = b(t)e^{-\int_0^t a(t')dt'}. \tag{B.21}$$

Integrating this expression gives:

$$u(t) = u(0) + \int_0^t b(t')e^{-\int_0^{t'} a(t'')dt''}dt'. \tag{B.22}$$

Using (B.20) gives:

$$x(t)e^{-\int_0^t a(t')dt'} = x(0) + \int_0^t b(t')e^{-\int_0^{t'} a(t'')dt''} dt', \tag{B.23}$$

or

$$x(t) = x(0)e^{\int_0^t a(t')dt'} + e^{\int_0^t a(t')dt'} \int_0^t b(t')e^{-\int_0^{t'} a(t'')dt''} dt'. \tag{B.24}$$

As in the previous examples, all that is required is for the integrals involving $a(t)$ and $b(t)$ to be well-defined. Continuity is a sufficient condition.

The previous examples were all one-dimensional. Now we will consider two n-dimensional examples.

Example 36. Consider the n-dimensional autonomous linear vector field:

$$\dot{x} = Ax, \quad x \in \mathbb{R}^n, \tag{B.25}$$

where A is a $n \times n$ matrix of real numbers. We make the following transformation of variables (compare with):

$$x = e^{At}u. \tag{B.26}$$

Differentiating this expression with respect to t, and using (B.25), gives:

$$\dot{u} = 0. \tag{B.27}$$

Integrating this expression gives:

$$u(t) = u(0). \tag{B.28}$$

Using (B.26) with (B.28) gives:

$$u(t) = e^{-At}x(t) = u(0) = x(0), \tag{B.29}$$

from which it follows that:

$$x(t) = e^{At}x(0). \tag{B.30}$$

Example 37. Consider the n-dimensional inhomogeneous nonautonomous linear vector field:

$$\dot{x} = Ax + g(t), \quad x \in \mathbb{R}^n, \tag{B.31}$$

where $g(t)$ is a vector valued function of t whose required properties will be considered later on. We use the same transformation as in the previous example:

$$x = e^{At}u, \tag{B.32}$$

Differentiating this expression with respect to t, and using (B.31), gives:

$$\dot{u} = e^{-At}g(t), \tag{B.33}$$

from which it follows that:

$$u(t) = u(0) + \int_0^t e^{-At'}g(t')dt', \tag{B.34}$$

or, using (B.32):

$$x(t) = e^{At}x(0) + e^{At}\int_0^t e^{-At'}g(t')dt'. \tag{B.35}$$

C

Solutions of Some Second Order ODEs Arising in Applications: Newton's Equations*

In this appendix we will illustrate some of the issues associated with the solutions of second order ODEs in the setting of a class of second order ODEs arising in a familiar physical setting — the classical mechanical motion of a particle of constant mass m in one dimension as described by Newton's equations. By "the motion of a particle in one dimension" we mean that the particle is restricted to lie on a (one-dimensional) curve. If you think hard about this statement you might think that the motion of all particles is one-dimensional since the motion traces out a curve in space. Actually, this is true. However, in order to take advantage of this you must know the curve, i.e. know the motion. In practice, this is usually what we are trying to find out. The situations we have in mind here are when the forces on a particle are always acting so that at each point of a *known curve* they are tangential to that curve. In particular, the "curve" of interest may be a straight line. In this case, the forces act only in the direction of this straight line. We could also consider particles constrained to move in a circle, as well as other curves.

*The material in this appendix is adapted from *Elementary Classical Mechanics* (World Scientific, 2023).

We will denote the generic coordinate describing the position of the particle by "s". For a particle moving horizontally in the x direction, s would just be x. For a particle falling under the influence of gravity s would be the typical vertical coordinate, z. For a simple pendulum the particle would be constrained to move in a circle and s would be an angular coordinate, where the angle is measured from some fixed (say vertical) line.

Assuming constant mass, Newton's second law of motion becomes:

$$m\frac{d^2 s}{dt^2} = F. \tag{C.1}$$

This is an example of a second order ODE, and all we now need to do is "solve it" to obtain the motion, $s(t)$. However, we are getting a bit ahead of ourselves, because there is the problem of specifying the right-hand-side of (C.1), the F. This requires an understanding of the physics of the system under consideration. But first, we want to consider some strictly mathematical issues associated with (C.1).

In order to specify a solution of a second order ODE uniquely we require two constants (which generally will have physical meaning). Why is this? Here is an argument that is not generally found in textbooks, but it is pretty straightforward. Suppose $s(t)$ is a solution of (C.1) and we are interested in the solution near $t = t_0$. So we could consider a Taylor expansion near t_0:

$$s(t + t_0) = s(t_0) + \frac{ds}{dt}(t_0)t + \frac{1}{2}\frac{d^2 s}{dt^2}(t_0)t^2 + \frac{1}{6}\frac{d^3 s}{dt^3}(t_0)t^3 + \cdots, \tag{C.2}$$

and the Taylor coefficients are just the derivatives of $s(t)$ evaluated at t_0. Now in what sense does Eq. (C.1) provide the information to completely specify (C.2)? Here, "completely specify" would mean to completely specify all of the Taylor coefficients. Equation (C.1) provides us with the second derivative of $s(t)$, we can repeatedly differentiate it to obtain all the higher order derivatives (assuming all of the derivatives that we require exist). However, it tells us nothing about $s(t_0)$ and $\frac{ds}{dt}(t_0)$. These we must specify separately. If t_0 is the *initial time* we say that we must specify *initial conditions*; the initial position, $s(t_0)$, and the initial velocity, $\dot{s}(t_0)$.[1]

Now for what type of forces can we actually solve (C.1) analytically (as opposed to using a computer)? This depends on the "functional form"

[1] In this appendix we will assume that all quantities have sufficient regularity properties, e.g. continuity, differentiability, in order for the necessary mathematical operations, e.g. differentiation, integration, to be valid.

Table C.1. Examples of mathematical forms of forces for Newton's equations in one dimension.

Force function	Solvability of Newton's equations
F = constant	yes
F = F(t)	yes
F = F($\dot{s}$)	yes
F = F(s)	yes
F = F($\dot{s}$, t)	generally no, except for special cases
F = F(s, t)	generally no, except for special cases
F = F(s, $\dot{s}$, t)	generally no, except for special cases

of the force, i.e. how it depends on s, $\dot{s}$ and t. We summarize the results in Table C.1, and then provide some detailed calculations backing up the claims.

Everything was going fine until the last three rows of Table C.1; and that requires some explanation.

There is an important distinction between linear and nonlinear ODEs. The distinction being that solutions of linear ODEs are fairly simple, while the solutions of nonlinear ODEs *may* be extremely complicated (in ways that can be made mathematically precise).

For these reasons it is important to understand from the outset whether you are dealing with a linear or nonlinear ODE.

First, recall what we mean by a linear ODE. This means that $F(s, \dot{s}, t)$ is a linear function of s and $\dot{s}$, plus a function solely of t (which could be constant), i.e.

$$F(s, \dot{s}, t) = (a_0 + a_1(t))s + (b_0 + b_1(t))\dot{s} + c_0 + c_1(t), \qquad \text{(C.3)}$$

where a_0, b_0, c_0 are constants, and $a_1(t)$, $b_1(t)$, $c_1(t)$ are functions of t.

So are Newton's equations solvable with a force of the form of Eq. (C.3)? No. The problem comes from the coefficients on the s and $\dot{s}$ terms. If they are *constant* (i.e. $a_1(t) = b_1(t) = 0$), then Newton's equations can always be solved.

So what is a nonlinear ODE? It is one that is not linear, according to our definition above.

Now let us turn to justifying Table C.1.

F = constant. Newton's equations are:

$$m\frac{d^2 s}{dt^2} = F = \text{constant}, \quad s(t_0) = s_0, \ \ \dot{s}(t_0) = v_0,$$

and this is about the easiest differential equation that you could be given to solve. To solve it, you just "integrate twice", as we now show.

$$m \int_{t_0}^{t} \frac{d}{d\tau}\left(\frac{ds}{d\tau}(\tau)\right) d\tau = \int_{t_0}^{t} F d\tau,$$

Performing the integrals gives:

$$\frac{ds}{dt}(t) = v_0 + \frac{F}{m}(t - t_0).$$

Integrating this equation gives:

$$\int_{t_0}^{t} \frac{ds}{d\tau}(\tau) d\tau = v_0 \int_{t_0}^{t} d\tau + \frac{F}{m} \int_{t_0}^{t} (\tau - t_0) d\tau.$$

Performing the integrals gives:

$$s(t) = s_0 + v_0(t - t_0) + \frac{F}{2m}(t - t_0)^2.$$

$\mathbf{F} = \mathbf{F}(\mathbf{t})$. Newton's equations are:

$$m\frac{d^2 s}{dt^2} = F(t), \quad s(t_0) = s_0, \ \dot{s}(t_0) = v_0,$$

and this is about the second easiest differential equation that you could be given to solve. To solve it, you also integrate twice.

$$m \int_{t_0}^{t} \frac{d}{d\tau}\left(\frac{ds}{d\tau}(\tau)\right) d\tau = \int_{t_0}^{t} F(\tau) d\tau,$$

Performing the integrals gives:

$$\frac{ds}{dt}(t) = v_0 + \frac{1}{m} \int_{t_0}^{t} F(\tau) d\tau.$$

Integrating this expression gives:

$$\int_{t_0}^{t} \frac{ds}{d\tau'}(\tau') d\tau' = v_0 \int_{t_0}^{t} d\tau' + \frac{1}{m} \int_{t_0}^{t} \int_{t_0}^{\tau'} F(\tau) d\tau d\tau',$$

or

$$s(t) = s_0 + v_0(t - t_0) + \frac{1}{m} \int_{t_0}^{t} \int_{t_0}^{\tau'} F(\tau) d\tau d\tau'.$$

$\mathbf{F} = \mathbf{F}(\dot{\mathbf{s}})$. Newton's equations are:

$$m\frac{d^2 s}{dt^2} = F\left(\frac{ds}{dt}\right), \quad s(t_0) = s_0,\ \dot{s}(t_0) = v_0.$$

To solve this equation let

$$u = \frac{ds}{dt}, \tag{C.4}$$

then Newton's equations become:

$$m\frac{du}{dt} = F(u). \tag{C.5}$$

We can solve this equation for $u(t)$ (*provided* we can do the integrals that arise), and then integrate $u(t)$ with respect to t to get $s(t)$.

More precisely, we solve for $u(t)$ by rewriting (C.5) in the following form:

$$m\int_{u(t_0)=v_0}^{u(t)} \frac{du}{F(u)} = \int_{t_0}^{t} d\tau.$$

If this integral can be performed (which will depend on $F(u)$), then we may be able to obtain $u(t) = \frac{ds}{dt}(t)$. We integrate this expression from t_0 to t to obtain $s(t)$.

$\mathbf{F} = \mathbf{F}(\mathbf{s})$. Newton's equations are given by:

$$m\frac{d^2 s}{dt^2} = F(s), \quad s(t_0) = s_0,\ \dot{s}(t_0) = v_0. \tag{C.6}$$

Solving this equation requires a certain insight that, fortunately, others have had earlier.

Define the function:

$$V(s) = -\int_{c}^{s} F(s')ds', \tag{C.7}$$

where c is any constant. Then Eq. (C.6) becomes:

$$m\frac{d^2 s}{dt^2} = -\frac{dV}{ds}(s), \quad s(t_0) = s_0,\ \dot{s}(t_0) = v_0. \tag{C.8}$$

Notice that:

$$\frac{d}{dt}\left(\frac{m}{2}\dot{s}^2 + V(s)\right) = \dot{s}\left(m\ddot{s} + \frac{dV}{ds}(s)\right) = 0. \tag{C.9}$$

Now it is important to interpret this equation correctly. It says that a solution of $m\ddot{s} + \frac{dV}{ds}(s) = 0$ must also satisfy:

$$\frac{m}{2}\dot{s}^2 + V(s) = \text{constant}. \tag{C.10}$$

The constant is determined by the initial conditions. This is a very important expression and it is related to "energy" (see also Appendix F).

$\mathbf{F} = \mathbf{F}(\dot{\mathbf{s}}, \mathbf{t})$. If we let

$$\dot{s} = u,$$

then Newton's equations become first order equations for the velocity (u):

$$m\dot{u} = F(u, t), \quad u(t_0) = v_0.$$

If these equations could be solved then the velocity could be integrated to give the position. Unfortunately, even though they are first order, the general equation cannot be solved explicitly (although many "special cases" are known that can be solved). However, it is true that all *linear first order equations* can be solved, i.e. equations of the form:

$$m\dot{u} = a(t)u + b(t), \quad u(t_0) = v_0, \tag{C.11}$$

where $a(t)$ and $b(t)$ are continuous functions of t (we show this in Appendix B). Before demonstrating how this can be solved, let us first simplify this equation by getting rid of the annoying mass term (we can restore it later). We do this by defining:

$$\bar{a}(t) \equiv \frac{a(t)}{m}, \quad \bar{b}(t) \equiv \frac{b(t)}{m}.$$

Then (C.11) becomes:

$$\dot{u} = \bar{a}(t)u + \bar{b}(t), \quad u(t_0) = v_0, \tag{C.12}$$

Now the "trick" to solving this equation is due to Johann Bernoulli. He proposed to write the solution of (C.12) in the form $u(t) = n(t)m(t)$. Substituting this into (C.12) gives:

$$\dot{n}m + \dot{m}n = \bar{a}(t)nm + \bar{b}(t), \quad u(t_0) = n(t_0)m(t_0) = v_0. \tag{C.13}$$

The solution of this equation can be obtained by breaking it into two pieces, and solving each piece separately (why?):

$$\dot{n} = \bar{a}(t)n, \quad \text{solve for } n, \tag{C.14}$$

$$\dot{m} = \frac{\bar{b}(t)}{n}, \quad \text{integrate to get } m, \text{ with } n \text{ substituted in from above} \tag{C.15}$$

The solution of (C.14) is given by:

$$n(t) = n(t_0)e^{\int_{t_0}^{t} \bar{a}(\tau)d\tau}. \tag{C.16}$$

This is then substituted into (C.15), and integrated, to obtain:

$$m(t) = m(t_0) + \frac{1}{n(t_0)} \int_{t_0}^{t} \bar{b}(\tau)e^{-\int_{t_0}^{\tau} \bar{a}(\tau'')d\tau''} d\tau. \tag{C.17}$$

To obtain $u(t)$, we multiply $n(t)$ and $m(t)$ to obtain:

$$u(t) = n(t)m(t) = n(t_0)m(t_0)e^{\int_{t_0}^{t} \bar{a}(\tau)d\tau} + e^{\int_{t_0}^{t} \bar{a}(\tau)d\tau} \int_{t_0}^{t} \bar{b}(\tau)e^{-\int_{t_0}^{\tau} \bar{a}(\tau'')d\tau''} d\tau, \tag{C.18}$$

and remember that

$$u(t_0) = n(t_0)m(t_0).$$

This expression illustrates the fact that the general solution of (C.12) is the sum of a solution to the homogeneous equation:

$$\dot{u} = \bar{a}(t)u,$$

which is the first term in the solution (C.18), and a particular solution (the second term in the solution (C.18). A particular solution is just any solution of:

$$\dot{u} = \bar{a}(t)u + \bar{b}(t),$$

where you do not worry about the initial conditions. The initial conditions are then satisfied for the *sum* of the homogeneous plus particular solution.

$\mathbf{F} = \mathbf{F}(\mathbf{s}, \mathbf{t})$. Newton's equations are:

$$m\frac{d^2 s}{dt^2} = F(s,t), \quad s(t_0) = s_0, \, \dot{s}(t_0) = v_0.$$

Equations of this type cannot generally be solved analytically, *even if they are linear.* However, there is one class of systems of this form which always have a solution: the linear, *constant coefficient* systems, i.e. systems of the form:

$$m\frac{d^2s}{dt^2} = as + b(t),$$

where a is a real number and $b(t)$ is a continuous function of t.

$\mathbf{F} = \mathbf{F}(\mathbf{s}, \dot{\mathbf{s}}, \mathbf{t})$. Newton's equations are:

$$m\frac{d^2s}{dt^2} = F(s, \dot{s}, t), \quad s(t_0) = s_0, \; \dot{s}(t_0) = v_0.$$

Equations of this type cannot generally be solved analytically, *even if they are linear.* However, there is one class of systems of this form which always have a solution: the linear, *constant coefficient* systems, i.e. systems of the form:

$$m\frac{d^2s}{dt^2} = a\dot{s} + bs + c(t),$$

where a and b are real numbers and $c(t)$ is a continuous function of t.

D

Finding Lyapunov Functions

Lyapunov's method and the LaSalle invariance principle are very powerful techniques, but the obvious question always arises, "how do I find *the* Lyapunov function? The unfortunate answer is that given an arbitrary ODE there is no general method to find a Lyapunov function appropriate for a given ODE for the application of these methods.

In general, to determine a Lyapunov function appropriate for a given ODE, the ODE must have a structure that lends itself to the construction of the Lyapunov function. Therefore the next question is "what is this structure?" If the ODE arises from physical modeling there may be an "energy function" that is "almost conserved". What this means is that when certain terms of the ODE are neglected, the resulting ODE has a conserved quantity, i.e. a scalar valued function whose time derivative along trajectories is zero, and this conserved quantity may be a candidate for a Lyapunov function. If that sounds vague it is because the construction of Lyapunov functions often requires a bit of "mathematical artistry". We will consider this procedure with some examples. Energy methods are important techniques for understanding stability issues in science and engineering; see, for example see the book by Langhaar[1] and the article by Maschke.[2]

[1] H. L. Langhaar. *Energy Methods in Applied Mechanics.* John Wiley & Sons Inc, 1962.

[2] B. Maschke, R. Ortega, and A. J. Van Der Schaft. Energy-based Lyapunov functions for forced Hamiltonian systems with dissipation. *IEEE Transactions on Automatic Control* 45(8):1498–1502, 2000.

To begin, we consider Newton's equations for the motion of a particle of mass m under a conservative force in one dimension:

$$m\ddot{x} = -\frac{d\Phi}{dx}(x), \quad x \in \mathbb{R}. \tag{D.1}$$

Writing this as a first order system gives:

$$\begin{aligned} \dot{x} &= y, \\ \dot{y} &= -\frac{1}{m}\frac{d\Phi}{dx}(x). \end{aligned} \tag{D.2}$$

It is easy to see that the time derivative of the following function is zero

$$E = \frac{my^2}{2} + \Phi(x), \tag{D.3}$$

since

$$\begin{aligned} \dot{E} &= my\dot{y} + \frac{d\Phi}{dx}(x)\dot{x}, \\ &= -y\frac{d\Phi}{dx}(x) + y\frac{d\Phi}{dx}(x) = 0. \end{aligned} \tag{D.4}$$

In terms of dynamics, the function (D.3) has the interpretation as the conserved kinetic energy associated with (D.1).

Now we will consider several examples. In all cases we will simplify matters by taking $m = 1$.

Example 38. Consider the following autonomous vector field on $\mathbb{R}^2$:

$$\begin{aligned} \dot{x} &= y, \\ \dot{y} &= -x - \delta y, \quad \delta \geq 0, \quad (x, y) \in \mathbb{R}^2. \end{aligned} \tag{D.5}$$

For $\delta = 0$ (D.5) has the form of (D.1):

$$\begin{aligned} \dot{x} &= y, \\ \dot{y} &= -x, \quad (x, y) \in \mathbb{R}^2. \end{aligned} \tag{D.6}$$

with

$$E = \frac{y^2}{2} + \frac{x^2}{2}. \tag{D.7}$$

It is easy to verify that $\frac{dE}{dt} = 0$ along the trajectories of (D.6).

Now we differentiate E along trajectories of (D.5) and obtain:

$$\frac{dE}{dt} = -\delta y^2. \tag{D.8}$$

(D.6) has only one equilibrium point located at the origin. E is clearly positive everywhere, except for the origin, where it is zero. Using E as a Lyapunov function we can conclude that the origin is Lyapunov stable. If we use E to apply the LaSalle invariance principle, we can conclude that the origin is asymptotically stable. Of course, in this case we can linearize and conclude that the origin is a hyperbolic sink for $\delta > 0$.

Example 39. Consider the following autonomous vector field on $\mathbb{R}^2$:

$$\begin{aligned} \dot{x} &= y, \\ \dot{y} &= x - x^3 - \delta y, \quad \delta \geq 0, \quad (x, y) \in \mathbb{R}^2. \end{aligned} \tag{D.9}$$

For $\delta = 0$ (D.9) has the form of (D.1):

$$\begin{aligned} \dot{x} &= y, \\ \dot{y} &= x - x^3, \quad (x, y) \in \mathbb{R}^2. \end{aligned} \tag{D.10}$$

with

$$E = \frac{y^2}{2} - \frac{x^2}{2} + \frac{x^4}{4}. \tag{D.11}$$

It is easy to verify that $\frac{dE}{dt} = 0$ along trajectories of (D.10).

The question now is how we will use E to apply Lyapunov's method or the LaSalle invariance principle? (D.9) has three equilibrium points, a hyperbolic saddle at the origin for $\delta \geq 0$ and hyperbolic sinks at $(x, y) = (\pm 1, 0)$ for $\delta > 0$ and centers for $\delta = 0$. So linearization gives us complete information for $\delta > 0$. For $\delta = 0$, linearization is sufficient to allow us to conclude that the origin is a saddle. The equilibria $(x, y) = (\pm 1, 0)$ are Lyapunov stable for $\delta = 0$, but an argument involving the function E would be necessary in order to conclude this. Linearization allows us to conclude that the equilibria $(x, y) = (\pm 1, 0)$ are asymptotically stable for $\delta > 0$.

The function E can be used to apply the LaSalle invariance principle to conclude that for $\delta > 0$ all trajectories approach one of the three equilibria as $t \to \infty$.

E

Center Manifolds Depending on Parameters

In this appendix we describe the situation of center manifolds that depend on a parameter. The theoretical framework plays an important role in bifurcation theory.

As when we developed the theory earlier, we begin by describing the set-up. As before, it is important to realize that when applying these results to a vector field, it must be in the following form.

$$\begin{aligned} \dot{x} &= Ax + f(x, y, \mu), \\ \dot{y} &= By + g(x, y, \mu), \quad (x, y, \mu) \in \mathbb{R}^c \times \mathbb{R}^s \times \mathbb{R}^p, \end{aligned} \tag{E.1}$$

where $\mu \in \mathbb{R}^p$ is a vector of parameters and the matrices A and B have the following properties:

$$\begin{aligned} &A- \ c \times c \quad \text{matrix of real numbers} \\ &\qquad \text{having eigenvalues with zero real parts,} \\ &B- \ s \times s \quad \text{matrix of real numbers} \\ &\qquad \text{having eigenvalues with negative real parts,} \end{aligned}$$

and f and g are nonlinear functions. That is, they are of order two or higher in x, y and μ, as expressed in the following properties:

$$\begin{aligned} f(0,0,0) &= 0, \qquad Df(0,0,0) = 0, \\ g(0,0,0) &= 0, \qquad Dg(0,0,0) = 0, \end{aligned} \tag{E.2}$$

and they are C^r, r as large as is required to compute an adequate approximation the center manifold. With this set-up $(x, y, \mu) = (0, 0, 0)$ is a fixed point for (E.1) and we are interested in its stability properties.

The conceptual "trick" that reveals the nature of the parameter dependence of center manifolds is to include the parameter μ as a new *dependent* variable[1]:

$$\begin{aligned} \dot{x} &= Ax + f(x, y, \mu), \\ \dot{\mu} &= 0, \\ \dot{y} &= By + g(x, y, \mu), \quad (x, y, \mu) \in \mathbb{R}^c \times \mathbb{R}^s \times \mathbb{R}^p, \end{aligned} \tag{E.3}$$

The linearization of (E.3) about the fixed point is given by:

$$\begin{aligned} \dot{x} &= Ax, \\ \dot{\mu} &= 0, \\ \dot{y} &= By, \quad (x, y, \mu) \in \mathbb{R}^c \times \mathbb{R}^s \times \mathbb{R}^p. \end{aligned} \tag{E.4}$$

Even after increasing the dimension of the phase space by p dimensions by including the parameters as new dependent variables, the fixed point $(x, y, \mu) = (0, 0, 0)$ remains a nonhyperbolic fixed point. It has a $(c + p)$-dimensional invariant center subspace and a s-dimensional invariant stable subspace given by:

$$E^c = \{(x, y, \mu) \in \mathbb{R}^c \times \mathbb{R}^s \times \mathbb{R}^p \,|\, y = 0\}, \tag{E.5}$$

$$E^s = \{(x, y, \mu) \in \mathbb{R}^c \times \mathbb{R}^s \times \mathbb{R}^p \,|\, x = 0, \; \mu = 0\}, \tag{E.6}$$

respectively.

It should be clear that center manifold theory, as we have already developed, applies to (E.3). Including the parameters, μ as additional dependent variables has the effect of increasing the dimension of the "center variables", but there is also an important consequence. Since μ are now *dependent* variables they enter into the determination of the nonlinear terms in the equations. In particular, terms of the form

$$x_i^\ell \mu_j^m y_k^n,$$

[1] At this point it may be useful to go back to the first chapter and recall how nonlinearity of an ODE is defined in terms of the dependent variable.

now are interpreted as *nonlinear terms*, when $\ell+m+n > 1$, for nonnegative integers ℓ, m, n. We will see this in the example below.

Now we consider the information that center manifold theory provides us near the origin of (E.3).

1. In a neighborhood of the origin there exists a C^r center manifold that is represented as the graph of a function over the center variables, $h(x, \mu)$, it passes through the origin ($h(0,0) = 0$) and is tangent to the center subspace at the origin ($Dh(0,0) = 0$)
2. All solutions sufficiently close to the origin are attracted to a trajectory in the center manifold at an exponential rate.
3. The center manifold can be approximated by a power series expansion.

It is significant that the center manifold is defined in a neighborhood of the origin in both the x *and* μ coordinates since $\mu = 0$ is a *bifurcation value*. This means that *all bifurcating solutions* are contained in the center manifold. This is why, for example, that without loss of generality bifurcations from a single zero eigenvalue can be described by a parametrized family of *one*-dimensional vector fields.[2]

Example 40. We now consider an example which was Exercise 1b from Problem Set 8.

$$\begin{aligned}
\dot{x} &= \mu x + 10x^2, \\
\dot{\mu} &= 0, \\
\dot{y} &= x - 2y, \quad (x, y) \in \mathbb{R}^2, \ \mu \in \mathbb{R}.
\end{aligned} \tag{E.7}$$

The Jacobian associated with the linearization about $(x, \mu, y) = (0, 0, 0)$ is given by:

$$\begin{pmatrix} 0 & 0 & 0 \\ 0 & 0 & 0 \\ 1 & 0 & -2 \end{pmatrix}. \tag{E.8}$$

It is easy to check that the eigenvalues of this matrix are 0, 0, and -2 (as we would have expected). However, (E.7) is not quite of the form of (E.3) as a

[2]This is a very significant statement and it explains why "bifurcation problems" are amenable to dimensional reduction. In particular, an understanding of the nature of bifurcation of equilibria for autonomous vector fields can be reduced to a lower dimensional problem, where the dimension of the problem is equal to the number of eigenvalues with zero real part.

result of there being a 1 in the lower left-hand entry of the Jacobian (E.8). Nevertheless, this can be dealt with via a linear transformation obtained from the eigenvalues and eigenvectors of the Jacobian (E.8). It can be verified that each of its eigenvalues has an eigenvector. It is easily checked that two eigenvectors corresponding to the eigenvalue 0 are given by:

$$\begin{pmatrix} 2 \\ 0 \\ 1 \end{pmatrix}, \tag{E.9}$$

$$\begin{pmatrix} 0 \\ 1 \\ 0 \end{pmatrix}, \tag{E.10}$$

and an eigenvector corresponding to the eigenvalue -2 is given by:

$$\begin{pmatrix} 0 \\ 0 \\ 1 \end{pmatrix}. \tag{E.11}$$

From these eigenvectors, we form the transformation matrix

$$T = \begin{pmatrix} 2 & 0 & 0 \\ 0 & 1 & 0 \\ 1 & 0 & 1 \end{pmatrix} \tag{E.12}$$

with inverse

$$T^{-1} = \begin{pmatrix} \frac{1}{2} & 0 & 0 \\ 0 & 1 & 0 \\ -\frac{1}{2} & 0 & 1 \end{pmatrix}. \tag{E.13}$$

The transformation matrix, T, defines the following transformation of the dependent variables of (E.7):

$$\begin{pmatrix} x \\ \mu \\ y \end{pmatrix} = T \begin{pmatrix} u \\ \mu \\ v \end{pmatrix} = \begin{pmatrix} 2 & 0 & 0 \\ 0 & 1 & 0 \\ 1 & 0 & 1 \end{pmatrix} \begin{pmatrix} u \\ \mu \\ v \end{pmatrix} = \begin{pmatrix} 2u \\ \mu \\ u+v \end{pmatrix}. \tag{E.14}$$

It then follows that the transformed vector field has the form:

$$\begin{pmatrix} \dot{u} \\ \dot{\mu} \\ \dot{v} \end{pmatrix} = \begin{pmatrix} 0 & 0 & 0 \\ 0 & 0 & 0 \\ 0 & 0 & -2 \end{pmatrix} \begin{pmatrix} u \\ \mu \\ v \end{pmatrix} + T^{-1} \begin{pmatrix} \mu(2u) + 10(2u)^2 \\ 0 \\ 0 \end{pmatrix}, \tag{E.15}$$

or

$$\begin{pmatrix} \dot{u} \\ \dot{\mu} \\ \dot{v} \end{pmatrix} = \begin{pmatrix} 0 & 0 & 0 \\ 0 & 0 & 0 \\ 0 & 0 & -2 \end{pmatrix} \begin{pmatrix} u \\ \mu \\ v \end{pmatrix} + \begin{pmatrix} \frac{1}{2} & 0 & 0 \\ 0 & 1 & 0 \\ -\frac{1}{2} & 0 & 1 \end{pmatrix} \begin{pmatrix} 2\mu u + 40u^2 \\ 0 \\ 0 \end{pmatrix}, \quad \text{(E.16)}$$

or

$$\begin{aligned} \dot{u} &= \mu u + 20u^2, \\ \dot{\mu} &= 0, \\ \dot{v} &= -2v - \mu u - 20u^2, \end{aligned} \quad \text{(E.17)}$$

from which one sees the form of the transcritical bifurcation that is captured on the center manifold.

F

Dynamics of Hamilton's Equations

In this appendix we give a brief introduction to some of the characteristics and results associated with Hamiltonian differential equations (or, Hamilton's equations or Hamiltonian vector fields). The Hamiltonian formulation of Newton's equations reveals a great deal of structure about dynamics and it also gives rise to a large amount of deep mathematics that is the focus of much contemporary research.

Our purpose here is not to derive Hamilton's equations from Newton's equations. Discussions of that can be found in many textbooks on mechanics (although it is often considered "advanced mechanics"). For example, a classical exposition of this topic can be found in the classic book of Landau,[1] and more modern expositions can be found in Abraham and Marsden[2] and Arnold.[3] Rather, our approach is to start with Hamilton's equations and to understand some simple aspects and consequences of the special structure associated with Hamilton's equations. Towards this end, our starting point will be Hamilton's equations. Keeping with the simple approach throughout these lectures, our discussion of Hamilton's equations will be for two-dimensional systems.

We begin with a scalar valued function defined on $\mathbb{R}^2$

$$H = H(q, p), \quad (q, p) \in \mathbb{R}^2. \tag{F.1}$$

[1]L. D. Landau and E. M. Lifshitz. *Classical Mechanics*, 1960.

[2]R. Abraham and J. E. Marsden. *Foundations of Mechanics*. Benjamin/Cummings Publishing Company Reading, Massachusetts, 1978.

[3]V. I. Arnol'd. *Mathematical Methods of Classical Mechanics*, Volume 60. Springer Science & Business Media, 2013.

This function is referred to as the *Hamiltonian*. From the Hamiltonian, Hamilton's equations take the following form:

$$\begin{aligned}\dot{q} &= \frac{\partial H}{\partial p}(q,p),\\ \dot{p} &= -\frac{\partial H}{\partial q}(q,p), \quad (q,p) \in \mathbb{R}^2.\end{aligned} \tag{F.2}$$

The form of Hamilton's equations implies that the Hamiltonian is constant on trajectories. This can be seen from the following calculation:

$$\begin{aligned}\frac{dH}{dt} &= \frac{\partial H}{\partial q}\dot{q} + \frac{\partial H}{\partial p}\dot{p},\\ &= \frac{\partial H}{\partial q}\frac{\partial H}{\partial p} - \frac{\partial H}{\partial p}\frac{\partial H}{\partial q} = 0.\end{aligned} \tag{F.3}$$

Furthermore, this calculation implies that the level sets of the Hamiltonian are invariant manifolds. We denote the level set of the Hamiltonian as:

$$H_E = \{(q,p) \in \mathbb{R}^2 \mid H(q,p) = E\}. \tag{F.4}$$

In general, the level set is a curve (or possibly an equilibrium point). Hence, in the two-dimensional case, the trajectories of Hamilton's equations are given by the level sets of the Hamiltonian.

The Jacobian of the Hamiltonian vector field (F.2), denoted J, is given by:

$$J(q,p) \equiv \begin{pmatrix} \frac{\partial^2 H}{\partial q \partial p} & \frac{\partial^2 H}{\partial p^2} \\ -\frac{\partial^2 H}{\partial q^2} & -\frac{\partial^2 H}{\partial p \partial q} \end{pmatrix}, \tag{F.5}$$

at an arbitrary point $(q,p) \in \mathbb{R}^2$. Note that the trace of $J(q,p)$, denoted $\mathrm{tr}J(q,p)$, is zero. This implies that the eigenvalues of $J(q,p)$, denoted by $\lambda_{1,2}$, are given by:

$$\lambda_{1,2} = \pm\sqrt{-\det J(q,p)}, \tag{F.6}$$

where $\det J(q,p)$ denotes the determinant of $J(q,p)$. Therefore, if (q_0,p_0) is an equilibrium point of (F.1) and $\det J(q_0,p_0) \neq 0$, then the equilibrium point is a center for $\det J(q_0,p_0) > 0$ and a saddle for $\det J(q_0,p_0) < 0$.[4]

[4]Constraints on the eigenvalues of the matrix associated with the linearization of a Hamiltonian vector field at a fixed point in higher dimensions are described in Abraham and Marsden or Wiggins.

R. Abraham and J. E. Marsden. *Foundations of Mechanics.* Benjamin/Cummings Publishing Company Reading, Massachusetts, 1978; and S. Wiggins. *Introduction to Applied Nonlinear Dynamical Systems and Chaos*, Volume 2. Springer Science & Business Media, 2003.

Next we describe some examples of two-dimensional, linear autonomous Hamiltonian vector fields.

Example 41 (The Hamiltonian Saddle). We consider the Hamiltonian.

$$H(q,p) = \frac{\lambda}{2}\left(p^2 - q^2\right) = \frac{\lambda}{2}(p-q)(p+q), \quad (q,p) \in \mathbb{R}^2 \tag{F.7}$$

with $\lambda > 0$. From this Hamiltonian, we derive Hamilton's equations:

$$\begin{aligned} \dot{q} &= \frac{\partial H}{\partial p}(q,p) = \lambda p, \\ \dot{p} &= -\frac{\partial H}{\partial q}(q,p) = \lambda q, \end{aligned} \tag{F.8}$$

or in matrix form:

$$\begin{pmatrix} \dot{q} \\ \dot{p} \end{pmatrix} = \begin{pmatrix} 0 & \lambda \\ \lambda & 0 \end{pmatrix} \begin{pmatrix} q \\ p \end{pmatrix}. \tag{F.9}$$

The origin is a fixed point, and the eigenvalues associated with the linearization are given by $\pm\lambda$. Hence, the origin is a saddle point. The value of the Hamiltonian at the origin is zero. We also see from (F.7) that the Hamiltonian is zero on the lines $p - q = 0$ and $p + q = 0$. These are the unstable and stable manifolds of the origin, respectively. The phase portrait is illustrated in Fig. F.1.

The flow generated by this vector field is given in Chapter 2, Problem Set 2, Problem 6.

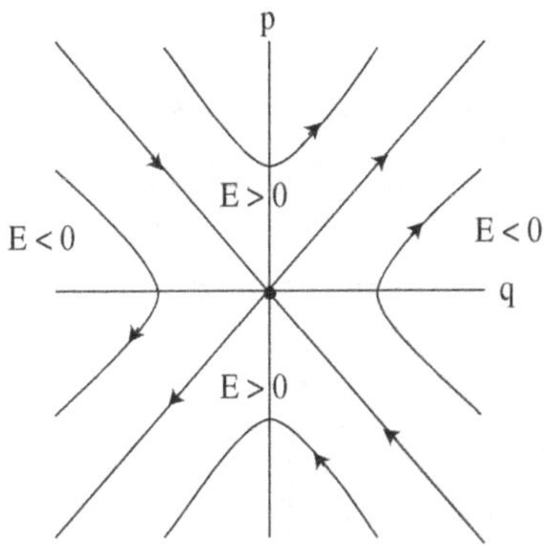

Fig. F.1. The phase portrait of the linear Hamiltonian saddle. The stable manifold of the origin is given by $p - q = 0$ and the unstable manifold of the origin is given by $p + q = 0$.

Example 42 (The Hamiltonian Center). We consider the Hamiltonian:

$$H(q,p) = \frac{\omega}{2}\left(p^2 + q^2\right), \quad (q,p) \in \mathbb{R}^2 \tag{F.10}$$

with $\omega > 0$. From this Hamiltonian, we derive Hamilton's equations:

$$\begin{aligned} \dot{q} &= \frac{\partial H}{\partial p}(q,p) = \omega p, \\ \dot{p} &= -\frac{\partial H}{\partial q}(q,p) = -\omega q, \end{aligned} \tag{F.11}$$

or, in matrix form:

$$\begin{pmatrix} \dot{q} \\ \dot{p} \end{pmatrix} = \begin{pmatrix} 0 & \omega \\ -\omega & 0 \end{pmatrix} \begin{pmatrix} q \\ p \end{pmatrix} \tag{F.12}$$

The level sets of the Hamiltonian are circles, and are illustrated in Fig. F.2.

The flow generated by this vector field is given in Chapter 2, Problem Set 2, Problem 5.

We will now consider two examples of bifurcation of equilibria in two-dimensional Hamiltonian systems. Bifurcation associated with one zero eigenvalue (as we studied in Chapter 8) is not possible since, following (F.6), if there is one zero eigenvalue, the other eigenvalue must also be zero. We will consider examples of the Hamiltonian saddle-node and

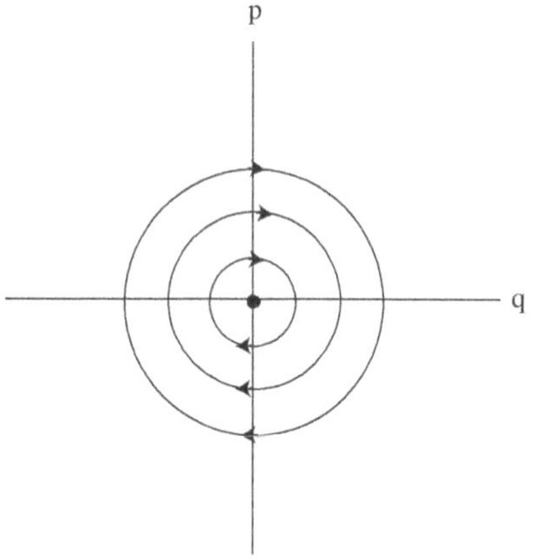

Fig. F.2. The phase portrait for the linear Hamiltonian center.

Hamiltonian pitchfork bifurcations. Discussions of the Hamiltonian versions of these bifurcations can also be found in Golubitsky *et al.*[5]

Example 43 (Hamiltonian saddle-node bifurcation). We consider the Hamiltonian:

$$H(q,p) = \frac{p^2}{2} - \lambda q + \frac{q^3}{3}, \quad (q,p) \in \mathbb{R}^2. \tag{F.13}$$

where λ is considered to be a parameter that can be varied. From this Hamiltonian, we derive Hamilton's equations:

$$\begin{aligned} \dot{q} &= \frac{\partial H}{\partial p} = p, \\ \dot{p} &= -\frac{\partial H}{\partial q} = \lambda - q^2. \end{aligned} \tag{F.14}$$

The fixed points for (F.14) are:

$$(q,p) = (\pm\sqrt{\lambda}, 0), \tag{F.15}$$

from which it follows that there are no fixed points for $\lambda < 0$, one fixed point for $\lambda = 0$, and two fixed points for $\lambda > 0$. This is the scenario for a saddle-node bifurcation.

Next we examine stability of the fixed points. The Jacobian of (F.14) is given by:

$$\begin{pmatrix} 0 & 1 \\ -2q & 0 \end{pmatrix}. \tag{F.16}$$

The eigenvalues of this matrix are:

$$\lambda_{1,2} = \pm\sqrt{-2q}.$$

Hence $(q,p) = (-\sqrt{\lambda}, 0)$ is a saddle, $(q,p) = (\sqrt{\lambda}, 0)$ is a center, and $(q,p) = (0,0)$ has two zero eigenvalues. The phase portraits are shown in Fig. F.3.

[5] M. Golubitsky, I. Stewart, and J. Marsden. Generic bifurcation of Hamiltonian systems with symmetry. *Physica D: Nonlinear Phenomena*, 24(1–3):391–405, 1987.

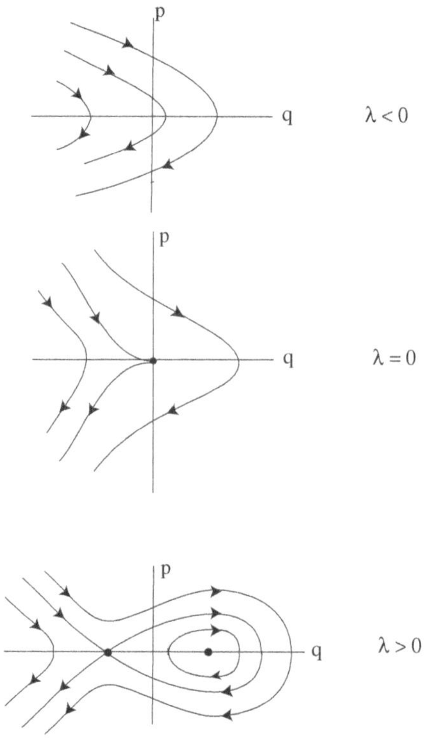

Fig. F.3. The phase portraits for the Hamiltonian saddle-node bifurcation.

Example 44 (Hamiltonian pitchfork bifurcation). We consider the Hamiltonian:

$$H(q,p) = \frac{p^2}{2} - \lambda\frac{q^2}{2} + \frac{q^4}{4}, \tag{F.17}$$

where λ is considered to be a parameter that can be varied. From this Hamiltonian, we derive Hamilton's equations:

$$\begin{aligned} \dot{q} &= \frac{\partial H}{\partial p} = p, \\ \dot{p} &= -\frac{\partial H}{\partial q} = \lambda q - q^3. \end{aligned} \tag{F.18}$$

The fixed points for (F.18) are:

$$(q,p) = (0,0),\ (\pm\sqrt{\lambda},0), \tag{F.19}$$

from which it follows that there is one fixed point for $\lambda < 0$, one fixed point for $\lambda = 0$, and three fixed points for $\lambda > 0$. This is the scenario for a pitchfork bifurcation.

Next we examine stability of the fixed points. The Jacobian of (F.18) is given by:

$$\begin{pmatrix} 0 & 1 \\ \lambda - 3q^2 & 0 \end{pmatrix}. \tag{F.20}$$

The eigenvalues of this matrix are:

$$\lambda_{1,2} = \pm\sqrt{\lambda - 3q^2}.$$

Hence $(q, p) = (0, 0)$ is a center for $\lambda < 0$, a saddle for $\lambda > 0$ and has two zero eigenvalues for $\lambda = 0$. The fixed points $(q, p) = (\sqrt{\lambda}, 0)$ are centers for $\lambda > 0$. The phase portraits are shown in Fig. F.4.

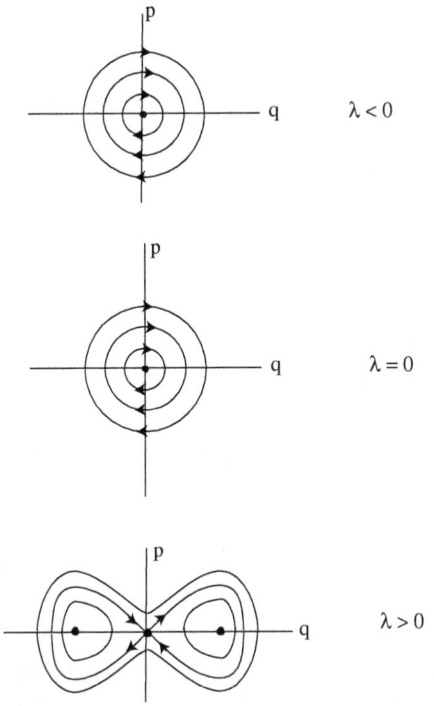

Fig. F.4. The phase portraits for the Hamiltonian pitchfork bifurcation.

We remark that, with a bit of thought, it should be clear that in *two dimensions* there is no analog of the Hopf bifurcation for Hamiltonian vector fields similar to the situation we analyzed earlier in the non-Hamiltonian context. There is a situation that is referred to as the Hamiltonian Hopf bifurcation, but this notion requires at least four dimensions, see Van Der Meer.[6]

In Hamiltonian systems a natural bifurcation parameter is the value of the level set of the Hamiltonian, or the "energy". From this point of view perhaps a more natural candidate for a Hopf bifurcation in a Hamiltonian system is described by the Lyapunov subcenter theorem, see Kelley.[7] The setting for this theorem also requires at least four dimensions, but the associated phenomena occur quite often in applications.

[6] J.-C. Van Der Meer. *The Hamiltonian Hopf Bifurcation.* Springer, 1985.

[7] A. Kelley. On the Liapounov subcenter manifold. *Journal of Mathematical Analysis and Applications*, 18(3):472–478, 1967.

G

A Brief Introduction to the Characteristics of Chaos

In this appendix we will describe some aspects of the phenomenon of chaos as arising in ODEs. Chaos is one of those notable topics that crosses disciplinary boundaries in mathematics, science, and engineering and captures the intrigue and curiousity of the general public. Numerous popularizations and histories of the topic, from different points of view, have been written; see, for example the books by Lorenz,[1] Diacu and Holmes,[2] Stewart,[3] and Gleick.[4]

Our goal here is to introduce some of the key characteristics of chaos based on notions that we have already developed so as to frame possible future directions of studies that the student might wish to pursue. Our discussion will be in the setting of a flow generated by an autonomous vector field.

The phrase "chaotic behavior" calls to mind a form of randomness and unpredictability. But keep in mind, we are working in the setting of, i.e. our ODE satisfies the criteria for existence and uniqueness of solutions. Therefore specifying the initial condition *exactly* implies that the

[1] E. N. Lorenz. *The Essence of Chaos.* University of Washington Press, Seattle, 1993.
[2] F. Diacu and P. Holmes. *Celestial Encounters: The Origins of Chaos and Stability.* Princeton University Press, 1996.
[3] I. Stewart. *Does God Play Dice?: The New Mathematics of Chaos.* Penguin UK, 1997.
[4] J. Gleick. *Chaos: Making a New Science (Enhanced Edition).* Open Road Media, 2011.

future evolution is uniquely determined, i.e. there is no "randomness or unpredictability". The key here is the word "exactly". Chaotic systems have an intrinsic property in their dynamics that can result in slight perturbations of the initial conditions leading to behavior, over time, that is unlike the behavior of the trajectory through the original initial condition. Often it is said that a chaotic system exhibits sensitive dependence on initial conditions. Now this is a lot of words for a mathematics course. Just like when we studied stability, we will give a mathematical definition of sensitive dependence on initial conditions, and then consider the meaning of the definition in the context of specific examples.

As mentioned above, we consider an autonomous, C^r, $r \geq 1$ vector field on $\mathbb{R}^n$:

$$\dot{x} = f(x), \quad x \in \mathbb{R}^n, \tag{G.1}$$

and we denote the flow generated by the vector field by $\phi_t(\cdot)$, and we assume that it exists for all time. We let $\Lambda \subset \mathbb{R}^n$ denote an invariant set for the flow. Then we have the following definition.

Definition 22 (Sensitive dependence on initial conditions). The flow $\phi_t(\cdot)$ is said to have sensitive dependence on initial conditions on Λ if there exists $\epsilon > 0$ such that, for any $x \in \Lambda$ and any neighborhood $U \subset \Lambda$ of x there exists $y \in U$ and $t > 0$ such that $|\phi_t(x) - \phi_t(y)| > \epsilon$.

Now we consider an example and analyze whether or not sensitive dependence on initial conditions is present in the example.

Example 45. Consider the autonomous linear vector field on $\mathbb{R}^2$:

$$\begin{aligned} \dot{x} &= \lambda x, \\ \dot{y} &= -\mu y, \qquad (x, y) \in \mathbb{R}^2. \end{aligned} \tag{G.2}$$

with λ, $\mu > 0$. This is just a standard "saddle point". The origin is a fixed point of saddle type with its stable manifold given by the y-axis (i.e. $x = 0$) and its unstable manifold given by the x-axis (i.e. $y = 0$). The flow generated by this vector field is given by:

$$\phi_t(x_0, y_0) = \left(x_0 e^{\lambda t},\, y_0 e^{-\mu t}\right). \tag{G.3}$$

Following the definition, sensitive dependence on initial conditions is defined with respect to invariant sets. Therefore, we must identify the invariant sets for which we want to determine whether or not they possess the property of sensitive dependence on initial condition.

The simplest invariant set is the fixed point at the origin. However, that invariant set clearly does not exhibit sensitive dependence on initial conditions.

Then we have the one-dimensional stable (y-axis) and unstable manifolds (x-axis). We can consider the issue of sensitive dependence on initial conditions on these invariant sets. The stable and unstable manifolds divide the plane into four quadrants. Each of these is an invariant set (with a segment of the stable and unstable manifold forming part of their boundary), and the entire plane (i.e. the entire phase space) is also an invariant set.

We consider the unstable manifold, $y = 0$. The flow restricted to the unstable manifold is given by

$$\phi_t(x_0, 0) = \left(x_0 e^{\lambda t},\, 0\right). \tag{G.4}$$

It should be clear that the unstable manifold is an invariant set that exhibits sensitive dependence on initial conditions. Choose an arbitrary point on the unstable manifold, $\bar{x}_1$. Consider another point arbitrarily close to $\bar{x}_1$, $\bar{x}_2$. Now consider any $\epsilon > 0$. We have

$$|\phi_t(\bar{x}_1, 0) - \phi_t(\bar{x}_2, 0)| = |\bar{x}_1 - \bar{x}_2| e^{\lambda t}. \tag{G.5}$$

Now since $|\bar{x}_1 - \bar{x}_2|$ is a fixed constant, we can clearly find a $t > 0$ such that

$$|\bar{x}_1 - \bar{x}_2| e^{\lambda t} > \epsilon. \tag{G.6}$$

Therefore, the invariant unstable manifold exhibits sensitive dependence on initial conditions. Of course, this is not surprising because of the $e^{\lambda t}$ term in the expression for the flow since this term implies exponential growth in time of the x component of the flow.

The stable manifold, $x = 0$, does not exhibit sensitive dependence on initial conditions since the restriction to the stable manifold is given by:

$$\phi_t(0, y_0) = \left(0,\, y_0 e^{-\mu t}\right), \tag{G.7}$$

which implies that neighboring points actually get closer together as t increases.

Moreover, the term $e^{\lambda t}$ implies that the four quadrants separated by the stable and unstable manifolds of the origin also exhibit sensitive dependence on initial conditions.

Of course, we would not consider a linear autonomous ODE on the plane having a hyperbolic saddle point to be a chaotic dynamical system, even though it exhibits sensitive dependence on initial conditions. Therefore, there must be something more to "chaos", and we will explore this through more examples.

Before we consider the next example we point out two features of this example that we will consider in the context of other examples.

1. The invariant sets that we considered (with the exception of the fixed point at the origin) were unbounded. This was a feature of the linear nature of the vector field.
2. The "separation of trajectories" occurred at an exponential rate. There was no requirement on the "rate of separation" in the definition of sensitive dependence on initial conditions.
3. Related to these two points is the fact that trajectories continue to separate for all time, i.e. they never again "get closer" to each other.

Example 46. Consider the autonomous vector field on the cylinder:

$$\begin{aligned} \dot{r} &= 0, \\ \dot{\theta} &= r, \quad (r, \theta) \in \mathbb{R}^+ \times S^1. \end{aligned} \tag{G.8}$$

The flow generated by this vector field is given by:

$$\phi_t(r_0, \theta_0) = (r_0,\, r_0\, t + \theta_0). \tag{G.9}$$

Note that r is constant in time. This implies that any annulus is an invariant set. In particular, choose any $r_1 < r_2$. Then the annulus

$$\mathcal{A} \equiv \{(r, \theta) \in \mathbb{R}^+ \times S^1 \,|\, r_1 \le r \le r_2,\, \theta \in S^1\}, \tag{G.10}$$

is a bounded invariant set.

Now choose initial conditions in $\mathcal{A}$, (r_1', θ_1), (r_2', θ_2), with $r_1 \le r_1' < r_2' \le r_2$. Then we have that:

$$\begin{aligned} |\phi_t(r_1', \theta_1) - \phi_t(r_2', \theta_2)| &= |\, (r_1',\, r_1'\, t + \theta_1) - (r_2',\, r_2'\, t + \theta_2)\,|, \\ &= (r_1' - r_2',\, (r_1' - r_2')t + (\theta_1 - \theta_2)). \end{aligned}$$

Hence we see that the distance between trajectories will grow linearly in time, and therefore trajectories exhibit sensitive dependence on initial conditions. However, the distance between trajectories will not grow unboundedly (as in the previous example). This is because θ is on the circle. Trajectories will move apart (in θ, but their r values will remain constant) and then come close, then move apart, etc. Nevertheless, this is not an example of a chaotic dynamical system.

Example 47. Consider the following autonomous vector field defined on the two-dimensional torus (i.e. each variable is an angular variable):

$$\begin{aligned}\dot{\theta}_1 &= \omega_1,\\ \dot{\theta}_2 &= \omega_2, \quad (\theta_1, \theta_2) \in S^1 \times S^1.\end{aligned} \tag{G.11}$$

This vector field is an example that is considered in many dynamical systems courses where it is shown that if $\frac{\omega_1}{\omega_2}$ is an irrational number, then the trajectory through *any* initial condition "densely fills out the torus". This means that given any point on the torus any trajectory will get arbitrarily close to that point at some time of its evolution, and this "close approach" will happen infinitely often. This is the classic example of an ergodic system, and this fact is proven in many textbooks, e.g. Arnold[5] or Wiggins.[6] This behavior is very different from the previous examples. For the case $\frac{\omega_1}{\omega_2}$ an irrational number, the natural invariant set to consider is the entire phase space (which is bounded).

Next we consider the issue of sensitive dependence on initial conditions. The flow generated by this vector field is given by:

$$\phi_t(\theta_1, \theta_2) = (\omega_1 t + \theta_1,\, \omega_2 t + \theta_2). \tag{G.12}$$

We choose two initial conditions, (θ_1, θ_2), (θ_1', θ_2'). Then we have

$$\begin{aligned}|\phi_t(\theta_1, \theta_2) - \phi_t(\theta_1', \theta_2')| &= |\,(\omega_1 t + \theta_1,\, \omega_2 t + \theta_2) - (\omega_1 t + \theta_1',\, \omega_2 t + \theta_2')\,|,\\ &= |(\theta_1 - \theta_1', \theta_2 - \theta_2')|,\end{aligned}$$

and therefore trajectories always maintain the same distance from each other during the course of their evolution.

Sometimes it is said that chaotic systems contain an infinite number of unstable periodic orbits. We consider an example.

Example 48. Consider the following two-dimensional autonomous vector field on the cylinder:

$$\begin{aligned}\dot{r} &= \sin\frac{\pi}{r},\\ \dot{\theta} &= r, \qquad (r, \theta) \in \mathbb{R}^+ \times S^1.\end{aligned}$$

---

[5]V. I. Arnold. *Ordinary Differential Equations.* M.I.T. Press, Cambridge, 1973. ISBN 0262010372.

[6]S. Wiggins. *Introduction to Applied Nonlinear Dynamical Systems and Chaos*, Volume 2. Springer Science & Business Media, 2003.

Equilibrium points of the $\dot{r}$ component of this vector field correspond to periodic orbits. These equilibrium points are given by

$$r = \frac{1}{n}, \quad n = 0, 1, 2, 3, \ldots. \tag{G.13}$$

Stability of the periodic orbits can be determined by computing the Jacobian of the $\dot{r}$ component of the equation and evaluating it on the periodic orbit. This is given by:

$$-\frac{\pi}{r^2} \cos \frac{\pi}{r},$$

and evaluating this on the periodic orbits gives;

$$-\frac{\pi}{n^2}(-1)^n.$$

Therefore all of these periodic orbits are hyperbolic and stable for n even and unstable for n odd. This is an example of a two-dimensional autonomous vector field that contains an infinite number of unstable hyperbolic periodic orbits in a bounded region, yet it is not chaotic.

Now we consider what we have learned from these four examples. In Example 45 we identified invariant sets on which the trajectories exhibited sensitive dependence on initial conditions (i.e. trajectories separated at an exponential rate), but those invariant sets were unbounded, and the trajectories also became unbounded. This illustrates why boundedness is part of the definition of invariant set in the context of chaotic systems.

In Example 46 we identified an invariant set, $\mathcal{A}$, on which all trajectories were bounded and they exhibited sensitive dependence on initial conditions, although they only separated linearly in time. However, the r coordinates of all trajectories remained constant, indicating that trajectories were constrained to lie on circles ("invariant circles") within $\mathcal{A}$.

In Example 47, for $\frac{\omega_1}{\omega_2}$ an irrational number, *every* trajectory densely fills out the entire phase space, the torus (which is bounded). However, the trajectories did not exhibit sensitive dependence on initial conditions.

Finally, in Example 48, we gave an example having an infinite number of unstable hyperbolic orbits in a bounded region of the phase space. We did not explicitly examine the issue of sensitive dependence on initial conditions for this example.

So what characteristics would we require of a chaotic invariant set? A combination of Examples 45 and 47 would capture many manifestations of "chaotic invariant sets":

1. the invariant set is bounded,
2. every trajectory comes arbitrarily close to every point in the invariant set during the course of its evolution in time, and
3. every trajectory has sensitive dependence on initial condition.

While simple to state, developing a unique mathematical framework that makes these three criteria mathematically rigorous, and provides a way to verify them in particular examples, is not so straightforward.

Property 1 is fairly straightforward, once we have identified a candidate invariant set (which can be very difficult in explicit ODEs). If the phase space is equipped with a norm, then we have a way of verifying whether or not the invariant set is bounded.

Property 2 is very difficult to verify, as well as to develop a universally accepted definition amongst mathematicians as to what it means for "every trajectory to come arbitrarily close to every point in phase space during the course of its evolution". Its definition is understood within the context of recurrence properties of trajectories. Those can be studied from either the topological point of view (see Akin[7]). or from the point of view of ergodic theory (see Katok and Hasselblatt[8] or Brin and Stuck[9]). The settings for both of these points of view utilize different mathematical structures (topology in the former case, measure theory in the latter case). A book that describes how both of these points of view are used in the application of mixing is Sturman *et al.*[10]

Verifying that Property 3 holds for *all* trajectories is also not straightforward. What "all" means is different in the topological setting ("open set", Baire category) and the ergodic theoretic setting (sets of "full measure").

---

[7] E. Akin. *The General Topology of Dynamical Systems*, Volume 1. American Mathematical Soc., 2010.

[8] A. Katok and B. Hasselblatt. *Introduction to the Modern Theory of Dynamical Systems*, Volume 54. Cambridge University Press, 1997.

[9] M. Brin and G. Stuck. *Introduction to Dynamical Systems*. Cambridge University Press, 2002.

[10] R. Sturman, J. M. Ottino, and S. Wiggins. *The Mathematical Foundations of Mixing: The Linked Twist Map as a Paradigm in Applications: Micro to Macro, Fluids to Solids*, Volume 22. Cambridge University Press, 2006.

What "sensitive dependence on initial conditions" means is also different in each setting. The definition we gave above was more in the spirit of the topological point of view (no specific "rate of separation" was given) and the ergodic theoretic framework focuses on Lyapunov exponents ("Lyapunov's first method") and exponential rate of separation of trajectories.

Therefore we have not succeeded in giving a specific example of an ODE whose trajectories behave in a chaotic fashion. We have been able to describe some of the issues, but the details will be left for other courses (which could be either courses in dynamical systems theory or ergodic theory or, ideally, a bit of both). But we have illustrated just how difficult it can be to formulate mathematically precise definitions that can be verified in specific examples.

All of our examples above were two-dimensional, autonomous vector fields. The type of dynamics that can be exhibited by such systems is very limited, according to the Poincaré–Bendixson theorem (see Hirsch *et al.*[11] or Wiggins[12]). There are a number of variations of this theorem, so we will leave the exploration of this theorem to the interested student.

[11]M. W. Hirsch, S. Smale, and R. L. Devaney. *Differential Equations, Dynamical Systems, and an Introduction to Chaos.* Academic Press, 2012.

[12]S. Wiggins. *Introduction to Applied Nonlinear Dynamical Systems and Chaos,* Volume 2. Springer Science & Business Media, 2003.

Bibliography

R. Abraham and J. E. Marsden. *Foundations of Mechanics*. Benjamin/ Cummings Publishing Company Reading, Massachusetts, 1978.

E. Akin. *The General Topology of Dynamical Systems*, Volume 1. American Mathematical Soc., 2010.

V. I. Arnold. *Ordinary Differential Equations*. M.I.T. Press, Cambridge, 1973. ISBN 0262010372.

V. I. Arnol'd. *Mathematical Methods of Classical Mechanics*, Volume 60. Springer Science & Business Media, 2013.

I. Barkana. Defending the beauty of the invariance principle. *International Journal of Control*, 87(1):186–206, 2014.

M. Brin and G. Stuck. *Introduction to Dynamical Systems*. Cambridge University Press, 2002.

J. Carr. *Applications of Centre Manifold Theory*, Volume 35. Springer Science & Business Media, 2012.

C. Chicone. *Ordinary Differential Equations with Applications*. Springer, 2000.

E. A. Coddington and N. Levinson. *Theory of Ordinary Differential Equations*. Krieger, 1984.

F. Diacu and P. Holmes. *Celestial Encounters: The Origins of Chaos and Stability*. Princeton University Press, 1996.

W. Fulton. *Algebraic Topology: A First Course*, Volume 153. Springer Science & Business Media, 2013.

J. Gleick. *Chaos: Making a New Science (Enhanced Edition)*. Open Road Media, 2011.

M. Golubitsky, I. Stewart, and J. Marsden. Generic bifurcation of Hamiltonian systems with symmetry. *Physica D: Nonlinear Phenomena*, 24(1–3):391–405, 1987.

J. Guckenheimer and P. J. Holmes. *Nonlinear Oscillations, Dynamical Systems, and Bifurcations of Vector Fields*, Volume 42. Springer Science & Business Media, 2013.

V. Guillemin and A. Pollack. *Differential Topology*, Volume 370. American Mathematical Soc., 2010.

J. K. Hale. *Ordinary Differential Equations.* Dover, 2009.

P. Hartman. *Ordinary Differential Equations.* Society for Industrial and Applied Mathematics, 2002.

B. D. Hassard, N. D. Kazarinoff, and Y.-H. Wan. *Theory and Applications of Hopf Bifurcation*, Volume 41. CUP Archive, 1981.

D. Henry. *Geometric Theory of Semilinear Parabolic Equations.* Lecture Notes in Mathematics. Springer Berlin Heidelberg, 1993.

M. W. Hirsch, S. Smale, and R. L. Devaney. *Differential Equations, Dynamical Systems, and an Introduction to Chaos.* Academic Press, 2012.

A. Katok and B. Hasselblatt. *Introduction to the Modern Theory of Dynamical Systems*, Volume 54. Cambridge University Press, 1997.

A. Kelley. On the Liapounov subcenter manifold. *Journal of Mathematical Analysis and Applications*, 18(3):472–478, 1967.

L. D. Landau and E. M. Lifshitz. *Classical Mechanics*, 1960.

H. L. Langhaar. *Energy Methods in Applied Mechanics.* John Wiley & Sons Inc, 1962.

J. P. LaSalle. *The Stability of Dynamical Systems*, Volume 25. SIAM, 1976.

E. N. Lorenz. *The Essence of Chaos.* University of Washington Press, Seattle, 1993.

A. M. Lyapunov. *General Problem of the Stability of Motion.* Control Theory and Applications Series. Taylor & Francis, 1992.

J. E. Marsden and M. McCracken. *The Hopf Bifurcation and Its Applications.* Applied Mathematical Sciences. Springer-Verlag, 1976.

B. Maschke, R. Ortega, and A. J. Van Der Schaft. Energy-based Lyapunov functions for forced Hamiltonian systems with dissipation. *IEEE Transactions on Automatic Control* 45(8):1498–1502, 2000.

P. C. Parks. A. M. Lyapunov's stability theory — 100 years on. *IMA Journal of Mathematical Control and Information*, 9(4):275–303, 1992.

I. Stewart. *Does God Play Dice?: The New Mathematics of Chaos.* Penguin UK, 1997.

R. Sturman, J. M. Ottino, and S. Wiggins. *The Mathematical Foundations of Mixing: The Linked Twist Map as a Paradigm in Applications: Micro to Macro, Fluids to Solids*, Volume 22. Cambridge University Press, 2006.

J.-C. Van Der Meer. *The Hamiltonian Hopf Bifurcation.* Springer, 1985.

S. Wiggins. *Introduction to Applied Nonlinear Dynamical Systems and Chaos*, Volume 2. Springer Science & Business Media, 2003.

Index

Other World Scientific Titles by the Author

Infinite-Dimensional Dynamical Systems in Atmospheric and Oceanic Science
ISBN: 978-981-4590-37-2

Fractional Partial Differential Equations and Their Numerical Solutions
ISBN: 978-981-4667-04-3

Quantum Hydrodynamic Equation and Its Mathematical Theory

Quantum Hydrodynamic Equation and Its Mathematical Theory

BOLING GUO

Institute of Applied Physics and Computational Mathematics, China

World Scientific

NEW JERSEY • LONDON • SINGAPORE • BEIJING • SHANGHAI • HONG KONG • TAIPEI • CHENNAI • TOKYO

Published by

World Scientific Publishing Co. Pte. Ltd.
5 Toh Tuck Link, Singapore 596224
USA office: 27 Warren Street, Suite 401-402, Hackensack, NJ 07601
UK office: 57 Shelton Street, Covent Garden, London WC2H 9HE

Library of Congress Cataloging-in-Publication Data
Names: Guo, Boling, author.
Title: Quantum hydrodynamic equation and its mathematical theory / Boling Guo.
Description: New Jersey : World Scientific, [2023] | Includes bibliographical references.
Identifiers: LCCN 2022018511 | ISBN 9789811260834 (hardcover) |
ISBN 9789811260841 (ebook for institutions) | ISBN 9789811260858 (ebook for individuals)
Subjects: LCSH: Hydrodynamics--Mathematical models. | Quantum theory--Mathematics.
Classification: LCC QA911 .G785 2023 | DDC 530.12--dc23/eng20220722
LC record available at https://lccn.loc.gov/2022018511

British Library Cataloguing-in-Publication Data
A catalogue record for this book is available from the British Library.

《可压缩量子流体力学方程及其数学理论》
Originally published in Zhejiang Science and Technology Publishing House

For any available supplementary material, please visit
https://www.worldscientific.com/worldscibooks/10.1142/12978#t=suppl

Typeset by Stallion Press
Email: enquiries@stallionpress.com

Preface

Quantum mechanics is a branch of physics which is the fundamental theory of nature at small scales and low energies of atoms and subatomic particles, which mainly concerns the structure and properties of atoms, molecules, condensed matter, atomic nucleus and elementary particles. Schrödinger equation is the fundamental equation of quantum mechanics. In the late 1920's, from Schrödinger equation $i\psi_t + \frac{h^2}{2}\Delta\psi = h(|\psi|^2)\psi$, $h'(\rho) = \frac{P'(\rho)}{\rho}$, Madelung deduced the inviscid quantum hydrodynamic equations, which can be regarded as the compressible Euler equation with Bohm potential. Quantum hydrodynamics comes from superfluid, superconductor, semiconductor and so on. Quantum hydrodynamic model describes Helium II superfluid, Bose–Einstein condensation in inert gas, dissipative perturbation of Homiton–Jacobi system, amplitude and dissipative perturbation of Eikonal quantum wave and so on. Recently, M. Gualdani and A. Jüngel deduced viscous quantum hydrodynamic equations from Wigner–Fokker–Planck equation by using method of momentum. Under some assumptions, S. Brull and F. M'ehats deduced the quantum hydrodynamic equations from Wigner equation by Chapman–Enskog expansion. Owing to the widely application of quantum hydrodynamic equations, the study of the quantum hydrodynamic equations has aroused the concern of more and more scholars.

Based on the above facts, recently, we collected and collated the data of quantum hydrodynamic equations, and studied the concerning mathematical problems. This book gives some newest results in this field, some of which are deduced by authors and coauthors. The main contents of this book are: The derivation and mathematical models of quantum hydrodynamic equations, global existence of weak solutions to the compressible quantum hydrodynamic equations, existence of finite energy weak solutions of inviscid quantum hydrodynamic equations, non-isentropic quantum Navier–Stokes equations with cold pressure, boundary problem of compressible quantum Euler–Poisson equations, asymptotic limit to the bipolar quantum hydrodynamic equations. We hope that the publishing of this book will help the readers to sort out issues that are not clarified in previous books and literatures, and enable the readers to carry out basic research by referring to the relevant references when they are interested in specific aspect of the problem.

Due to the limitation of the length of the book and author's knowledge level, there is still room for improvement. So we value the input of our readers who are quite welcomed to contact us for improvement of the work.

Boling Guo
19 May 2018, Beijing

Contents

Chapter 1

The Derivation and Mathematical Models of Quantum Hydrodynamic Equations

The quantum hydrodynamic phenomenon comes from superfluid, superconductor, semiconductor and so on. As it's well known, semiconductor is a kind of material whose conductivity is between the superconductor and insulator under room temperature. The semiconductor includes the element semiconductor and compound semiconductor. The element semiconductor is the semiconductor which consists of single element. On the other hand, we call the semiconductor which consists of two kinds or more elements as compound semiconductor. Silicon is the most common semiconductor used in the integrated circuit. Silicon and Germanium which are widely used and most early researched, are called the first generation semiconductor. The compound semiconductor has three forms: the first one, IIIA-VA group compounds, such as Gallium arsenide (GaAs), Indium phosphide (InP), Gallium nitride (GaN); the second one, IIB-VIA group compounds, such as Zinc sulfide (ZnS), Zinc oxide (ZnO); the third one, IVA group compounds, such as Silicon carbide (SiC). GaAs is the most commonly used compound semiconductor, which is called the second generation semiconductor material. Since 1980s, wide bandgap semiconductor material, especially GaN, has been paid more and more attention. Based on these, the blue diamond light diode and laser diode have been made. Wide bandgap semiconductor material on GaN as the representative is called the third generation semiconductor material.

According to the size of conductivity, solid can be divided into three parts: metal, semiconductor and insulator. The resistivity of the metal is $10^{-6}\Omega \cdot \text{cm}$, which has good conductivity. The resistivity of the insulator is above $10^{12}\Omega \cdot \text{cm}$, which is not conductive. The resistivity of semiconductor is about 10^{-3}–$10^{8}\Omega \cdot \text{cm}$ and its conductivity is between the metal and the insulator.

There are three kinds of energy band in the solid. The first one is that the electronic states in the band are empty, no electrons occupied. This band is called the empty band. The second one is that the electronic states in the band can be completely occupied by electrons, no empty state electron occupied. The band is called the full band. The third one is to be partially occupied by electrons. A part of the electronic states are filled by electrons in the band and there is the other part of electronic states which is empty. This band is called dissatisfaction zone or partially filled band.

Three solid energy band diagrams and filled conditions of metal, semiconductor and insulator by electrons are shown in Figure 1.1.

For semiconductor and insulator, at the absolute zero, the highest energy band occupied by electrons is a full band, while the adjacent band is empty. The empty band and the full band are separated by the band gap. Since there is no partially-occupied band, it does not conduct electricity. The band gap of insulator is very wide, and even when the temperature is high, it is difficult to excite the electrons from the full band to the empty band, so it is not conductive. The difference between semiconductor and insulator is only that the band gap of semiconductor is relatively narrow. So at a certain

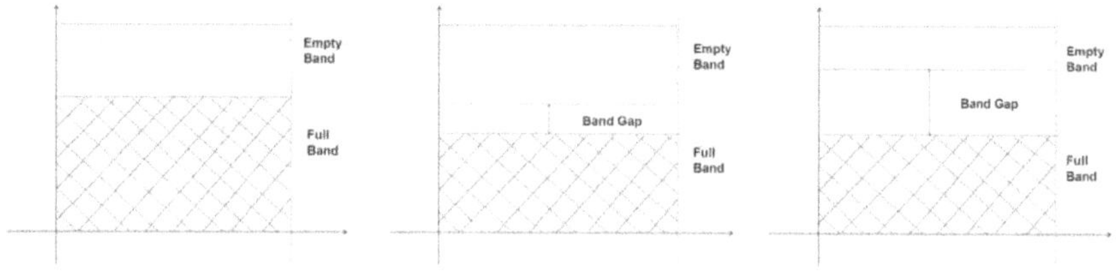

(a) Metal (b) Semiconductor (c) Insulator

Figure 1.1 Energy band diagram

temperature, the electrons are easy to go from the full band to the empty band. As a result, the original empty band which can be filled with a small amount of electrons, becomes the partially-occupied band and the original band which is full of electrons becomes the partially-occupied band. So the semiconductor has conductivity.

1.1 Isentropic quantum hydrodynamic model

Taking into account of one dimensional quantum plasma, a pure quantum state of electron, $\psi(t)$ is a wave function, satisfying the Schroödinger–Poisson system

$$\mathrm{i}^{h}\frac{\partial\psi}{\partial t} = -\frac{h^2}{2m}\frac{\partial^2\psi}{\partial x^2} - e\phi\psi, \tag{1.1.1}$$

$$\frac{\partial^2\phi}{\partial x^2} = \frac{e}{\varepsilon_0}(|\psi|^2 - n_0), \tag{1.1.2}$$

where h is the Planck constant, e is the charge of electron, m is the mass of electron, ϕ is the potential of electron, ε_0, n_0 are physical constants. Setting

$$\psi = A(x,t)\mathrm{e}^{i\boldsymbol{S}(x,t)}, \tag{1.1.3}$$

$$n = |\psi|^2, \quad \boldsymbol{u} = \frac{1}{m}\frac{\partial \boldsymbol{S}}{\partial x}, \tag{1.1.4}$$

substituting into the equations (1.1.1)–(1.1.2), separate the real part, the imaginary part of the equations satisfies

$$\frac{\partial n}{\partial t} + \frac{\partial n\boldsymbol{u}}{\partial x} = 0, \tag{1.1.5}$$

$$\frac{\partial \boldsymbol{u}}{\partial t} + \boldsymbol{u}\frac{\partial \boldsymbol{u}}{\partial x} = \frac{e}{m}\frac{\partial\phi}{\partial x} + \frac{h^2}{2m^2}\frac{\partial}{\partial x}\left(\frac{\frac{\partial^2(\sqrt{n})}{\partial x^2}}{\sqrt{n}}\right), \tag{1.1.6}$$

$$\frac{\partial^2\phi}{\partial x^2} = \frac{e}{\varepsilon_0}(n - n_0). \tag{1.1.7}$$

The initial data of equations (1.1.5)–(1.1.7) are

$$n|_{t=0} = n_0, \quad \boldsymbol{u}|_{t=0} = \boldsymbol{u}_0, \quad \phi|_{t=0} = 0. \tag{1.1.8}$$

Now considering the steady state of the equations (1.1.5)–(1.1.7), it holds

$$\frac{\mathrm{d}n\boldsymbol{u}}{\mathrm{d}x} = 0, \tag{1.1.9}$$

$$\boldsymbol{u}\frac{\mathrm{d}\boldsymbol{u}}{\mathrm{d}x} = \frac{e}{m}\frac{\mathrm{d}\phi}{\mathrm{d}x} + \frac{h^2}{2m^2}\frac{\mathrm{d}}{\mathrm{d}x}\left(\frac{\frac{\mathrm{d}^2(\sqrt{n})}{\mathrm{d}x^2}}{\sqrt{n}}\right). \tag{1.1.10}$$

The first integrals of equations (1.1.9)–(1.1.10) are

$$\boldsymbol{J} = n\boldsymbol{u}, \tag{1.1.11}$$

$$\Sigma = \frac{m\boldsymbol{u}^2}{2} - e\phi - \frac{h^2}{2m}\left(\frac{\frac{\mathrm{d}^2(\sqrt{n})}{\mathrm{d}x^2}}{\sqrt{n}}\right). \tag{1.1.12}$$

Let $\sqrt{n} = A$, eliminating $\boldsymbol{u}$, and using the Poisson equation, we have

$$h^2\frac{\mathrm{d}^2A}{\mathrm{d}x^2} = m\left(\frac{m\boldsymbol{J}^2}{A^3} - 2eA\phi\right), \tag{1.1.13}$$

$$\frac{\mathrm{d}^2\phi}{\mathrm{d}x^2} = \frac{e}{\varepsilon_0}(A^2 - n_0). \tag{1.1.14}$$

Setting $\boldsymbol{J} = n_0\boldsymbol{u}_0 \neq 0$, and

$$x^* = \frac{w_{\boldsymbol{p}}x}{\boldsymbol{u}_0}, \quad A^* = \frac{A}{\sqrt{n_0}}, \quad \phi^* = \frac{e\phi}{m\boldsymbol{u}_0^2}, \quad H = \frac{hw_{\boldsymbol{p}}}{m\boldsymbol{u}_0^2}, \tag{1.1.15}$$

where $w_{\boldsymbol{p}} = (\frac{n_0e^2}{m\varepsilon_0})^{1/2}$ is the frequency of the plasma, then from (1.1.13)–(1.1.14), we know

$$H^2\frac{\mathrm{d}^2A}{\mathrm{d}x^2} = -2\phi A + \frac{1}{A^3}, \tag{1.1.16}$$

$$\frac{\mathrm{d}^2\phi}{\mathrm{d}x^2} = A^2 - 1, \tag{1.1.17}$$

when $H = 0$, it will be the classical case. Since (1.1.16) holds $A^2 = \frac{1}{\sqrt{2\phi}}$, then we get

$$\frac{\mathrm{d}^2\phi}{\mathrm{d}x^2} = \frac{1}{\sqrt{2\phi}} - 1. \tag{1.1.18}$$

We can solve the equation (1.1.18) and the system (1.1.16)–(1.1.17).

1.2 Non-isentropic quantum hydrodynamic model

Firstly, we start from a collisional Wigner–Boltzman equation in the hydrodynamic scaling

$$\omega_t + \boldsymbol{p} \cdot \nabla_x \omega + \theta[V] = \frac{1}{\alpha}(M[\omega] - \omega) + \frac{\alpha}{\tau}(\Delta_p \omega + \mathrm{div}_p(\rho\omega)), \tag{1.2.1}$$

where $\omega(x, \boldsymbol{p}, t)$ is the Wigner function in the phase-space variables $(x, \boldsymbol{p}) \in \mathbf{R}^3 \times \mathbf{R}^3$, and $t > 0$, α is the scaled mean-free path (which is assumed to be small compared to one), τ is a relaxation time, and $\theta[V]$ is a non-local potential operator, involving the scaled Planck constant h. The electric potential V may be a given function or self-consistently coupled to the particle density through the Poisson equation. The collision operator on the right hand side of (1.2.1) consists of two terms. The dominant term is of relaxation type, namely the Wigner function having the tendency to tend to the quantum equilibrium Maxwellian $M[\omega]$. The second term is a Caldeira–Leggelt type collision operator.

Performing a Chapman–Enskog expansion $\omega = M[\omega] + \alpha g$ around the quantum Maxwellian gives non-local equations for the first momentum of ω, the particle density n, the momentum $n\boldsymbol{u}$, and the energy density ne, including viscous corrections of order α. Local expressions in density n, velocity $\boldsymbol{u}$ and temperature T are obtained by expanding the higher-order momentum in powers of h^2. Then assuming further that the temperature variations and the velocity $\boldsymbol{A}(\boldsymbol{u}) = \frac{1}{2}(\nabla \boldsymbol{u} - (\nabla \boldsymbol{u})^{\mathrm{T}})$ is of order h^2, $\nabla \log T = O(h^2)$ and $\boldsymbol{A}(\boldsymbol{u}) = O(\varepsilon^2)$ are the expansion up to order $O(\alpha^2 + \alpha h^2 + h^4)$. That is given as the following quantum Navier–Stokes system:

$$n_t + \mathrm{div}\,(n\boldsymbol{u}) = 0, \tag{1.2.2}$$

$$(n\boldsymbol{u})_t + \mathrm{div}\,(n\boldsymbol{u} \otimes \boldsymbol{u}) + \nabla(nT) - \frac{h^2}{12}\mathrm{div}\,(n\nabla^2 \log n) - n\nabla V$$
$$= \mathrm{div}\,\boldsymbol{S} - \frac{n\boldsymbol{u}}{\tau_0}, \tag{1.2.3}$$

$$
\begin{aligned}
&(ne)_t + \operatorname{div}((ne+nT)\boldsymbol{u}) - \frac{h^2}{12}\operatorname{div}((n\boldsymbol{\nabla}^2\log n)\boldsymbol{u}) + \operatorname{div} q_0 - n\boldsymbol{u}\boldsymbol{\nabla}V \\
&\quad = \operatorname{div}(\boldsymbol{S}\boldsymbol{u}) - \frac{2}{\tau_0}\left(ne - \frac{3}{2}n\right). \qquad (1.2.4)
\end{aligned}
$$

where $\boldsymbol{u}\otimes\boldsymbol{u}$ is a matrix with components $u_j u_k$, $\boldsymbol{\nabla}^2$ is the Hessian matrix, $\tau_0 = \frac{\tau}{\alpha}$. The energy density (up to terms of order $O(h^4)$) is given by

$$
ne = \frac{3}{2}nT + \frac{1}{2}n|\boldsymbol{u}|^2 - \frac{h^2}{24}n\Delta\log n. \qquad (1.2.5)
$$

The viscous stress tensor $\boldsymbol{S}$ and the total heat flux q_0 are defined by

$$
\begin{aligned}
\boldsymbol{S} &= 2\alpha nTD(\boldsymbol{u}) - \frac{2}{3}\alpha nT\operatorname{div}\boldsymbol{u}\boldsymbol{I}, \\
q_0 &= -\frac{5}{2}\alpha nT\boldsymbol{\nabla}T - \frac{h^2}{24}n(\Delta n + 2\boldsymbol{\nabla}\operatorname{div}\boldsymbol{u}), \qquad (1.2.6)
\end{aligned}
$$

where $D(\boldsymbol{u}) = \frac{1}{2}(\boldsymbol{\nabla}\boldsymbol{u} + (\boldsymbol{\nabla}\boldsymbol{u})^T)$ is the symmetric part of the velocity gradient, $\boldsymbol{I}$ is the unit matrix in $\mathbf{R}^3\times\mathbf{R}^3$.

Next we well derive the equations (1.2.2)–(1.2.4).

1.2.1 *Wigner–BGK equation*

From (1.2.1), we know the pseudo-differential operator

$$
\begin{aligned}
(\theta[V]\omega(x,\boldsymbol{p},t)) &= \frac{1}{(2\pi)^3}\int_{\mathbf{R}^3\times\mathbf{R}^3}(\delta V)(x,\eta,t)w(x,\boldsymbol{p}',t)\mathrm{e}^{\mathrm{i}(p-p')}\mathrm{d}\boldsymbol{p}'\mathrm{d}\eta, \\
(\delta V)(x,\eta,t) &= \frac{\mathrm{i}}{h}\left(V\left(x+\frac{h}{2}\eta,t\right) - V\left(x-\frac{h}{2}\eta,t\right)\right), \qquad (1.2.7)
\end{aligned}
$$

where $V = V(x,t)$ is a given function or self-consistently coupled to the Poisson equation. Here $h > 0$ denotes the scaled Planck constant. In the semiclassical limit $h \to 0$, the potential operator $\theta[V]\omega$ converges formally to its classical counterpart $\boldsymbol{\nabla}_x V\cdot\boldsymbol{\nabla}_{\boldsymbol{p}}\omega$. Let the quantum exponential and quantum logarithm function be defined by

$$\exp \omega = W(\exp W^{-1}(\omega)), \quad \log \omega = W(\log W^{-1}(\omega)),$$

where W is the Wigner transform, W^{-1} is inverse. Let

$$W(\rho)(x, \boldsymbol{p}) = \frac{1}{(2\pi)^2} \int_{\mathbf{R}^3} \tilde{\rho}\left(x + \frac{h}{2}\eta, x - \frac{h}{2}\eta\right) \mathrm{e}^{\mathrm{i}\eta \boldsymbol{p}} \mathrm{d}\eta,$$

set ρ as an operator in $L^2(\mathbf{R}^3)$, $\tilde{\rho}$ as the integral kernel, i.e.

$$\rho\phi(x) = \int_{\mathbf{R}^3} \tilde{\rho}(x, x')\phi(x')\mathrm{d}x', \quad \forall \phi \in L^2(\mathbf{R}^3).$$

The inverse W^{-1} is called the Weyl quantum, for any function $f(x, \boldsymbol{p})$, which is an operator defined on $L^2(\mathbf{R}^3)$:

$$(W^{-1}(f)\phi)(x, \boldsymbol{p}) = \int_{\mathbf{R}^3 \times \mathbf{R}^3} f\left(\frac{x+y}{2}\right) \phi(y) \mathrm{e}^{\mathrm{i}} \frac{\boldsymbol{p}(x-y)}{h} \mathrm{d}\boldsymbol{p}\mathrm{d}y,$$
$$\forall \phi \in L^2(\mathbf{R}^3). \quad (1.2.8)$$

The quantum free energy is given by

$$S(\omega) = \frac{-1}{(2\pi h)^3} \int_{\mathbf{R}^3 \times \mathbf{R}^3} \omega(x, \boldsymbol{p}, \cdot)$$
$$\times \left((\log W)(x, \boldsymbol{p}, \cdot) - 1 + \frac{\boldsymbol{p}^2}{2} - V(x, \cdot)\right) \mathrm{d}x\mathrm{d}\boldsymbol{p}. \quad (1.2.9)$$

For a given Wigner function ω, let $M[\omega]$ be the formal maximizer of $S(f)$, where f has the same momentum as ω, i.e.

$$\int_{\mathbf{R}^3} \omega \begin{pmatrix} 1 \\ \boldsymbol{p} \\ |\boldsymbol{p}|^2/2 \end{pmatrix} \mathrm{d}\boldsymbol{p} = \int_{\mathbf{R}^3} f \begin{pmatrix} 1 \\ \boldsymbol{p} \\ |\boldsymbol{p}|^2/2 \end{pmatrix} \mathrm{d}\boldsymbol{p}. \quad (1.2.10)$$

If such a solution exists, it has the form

$$M[\omega](x, \boldsymbol{p}, t) = \exp\left(A(x,t) - \frac{|\boldsymbol{p} - v(x,t)|^2}{2T(x,t)}\right), \quad (1.2.11)$$

where A, v and T are some Lagrange multipliers, it can be shown that there exists a unique S subject to a given local density $\int_R \frac{f\mathrm{d}p}{(2\pi h)^3}$ in the one dimensional setting. By definition, the quantum Maxwellian satisfies all function ω

$$\int_{\mathbf{R}^3} (M[\omega] - \omega) \begin{pmatrix} 1 \\ \boldsymbol{p} \\ |\boldsymbol{p}|^2/2 \end{pmatrix} \mathrm{d}\boldsymbol{p} = 0. \tag{1.2.12}$$

Physically, the equation (1.2.12) means that mass, momentum and energy of the collision operator $Q(\omega) = \frac{(M[\omega]-\omega)}{\alpha}$ is conserved.

1.2.2 *Non-local momentum equation*

We introduce the notation

$$\langle g(\boldsymbol{p}) \rangle = \frac{1}{(2\pi h)^3} \int_{\mathbf{R}^3} g(\boldsymbol{p}) \mathrm{d}\boldsymbol{p},$$

for given functions $g(\boldsymbol{p})$. Multiplying the Wigner equation (1.2.12) by 1, $\boldsymbol{p}$ and $\frac{|\boldsymbol{p}|^2}{2}$, respectively, integrating over $\boldsymbol{p} \in \mathbf{R}^3$, and employing (1.2.12), we obtain the momentum equations:

$$\begin{aligned}
&\partial_t \langle \omega \rangle + \mathrm{div}_x \langle \boldsymbol{p}\omega \rangle + \langle \theta[V]\omega \rangle = 0, \\
&\partial t \langle \boldsymbol{p}\omega \rangle + \mathrm{div}_x \langle \boldsymbol{p} \otimes \boldsymbol{p}\omega \rangle + \langle \boldsymbol{p}\theta[V]\omega \rangle = -\alpha\tau^{-1} \langle \boldsymbol{p}\omega \rangle, \\
&\partial t \left\langle \frac{1}{2} \boldsymbol{p}\omega \right\rangle + \mathrm{div}_x \left\langle \frac{1}{2} \boldsymbol{p}|\boldsymbol{p}|^2 \omega \right\rangle + \left\langle \frac{1}{2} |\boldsymbol{p}|^2 \theta[V]\omega \right\rangle \\
&\quad = -\alpha\tau^{-1} \langle |\boldsymbol{p}|^2 \omega - 3\omega \rangle,
\end{aligned}$$

where $\boldsymbol{p} \otimes \boldsymbol{p}$ denotes a matrix with components $p_j p_k (j, k = 1, 2, 3)$ governing the evolution of the particle density n, the momentum $n\boldsymbol{u}$ and the energy density ne are defined by

$$n = \langle \omega \rangle, \quad n\boldsymbol{u} = \langle \boldsymbol{p}\omega \rangle, \quad ne = \left\langle \frac{1}{2} |\boldsymbol{p}|^2 \omega \right\rangle.$$

The variable $\boldsymbol{u} = \frac{n\boldsymbol{u}}{n}$ is the macroscopic velocity, and $e = \frac{ne}{n}$ is the macroscopic energy. The right hand side of the momentum equations can be written as

$$\frac{-\alpha}{\tau} \langle \boldsymbol{p}\omega \rangle = -\frac{n\boldsymbol{u}}{\tau_0}, \quad -\frac{\alpha}{\tau} \langle |\boldsymbol{p}|^2 \omega - 3\omega \rangle = -\frac{2}{\tau_0} \left(ne - \frac{3}{2} n \right),$$

where $\tau_0 = \frac{\tau}{\alpha}$, and the momentum of the potential operator $\theta[V]$ can be simplified as shown in the following lemma.

Lemma 1.2.1. *The momentum of the potential operator $\theta[V]$ becomes*

$$\langle \theta[V] \rangle = 0, \quad \langle \boldsymbol{p}\theta[V]\omega \rangle = -n\boldsymbol{\nabla}_x V, \tag{1.2.13}$$

$$\langle \boldsymbol{p} \otimes \boldsymbol{p}\theta[V]\omega \rangle = -2n\boldsymbol{u} \otimes_s \boldsymbol{\nabla}_x V, \quad \left\langle \frac{1}{2}|\boldsymbol{p}|^2\theta[V]\omega \right\rangle = -n\boldsymbol{u} \cdot \boldsymbol{\nabla}_x V, \tag{1.2.14}$$

$$\left\langle \frac{1}{2}\boldsymbol{p}|\boldsymbol{p}|^2\theta[V]\omega \right\rangle = -(\langle \boldsymbol{p} \otimes \boldsymbol{p}\omega \rangle + ne\boldsymbol{I})\boldsymbol{\nabla}_x V + \frac{\varepsilon^2}{8} n\boldsymbol{\nabla}_x \Delta_x V, \tag{1.2.15}$$

where $\boldsymbol{u} \otimes_s \nabla_x V = \frac{1}{2}(\boldsymbol{u} \otimes \nabla_x V + \nabla_x V \otimes \boldsymbol{u})$ denotes the symmetrized tensor product, and $\boldsymbol{I}$ is the unit matrix in $R^3 \times R^3$.

Proof. The momentum (1.2.13) is computed in [1]–[3]. (1.2.15) is shown in [3], and (1.2.14) is proved in [4].

In order to calculate the momentum $\langle \boldsymbol{p} \otimes \boldsymbol{p}\omega \rangle$ and $\langle \frac{1}{2}\boldsymbol{p}|\boldsymbol{p}|^2\omega \rangle$, which appear in the momentum equations, we use the Chapman–Enskog expansion,

$$\omega = M[\omega] + \alpha g,$$

which defines the function g. The momentum can be written as

$$\begin{aligned} \langle \boldsymbol{p} \otimes \boldsymbol{p}\omega \rangle &= \langle \boldsymbol{p} \otimes \boldsymbol{p}M[\omega] \rangle + \alpha \langle \boldsymbol{p} \otimes \boldsymbol{p}g \rangle, \\ \left\langle \frac{1}{2}\boldsymbol{p}|\boldsymbol{p}|^2\omega \right\rangle &= \left\langle \frac{1}{2}\boldsymbol{p}|\boldsymbol{p}|^2 M[\omega] \right\rangle + \alpha \left\langle \frac{1}{2}\boldsymbol{p}|\boldsymbol{p}|^2 g \right\rangle. \end{aligned} \tag{1.2.16}$$

Inserting the Chapman–Enskog expansion in the Wigner equation (1.2.1), we obtain an explicit expression for g

$$\begin{aligned} g &= -\frac{1}{\alpha}(M[\omega] - \omega) \\ &= -\omega_t - \boldsymbol{p} \cdot \boldsymbol{\nabla}_x \omega - \theta[V]\omega + \alpha\tau^{-1}(\Delta_{\boldsymbol{p}}\omega + \mathrm{div}_{\boldsymbol{p}}(\boldsymbol{p}\omega)) \\ &= -M[\omega]_t - \boldsymbol{p} \cdot \boldsymbol{\nabla}_x M[\omega] - \theta[V]M[\omega] + O(\alpha). \end{aligned} \tag{1.2.17}$$

We introduce the quantum stress tensor $\boldsymbol{P}$ and quantum heat flux q,

$$\boldsymbol{P} = \langle (\boldsymbol{p}-\boldsymbol{u})\otimes(\boldsymbol{p}-\boldsymbol{u})M[\omega]\rangle, \quad q = \left\langle \frac{1}{2}(\boldsymbol{p}-\boldsymbol{u})|\boldsymbol{p}-\boldsymbol{u}|^2 M[\omega]\right\rangle.$$

Thus, using the relations $\langle M[\omega]\rangle = \langle \omega\rangle = n$, $\langle \boldsymbol{p}M[\omega]\rangle = n\boldsymbol{u}$, and $\langle \frac{1}{2}|\boldsymbol{p}|^2 M[\omega]\rangle = ne$, which follow from (1.2.12), we compute

$$\begin{aligned}
\boldsymbol{P} &= \langle \boldsymbol{p}\otimes \boldsymbol{p}M[\omega]\rangle - \boldsymbol{u}\otimes\langle \boldsymbol{p}M[\omega]\rangle - \langle \boldsymbol{p}M[\omega]\rangle\otimes \boldsymbol{u} + \boldsymbol{u}\otimes\boldsymbol{u}\langle M[\omega]\rangle \\
&= \langle \boldsymbol{p}\otimes \boldsymbol{p}M[\omega]\rangle - n\boldsymbol{u}\otimes\boldsymbol{u}, \\
q &= \left\langle \frac{1}{2}\boldsymbol{p}|\boldsymbol{p}|^2 M[\omega]\right\rangle - \left\langle \frac{1}{2}|\boldsymbol{p}|^2 M[\omega]\right\rangle \boldsymbol{u} + \frac{1}{2}|\boldsymbol{u}|^2\langle \boldsymbol{p}M[\omega]\rangle \\
&\quad - \frac{1}{2}\boldsymbol{u}|\boldsymbol{u}|^2\langle M[\omega]\rangle - \langle \boldsymbol{p}\otimes\boldsymbol{p}M[\omega]\rangle\boldsymbol{u} + \boldsymbol{u}\otimes\boldsymbol{u}\langle \boldsymbol{p}M[\omega]\rangle \\
&= \left\langle \frac{1}{2}\boldsymbol{p}|\boldsymbol{p}|^2 M[\omega]\right\rangle - (\boldsymbol{P} + ne^{\boldsymbol{I}})\boldsymbol{u}. \qquad (1.2.18)
\end{aligned}$$

From (1.2.16), we know that

$$\begin{aligned}
\langle \boldsymbol{p}\otimes\boldsymbol{p}\omega\rangle &= \boldsymbol{P} + n\boldsymbol{u}\otimes\boldsymbol{u} + \alpha\langle \boldsymbol{p}\otimes\boldsymbol{p}g\rangle, \\
\left\langle \frac{1}{2}\boldsymbol{p}|\boldsymbol{p}|^2\omega\right\rangle &= (\boldsymbol{P} + ne\boldsymbol{I})\boldsymbol{u} + q + \alpha\left\langle \frac{1}{2}\boldsymbol{p}|\boldsymbol{p}|^2 g\right\rangle.
\end{aligned}$$

□

Thus we show the following lemma.

Lemma 1.2.2. *Up to terms of $O(\alpha^2)$, the momentum equations of the Wigner equation read as follows*

$$n_t + \mathrm{div}_x(n\boldsymbol{u}) = 0 \qquad (1.2.19)$$

$$(n\boldsymbol{u})_t + \mathrm{div}_x(\boldsymbol{P} + n\boldsymbol{u}\otimes\boldsymbol{u}) - n\boldsymbol{\nabla}_x\boldsymbol{u} + \alpha\,\mathrm{div}_x S_1 - \frac{n\boldsymbol{u}}{\tau_0}, \qquad (1.2.20)$$

$$\begin{aligned}
&(ne)_t + \mathrm{div}_x((\boldsymbol{P} + ne\boldsymbol{I})\boldsymbol{u}) + \mathrm{div}_x q - n\boldsymbol{u}\cdot\boldsymbol{\nabla}_x V \\
&\quad = \alpha\,\mathrm{div}_x S_2 - \frac{2}{\tau_0}\left(ne - \frac{3}{2}n\right), \qquad (1.2.21)
\end{aligned}$$

where $\tau_0 = \frac{\tau}{\alpha_0}$, *and* S_1, S_2 *are given by*

$$S_1 = \partial_t \langle \boldsymbol{p} \otimes \boldsymbol{p} M[\omega] \rangle + \operatorname{div}_x \langle \boldsymbol{p} \otimes \boldsymbol{p} \otimes \boldsymbol{p} M[\omega] \rangle + \langle \boldsymbol{p} \otimes \boldsymbol{p} - [V] M[\omega] \rangle,$$

$$S_2 = \partial_t \left\langle \frac{1}{2} \boldsymbol{p} |\boldsymbol{p}|^2 M[\omega] \right\rangle + \operatorname{div}_x \left\langle \frac{1}{2} \boldsymbol{p} \otimes \boldsymbol{p} |\boldsymbol{p}|^2 M[\omega] \right\rangle + \left\langle \frac{1}{2} \boldsymbol{p} |\boldsymbol{p}|^2 \theta[V] M[\omega] \right\rangle.$$

These equations can be interpreted as a non-local quantum Navier–Stokes system. By expanding the quantum Maxwellian $M[\omega]$ in powers of $O(h^2)$, we derive a local version of this system. Such an expansion has been carried out in [5] for $\boldsymbol{P}, q, ne$.

$$\boldsymbol{P} = nT\boldsymbol{I} - \frac{h^2}{12} \boldsymbol{\nabla}_x^2 \log n + O(h^4), \tag{1.2.22}$$

$$q = -\frac{h^2}{24} n(\Delta_x \boldsymbol{u} + 2\boldsymbol{\nabla}_x \operatorname{div}_x \boldsymbol{u}) + O(h^4), \tag{1.2.23}$$

$$ne = \frac{3}{2} nT + \frac{1}{2} n|\boldsymbol{u}|^2 - \frac{h^2}{24} n\Delta_x \log n + O(h^4). \tag{1.2.24}$$

Under the assumption that the temperature varies slowly and the vorticity tensor $\boldsymbol{A}(\boldsymbol{u})$ is small, we know that $\boldsymbol{\nabla}_x \log T = O(h^2)$, and $\boldsymbol{A}(\boldsymbol{u}) = 1\frac{1}{2}(\boldsymbol{\nabla u} - (\boldsymbol{\nabla u})^{\mathrm{T}}) = O(h^2)$. It is shown in [6].

Moreover, the Lagrange multiplier $\boldsymbol{v}$ is related to the macroscopic velocity $\boldsymbol{u}$ by [7]

$$n\boldsymbol{u} = n\boldsymbol{v} + O(h^2).$$

Then we can derive local expressions for the viscous terms S_1, S_2 up to order $O(\alpha h^2 + \alpha^2)$.

1.2.3 *Calculation of* S_1

We set $S_1 = S_{11} + S_{12} + S_{13}$, where

$$S_{11} = \partial_t \langle \boldsymbol{p} \otimes \boldsymbol{p} M[\omega] \rangle, \quad S_{12} = \operatorname{div}_x \langle \boldsymbol{p} \otimes \boldsymbol{p} \otimes \boldsymbol{p} M[\omega] \rangle,$$
$$S_{13} = \langle \boldsymbol{p} \otimes \boldsymbol{p} \theta[V] M[\omega] \rangle.$$

It is sufficient to compute these expressions up to terms of order $O(\alpha)$ or $O(h^2)$, since the expressions $\alpha\,\mathrm{div}_x S_1$ and $\alpha\,\mathrm{div}_x S_2$ are already of order $O(\alpha)$. By Lemma 1.2.1, we have

$$S_{13} = -2n\boldsymbol{u} \otimes_s \boldsymbol{\nabla}_x V. \tag{1.2.25}$$

For the calculations of S_{12}, we set $\boldsymbol{s} = \frac{(\boldsymbol{p}-\boldsymbol{v})}{\sqrt{T}}$, and employ the following expansions, see [8].

Lemma 1.2.3.

$$\langle \boldsymbol{s}M[\omega]\rangle = O(h^2), \quad \langle \boldsymbol{s}\otimes\boldsymbol{s}\otimes\boldsymbol{s}M[\omega]\rangle = O(h^2),$$
$$\langle \boldsymbol{s}\otimes\boldsymbol{s}M[\omega]\rangle = n\boldsymbol{I} + O(h^2), \quad \langle \boldsymbol{s}\otimes\boldsymbol{s}|\boldsymbol{s}|^2M[\omega]\rangle = 5n\boldsymbol{I} + O(h^2).$$

Taking into account (1.2.21) and Lemma 1.2.3, we infer that

$$\begin{aligned}
&\langle p_j p_k p_l M[\omega]\rangle \\
&\quad = \langle (v_j + \sqrt{T}s_j)(v_k + \sqrt{T}s_k)(v_l + \sqrt{T}s_l)M[\omega]\rangle \\
&\quad = \langle v_j v_k v_l M[\omega]\rangle + T^{\frac{3}{2}}\langle s_j s_k s_l M[\omega]\rangle \\
&\qquad + \sqrt{T}(v_k v_l\langle s_j M[\omega]\rangle + v_l v_j\langle s_l M[\omega]\rangle + v_j v_l\langle s_k M[\omega]\rangle) \\
&\qquad + T(v_j\langle s_k s_l M[\omega]\rangle + v_k\langle s_j s_l M[\omega]\rangle + v_l\langle s_j s_k\rangle M[\omega]) \\
&\quad = n u_j u_k u_l + nT(u_j\delta_{kl} + u_k\delta_{jl} + u_l\delta_{jk}) + O(h^2).
\end{aligned} \tag{1.2.26}$$

Hence,

$$\begin{aligned}
S_{12} &= \mathrm{div}_x(n\boldsymbol{u}\otimes\boldsymbol{u}\otimes\boldsymbol{u}) + \mathrm{div}_x(nT\boldsymbol{u})\boldsymbol{I} + 2\boldsymbol{\nabla}_x(nT)\otimes_s\boldsymbol{u} \\
&\quad + 2nTD(\boldsymbol{u}) + O(h^2),
\end{aligned} \tag{1.2.27}$$

where we recall that $D(\boldsymbol{u}) = \frac{1}{2}(\boldsymbol{\nabla}\boldsymbol{u} + (\boldsymbol{\nabla}\boldsymbol{u})^{\mathrm{T}})$.

The calculation of S_{11} is more involved. By (1.2.18) and (1.2.22), we have

$$\langle \boldsymbol{p}\otimes\boldsymbol{p}M[\omega]\rangle = \boldsymbol{P} + n\boldsymbol{u}\otimes\boldsymbol{u} = nT\boldsymbol{I} + n\boldsymbol{u}\otimes\boldsymbol{u} + O(h^2),$$

and thus, the time derivative becomes

$$\begin{aligned}
S_{11} &= (nT\boldsymbol{I} + n\boldsymbol{u}\otimes\boldsymbol{u})_t + O(h^2) \\
&= \frac{2}{3}(ne)_t\boldsymbol{I} - \frac{1}{3}(n|\boldsymbol{u}|^2)_t\boldsymbol{I} + (n\boldsymbol{u}\otimes\boldsymbol{u})_t + O(h^2).
\end{aligned}$$

Elementary computations show that

$$(n|\boldsymbol{u}|^2)_t = 2(n\boldsymbol{u})_t\boldsymbol{u} + \operatorname{div}_x(n\boldsymbol{u})|\boldsymbol{u}|^2,$$
$$(n\boldsymbol{u}\otimes\boldsymbol{u})_t = 2(n\boldsymbol{u})_t \otimes_s \boldsymbol{u} + \operatorname{div}_x(n\boldsymbol{u})\boldsymbol{u}\otimes\boldsymbol{u}. \tag{1.2.28}$$

Then, employing (1.2.21) and the expression for $\boldsymbol{P}, ne, q$, i.e.

$$(ne)_t = -\operatorname{div}_x\left(\frac{5}{2}nT\boldsymbol{u} + \frac{1}{2}n\boldsymbol{u}|\boldsymbol{u}|^2\right) + n\boldsymbol{u}\cdot\boldsymbol{\nabla}_x V + O(h^2+\alpha),$$

we find that

$$\begin{aligned} S_{11} &= \frac{2}{3}(ne)_t\boldsymbol{I} - \frac{2}{3}(n\boldsymbol{u})_t\boldsymbol{u}\boldsymbol{I} + 2(n\boldsymbol{u})_t\otimes_s\boldsymbol{u} + \operatorname{div}_x(n\boldsymbol{u}) \\ &\quad + \left(\boldsymbol{u}\otimes\boldsymbol{u} - \frac{1}{3}|\boldsymbol{u}|^2\boldsymbol{I}\right) + O(h^2) \\ &= -\frac{2}{3}\operatorname{div}_x\left(n\boldsymbol{u}\left(\frac{5}{2}T + \frac{1}{2}|\boldsymbol{u}|^2\right)\right)\boldsymbol{I} + \frac{2}{3}n\boldsymbol{u}\cdot\boldsymbol{\nabla}_x V\boldsymbol{I} \\ &\quad + \frac{2}{3}\operatorname{div}_x(n\boldsymbol{u}\otimes\boldsymbol{u})\cdot\boldsymbol{u}\boldsymbol{I} + \frac{2}{3}\boldsymbol{\nabla}_x(nT)\cdot\boldsymbol{u}\boldsymbol{I} - \frac{2}{3}n\boldsymbol{u}\cdot\boldsymbol{\nabla}_x\boldsymbol{u}\boldsymbol{I} \\ &\quad - 2\operatorname{div}(n\boldsymbol{u}\otimes_s\boldsymbol{u})\otimes\boldsymbol{u} - 2\boldsymbol{\nabla}_x(nT)\otimes_s\boldsymbol{u} + 2n\boldsymbol{\nabla}_x V\otimes_s\boldsymbol{u} \\ &\quad + \operatorname{div}_x(n\boldsymbol{u})\left(\boldsymbol{u}\otimes\boldsymbol{u} - \frac{1}{3}|\boldsymbol{u}|^2\boldsymbol{I}\right) + O(h^2+\alpha). \end{aligned}$$

The term involving T simplifies to

$$-\operatorname{div}_x(nT\boldsymbol{u})\boldsymbol{I} - \frac{2}{3}nT\operatorname{div}_x\boldsymbol{u}\boldsymbol{I} - 2\boldsymbol{\nabla}_x(nT)\otimes_s\boldsymbol{u}.$$

The term involving third powers of $\boldsymbol{u}$ sums up to $-\operatorname{div}_x(n\boldsymbol{u}\otimes\boldsymbol{u}\otimes\boldsymbol{u})$. Hence,

$$\begin{aligned} S_{11} &= -\operatorname{div}_x(nT\boldsymbol{u})\boldsymbol{I} - \frac{2}{3}nT\operatorname{div}_x\boldsymbol{u}\boldsymbol{I} - 2\boldsymbol{\nabla}_x(nT)\otimes_{\boldsymbol{s}}\boldsymbol{u} \\ &\quad - \operatorname{div}_x(n\boldsymbol{u}\otimes\boldsymbol{u}\otimes\boldsymbol{u}) + 2n\boldsymbol{\nabla}_x V\otimes_s\boldsymbol{u} + O(h^2+\alpha). \end{aligned}$$

In view of (1.2.27) and (1.2.28), we conclude that

$$S_1 = 2nTD(\boldsymbol{u}) - \frac{2}{3}nT\operatorname{div}_x\boldsymbol{u}\boldsymbol{I} + O(h^2+\alpha).$$

1.2.4 Calculation of S_2

$$S_2 = S_{21} + S_{22} + S_{23},$$

$$S_{21} = \partial_t \left(\frac{1}{2}\boldsymbol{p}|\boldsymbol{p}|^2 M[\omega]\right), \quad S_{22} = \operatorname{div}_x \left(\frac{1}{2}\boldsymbol{p}\otimes \boldsymbol{p}|\boldsymbol{p}|^2 M[\omega]\right),$$

$$S_{23} = \left\langle \frac{1}{2}\boldsymbol{p}|\boldsymbol{p}|^2\theta[V]M[\omega]\right\rangle.$$

The expressions (1.2.14), (1.2.22), and (1.2.23) show that

$$\begin{aligned} S_{23} &= -(\boldsymbol{P} + n\boldsymbol{u}\otimes \boldsymbol{u} + ne\boldsymbol{I})\boldsymbol{\nabla}_x V + O(h^2) \\ &= \left(\frac{5}{2}nT\boldsymbol{I} + \frac{1}{2}n|\boldsymbol{u}|^2\boldsymbol{I} + n\boldsymbol{u}\otimes\boldsymbol{u}\right)\boldsymbol{\nabla}_x V + O(h^2). \end{aligned}$$

Next, we calculate the fourth momentum $\langle\frac{1}{2}\boldsymbol{p}\otimes\boldsymbol{p}|\boldsymbol{p}|M[\omega]\rangle$. Setting $\boldsymbol{s} = \frac{(\boldsymbol{p}-\boldsymbol{v})}{\sqrt{T}}$, and employing Lemma 1.2.3 and (1.2.24), we obtain

$$\begin{aligned} &\left\langle \frac{1}{2}\boldsymbol{p}\otimes\boldsymbol{p}|\boldsymbol{p}|^2 M[\omega]\right\rangle \\ &= \left\langle \frac{1}{2}(\boldsymbol{v}+\sqrt{T}\boldsymbol{s})\otimes\langle\boldsymbol{v}+\sqrt{T}\boldsymbol{s}\rangle|\boldsymbol{v}+\sqrt{T}\boldsymbol{s}|^2 M[\omega]\right\rangle \\ &= \left\langle \frac{1}{2}\boldsymbol{v}\otimes\boldsymbol{v}|\boldsymbol{v}|^2 M[\omega]\right\rangle + \sqrt{T}(\boldsymbol{v}\otimes_{\boldsymbol{s}}\langle \boldsymbol{s}M[\omega] > |\boldsymbol{v}^2| \\ &\quad + (\boldsymbol{v}\otimes\boldsymbol{v})\boldsymbol{v}\cdot\langle\boldsymbol{s}M[\omega]\rangle) + T\left(\frac{1}{2}\langle\boldsymbol{s}\otimes\boldsymbol{s}M[\omega]\rangle|\boldsymbol{v}|^2\right. \\ &\quad \left. + \boldsymbol{v}\otimes\boldsymbol{v}\left\langle\frac{1}{2}|\boldsymbol{s}|^2 M[\omega]\right\rangle + 2\langle(\boldsymbol{v}\otimes_{\boldsymbol{s}}\boldsymbol{s})\boldsymbol{v}\cdot\boldsymbol{s}M[\omega]\rangle\right) \\ &\quad + T^{\frac{3}{2}}(\boldsymbol{v}\otimes_s\langle\boldsymbol{s}|\boldsymbol{s}|^2 M[\omega]\rangle + \langle\boldsymbol{s}\otimes\boldsymbol{s}\otimes\boldsymbol{s}M[\omega]\rangle\cdot\boldsymbol{v}) \\ &\quad + T^2\left\langle\frac{1}{2}\boldsymbol{s}\otimes\boldsymbol{s}|\boldsymbol{s}|^2 M[\omega]\right\rangle \\ &= \frac{1}{2}n\boldsymbol{u}\otimes\boldsymbol{u}|\boldsymbol{u}|^2 + \frac{1}{2}T|\boldsymbol{u}|^2\langle\boldsymbol{s}\otimes\boldsymbol{s}M[\omega]\rangle + \frac{1}{2}T\boldsymbol{u}\otimes\boldsymbol{u}\langle|\boldsymbol{s}|^2 M[\omega]\rangle \\ &\quad + 2T\boldsymbol{u}\otimes_{\boldsymbol{s}}\langle\boldsymbol{s}(\boldsymbol{u}\cdot\boldsymbol{s})M[\omega]\rangle + \frac{1}{2}T^2\langle\boldsymbol{s}\otimes\boldsymbol{s}|\boldsymbol{s}|^2 M[\omega]\rangle + O(h^2) \\ &= \frac{1}{2}n\boldsymbol{u}\otimes\boldsymbol{u}|\boldsymbol{u}|^2 + \frac{1}{2}nT|\boldsymbol{u}|^2\boldsymbol{I} + \frac{7}{2}nT(\boldsymbol{u}\otimes\boldsymbol{u}) + \frac{5}{2}nT^2\boldsymbol{I} + O(h^2). \end{aligned}$$

Therefore

$$
\begin{aligned}
S_{22} &= \frac{1}{2}\operatorname{div}_x(n\boldsymbol{u}\otimes\boldsymbol{u}|\boldsymbol{u}|^2) + \frac{1}{2}\boldsymbol{\nabla}_x(nT|\boldsymbol{u}|^2) \\
&\quad + \frac{7}{2}\operatorname{div}_x(nT\boldsymbol{u}\otimes\boldsymbol{u})\frac{5}{2}\boldsymbol{\nabla}_x(nT^2) + O(h^2).
\end{aligned}
$$

It remains to calculate the time derivative of $\langle\frac{1}{2}\boldsymbol{p}|\boldsymbol{p}|^2M[\omega]\rangle$. Setting $k=l$ in (1.2.26) and summing over j, we find that

$$
\left\langle \frac{1}{2}\boldsymbol{p}|\boldsymbol{p}|^2M[\omega]\right\rangle = \frac{5}{2}nT\boldsymbol{u} + \frac{1}{2}n\boldsymbol{u}|\boldsymbol{u}|^2 + O(h^2).
$$

Hence,

$$
\begin{aligned}
S_{21} &= \partial_t\left(\frac{5}{2}nT\boldsymbol{u} + \frac{1}{2}n\boldsymbol{u}|\boldsymbol{u}|^2\right) + O(h^2) \\
&= \partial_t\left(\frac{5}{3}ne\boldsymbol{u} - \frac{1}{3}n\boldsymbol{u}|\boldsymbol{u}|^2\right) + O(h^2) \\
&= \frac{5}{3}(ne)_t\boldsymbol{u} + \frac{5}{3}\left(\frac{3}{2}T + \frac{1}{2}|\boldsymbol{u}^2|\right)n\boldsymbol{u}_t - \frac{1}{3}(n|\boldsymbol{u}|^2)_t\boldsymbol{u} \\
&\quad - \frac{1}{3}|\boldsymbol{u}|^2n\boldsymbol{u}_t + O(h^2) \\
&= \frac{5}{3}(ne)_t\boldsymbol{u} + \left(\frac{5}{2}T + \frac{1}{2}|\boldsymbol{u}|^2\right)n\boldsymbol{u}_t - \frac{1}{3}(n|\boldsymbol{u}|^2)_t\boldsymbol{u} + O(h^2).
\end{aligned}
$$

By (1.2.28), and $n\boldsymbol{u}_t = (n\boldsymbol{u})_t + \operatorname{div}_x(n\boldsymbol{u})\boldsymbol{u}$, it follows that

$$
\begin{aligned}
S_{21} &= \frac{5}{3}(ne)_t\boldsymbol{u} + \left(\frac{5}{2}T + \frac{1}{2}|\boldsymbol{u}|^2\right)(n\boldsymbol{u})_t + \left(\frac{5}{2}T + \frac{1}{2}|\boldsymbol{u}|^2\right)\operatorname{div}_x(n\boldsymbol{u})\boldsymbol{u} \\
&\quad - \frac{1}{3}\operatorname{div}_x(n\boldsymbol{u})|\boldsymbol{u}|^2\boldsymbol{u} - \frac{2}{3}n\boldsymbol{u}\cdot(n\boldsymbol{u})_t + O(h^2) \\
&= \frac{5}{3}(ne)_t\boldsymbol{u} + \left(\frac{5}{2}T + \frac{1}{2}|\boldsymbol{u}|^2\right)(n\boldsymbol{u})_t + \left(\frac{5}{2}T + \frac{1}{6}|\boldsymbol{u}|^2\right)\operatorname{div}_x(n\boldsymbol{u})\boldsymbol{u} \\
&\quad - \frac{2}{3}n\boldsymbol{u}\cdot(n\boldsymbol{u})_t + O(h^2).
\end{aligned}
$$

From (1.2.20) and (1.2.21), we have

$$\begin{aligned}S_{21} &= -\frac{25}{6}\mathrm{div}_x(nT\boldsymbol{u})\boldsymbol{u} - \frac{5}{6}\mathrm{div}_x(n\boldsymbol{u}|\boldsymbol{u}|^2)\boldsymbol{u} + \frac{5}{3}n\boldsymbol{u}\boldsymbol{u}\cdot\boldsymbol{\nabla}_x V\\
&\quad - \left(\frac{5}{2}T + \frac{1}{2}|\boldsymbol{u}|^2\right)\mathrm{div}_x(nT\boldsymbol{I} + n\boldsymbol{u}\otimes\boldsymbol{u}) + \left(\frac{5}{2}T + \frac{1}{2}|\boldsymbol{u}|^2\right)n\boldsymbol{\nabla}_x V\\
&\quad + \left(\frac{5}{2}T + \frac{1}{6}|\boldsymbol{u}|^2\right)\mathrm{div}(n\boldsymbol{u})\cdot\boldsymbol{u} + \frac{2}{3}\mathrm{div}_x(nT\boldsymbol{I} + n\boldsymbol{u}\otimes\boldsymbol{u})\cdot n\boldsymbol{u}\\
&\quad - \frac{2}{3}n\boldsymbol{u}\boldsymbol{u}\cdot\boldsymbol{\nabla}_x V + O(h^2+\alpha).\end{aligned}$$

The terms involving the potential V sum up to

$$n\boldsymbol{u}\boldsymbol{u}\cdot\boldsymbol{\nabla}_x V + \frac{5}{2}nT\boldsymbol{\nabla}_x V + \frac{1}{2}n|\boldsymbol{u}|^2\boldsymbol{\nabla}_x V,$$

which equals $-S_{23}$ up to terms of order $O(h^2)$. We simplify the terms involving T:

$$\begin{aligned}&-\frac{7}{2}\mathrm{div}_x(nT\boldsymbol{u}\otimes\boldsymbol{u}) - \frac{2}{3}nT\boldsymbol{u}\,\mathrm{div}_x\boldsymbol{u} + nT(\boldsymbol{\nabla}\boldsymbol{u} + (\boldsymbol{\nabla}\boldsymbol{u})^{\mathrm{T}})\boldsymbol{u}\\
&\quad - \frac{5}{2}T\boldsymbol{\nabla}_x(nT) - \frac{1}{2}\boldsymbol{\nabla}_x(nT|\boldsymbol{u}|^2).\end{aligned}$$

Finally, we summarize all remaining terms involving fourth powers of $\boldsymbol{u}$:

$$\begin{aligned}&-\frac{1}{2}|\boldsymbol{u}|^2\boldsymbol{u}\,\mathrm{div}_x(n\boldsymbol{u})\\
&\quad -n\boldsymbol{u}(\boldsymbol{u}\boldsymbol{\nabla}n\boldsymbol{u}) - \frac{1}{2}|\boldsymbol{u}|^2(\boldsymbol{u}\cdot\boldsymbol{\nabla})\boldsymbol{u} = -\frac{1}{2}\mathrm{div}_x(n\boldsymbol{u}\otimes\boldsymbol{u}|\boldsymbol{u}|^2).\end{aligned}$$

We conclude that

$$\begin{aligned}S_{21} &= \left(\frac{5}{2}nT\boldsymbol{I} + \frac{1}{2}n|\boldsymbol{u}|^2\boldsymbol{I} + n\boldsymbol{u}\otimes\boldsymbol{u}\right)\boldsymbol{\nabla}_x V\\
&\quad + 2nTD(\boldsymbol{u})\boldsymbol{u} - \frac{2}{3}nT\boldsymbol{u}\,\mathrm{div}_x\boldsymbol{u}\\
&\quad - \frac{1}{2}\boldsymbol{\nabla}_x(nT|\boldsymbol{u}|^2) - \frac{7}{2}\mathrm{div}_x(nT\boldsymbol{u}\otimes\boldsymbol{u}) - \frac{5}{2}T\boldsymbol{\nabla}_x(nT)\\
&\quad - \frac{1}{2}\mathrm{div}_x(n\boldsymbol{u}\otimes\boldsymbol{u}|\boldsymbol{u}|^2) + O(h^2+\alpha).\end{aligned}$$

Summing S_{21}, S_{22}, S_{23}, cancelling some terms, and we end up with

$$S_2 = \frac{5}{2} nT\boldsymbol{\nabla}_x T + 2nTD(\boldsymbol{u})\boldsymbol{u} - \frac{2}{3} nT\boldsymbol{u}\,\mathrm{div}_x \boldsymbol{u} + O(h^2 + \alpha).$$

Theorem 1.2.1. *Assume that* $\boldsymbol{A}(\boldsymbol{u}) = \frac{\boldsymbol{\nabla u} - (\boldsymbol{\nabla u})^{\mathrm{T}}}{2} = O(h^2)$, $\boldsymbol{\nabla} \log T = O(h^2)$, *then the momentum equations of the Wigner–Boltzman equation* (1.2.1) *are given by* (1.2.2)–(1.2.4), *up to terms of order* $O(\alpha^2 + \alpha h + h^4)$, *where the energy density* ne, *the viscous stress tensor* $\boldsymbol{S}$, *and the total heat flux* q_0 *are given by* (1.2.5) *and* (1.2.6).

1.2.5 *Energy and entropy estimates*

In this section, we prove energy and entropy estimates for the quantum Navier–Stokes system (1.2.2)–(1.2.4). Let the electric potential V solve the Poisson equation

$$\lambda^2 \Delta V = n - C(x), \tag{1.2.29}$$

where λ is the scaled Debye length, and C is models fixed background ions.

Proposition 1.2.1. *Let* $(n, n\boldsymbol{u}, ne)$ *be a smooth solution to* (1.2.2)–(1.2.4). *Let* V *be the solution to* (1.2.29). *We assume that these variables decay sufficiently fast to zero as* $|x| \to \infty$ *uniformly in* t. *Then the total mass* $N(t) = \int_{\mathbf{R}^3} n(x,t)\mathrm{d}x$ *is conserved, i.e.* $\frac{\mathrm{d}}{\mathrm{d}t}N(t) = 0$, *and the total energy*

$$E(t) = \int_{\mathbf{R}^3} \left(ne + \frac{\lambda^2}{2} |\boldsymbol{\nabla} V|^2 \right) \mathrm{d}x$$

is dissipated according to

$$\frac{\mathrm{d}E(t)}{\mathrm{d}t} + \frac{2}{\tau_0} \int_{\mathbf{R}^3} \left(\frac{3}{2} n(T-1) + \frac{1}{2} n|\boldsymbol{u}|^2 + \frac{h^2}{6} |\boldsymbol{\nabla}\sqrt{n}|^2 \right) \mathrm{d}x = 0.$$

In particular, without relaxation $\frac{1}{\tau_0} = 0$, *the total energy is conserved. Furthermore, the total energy can be written as*

$$E(t) = \int_{\mathbf{R}^3} \left(\frac{3}{2} nT + \frac{1}{2} n|\boldsymbol{u}|^2 + \frac{h^2}{6} |\boldsymbol{\nabla}\sqrt{n}|^2 + \frac{\lambda^2}{2} |\boldsymbol{\nabla} V|^2 \right) \mathrm{d}x \geq 0.$$

The total energy is the sum of the thermal, kinetic, quantum and electric energy.

Proof.

$$\begin{aligned}
\frac{\mathrm{d}E}{\mathrm{d}t} &= \int_{\mathbf{R}^3}((ne)_t + \lambda^2\nabla V\cdot\nabla V_t)\mathrm{d}x \\
&= \int_{\mathbf{R}^3}(n\boldsymbol{u}\cdot\nabla V - \lambda^2 V(\Delta V)_t)\mathrm{d}x - \frac{2}{\tau_0}\int_{\mathbf{R}^3}\left(ne - \frac{3}{2}\boldsymbol{u}\right)\mathrm{d}x \\
&= \int_{\mathbf{R}^3}(-\mathrm{div}(n\boldsymbol{u})V - Vn_t)\mathrm{d}x \\
&\quad - \frac{2}{\tau_0}\int_{\mathbf{R}^3}\left(\frac{3}{2}n(T-1) + \frac{1}{2}n|\boldsymbol{u}|^2 + \frac{\varepsilon^2}{24}\frac{|\nabla n|^2}{n}\right)\mathrm{d}x \\
&\quad - \frac{2}{\tau_0}\int_{\mathbf{R}^3}\left(\frac{3}{2}n(T-1) + \frac{1}{2}n|\boldsymbol{u}|^2 + \frac{\varepsilon^2}{6}|\nabla\sqrt{n}|^2\right)\mathrm{d}x.
\end{aligned}$$

Next we calculate the evolution of the thermal energy $\frac{3}{2}nT$. By (1.2.5), we find that

$$\left(\frac{3}{2}n|T\right)_t = (ne)_t - \frac{1}{2}(n|\boldsymbol{u}|^2)_t\frac{h^2}{24}(n\Delta\log n)_t. \tag{1.2.30}$$

We employ (1.2.4) and (1.2.28) to reformulate the first two terms on the right hand side of (1.2.30):

$$\begin{aligned}
(ne)_t = &\frac{5}{2}\mathrm{div}(-nT\boldsymbol{u} + \alpha nT\nabla T) - \frac{1}{2}\mathrm{div}\,(n\boldsymbol{u}|\boldsymbol{u}|^2) \\
&+ \mathrm{div}\,(\boldsymbol{S}\boldsymbol{u}) + n\boldsymbol{u}\cdot\nabla\boldsymbol{u} \\
&+ \frac{h^2}{24}\mathrm{div}\,(n\boldsymbol{u}\Delta\log n + 2n(\nabla^2\log n)\boldsymbol{u} + n\Delta\boldsymbol{u} + 2n\nabla\mathrm{div}\,\boldsymbol{u}) \\
&- \frac{2}{\tau_0}\left(ne - \frac{3}{2}n\right),
\end{aligned}$$

$$\begin{aligned}
-\frac{1}{2}(n|\boldsymbol{u}|^2)_t &= -(n\boldsymbol{u})_t\cdot\boldsymbol{u} - \frac{1}{2}\mathrm{div}\,(n\boldsymbol{u})\cdot|\boldsymbol{u}|^2 \\
&= \mathrm{div}\,(n\boldsymbol{u}\otimes\boldsymbol{u})\cdot\boldsymbol{u} - \frac{1}{2}\mathrm{div}\,(n\boldsymbol{u})|\boldsymbol{u}|^2 + \nabla(nT)\boldsymbol{u} \\
&\quad - \frac{h^2}{12}\mathrm{div}\,(n\nabla^2\log n)\boldsymbol{u} - n\boldsymbol{u}\nabla\boldsymbol{u} - (\mathrm{div}\,\boldsymbol{S})\boldsymbol{u} - \frac{n}{\tau_0}|u|^2.
\end{aligned}$$

Since, the third-order terms in $\boldsymbol{u}$ sum up to zero

$$-\frac{1}{2}\mathrm{div}\,(n\boldsymbol{u}|\boldsymbol{u}|^2) + \mathrm{div}\,(n\boldsymbol{u}\otimes\boldsymbol{u})\boldsymbol{u} - \frac{1}{2}\mathrm{div}\,(n\boldsymbol{u})|\boldsymbol{u}|^2 = 0.$$

The sum $(ne)_t - \frac{1}{2}(n|\boldsymbol{u}|^2)_t$ simplifies to

$$\begin{aligned}
&(ne)_t - \frac{1}{2}(n|\boldsymbol{u}|^2)_t \\
&\quad = \mathrm{div}\left(-\frac{3}{2}nT\boldsymbol{u} + \frac{5}{2}\alpha nT\boldsymbol{\nabla}T\right) - nT\mathrm{div}\,\boldsymbol{u} - \boldsymbol{S} : \boldsymbol{\nabla}\boldsymbol{u} \\
&\quad\quad - \frac{h^2}{12}\mathrm{div}\,(n\boldsymbol{\nabla}^2\log n)\boldsymbol{u} + \frac{h^2}{24}\mathrm{div}\,\big(n\boldsymbol{u}\Delta\log n + 2n(\boldsymbol{\nabla}^2\log n)\boldsymbol{u} \\
&\quad\quad + n\Delta\boldsymbol{u} + \alpha n\boldsymbol{\nabla}\mathrm{div}\,\boldsymbol{u}\big) - \frac{2}{\tau_0}\left(\frac{3}{2}n(T-1) - \frac{h^2}{24}n\Delta\log n\right),
\end{aligned} \tag{1.2.31}$$

where ‘:’ means summation of both matrix indices. We notice

$$\begin{aligned}
\boldsymbol{S} : \boldsymbol{\nabla}\boldsymbol{u} &= \alpha nT\left((\boldsymbol{\nabla}\boldsymbol{u} + (\boldsymbol{\nabla}\boldsymbol{u})^{\mathrm{T}}) : \boldsymbol{\nabla}\boldsymbol{u} - \frac{3}{2}(\mathrm{div}\,\boldsymbol{u})^2\right) \\
&= \alpha nT\left(\frac{1}{2}(\boldsymbol{\nabla}\boldsymbol{u} + (\boldsymbol{\nabla}\boldsymbol{u})^{\mathrm{T}}) : (\boldsymbol{\nabla}\boldsymbol{u} + (\boldsymbol{\nabla}\boldsymbol{u})^{\mathrm{T}}) - \frac{3}{2}(\mathrm{div}\,\boldsymbol{u})^2\right) \\
&= \alpha nT\left(2|D(\boldsymbol{u})|^2 - \frac{3}{2}(\mathrm{div}\,\boldsymbol{u})^2\right).
\end{aligned}$$

The remaining term derivative becomes

$$(n\Delta\log n)_t = -\mathrm{div}\,(n\boldsymbol{u})\Delta\log n - n\Delta(\boldsymbol{\nabla}\log n\cdot\boldsymbol{u} + \mathrm{div}\,\boldsymbol{u})$$

by (1.2.2), since

$$\begin{aligned}
&n\Delta(\boldsymbol{\nabla}\log n\cdot\boldsymbol{u} + \mathrm{div}\,\boldsymbol{u}) \\
&\quad = n\boldsymbol{\nabla}\Delta\log n\cdot\boldsymbol{u} + 2n\boldsymbol{\nabla}^2\log n : \boldsymbol{\nabla}\boldsymbol{u} + \Delta\boldsymbol{u}\cdot\boldsymbol{\nabla}n + n\Delta\mathrm{div}\,\boldsymbol{u} \\
&\quad = \mathrm{div}\,(n\boldsymbol{u}\Delta\log n + 2n(\boldsymbol{\nabla}^2\log n)\boldsymbol{u} + n\Delta\boldsymbol{u}) - \mathrm{div}\,(n\boldsymbol{u})\Delta\log n \\
&\quad\quad - 2\mathrm{div}\,(n\boldsymbol{\nabla}^2\log n)\cdot\boldsymbol{u}.
\end{aligned}$$

It follows that

$$\frac{h^2}{24}(n\Delta\log n)_t = -\frac{h^2}{24}\mathrm{div}\,(n\boldsymbol{u}\Delta\log n + 2n(\boldsymbol{\nabla}^2\log n)\cdot\boldsymbol{u} + n\Delta\boldsymbol{u})$$
$$+\frac{h^2}{12}\mathrm{div}\,(n\boldsymbol{\nabla}^2\log n)\cdot n.$$

From (1.2.30) and (1.2.31), we have

$$\left(\frac{3}{2}nT\right)_t + \mathrm{div}\left(\frac{3}{2}nT\boldsymbol{u} - \frac{5}{2}\alpha nT\boldsymbol{\nabla}T - \frac{h^2}{12}n\nabla\,\mathrm{div}\,\boldsymbol{u}\right)$$
$$= -nT\,\mathrm{div}\,\boldsymbol{u} + \alpha nT(2|D(\boldsymbol{u})|^2 - \frac{2}{3}(\mathrm{div}\,\boldsymbol{u})^2)$$
$$-\frac{2}{\tau_0}\left(\frac{3}{2}n(T-1) - \frac{h^2}{24}n\Delta\log n\right). \tag{1.2.32}$$

This formulation allows us to formulate the evolutions of the thermal energy and entropy. □

Proposition 1.2.2. *Let $(n, n\boldsymbol{u}, ne)$ be a smooth solution to (1.2.2)–(1.2.4). We assume that these variables decay sufficiently fast to zero as $|x|\to\infty$ uniformly in t. Then*

$$\frac{3}{2}\frac{\mathrm{d}}{\mathrm{d}t}\int_{R^3} nT\mathrm{d}x$$
$$= \int_{R^3}\boldsymbol{\nabla}(nT)\cdot\boldsymbol{u}\mathrm{d}x + \alpha\int_{R^3} nT(2|D(\boldsymbol{u})|^2$$
$$-\frac{3}{2}(\mathrm{div}\,\boldsymbol{u})^2)dx - \frac{2}{\tau_0}\int_{R^3}\left[\frac{3}{2}n(T-1) + \frac{h^2}{6}|\boldsymbol{\nabla}\sqrt{n}|^2\right]\mathrm{d}x, \tag{1.2.33}$$

$$\frac{\mathrm{d}}{\mathrm{d}t}\int_{R^3}\left(n\log\frac{n}{T^{3/2}} + \frac{3}{2}nT\right)\mathrm{d}x$$
$$= \int_{R^3}\left(\frac{5}{2}\alpha\frac{n}{T}|\boldsymbol{\nabla}T|^2 + \alpha n(2|D(\boldsymbol{u})|^2 - \frac{2}{3}(\mathrm{div}\,\boldsymbol{u})^2\right)$$

$$+\frac{2}{\tau_0}\frac{n}{T}(T-1)^2)\mathrm{d}x - \frac{h^2}{12}\int_{R^3}\left(\frac{n}{T^2}\nabla T\cdot\nabla\mathrm{div}\,\boldsymbol{u}\right.$$
$$\left.-\frac{1}{\tau_0}\frac{n}{T}(T-1)\Delta\log n\right)\mathrm{d}x. \tag{1.2.34}$$

The first integral term on the right hand side of (1.2.33) expresses the energy change due to the work of compression. In incompressible fluids, this term vanishes. The second integral term is non-negative since

$$2|D(\boldsymbol{u})|^2 - \frac{2}{3}(\mathrm{div}\,\boldsymbol{u})^2 = \frac{1}{2}\sum_{j\neq k}\left(\left|\frac{\partial u_j}{\partial x_k}+\frac{\partial u_k}{\partial x_j}\right|^2+\frac{4}{3}\right)(\mathrm{div}\,\boldsymbol{u})^2 \geq 0.$$

The third integral term represents the dissipation due to relaxation. The first integral term on the right hand side of (1.2.34) expresses the dissipation due to temperature variations, viscosity and relaxation. The second integral term vanishes for incompressible fluids without relaxation. In this situation, the entropy is non-decreasing.

Proof. The first identity (1.2.33) follows immediately from (1.2.32) after integration over $x \in \mathbf{R}^3$ and integrating by parts in the terms nT div $\boldsymbol{u}$ and $n\Delta\log n$. For the proof of the second identity (1.2.34), we employ (1.2.2) and (1.2.32) to obtain

$$\begin{aligned}
&\frac{\mathrm{d}}{\mathrm{d}t}\int_{R^3}\left(n\log\frac{n}{T^{3/2}}+\frac{3}{2}nT\right)dx \\
&= \frac{\mathrm{d}}{\mathrm{d}t}\int_{R^3}\left(\frac{5}{2}n\log n - \frac{3}{2}n\log(nT)+\frac{3}{2}nT\right)\mathrm{d}x \\
&= \int_{R^3}\left(n_t\left(\log n - \frac{3}{2}\log T\right) - \frac{5}{2}n_t - \frac{3}{2T}(nT)_t\right)(1-T)\mathrm{d}x.
\end{aligned}$$

□

1.3 Quantum electron-magnetic model in plasma

Quantum magnetohydrodynamic model

In the laser, ultra electronics, astrophysical plasmas, the particle motion in the plasma are paid more attention. Taking into account about N particles, wave functions $\psi_\alpha = \psi_\alpha(r,t)$, which have the probabilities $P_\alpha, \alpha = 1,\ldots,N$. $P\alpha \geq 0$, $\sum_{\alpha=1}^{N} P_\alpha = 1$, satisfy the Schrödinger equation

$$\frac{1}{2m}(ih\boldsymbol{\nabla} - q\boldsymbol{A})^2\psi_\alpha + q\phi\psi_\alpha = \mathrm{i}h\frac{\partial\psi_\alpha}{\partial t}, \tag{1.3.1}$$

where ϕ, $\boldsymbol{A}$ are scaled potential and vector potential, respectively. m is the mass of electron. q is the charge of electron. Set Coulomp gauge field $\boldsymbol{\nabla}\cdot\boldsymbol{A} = 0$, and ψ_α satisfies the Schrödinger equation (1.3.1). Let the Wigner function be

$$f(r,\boldsymbol{p},t) = \frac{1}{(2\pi h)^3}\sum_{\alpha=1}^{N} P_\alpha \int \psi_\alpha^* \mathrm{e}^{\mathrm{i}\boldsymbol{p}s/h}\psi_\alpha\left(r - \frac{s}{2}\right)\mathrm{d}s. \tag{1.3.2}$$

Then $f = f(r,\boldsymbol{p},t)$ satisfies the transport equation

$$\begin{aligned}
&\frac{\partial f}{\partial t} + \frac{\boldsymbol{p}}{m}\cdot\boldsymbol{\nabla} f \\
&= \frac{\mathrm{i}q}{h(2\pi h)^3}\iint \mathrm{e}^{\mathrm{i}(\boldsymbol{p}-\boldsymbol{p}')s/h}\left[\phi\left(r+\frac{s}{2}\right) - \phi\left(r - \frac{s}{2}\right)\right] f(r,\boldsymbol{p}',t)\mathrm{d}s\mathrm{d}\boldsymbol{p}' \\
&\quad + \frac{\mathrm{i}q^2}{2hm(2\pi h)^3}\boldsymbol{\nabla}\iint \mathrm{e}^{\mathrm{i}(\boldsymbol{p}-\boldsymbol{p}')s/h}\left(\boldsymbol{A}^2\left(r+\frac{s}{2}\right) - \boldsymbol{A}^2\left(r-\frac{s}{2}\right)\right) \\
&\quad \times f(r,\boldsymbol{p}',t)\mathrm{d}s\mathrm{d}\boldsymbol{p}' + \frac{q}{hm(2\pi h)^3}\boldsymbol{p}\cdot\iint \mathrm{e}^{\mathrm{i}(\boldsymbol{p}-\boldsymbol{p}')s/h} \\
&\quad \times \left(\boldsymbol{A}\left(r+\frac{s}{2}\right) - \boldsymbol{A}\left(r-\frac{s}{2}\right)\right) f(r,\boldsymbol{p}',t)\mathrm{d}s\mathrm{d}\boldsymbol{p}'.
\end{aligned} \tag{1.3.3}$$

When $h \to 0$, the Wigner equation (1.3.3) tends to the Vlasov equation

$$\frac{\partial f}{\partial t} + \boldsymbol{v}\cdot\boldsymbol{\nabla} f + \frac{q}{m}(\boldsymbol{E} + \boldsymbol{v}\times\boldsymbol{B})\cdot\frac{\partial f}{\partial \boldsymbol{v}} = 0, \tag{1.3.4}$$

where $\boldsymbol{v} = \frac{\boldsymbol{p}-q\boldsymbol{A}}{m}$, $\boldsymbol{E} = -\boldsymbol{\nabla}\phi - \frac{\partial \boldsymbol{A}}{\partial t}$, $\boldsymbol{B} = \boldsymbol{\nabla}\times\boldsymbol{A}$.

To simplify the equation, we set

$$\text{density}\quad n = \int f \mathrm{d}\boldsymbol{p}, \tag{1.3.5}$$

$$\text{velocity}\quad \boldsymbol{u} = \frac{1}{mn}\int (\boldsymbol{p} - q\boldsymbol{A}) f \mathrm{d}\boldsymbol{p}, \tag{1.3.6}$$

$$\text{pressure}\quad \boldsymbol{p} = \frac{1}{m^2}\int (\boldsymbol{p} - q\boldsymbol{A}) \otimes (\boldsymbol{p} - q\boldsymbol{A}) f - n\boldsymbol{u}\otimes\boldsymbol{u}\mathrm{d}\boldsymbol{p}. \tag{1.3.7}$$

Make all kinds of momentum in the Wigner equation (1.3.3), and by the definition (1.3.5)–(1.3.7), we can get the quantum hydrodynamic model

$$\frac{\partial n}{\partial t} + \nabla \cdot (n\boldsymbol{u}) = 0, \tag{1.3.8}$$

$$\frac{\partial n}{\partial t} + \boldsymbol{u}\cdot\nabla\boldsymbol{u} = -\frac{1}{n}\nabla\boldsymbol{p} + \frac{q}{m}(\boldsymbol{E} + \boldsymbol{u}\times\boldsymbol{B}). \tag{1.3.9}$$

Because of

$$\psi_\alpha = \sqrt{n_\alpha}\mathrm{e}^{\mathrm{i}s_\alpha/h}, \tag{1.3.10}$$

let

$$\boldsymbol{p} = \boldsymbol{p}^C + \boldsymbol{p}^Q, \tag{1.3.11}$$

where

$$\begin{aligned}\boldsymbol{p}^C &= m\sum_{\alpha=1}^{N} P_\alpha n_\alpha(\boldsymbol{u}_\alpha - \boldsymbol{u})\otimes(\boldsymbol{u}_\alpha - \boldsymbol{u})\\ &\quad + m\sum_{\alpha=1}^{N} P_\alpha n_\alpha(\boldsymbol{u}_\alpha^0 - \boldsymbol{u}^0)\otimes(\boldsymbol{u}_\alpha^0 - \boldsymbol{u}^0),\end{aligned} \tag{1.3.12}$$

$$\boldsymbol{p}^Q = -\frac{h^2 n}{4m}\nabla\otimes\nabla\log n. \tag{1.3.13}$$

$\boldsymbol{p}^C$ is the classical pressure, $\boldsymbol{p}^Q$ is the quantum pressure. Set

$$\boldsymbol{u}_\alpha = \frac{\nabla s_\alpha}{m}, \tag{1.3.14}$$

$$\boldsymbol{u} = \sum_{\alpha=1}^{N} \frac{P_\alpha n_\alpha}{n} \boldsymbol{u}_\alpha, \tag{1.3.15}$$

$$\boldsymbol{u}_\alpha^0 = \frac{h}{2m} \frac{\boldsymbol{\nabla} n_\alpha}{n_\alpha}, \tag{1.3.16}$$

$$\boldsymbol{u}^0 = \sum_{\alpha=1}^{N} \frac{P_\alpha n_\alpha}{n} \boldsymbol{u}_\alpha^0, \tag{1.3.17}$$

$$n = \sum_{\alpha=1}^{N} P_\alpha n_\alpha. \tag{1.3.18}$$

If $\boldsymbol{p}_{ij} = \delta_{ij}\boldsymbol{p}$, $\boldsymbol{p} = \boldsymbol{p}(n)$, we can get the transport equation (1.3.19),

$$\frac{\partial \boldsymbol{u}}{\partial t} + u \cdot \boldsymbol{\nabla} u = -\frac{1}{mn}\boldsymbol{\nabla}_{\boldsymbol{p}} + \frac{q}{m}(\boldsymbol{E} + \boldsymbol{u} \times \boldsymbol{B}) + \frac{h^2}{2m^2}\boldsymbol{\nabla}\left(\frac{\boldsymbol{\nabla}^2\sqrt{n}}{\sqrt{n}}\right). \tag{1.3.19}$$

1.4 Bipolar quantum hydrodynamic model

$$\frac{\partial n_e}{\partial t} + \boldsymbol{\nabla} \cdot (n_e \boldsymbol{u}_e) = 0, \tag{1.4.1}$$

$$\frac{\partial n_i}{\partial t} + \boldsymbol{\nabla} \cdot (n_i \boldsymbol{u}_i) = 0, \tag{1.4.2}$$

$$\begin{aligned} \frac{\partial \boldsymbol{u}_e}{\partial t} + \boldsymbol{u}_e \cdot \boldsymbol{\nabla} \boldsymbol{u}_e = {} & -\frac{\boldsymbol{\nabla} P_e}{m_e n_e} - \frac{e}{m_e}(\boldsymbol{E} + \boldsymbol{u}_e \times \boldsymbol{B}) \\ & + \frac{h^2}{2m_e^2}\boldsymbol{\nabla}\left(\frac{\boldsymbol{\nabla}^2\sqrt{n_e}}{\sqrt{n_e}}\right) - \nu_{ei}(\boldsymbol{u}_e - \boldsymbol{u}_i), \end{aligned} \tag{1.4.3}$$

$$\begin{aligned} \frac{\partial \boldsymbol{u}_i}{\partial t} + \boldsymbol{u}_i \cdot \boldsymbol{\nabla} \boldsymbol{u}_i = {} & -\frac{\boldsymbol{\nabla} P_i}{m_i n_i} - \frac{e}{m_i}(\boldsymbol{E} + \boldsymbol{u}_i \times \boldsymbol{B}) + \frac{h^2}{2m_i^2}\boldsymbol{\nabla}\left(\frac{\boldsymbol{\nabla}^2\sqrt{n_i}}{\sqrt{n_i}}\right) \\ & - \nu_{ie}(\boldsymbol{u}_i - \boldsymbol{u}_e). \end{aligned} \tag{1.4.4}$$

where the subscript i is respect to ions, and the subscript e is respect to electrons, ν_{ei}, ν_{ie} denote the interaction constants respectively.

Supplying the Maxwellian equation

$$\nabla \cdot \boldsymbol{E} = \frac{\rho}{\varepsilon_0}, \quad \Delta\psi = \frac{e}{\varepsilon_0}(n_i - n_e), \quad \boldsymbol{E} = \nabla\psi, \tag{1.4.5}$$

$$\nabla \cdot \boldsymbol{B} = 0, \tag{1.4.6}$$

$$\nabla \times \boldsymbol{E} = -\frac{\partial \boldsymbol{B}}{\partial t}, \tag{1.4.7}$$

$$\nabla \times \boldsymbol{B} = \left(\mu_0 \boldsymbol{J} + \mu_0\varepsilon_0 \frac{\partial \boldsymbol{E}}{\partial t}\right), \tag{1.4.8}$$

$$\rho = e(n_i - n_e), \quad \boldsymbol{J} = e(n_i\boldsymbol{u}_i - n_e\boldsymbol{u}_e), \tag{1.4.9}$$

where μ_0, ε_0 are the physical constants.

In order to establish a similar model to the classical electromagnetic fluid dynamics model, we set the total mass

$$\rho_m = m_e n_e + m_i n_i \tag{1.4.10}$$

and the whole velocity of the fluid

$$\boldsymbol{U} = \frac{m_e n_e \boldsymbol{u}_e + m_i n_i \boldsymbol{u}_i}{m_e n_e + m_i n_i}, \tag{1.4.11}$$

where the densities of the electrons and ions are

$$n_e = \frac{1}{m_i + m_e}\left(\rho_m - \frac{m_i}{e}\rho\right), \tag{1.4.12}$$

$$n_i = \frac{1}{m_i + m_e}\left(\rho_m + \frac{m_e}{e}\rho\right). \tag{1.4.13}$$

Then we can obtain the hydrodynamic model

$$\frac{\partial \rho_m}{\partial t} + \nabla \cdot (\rho_m \boldsymbol{U}) = 0, \tag{1.4.14}$$

$$\rho_m\left(\frac{\partial \boldsymbol{U}}{\partial t} + \boldsymbol{U} \cdot \nabla \boldsymbol{U}\right) = -\nabla \cdot \Pi + \boldsymbol{J} \times \boldsymbol{B} + \frac{h^2 \rho_m}{2m_e m_i} \nabla \left(\frac{\nabla^2 \sqrt{\rho_m}}{\sqrt{\rho_m}}\right), \tag{1.4.15}$$

where

$$\Pi = P\boldsymbol{I} + \frac{m_e m_i n_e n_i}{\rho_m}(\boldsymbol{u}_e - \boldsymbol{u}_i) \otimes (\boldsymbol{u}_e - \boldsymbol{u}_i), \tag{1.4.16}$$

$P = P_e + P_i$, and $\boldsymbol{I}$ is the unit matrix. If we set $P_e = P_i = \frac{P}{2}$ in (1.4.16) and neglect the last term in the right hand side of equation (1.4.16), we can get

$$\frac{\partial \boldsymbol{U}}{\partial t} + \boldsymbol{U} \cdot \boldsymbol{\nabla} \boldsymbol{U} = -\frac{1}{\rho_m} \boldsymbol{\nabla} P + \frac{1}{\rho_m} \boldsymbol{J} \times \boldsymbol{B} + \frac{h^2}{2m_e m_i} \boldsymbol{\nabla} \left(\frac{\boldsymbol{\nabla}^2 \sqrt{\rho m}}{\sqrt{\rho m}} \right). \tag{1.4.17}$$

Under the assumption of the neutral, $m_e \ll m_i (m_i \approx 1836 m_e)$, we can obtain the equation about $\boldsymbol{J}$

$$\begin{aligned} \frac{m_e m_i}{\rho_m e^2} \frac{\partial \boldsymbol{J}}{\partial t} - \frac{m_i \boldsymbol{\nabla} P_e}{\rho_m e} &= \boldsymbol{E} + \boldsymbol{U} \times \boldsymbol{B} - \frac{m_i}{\rho_m e} \boldsymbol{J} \times \boldsymbol{B} \\ &\quad - \frac{h^2}{2 e m_e} \boldsymbol{\nabla} \left(\frac{\boldsymbol{\nabla}^2 \sqrt{\rho m}}{\sqrt{\rho m}} \right) - \frac{1}{\sigma} \boldsymbol{J}, \end{aligned} \tag{1.4.18}$$

where $\sigma = \frac{\rho_m e^2}{m_e m_i \nu_{ei}}$.

Making a further simplification for the equation (1.4.18), then we can get the quantum magnetohydrodynamic model:

$$\frac{\partial \rho_m}{\partial t} + \boldsymbol{\nabla} \cdot (\rho_m \boldsymbol{U}) = 0, \tag{1.4.19}$$

$$\frac{\partial \boldsymbol{U}}{\partial t} + \boldsymbol{U} \cdot \boldsymbol{\nabla} \boldsymbol{u} = -\frac{1}{\rho_m} \boldsymbol{\nabla} P + \frac{1}{\rho_m} \boldsymbol{J} \times \boldsymbol{B} + \frac{h^2}{2m_e m_i} \boldsymbol{\nabla} \left(\frac{\boldsymbol{\nabla}^2 \sqrt{\rho m}}{\sqrt{\rho m}} \right), \tag{1.4.20}$$

$$\boldsymbol{\nabla} P = V_s^2 \boldsymbol{\nabla} \rho_m, \tag{1.4.21}$$

$$\boldsymbol{\nabla} \times \boldsymbol{E} = -\frac{\partial B}{\partial t}, \tag{1.4.22}$$

$$\boldsymbol{\nabla} \times \boldsymbol{B} = \mu_0 \boldsymbol{J}, \tag{1.4.23}$$

$$\begin{aligned} \boldsymbol{J} = \sigma \Bigg[& \boldsymbol{E} + \boldsymbol{U} \times \boldsymbol{B} - \frac{m_i}{\rho_m e} \boldsymbol{J} \times \boldsymbol{B} \\ & - \frac{h^2}{2 e m_e} \boldsymbol{\nabla} \left(\frac{\boldsymbol{\nabla}^2 \sqrt{\rho_m}}{\sqrt{\rho_m}} \right) \Bigg]. \end{aligned} \tag{1.4.24}$$

In equation (1.4.21), V_s is the adiabatic sonic of the fluid. There are 13 variables (ρ_m, $\boldsymbol{U}, \boldsymbol{J}, \boldsymbol{B}, \boldsymbol{E}$), and 13 equations.

In the ideal magnetohydrodynamic model, setting $\sigma = \infty$ in (1.4.24), and neglecting $\boldsymbol{J} \times \boldsymbol{B}$, we have

$$\rho_m \left(\frac{\partial \boldsymbol{U}}{\partial t} + \boldsymbol{U} \cdot \boldsymbol{\nabla U} \right) = -\boldsymbol{\nabla} P + \frac{1}{\mu_0} (\boldsymbol{\nabla} \times B) \times B + \frac{h^2 \rho_m}{2 m_e m_i} \boldsymbol{\nabla} \left(\frac{\boldsymbol{\nabla}^2 \sqrt{\rho_m}}{\sqrt{\rho_m}} \right), \tag{1.4.25}$$

$$\frac{\partial B}{\partial t} = \boldsymbol{\nabla} \times (\boldsymbol{U} \times \boldsymbol{B}), \tag{1.4.26}$$

$$\boldsymbol{E} = -\boldsymbol{U} \times \boldsymbol{B} + \frac{h^2}{2 e m_e} \boldsymbol{\nabla} \left(\frac{\boldsymbol{\nabla}^2 \sqrt{\rho_m}}{\sqrt{\rho_m}} \right). \tag{1.4.27}$$

1.5 Some plasma equations with quantum effect

1.5.1 *Quantum KdV equation*

By using singular perturbation method, we can obtain the following quantum KdV equation:

$$\boldsymbol{u}_t + 2\boldsymbol{u}\boldsymbol{u}_x + \frac{1}{2} \left(1 - \frac{H^2}{4} \right) \boldsymbol{u}_{xxx} = 0, \tag{1.5.1}$$

where

$$H = \frac{h\omega pe}{k_b T_{Fe}}, \tag{1.5.2}$$

h is the Planck constant, ω_{pe} is the electronic plasma frequency, and T_{Fe} is the Femi temperature.

1.5.2 *Quantum Zakharov equation*

By using the fast, slow decomposition of electromagnetic fluid, we have the quantum Zakharov equation

$$\begin{cases} \mathrm{i}\boldsymbol{E}_t + \boldsymbol{E}_{xx} - H^2 \boldsymbol{E}_{xxxx} = n\boldsymbol{E}, \\ n_{tt} - n_{xx} + H^2 n_{xxxx} = |\boldsymbol{E}|^2_{xx}. \end{cases} \tag{1.5.3}$$

One dimensional Schrödinger equation with quantum effect

$$\mathrm{i}\boldsymbol{E}_t + \boldsymbol{E}_{xx} + |\boldsymbol{E}|^2\boldsymbol{E} = H^2(\boldsymbol{E}_{xxxx} - \boldsymbol{E}|\boldsymbol{E}|^2_{xx}). \tag{1.5.4}$$

Three dimensional quantum Zakharov equation

$$\mathrm{i}\varepsilon_t - \frac{5c^2}{3V_{Fe}^2}\boldsymbol{\nabla} \times (\boldsymbol{\nabla} \times \varepsilon) + \boldsymbol{\nabla}(\boldsymbol{\nabla} \cdot \varepsilon) = n\varepsilon + H\boldsymbol{\nabla}[\boldsymbol{\nabla}^2(\boldsymbol{\nabla} \cdot \varepsilon)], \tag{1.5.5}$$

$$n_{tt} - \Delta n + \Delta|\varepsilon|^2 + H\delta^2 n = 0. \tag{1.5.6}$$

Three dimensional quantum non-linear Schrödinger equation

$$\mathrm{i}\varepsilon_t + \boldsymbol{\nabla}(\boldsymbol{\nabla} \cdot \varepsilon) - \frac{5c^2}{3V_{Fe}^2}\boldsymbol{\nabla} \times (\boldsymbol{\nabla} \times \varepsilon) + |\varepsilon|^2\varepsilon = H\boldsymbol{\nabla}[\boldsymbol{\nabla}^2(\boldsymbol{\nabla} \cdot \varepsilon)] - H\varepsilon\nabla^2(|\varepsilon^2|). \tag{1.5.7}$$

Chapter 2

Global Existence of Weak Solutions to the Compressible Quantum Hydrodynamic Equations

2.1 Global existence of weak solutions to one-dimensional compressible quantum hydrodynamic equations

The aim of this section is to study global existence of the weak solution to the following viscous Korteweg system:

$$\rho_t + (\rho u)_x = \nu \rho_{xx}, \tag{2.1.1}$$

$$\begin{aligned}(\rho u)_t + (\rho u^2 + p(\rho))_x &- \delta^2 \rho((\varphi(\rho))_{xx}\varphi'(\rho))_x \\ &= \nu(\rho u)_{xx} + \varepsilon u_{xx} - \frac{\rho u}{\tau}.\end{aligned} \tag{2.1.2}$$

where ρ is the density, u is the velocity, ν is the viscosity depending also on the parameter δ, ε accounts for classical mass-conservative viscous effects, τ is the relaxation time, δ is a positive parameter, $p(\rho)$ denotes the pressure, the dispersion term $\varphi(\rho) = \rho^\alpha, 0 < \alpha \leq 1$. We consider the initial value problem of the system (2.1.1)–(2.1.2) in one-dimensional torus $\mathbb{T}$ with initial data:

$$\rho(x,0) = \rho_0(x), \rho u(x,0) = \rho_0 u_0. \tag{2.1.3}$$

For $0 < \alpha \leq 1$, there exists a global existence weak solution to the above Korteweg system. In addition, we perform the limit $\varepsilon \to 0$ with respect to $0 < \alpha \leq \frac{1}{2}$, then the global existence of the weak solution to the corresponding viscous quantum hydrodynamic (QHD) equations ($\alpha = \frac{1}{2}$) is obtained. Therefore, by virtue of the effective velocity transformation introduced in Section 2.2, we can deduce a companion result. One can refer to [1]–[23].

Theorem 2.1.1. *Let $T > 0$, $\varepsilon > 0$, and $0 < \alpha \leq 1$. Assume that the function $p \in C^1([0, \infty))$ is monotone and the primitive H satisfies $H(y) \geq -h_0$ for some $h_0 > 0$. Assume that the initial data $(\rho_0, u_0) \in H^1(\mathbb{T}) \times L^\infty(\mathbb{T})$, $\rho_0(x) \geq \eta_0 > 0$ for $x \in T$ and for some $\eta_0 > 0$, and $E(\rho_0, u_0) < \infty$. Then, there exists a constant $\eta > 0$, and a weak solution (ρ, u) to (2.1.1)–(2.1.2) with the regularity*

$$\begin{aligned}
&\rho(x,t) \geq \eta > 0, t > 0, x \in \mathbb{T},\\
&\rho_t \in L^2(0,T; L^2(\mathbb{T})), (\rho u)t \in L^2(0,T; H^{-2}(\mathbb{T})),\\
&\varphi(\rho) \in L^\infty(0,T; H^1(\mathbb{T})) \cap L^2(0,T; H^2(\mathbb{T})),\\
&u \in L^2(0,T; H^1(\mathbb{T})) \cap L^\infty(0,T; L^2(\mathbb{T})),
\end{aligned}$$

satisfying (2.1.1) pointwise, and for all smooth test functions with $\phi(\cdot, T) = 0$,

$$\begin{aligned}
&\int_{\mathbb{T}} \rho u \phi_t \mathrm{d}x + \int_{\mathbb{T}} (\rho u^2 + p(\rho))\phi_x \mathrm{d}x - \int_{\mathbb{T}} \delta^2 (\varphi(\rho)_{xx}\varphi'(\rho))(\rho\phi)_x \mathrm{d}x\\
&\quad - \int_{\mathbb{T}} \nu(\rho u)_x \phi_x \mathrm{d}x - \varepsilon(u)_x\phi_x - \int_{\mathbb{T}} \frac{\rho u \phi}{\tau} \mathrm{d}x = \int_{\mathbb{T}} \rho_0 u_0 \phi(\cdot, 0) \mathrm{d}x,
\end{aligned} \tag{2.1.4}$$

the lower bound $\eta > 0$ depends on ε, and the initial conditions (2.1.3) are satisfied in the sense of H^{-2}.

Assume that $(\rho_\varepsilon, u_\varepsilon)$ is a weak solution to (2.1.1)–(2.1.2). Then, we perform the limit $\varepsilon \to 0$, the lower bound of ρ_ε is useless as it depends on ε. Moreover, because we have only weak convergence of $\sqrt{\rho_\varepsilon} u_\varepsilon$, it's not easy to deal with the term $\rho_\varepsilon u_\varepsilon^2$. Furthermore, we can only obtain the strong convergence of $\rho_\varepsilon u_\varepsilon$ to J. By employing the test function $\rho_\varepsilon^{3/2}\phi$ in the momentum equation, we can overcome these above difficulties and deduce for the convective term that

$$\int_{\mathbb{T}} \rho_\varepsilon^{3/2}(\rho_\varepsilon u_\varepsilon^2)_x \phi \mathrm{d}x = -\int_{\mathbb{T}} (\rho_\varepsilon^{1/2-\alpha}\varphi(\rho_\varepsilon)\phi_x + \frac{3}{2\alpha}{\rho_\varepsilon}^{1/2-\alpha}(\varphi(\rho_\varepsilon))_x\phi)(\rho_\varepsilon u_\varepsilon)^2 \mathrm{d}x.$$

Then we have it as follows.

Theorem 2.1.2. *Let $T > 0$, and assume that the assumptions of Theorem 2.1.1 hold. Then, for $\varepsilon = 0$, $0 < \alpha \leq \frac{1}{2}$, there exists a weak solution (ρ, J) to (2.1.1)–(2.1.2) satisfying*

$$\begin{aligned}
&\rho(x,t) \geq 0 \,\text{for}\, t > 0, x \in \mathbb{T},\\
&\rho_t \in L^2(0,T;L^2(\mathbb{T})), (\rho^{3/2}J)_t \in L^2(0,T;H^{-1}(\mathbb{T})),\\
&\varphi(\rho) \in L^\infty(0,T;H^1(\mathbb{T})) \cap L^2(0,T;H^2(\mathbb{T})),\\
&J \in L^2(0,T;H^1(\mathbb{T})),
\end{aligned}$$

where ρ and J satisfy $\rho_t + J_x = \nu\rho_{xx}$ almost everywhere in $(0,T)\times\mathbb{T}$, and for all test functions $\phi \in L^\infty(0,T;H^1(\mathbb{T}))$, we have

$$\begin{aligned}
&\int_0^T \langle(\rho^{3/2}J)_t, \phi\rangle H^{-1}, H^{1\mathrm{d}t} - \frac{3}{2}\int_0^T \sqrt{\rho}\rho_t J\phi \mathrm{d}x\mathrm{d}t\\
&\quad - \int_0^T\int_{\mathbb{T}} J^2(3(\sqrt{\rho})_x\phi + \sqrt{\rho}\phi_x)\mathrm{d}x\mathrm{d}t + \int_0^T\int_{\mathbb{T}} (p(\rho))_x\rho^{3/2}\phi \mathrm{d}x\mathrm{d}t\\
&\quad + \delta^2\int_0^T\int_{\mathbb{T}} ((\varphi(\rho)\,)_{xx}(\alpha\rho^{\alpha+3/2}\phi_x + \frac{5}{2}\rho^{3/2}(\varphi(\rho))_x\phi))\mathrm{d}x\mathrm{d}t\\
&= -\nu\int_0^T\int_{\mathbb{T}} J_x\rho(3(\sqrt{\rho})_x\phi + \sqrt{\rho}\phi_x)\mathrm{d}x\mathrm{d}t - \frac{1}{\tau}\int_0^T\int_{\mathbb{T}} \rho^{3/2}J\phi \mathrm{d}x\mathrm{d}t,
\end{aligned} \tag{2.1.5}$$

where ρ and $\rho^{3/2}$ J satisfy the following initial conditions:

$$\rho(\cdot,0) = \rho_0 \,\text{in}\, L^2(\mathbb{T}), \quad (\rho^{3/2}J)(\cdot,0) = \rho_0^{5/2}u_0 \,\text{in}\, H^{-1}(\mathbb{T}).$$

2.1.1 *Faedo–Galerkin approximation*

In this section, we will prove the local existence of the solution to the approximate viscous quantum equations. Here, we proceed similarly as in [9].

Let $T > 0$, and let $(\boldsymbol{e}_n)$ be an orthonormal basis of $L^2(\mathbb{T})$. Introduce the finite-dimensional space $X_n = \operatorname{span}(\boldsymbol{e}_1, \ldots, \boldsymbol{e}_n)$, $n \in \mathbb{N}$. Let $(\rho_0, u_0) \in C^\infty(\mathbb{T})^2$ be some initial data satisfying $\rho_0(x) \geq \eta_0 > 0$, $x \in \mathbb{T}$. For given velocity $v \in C^0(0, T; X_n)$, we obtain

$$v(x,t) = \sum_{i=1}^{n} \lambda_i(t)\boldsymbol{e}_i(x), \quad t \in [0,T], x \in \mathbb{T},$$

for some $\lambda_i(t)$, and we have

$$\|v\|_{C^0(0,T;X_n)} = \max_{t\in[0,T]} \sum_{i=1}^{n} |\lambda_i(t)|.$$

Consequently, v can be bounded in $C^0(0, T; C^k(\mathbb{T}))$, and there exists a constant C depending on k such that

$$\|v\|_{C^0(0,T;C^k(\mathbb{T}))} \leq C\|v\|_{C^0(0,T;L^2(\mathbb{T}))}.$$

Now, we define the approximate system. Let ρ be the classical solution to

$$\rho_t + (\rho v)_x = \nu\rho_{xx}, x \in \mathbb{T}, t > 0, \tag{2.1.6}$$

$$\rho(x,0) = \rho_0(x), \quad x \in \mathbb{T}. \tag{2.1.7}$$

The solution $\rho \in C^0(0, T; C^k(\mathbb{T}))$ for any $k \in \mathbb{N}$, and it holds that $\int_{\mathbb{T}} \rho \mathrm{d}x = \int_{\mathbb{T}} \rho_0(x)\mathrm{d}x$. We set the operator $S : C^0(0,T;X_n) \to C^0(0,T;C^3(\mathbb{T}))$ by $S(v) = \rho$. Because v is smooth, $\rho = S(v)$ is bounded from above and below by the maximum principle. That is, for $\|v\|_{C^0(0,T;L^2(\mathbb{T}))}$, there exist positive constants $K_0(c)$ and $K_1(c)$

such that

$$0 < K_0(c) \leq (S(v))(t,x) \leq K_1(c), t \in [0,T], x \in \mathbb{T}. \tag{2.1.8}$$

Moreover, there exists a positive constant K_2 depending on k and n such that for any $v_1, v_2 \in C^0(0,T;X_n)$,

$$\|S(v_1) - S(v_2)\|_{C^0(0,T;C^k(\mathbb{T}))} \leq K_2\|v_1 - v_2\|_{C^0(0,T;L^2(\mathbb{T}))}. \tag{2.1.9}$$

We claim that the lower bound of $S(v)$ only depends on $\|v_x\|_{L^2(0,T;L^2(T))}$,

$$\rho = S(v) \geq \eta = \eta(\|v_x\|_{L^2(0,T;L^2(\mathbb{T}))}) > 0 \quad \text{in } [0,T] \times \mathbb{T}. \tag{2.1.10}$$

This claim is a consequence of the following lemma, see [14] for the detailed proof.

Lemma 2.1.1. *Let $T > 0$ and $v \in L^2(0,T;H^1(\mathbb{T}))$. Assume that ρ is a solution to (2.1.6)–(2.1.7) with initial data $\rho_0 \in L^\infty(\mathbb{T})$ satisfying $\rho_0(x) \geq \eta_0 \geq 0$. Then, there exists a constant $\eta > 0$ only depending on v, ρ_0, and the $L^2(0,T;L^2(\mathbb{T}))$ norm of v_x such that*

$$\rho(x,t) \geq \eta > 0, x \in \mathbb{T}, t \in [0,T]. \tag{2.1.11}$$

Then, for $\rho = S(v)$, we want to solve the following problem for u_n on X_n:

$$\begin{aligned}(\rho u_n)_t + (\rho v u_n + p(\rho))_x - \delta^2\rho((\varphi(\rho))_{xx}\varphi'(\rho))_x \\ = v(\rho u_n)_{xx} + \varepsilon(u_n)_{xx} - \frac{\rho u_n}{\tau}.\end{aligned} \tag{2.1.12}$$

For all test functions $\phi \in C^1(0,T;X_n)$ with $\phi(\cdot,T) = 0$, we are looking for a function $u_n \in C^1(0,T;X_n)$ satisfying

$$\begin{aligned}&\int_{\mathbb{T}} \rho u_n \phi_t \mathrm{d}x + \int_{\mathbb{T}} (\rho v u_n + p(\rho))\phi_x \mathrm{d}x - \int_{\mathbb{T}} \delta^2(\varphi(\rho)_{xx}\varphi'(\rho))(\rho\phi)_x \mathrm{d}x \\ &- \int_{\mathbb{T}} v(\rho u_n)_x \phi_x \mathrm{d}x - \varepsilon(u_n)_x\phi_x - \int_{\mathbb{T}} \frac{\rho u_n \phi}{\tau}\mathrm{d}x = \int_{\mathbb{T}} \rho_0 u_0 \phi(\cdot,0)\mathrm{d}x.\end{aligned}$$

For given $\rho \in K_\eta = \{L^1(\mathbb{T}) : \inf_{x\in\mathbb{T}}\rho \geq \eta > 0\}$, we introduce the following family of operators [9]:

$$M[\rho] : X_n \to X_n^*, \quad (M[\rho]u, w) = \int_{\mathbb{T}} \rho u w \mathrm{d}x, u, w \in X_n.$$

These operators are symmetric and positive definite with the smallest eigenvalue

$$\inf_{\|w\|_{L^2(\mathbb{T})=1}} \langle M(\rho)w, w\rangle = \inf_{\|w\|_{L^2(\mathbb{T})=1}} \int_{\mathbb{T}} \rho w^2 \mathrm{d}x \geq \inf_{x\in\mathbb{T}} \rho(x) > \eta.$$

Therefore, because X_n is finite-dimensional, the operators are invertible:

$$\|M_1^-(\rho)\|_{L(X_n^*, X_n)} \leq \eta^{-1},$$

where $L(X_n^*, X_n)$ is the set of continuous mappings from X_n^* to X_n. Furthermore, M^{-1} is Lipschitz continuous:

$$\|M^{-1}(\rho_1) - M^{-1}(\rho_2)\|_{L(X_n, X_n^*)} \leq K(n,\eta)\|\rho_1 - \rho_2\|_{L^1(\mathbb{T})}, \quad (2.1.13)$$

for all $\rho 1, \rho 2 \in K\eta$. Then, we can replace the problem (2.1.12) as an ordinary differential equation on the finite-dimensional space X_n:

$$\frac{\mathrm{d}}{\mathrm{d}t}(M[\rho(t)]u_n(t)) = N[v, u_n(t)], t > 0, M[\rho_0]u_n(0) = M[\rho_0]u_0, \quad (2.1.14)$$

where

$$(N[v, u_n(t)], \phi) = \int_{\mathbb{T}} (-(\rho v u_n + p(\rho))x + \delta^2 \rho((\varphi(\rho))_{xx}\varphi'(\rho))_x$$
$$+ \nu(\rho u_n)_{xx} + \varepsilon(u_n)_{xx} - \frac{\rho u_n}{\tau})\phi \mathrm{d}x.$$

The above integral is well defined as $\rho = S(v)$ is bounded from below, and for any $t \in [0,T]$, the operator $N[v,\cdot]$ from X_n to X_n^* is continuous in time. Therefore, standard theory for systems of ordinary differential equations provides the existence of a unique solution to (2.1.14). That is, there exists a unique solution $u_n \in C^1(0,T;X_n)$ to (2.1.12) for given $v \in C^0(0,T;X_n)$.

2.1.2 *Existence of the approximate solutions*

Now, we will establish the global existence of the solution to the approximate problem (2.1.6)–(2.1.7) and (2.1.12).

Proposition 2.1.1. *Let the assumptions of Theorem* 2.1.1 *hold, and the initial data be smooth with positive density. Then, there exists a solution* $(\rho_n, u_n) \in C^0(0,T;C^3(\mathbb{T})) \times C^1(0,T;X_n)$ *to* (2.1.6)–(2.1.7) *and* (2.1.12), *with* $v = u_n$, *and* $\rho = \rho_n = S(u_n)$, *satisfying the following estimates*:

$$\rho_n(x,t) \geq \eta(\varepsilon) > 0 \, \text{for} \, t \in [0,T], x \in \mathbb{T}, \tag{2.1.15}$$

$$\|\sqrt{\rho_n} u_n\|_{L^\infty(0,T;L^2(\mathbb{T}))} + \|\sqrt{\rho_n}(u_n)_x\|_{L^2(0,T;L^2(\mathbb{T}))} \leq K, \tag{2.1.16}$$

$$\|\varphi(\rho_n)\|_{L^\infty(0,T;H^1(\mathbb{T}))} + \|\varphi(\rho_n)\|_{L^2(0,T;H^2(T))} \leq K, \tag{2.1.17}$$

$$\varepsilon\|(u_n)_x\|_{L^2(0,T;L^2(\mathbb{T}))} \leq K, \tag{2.1.18}$$

where $\eta(\varepsilon) > 0$ *depends on* ε *and the initial data*, K *depends only on* ν *and the initial data.*

Proof. Integrating (2.1.14) over $(0,\ t)$, we obtain

$$u_n(t) = M^{-1}[(S(u_n))(t)](M[\rho_0](u_0) + \int_0^t N[u_n, u_n(s)]\mathrm{d}s) \, \text{in} \, X_n.$$

Using (2.1.9) and (2.1.13), we can solve this equation with the fixed-point theorem of Banach, at least on a short time interval $[0,T']$ with $T' \leq T$ in the space $C^0(0,T';X_n)$. In fact, we have $u_n \in C^1(0,T';X_n)$. In order to prove that u_n is bounded in X_n on the whole time $[0,T]$, we can employ the energy estimate. Multiply (2.1.6) by $\phi = h(\rho_n) - \frac{u_n^2}{2} - \delta^2(\varphi(\rho_n))_{xx}\varphi'(\rho_n)$, using the test function u_n in (2.1.12) with $v = u_n, \rho = \rho_n$, and summing both equations, we have

$$0 = \int_{\mathbb{T}} ((\rho_n)_t h(\rho_n) - (\rho_n)_t \frac{u_n^2}{2} + (\rho_n u_n)_t u_n)\mathrm{d}x$$

$$- \delta^2 \int_{\mathbb{T}} ((\rho_n u_n)_x (\varphi(\rho_n))_{xx}\varphi'(\rho_n) + \rho_n u_n ((\varphi(\rho_n))_{xx}\varphi'(\rho_n))_x$$

$$
\begin{aligned}
&+(\rho_n)_t(\varphi(\rho_n))_{xx}\varphi'(\rho_n))\mathrm{d}x+\int_{\mathbb{T}}((\rho_n u_n)_x h(\rho_n)+(p(\rho_n))_x u_n)\mathrm{d}x\\
&+\nu\int_{\mathbb{T}}(-(\rho_n)_{xx}h(\rho_n)+\frac{1}{2}(\rho_n)_{xx}u_n^2+\delta^2(\rho_n)_{xx}(\varphi(\rho_n))_{xx}\varphi'(\rho_n)\\
&-(\rho_n u_n)_{xx}u_n)\mathrm{d}x\\
&+\int_{\mathbb{T}}((\rho_n u_n^2)_x u_n-\frac{1}{2}(\rho_n u_n)_x {u_n}^2)\mathrm{d}x\\
&-\varepsilon\int_{\mathbb{T}}u_n(u_n)_{xx}\mathrm{d}x+\frac{1}{\tau}\int_{\mathbb{T}}\rho_n u_n^2\mathrm{d}x\\
=:&\ I_1+\cdots+I_7.
\end{aligned}
\tag{2.1.19}
$$

Here,

$$
\begin{aligned}
I_1&=\partial_t\int_{\mathbb{T}}(H(\rho_n)+\frac{1}{2}\rho_n u_n^2)\mathrm{d}_x,\\
I_2&=-\delta^2\int_{\mathbb{T}}(\varphi(\rho_n))_t(\varphi(\rho_n))_{xx}\mathrm{d}x=\frac{\delta^2}{2}\partial_t\int_{\mathbb{T}}(\varphi(\rho_n))_x^2\mathrm{d}x.
\end{aligned}
$$

Using $\mathrm{p}'(\rho_n)=\rho_n h'(\rho_n)$, and integrating by parts, we obtain

$$
I_3=\int_{\mathbb{T}}(-\rho_n u_n h'(\rho_n)(\rho_n)x+p'(\rho_n)(\rho_n)_x u_n)\mathrm{d}x=0,
$$

and

$$
\begin{aligned}
I_4&=\nu\int_{\mathbb{T}}\Big(h'(\rho_n)(\rho_n)_x^2-(\rho_n)_x u_n(u_n)_x+\delta^2(\varphi(\rho_n))_{xx}^2\\
&\quad+\frac{\delta^2(1-\alpha)}{3\alpha\varphi(\rho_n)}((\varphi(\rho_n))_x^3)_x+(\rho_n u_n)_x(u_n)_x)(u_n)_x)\mathrm{d}x\\
&=\nu\int_{\mathbb{T}}((G(\rho_n))_x^2+\rho_n(u_n)_x^2+\delta^2(\varphi(\rho_n))_{xx}^2\\
&\quad+\frac{16\delta^2(1-\alpha)}{3\alpha}(\sqrt{\varphi(\rho_n)}_x^4)\mathrm{d}x,
\end{aligned}
$$

$I_5 = 0, I_6 = \varepsilon \int_{\mathbb{T}} (u_n)_x^2 \mathrm{d}x$. In all, we obtain

$$
\begin{aligned}
0 = \partial_t & \int_{\mathbb{T}} \left(H(\rho_n) + \frac{1}{2}\rho_n u_n^2 + \frac{\delta^2}{2}(\varphi(\rho_n))_x^2 \right) dx \\
& + \nu \int_{\mathbb{T}} [(G(\rho_n))_x^2 + \rho_n (u_n)_x^2 + \delta^2 (\varepsilon(\rho_n))_{xx}^2 \\
& + \frac{16\delta^2(1-\alpha)}{3\alpha} (\sqrt{\varphi(\rho_n)})_x^4] dx \\
& + \varepsilon \int_{\mathbb{T}} (u_n)_x^2 \mathrm{d}x + \frac{1}{\tau} \int_{\mathbb{T}} \rho_n u_n^2 \mathrm{d}x.
\end{aligned}
$$

This estimate yields the bounds (2.1.16)–(2.1.18). By using (2.1.10) and (2.1.18), we obtain the bound (2.1.15). From the estimate (2.1.16), we obtain $\|u_n\|_{L^\infty(0,T;L^2(\mathbb{T}))} \le C(\varepsilon)$. Together with the estimates (2.1.9) and (2.1.13), we can apply the fixed-point theorem until $T' = T$. □

Lemma 2.1.2. *The following estimates hold:*

$$\|\partial_t \rho_n\|_{L^2(0,T;L^2(\mathbb{T}))} + \|\varphi(\rho_n)\|_{L^6(0,T;W^{1,6}(\mathbb{T}))} \le K, \tag{2.1.20}$$

$$\|\rho_n\|_{L^\infty(0,T;H^1(\mathbb{T}))} + \|\rho_n\|_{L^2(0,T;H^2(\mathbb{T}))} \le K, \tag{2.1.21}$$

$$\|\partial_t(\rho_n u_n)\|_{L^2(0,T;H^{-2}(\mathbb{T}))} \le K, \tag{2.1.22}$$

$$\|\rho_n^\beta \partial_t(\rho_n u_n)\|_{L^2(0,T;H^{-1}(\mathbb{T}))} \le K, \tag{2.1.23}$$

where $\beta \ge \frac{1}{2}, K > 0$ *are independent of* n *and* ε *for* $0 < \alpha \le \frac{1}{2}$.

Proof. By the Gagliardo–Nirenberg inequality, it follows that

$$
\begin{aligned}
\| (\varphi(\rho_n))_x \|_{L^6(0,T;L^6(\mathbb{T}))}^6 & \le K \int_0^T \|(\varphi(\rho_n))_x\|_{H^1(\mathbb{T})}^{6\theta} \|(\varphi(\rho_n))_x\|_{L^2(\mathbb{T})}^{6(1-\theta)} \mathrm{d}t \\
& \le K \|\varphi(\rho_n)\|_{L^\infty(0,T;H^1(\mathbb{T}))}^4 \\
& \quad \int_0^T \|\varphi(\rho_n)\|_{H^2(\mathbb{T})}^2 \mathrm{d}t \le K,
\end{aligned} \tag{2.1.24}
$$

where $\theta = \frac{1}{3}$. By using the bound (2.1.17), we obtain that $\varphi(\rho_n)$ is bounded in $L^6(0,T;W^{1,6}(\mathbb{T}))$. As the function ρ_n solving (2.1.6)–(2.1.7) with $v = u_n$, we have

$$\partial_t\rho_n = -\sqrt{\rho_n}\sqrt{\rho_n}(u_n)_x - \frac{\sqrt{\rho}_n u_n(\varphi(\rho_n))_x}{\sqrt{\rho_n}\varphi'(\rho_n)} + \nu\left[\frac{(\varphi(\rho_n))_{xx}}{\varphi'(\rho_n)} - \frac{\varphi''(\rho_n)(\rho_n)_x^2}{\varphi'(\rho_n)}\right].$$

By virtue of (2.1.16), (2.1.17) and (2.1.24), we have $\rho_n \in L^\infty(0,T;L^\infty(\mathbb{T}))$ and $\partial_t\rho_n \in L^2(0,T;L^2(T))$. Moreover, in view of $(\rho_n)_x = \frac{(\varphi(\rho_n))_x}{\varphi'(\rho_n)}$, and $(\rho_n)_{xx} = \frac{(\varphi(\rho_n))_{xx}}{\varphi'(\rho_n)} - \frac{\varphi''(\rho_n)(\rho_n)_x^2}{\varphi'(\rho n)} = \alpha\rho^{1-\alpha}(\varphi(\rho_n))_{xx} - \frac{\alpha-1}{\alpha^2}\rho_n^{1-2\alpha}(\varphi(\rho_n))_x^2$, we obtain $\rho_n \in L^\infty(0,T;H^1(\mathbb{T}))\cap L^2(0,T;H^2(\mathbb{T}))$.

Now, we claim that $\partial_t(\rho_n u_n)$ is bounded in $L^2(0,T;H^{-2}(\mathbb{T}))$. It's easy to verify that the terms $(p(\rho_n))_x, \varepsilon(u_n)_{xx}$ and $\frac{\rho_n u_n}{\tau}$ are bounded in $L^2(0,T;H^{-2}(\mathbb{T}))$. Furthermore, $\rho_n u_n^2 = (\sqrt{\rho_n}u_n)^2$ is bounded in $L^\infty(0,T;L^1(\mathbb{T}))$ so that $(\rho_n u_n^2)_x$ is bounded in $L^\infty(0,T;W^{-1,1}(\mathbb{T}))$, also in $L^\infty(0,T;H^{-2}(\mathbb{T})); \nu(\rho_n u_n)_{xx} = \nu(\sqrt{\rho_n}\sqrt{\rho_n}(u_n)_x + u_n\frac{(\varphi(\rho_n))_x}{\varphi'(\rho_n)})_x$ is bounded in $L(0,T;H^{-1}(\mathbb{T}))$;

$$\rho_n((\varphi(\rho_n))_{xx}\varphi'(\rho_n))_x = -\frac{1}{2}((\varphi(\rho_n))_x^2)_x + (\rho_n(\varphi(\rho_n))_{xx}\varphi'(\rho_n))_x$$

is bounded in $L^2(0,T;H^{-1}(\mathbb{T}))$, *also in* $L^2(0,T;H^{-2}(\mathbb{T}))$. Thus, we finish the claim.

Finally, we show that $\rho_n^\beta\partial_t(\rho_n u_n)$ is bounded in $L^2(0,T;H^{-1}(\mathbb{T}))$. The term

$$\rho_n^\beta(\rho_n u_n^2)_x = \frac{1}{\alpha}\rho_n^{\beta-\alpha}(\varphi(\rho_n))_x\rho_n u_n^2 + 2\rho_n^\beta\sqrt{\rho_n}u_n\sqrt{\rho_n}(u_n)_x$$

is bounded in $L^2(0,T;L^1(\mathbb{T})) \hookrightarrow L^2(0,T;H^{-1}(\mathbb{T}))$, where we have used that $\beta \geq \frac{1}{2}$ and $0 < \alpha \leq \frac{1}{2}$ to avoid the lower bound of ρ_n which depends on ε. Moreover, $\rho_n^\beta(p(\rho_n))_x$ and $\rho_n^{\beta+1}u_n/\tau$ are bounded in

$L^2(0,T;L^2(\mathbb{T}))$. The first term of

$$\rho_n^{\beta+1}(\varphi'(\rho)(\varphi(\rho))_{xx})_x - (\alpha\rho_n^{\beta+\alpha}(\varphi(\rho_n))_{xx})_x \\ - (\beta+1)\rho_n^{\beta}(\varphi(\rho_n))_x(\varphi(\rho_n))_{xx}$$

is bounded in $L^2(0,T;H^{-1}(\mathbb{T}))$, and the second term is bounded in $L^2(0,T;L^1(\mathbb{T}))$, also in $L^2(0,T;H^{-1}(\mathbb{T}))$. Similarly,

$$\varepsilon\rho_n^{\beta}(u_n)_{xx} = (\rho_n^{\beta}\varepsilon(u_n)_x)_x - \frac{\beta}{\alpha}\rho_n^{\beta-\alpha}(\varphi(\rho_n))_x\varepsilon(u_n)_x,$$

and

$$\rho_n^{\beta}(\rho_n u_n)_{xx} = (\rho_n^{\beta+1/2}\sqrt{\rho_n}(u_n)_x)_x - \frac{\beta-1}{\alpha}\rho_n^{\beta-\alpha+1/2}(\varphi(\rho_n))_x\sqrt{\rho_n}(u_n)_x \\ + \rho_n^{\beta-1/2}(\rho_n)_{xx}\sqrt{\rho_n}u_n$$

are bounded in $L^2(0,T;H^{-1}(\mathbb{T}))$. □

2.1.3 *Existence of weak solutions*

During this section, for fixed $\varepsilon > 0$, we perform the limit $n \to \infty$, in the system (2.1.6)– (2.1.7) and (2.1.12), with $\rho = \rho_n$, and $v = u_n$.

In view of (2.1.20) and (2.1.21), together with the compact embeddings of $H^1 \hookrightarrow L^\infty$ and $H^2 \hookrightarrow H^1$, using the Aubin's lemma, there exists a subsequence of (ρ_n) such that, as $n \to \infty$,

$$\begin{aligned} &\rho_n \to \rho \,\text{strongly in}\, L^2(0,T;H^1(\mathbb{T})) \,\text{and}\, L^\infty(0,T;L^\infty(\mathbb{T})), \\ &\rho_n \rightharpoonup \rho \,\text{weakly in}\, L^2(0,T;H^2(\mathbb{T})), \\ &\partial_t\rho_n \rightharpoonup \partial_t\rho \,\text{weakly in}\, L^2(0,T;L^2(\mathbb{T})). \end{aligned}$$

Because (ρ_n) is bounded from below, $(\varphi(\rho_n))_x$ converges weakly to $(\varphi(\rho_n))_x$ in $L^2(0,T;L^2(\mathbb{T}))$. Furthermore, for fixed $\varepsilon > 0$, by (2.1.16) and (2.1.18), u_n converges weakly to a function u in $L^2(0,T;H^1(\mathbb{T}))$.

Then, we have

$$\partial_t\rho_n+(\rho_n u_n)_x-\nu(\rho_n)_{xx} \rightharpoonup \partial_t\rho+(\rho u)_x-\nu\rho_{xx} \text{ weakly in } L^1(0,T;L^2(\mathbb{T})),$$

and

$$(p(\rho_n))_x - \delta^2\rho_n((\varphi(\rho_n))_{xx}\varphi'(\rho_n))_x - \nu(\rho_n u_n)_{xx} - \varepsilon(u_n)_{xx} + \frac{1}{\tau}\rho_n u_n$$

converges weakly in $L^1(0,T;H^{-1}(\mathbb{T}))$ to

$$(p(\rho))_x - \delta^2\rho((\varphi(\rho))_{xx}\varphi'(\rho))_x - \nu(\rho u)_{xx} - \varepsilon(u)_{xx} + \frac{1}{\tau}\rho u.$$

In order to pass the limit in the convection term. Firstly, $\rho_n u_n \rightharpoonup \rho u$ weakly* in $L^2(0,T;L^\infty(\mathbb{T}))$, because (ρ_n) converges strongly in $L^\infty(0,T;L^\infty(\mathbb{T}))$ and (u_n) converges weakly* in $L^2(0,T;L^\infty(\mathbb{T}))$. Besides, into account (2.1.22) and the bound take for $(\rho_n u_n)$ in $L^2(0,T;H^1(\mathbb{T}))$. By Aubin's lemma, we have $\rho_n u_n \to \rho u$ strongly in $L^2(0,T;L^\infty(\mathbb{T}))$.

Therefore,

$$\rho_n u_n^2 \rightharpoonup \rho u^2 \text{ weakly}^* \text{ in } L^1(0,T;L^\infty(\mathbb{T})).$$

By passing the limit $n\to\infty$ in (2.1.12) with $\rho=\rho_n$, and $v=u_n$, we show that (ρ,u) is a weak solution to (2.1.1)–(2.1.2) for $\varepsilon>0$ and thus finish the proof of Theorem 2.2.1.

2.1.4 *Vanishing viscosity limit* $\varepsilon \to 0$

Let $(\rho_\varepsilon, u_\varepsilon)$ be the solution to (2.1.1)–(2.1.2) constructed in the previous section, where $\varepsilon>0$, and $0<\alpha\le\frac{1}{2}$. In this section, we will perform the limit $\varepsilon\to 0$.

By virtue of estimates (2.1.20) and (2.1.21), we use the Aubin's lemma to obtain the existence of a subsequence of ρ_n such that

$$\begin{aligned}
&\rho_\varepsilon \to \rho \text{ strongly in } L^2(0,T;H^1(\mathbb{T})) \text{ and } L^\infty(0,T;L^\infty(\mathbb{T})),\\
&\rho_\varepsilon \rightharpoonup \rho \text{ weakly in } L^2(0,T;H^2(\mathbb{T})),\\
&\partial_t\rho_\varepsilon \rightharpoonup \partial_t\rho \text{ weakly in } L^2(0,T;L^2(\mathbb{T})).
\end{aligned}$$

Moreover, using (2.1.17), we have

$$\varphi(\rho_\varepsilon) \rightharpoonup \varphi(\rho) \text{ weakly}^* \text{ in } L^\infty(0,T;H^1(\mathbb{T}))$$
$$\text{and weakly in } L^2(0,T;H^2(\mathbb{T})).$$

By virtue of (2.1.16),

$$(\rho_\varepsilon u_\varepsilon)_x = \frac{1}{\alpha}\rho_\varepsilon^{1/2-\alpha}(\varphi(\rho_\varepsilon))_x\sqrt{\rho_\varepsilon}u_\varepsilon + \sqrt{\rho_\varepsilon}\sqrt{\rho_\varepsilon}(u_\varepsilon)_x$$

is bounded in $L^2(0,T;L^2(\mathbb{T}))$, hence, $\rho_\varepsilon u_\varepsilon$ is bounded in $L^2(0,T;H^1(\mathbb{T}))$. So

$$\rho_\varepsilon u_\varepsilon \rightharpoonup J \text{ weakly in } L^2(0,T;H^1(\mathbb{T})).$$

Therefore, using (2.1.20) and (2.1.21), the mass conservation equation (2.1.6) with $\rho = \rho_\varepsilon$ and $v = u_\varepsilon$ yields, as $\varepsilon \to 0$,

$$\rho_t + J_x = v\rho_{xx} \text{ in } L^2(0,T;H^1(\mathbb{T})).$$

Next, in order to perform $\varepsilon \to 0$ in the momentum conservation equation, we need to multiply (2.1.12) by $\rho_\varepsilon^{3/2}$ with the reason that we cannot control u_ε, but only $\rho_\varepsilon u_\varepsilon$, through which we can pass the limit in the convection term $(\rho_\varepsilon u_\varepsilon^2)_x$. By Aubin's lemma, (2.1.22), and the $L^2(0,T;H^1(\mathbb{T}))$ bound for $\rho_\varepsilon u_\varepsilon$, we have

$$\rho_\varepsilon u_\varepsilon \to J \text{ strongly in } L^2(0,T;L^\infty(\mathbb{T})).$$

Therefore, for any test function $\phi \in L^2(0,T;H^1(\mathbb{T}))$, as $\varepsilon \to 0$,

$$\begin{aligned}
&\int_{\mathbb{T}} \rho_\varepsilon^{3/2}(\rho_\varepsilon u_\varepsilon^2)_x\phi \mathrm{d}x \\
&\quad = -\int_{\mathbb{T}} (\rho_\varepsilon^{1/2-\alpha}\varphi(\rho_\varepsilon)\phi_x + \frac{3}{2\alpha}\rho_\varepsilon^{1/2-\alpha}(\varphi(\rho_\varepsilon))_x\phi)(\rho_\varepsilon u_\varepsilon)^2 \mathrm{d}x \\
&\quad \to -\int_{\mathbb{T}} (\rho^{1/2-\alpha}\varphi(\rho)\phi_x + \frac{3}{2\alpha}\rho^{1/2-\alpha}(\varphi(\rho))_x\phi)J^2 \mathrm{d}x.
\end{aligned}$$

By (2.1.23), $\rho_\varepsilon^{3/2}(\rho_\varepsilon u_\varepsilon)_t$ is bounded in $L^2(0,T;H^{-1}(\mathbb{T}))$. Hence,

$$(\rho_\varepsilon^{5/2}u_\varepsilon)_t = \rho_\varepsilon^{3/2}(\rho_\varepsilon u_\varepsilon)_t + \frac{3}{2}\rho_\varepsilon(\rho_\varepsilon)t(\sqrt{\rho_\varepsilon}u_\varepsilon)$$

is also bounded in $L^2(0,T;H^{-1}(\mathbb{T}))$, and we have

$$(\rho_\varepsilon^{5/2}u_\varepsilon)_t \rightharpoonup (\rho^{3/2}J)_t \text{ weakly in } L^2(0,T;H^{-1}(\mathbb{T})). \tag{2.1.25}$$

Choosing $\rho_\varepsilon^{3/2}\phi$ with $\phi \in L^\infty(0,T;H^1(\mathbb{T}))$ as a test function in the weak formulation of (2.1.2), we have

$$\begin{aligned}
0 = &\int_0^T\int_{\mathbb{T}} \rho_\varepsilon^{3/2}(\rho_\varepsilon u_\varepsilon)_t\phi \mathrm{d}x\mathrm{d}t - \int_0^T\int_{\mathbb{T}}(\rho_\varepsilon u_\varepsilon^2 + p(\rho_\varepsilon))(\rho_\varepsilon^{3/2}\phi)_x \mathrm{d}x\mathrm{d}t \\
&+ \delta^2\int_0^T\int_{\mathbb{T}}(\rho_\varepsilon^{5/2}\phi)_x(\varphi(\rho_\varepsilon))_{xx}\varphi'(\rho_\varepsilon)\mathrm{d}x\mathrm{d}t \\
&+ \nu\int_0^T\int_{\mathbb{T}}(\rho_\varepsilon u_\varepsilon)_x(\rho_\varepsilon^{3/2}\phi)_x \mathrm{d}x\mathrm{d}t \\
&+ \varepsilon\int_0^T\int_{\mathbb{T}}(\rho^{3/2}\phi)_x \mathrm{d}x\mathrm{d}t + \frac{1}{\tau}\int_0^T\int_{\mathbb{T}}\rho_\varepsilon^{5/2}u_\varepsilon\phi \mathrm{d}x\mathrm{d}t \\
=: &K_1 + K_2 + K_3 + K_4 + K_5 + K_6,
\end{aligned}$$

where

$$\begin{aligned}
K_1 &= \int_0^T \langle(\rho_\varepsilon^{5/2}u_\varepsilon)_t, \phi\rangle_{H^{-1},H^1}\mathrm{d}t - \frac{3}{2}\int_0^T\int_{\mathbb{T}}\rho_\varepsilon^{3/2}(\rho_\varepsilon)_t u_\varepsilon\phi \mathrm{d}x\mathrm{d}t \\
&\to \int_0^T \langle(\rho^{3/2}J)_t, \phi\rangle_{H^{-1},H^1}\mathrm{d}t - \frac{3}{2}\int_0^T\int_{\mathbb{T}}\sqrt{\rho}\rho_t J\phi \mathrm{d}x\mathrm{d}t,
\end{aligned}$$

and

$$\begin{aligned}
K_2 &= \int_0^T\int_{\mathbb{T}}((\rho_\varepsilon u_\varepsilon)^2 + \rho_\epsilon p(\rho_\epsilon))(3(\sqrt{\rho_\epsilon})_x\phi + \sqrt{\rho_\epsilon}\phi_x)\mathrm{d}x\mathrm{d}t \\
&= \int_0^T\int_{\mathbb{T}}((\rho_\varepsilon u_\varepsilon)^2 + \rho_\epsilon p(\rho_\varepsilon))(\rho_\varepsilon^{1/2-\alpha}\varphi(\rho_x)\phi_x
\end{aligned}$$

$$+\frac{3}{2\alpha}\rho_\varepsilon^{1/2-\alpha}(\varphi(\rho_\varepsilon))_x\,\phi)\,\mathrm{d}x\mathrm{d}t$$

$$\to\int_0^T\int_{\mathbb{T}}(J^2+\rho p(\rho))(\rho^{1/2-\alpha}\varphi(\rho)\phi_x+\frac{3}{2\alpha}\rho^{1/2-\alpha}(\varphi(\rho))_x\phi)\mathrm{d}x\mathrm{d}t$$

$$=\int_0^T\int_{\mathbb{T}}(J^2+\rho p(\rho))(3(\sqrt{\rho})_x\phi+\sqrt{\rho}\phi_x)\mathrm{d}x\mathrm{d}t,$$

because $(\rho_\varepsilon u_\varepsilon)^2\to J^2$ strongly in $L^1(0,T;L^\infty(\mathbb{T}))$, and $(\varphi(\rho_\varepsilon))_x\rightharpoonup(\varphi(\rho))_x$ weakly* in $L^\infty(0,T;L^2(\mathbb{T}))$.

For the third integral, we obtain

$$K_3=\delta^2\int_0^T\int_{\mathbb{T}}(\varphi(\rho_\varepsilon))_{xx}\left(\alpha\rho_\varepsilon^{\alpha+3/2}\phi_x+\frac{5}{2}\alpha\rho_\varepsilon^{\alpha+1/2}(\rho_\varepsilon)_x\phi\right)\mathrm{d}x\mathrm{d}t$$

$$\to\delta^2\int_0^T\int_{\mathbb{T}}(\varphi(\rho))_{xx}\left(\alpha\rho^{\alpha+3/2}\phi_x+\frac{5}{2}\rho^{3/2}(\varphi(\rho))_x\phi\right)\mathrm{d}x\mathrm{d}t,$$

where we have used the facts $\rho_\varepsilon\to\rho$ strongly in $L^\infty(0,T;L^\infty(\mathbb{T}))$ and $L^2(0,T;H^1(\mathbb{T}))$. For the fourth integral, $\rho_\varepsilon u_\varepsilon\rightharpoonup J$ in $L^2(0,T;H^1(\mathbb{T}))$, the strong convergence of ρ_ε in $L^\infty(0,T;L^\infty(\mathbb{T}))$, and the boundedness of $(\rho_\varepsilon)_x$ in $L^2(0,T;L^2(\mathbb{T}))$ imply that

$$K_4=\nu\int_0^T\int_{\mathbb{T}}(\rho_\varepsilon u_\varepsilon)_x\left(\frac{3}{2}\sqrt{\rho_\varepsilon}(\rho_\varepsilon)_x\phi+\rho_\varepsilon^{3/2}\phi_x\right)\mathrm{d}x\mathrm{d}t$$

$$\to\nu\int_0^T\int_{\mathbb{T}}J_x\left(\frac{3}{2}\sqrt{\rho}(\rho)_x\phi+\rho^{3/2}\phi_x\right)\mathrm{d}x\mathrm{d}t$$

$$=\nu\int_0^T\int_{\mathbb{T}}J_x(\rho^{3/2}\phi)_x\mathrm{d}x\mathrm{d}t,$$

and as $\varepsilon\to0$,

$$K_5=\varepsilon\int_0^T\int_{\mathbb{T}}\sqrt{\rho_\varepsilon}(u_\varepsilon)_x\left(\rho_\varepsilon\phi_x+\frac{3}{2}(\rho_\varepsilon)_x\phi\right)\mathrm{d}x\mathrm{d}t$$

$$\le\varepsilon\big(\|\rho_\varepsilon\|_{L^\infty(0,T;L^\infty(\mathbb{T}))}\|\phi_x\|_{L^2(0,T;L^2(\mathbb{T}))}$$

$$+\frac{3}{2}\|\phi\|_{L^\infty(0,T;L^\infty(\mathbb{T}))}\|(\rho_\varepsilon)_x\|_{L^2(0,T;L^2(\mathbb{T}))}\big)$$

$$\times\|\sqrt{\rho_\varepsilon}(u_\varepsilon)_x\|_{L^2(0,T;L^2(\mathbb{T}))}\to0.$$

Finally, we obtain

$$K_6 \to \frac{1}{\tau} \int_0^T \int_{\mathbb{T}} \rho^{3/2} J \phi \mathrm{d}x \mathrm{d}t.$$

Then, we have proved that $(\rho, \boldsymbol{J})$ solves the system (2.1.1) and (2.1.2) for $\varepsilon = 0$ and for smooth initial data. For the initial data $(\rho_0, u_0) \in H^1(\mathbb{T}) \times L^\infty(\mathbb{T})$ with positive density and finite energy, we can prove our main results by a standard approximation procedure. This completes the proof of Theorem 2.1.2.

2.2 Global existence of weak solutions to high dimensional compressible quantum hydrodynamic equations

In this section, we consider the following multi-dimensional quantum hydrodynamic equations:

$$\rho_t + \mathrm{div}(\rho \boldsymbol{u}) = 0, \tag{2.2.1}$$

$$(\rho \boldsymbol{u})_t + \mathrm{div}(\rho \boldsymbol{u} \otimes \boldsymbol{u}) + \boldsymbol{\nabla} p(\rho)$$
$$= 2h^2 \rho \boldsymbol{\nabla} \left(\frac{\Delta \sqrt{\rho}}{\sqrt{\rho}} \right) + 2\nu \mathrm{div}(\mu(\rho) D(\boldsymbol{u})) + \nu \boldsymbol{\nabla}(\lambda(\rho) \mathrm{div} \boldsymbol{u}), \tag{2.2.2}$$

$$\rho(\cdot, 0) = \rho_0, \rho \boldsymbol{u}(\cdot, 0) = \rho_0 \boldsymbol{u}_0 \,\mathrm{in}\, \mathbb{T}^d, \tag{2.2.3}$$

where $\rho, \boldsymbol{u}$ represent the mass density and velocity, $h (h > 0)$ and $\nu (\nu > 0)$ are Planck constant and viscosity coefficient, the pressure $p(\rho) = \rho^\gamma, \gamma \geq 1, \mathbb{T}^d$ denotes the d-dimensional period domain ($d \leq 3$). The expression $\frac{\Delta \sqrt{\rho}}{\sqrt{\rho}}$ represents the quantum Bohm potential. $\mu(\rho)$ and $\lambda(\rho)$ are two Lamé viscosity coefficients depending on the density ρ satisfying

$$\mu(\rho) \geq 0, 2\mu(\rho) + d\lambda(\rho) \geq 0, \tag{2.2.4}$$

where $\mu(\rho)$ is sometime called the shear viscosity coefficient of the fluid. While $\lambda(\rho)$ is usually referred to as the second viscosity coefficient, it is well known that the viscosity of a gas depends on the temperature, and thus on the density in the isentropic case. Here, we consider $\mu(\rho) = \rho, \lambda(\rho) = 0$. One can refer to [24]–[58].

First, we introduce the effective velocity transformation $\boldsymbol{\omega} = \boldsymbol{u} + \nu\boldsymbol{\nabla}\log\rho$. By calculation, the quantum Navier Stokes equation can be equivalently transformed into the following viscous quantum Euler equations

$$\rho_t + \mathrm{div}(\rho\boldsymbol{\omega}) = \nu\Delta\rho, \tag{2.2.5}$$

$$(\rho\boldsymbol{\omega})_t + \mathrm{div}(\rho\boldsymbol{\omega} \otimes \boldsymbol{\omega}) + \boldsymbol{\nabla}p(\rho) = 2h_0^2\rho\boldsymbol{\nabla}\left(\frac{\Delta\sqrt{\rho}}{\sqrt{\rho}}\right) + \nu\Delta(\rho\boldsymbol{\omega}), \tag{2.2.6}$$

$$\rho(\cdot, 0) = \rho_0, \rho\boldsymbol{\omega}(\cdot, 0) = \rho_0\boldsymbol{\omega}_0 \text{ in } \mathbb{T}^d, \tag{2.2.7}$$

specifically, we have the following lemma.

Lemma 2.2.1. *Let* $(\rho, \boldsymbol{u})$ *be a smooth solution to* (2.2.1)–(2.2.3), *then* $(\rho, \boldsymbol{\omega}) = (\rho, \boldsymbol{u} + \nu\boldsymbol{\nabla}\log\rho)$ *solves* (2.2.5)–(2.2.7). *Conversely, let* $(\rho, \boldsymbol{\omega})$ *be a smooth solution to* (2.2.5)–(2.2.7), *then* $(\rho, \boldsymbol{u}) = (\rho, \boldsymbol{\omega} - \nu\boldsymbol{\nabla}\log\rho)$ *solves* (2.2.1)–(2.2.3).

Proof. Let $(\rho, \boldsymbol{u})$ be a smooth solution to (2.2.1)–(2.2.3), take $\boldsymbol{\omega} = \boldsymbol{u} + c\boldsymbol{\nabla}\log\rho$, the mass equation transforms to

$$\rho_t + \mathrm{div}(\rho\boldsymbol{\omega}) - c\Delta\rho = \rho_t + \mathrm{div}(\rho(\boldsymbol{\omega} - c\boldsymbol{\nabla}\log\rho)) = \rho_t + \mathrm{div}(\rho\boldsymbol{u}) = 0.$$

Next, recalling the elementary identities

$$c(\rho\boldsymbol{\nabla}\log\rho)_t = \nu(\boldsymbol{\nabla}\rho)_t = -c\boldsymbol{\nabla}\mathrm{div}(\rho\boldsymbol{u}),$$

$$c^2(\rho\boldsymbol{\nabla}\log\rho \otimes \boldsymbol{\nabla}\log\rho) = c^2\Delta(\rho\boldsymbol{\nabla}\log\rho) - \nu^2\mathrm{div}(\rho\boldsymbol{\nabla}^2\log\rho)$$

$$= c^2\Delta(\rho\nabla\log\rho) - 2c^2\rho\nabla\left(\frac{\Delta\sqrt{\rho}}{\sqrt{\rho}}\right),$$

$$\text{cdiv}(\rho\nabla\log\rho \otimes \boldsymbol{u} + \rho\boldsymbol{u} \otimes \nabla\log\rho) = c\Delta(\rho\boldsymbol{u}) - 2c\text{div}(\rho D(\boldsymbol{u})) + c\nabla\text{div}(\rho\boldsymbol{u}),$$

then

$$\begin{aligned}
&(\rho\boldsymbol{\omega})_t + \text{div}(\rho\boldsymbol{\omega}\otimes\boldsymbol{\omega}) - c\Delta(\rho\boldsymbol{\omega}) \\
&\quad = -\nabla p(\rho) + 2(\nu - c)\text{div}(\rho D(\boldsymbol{\omega})) \\
&\qquad + 2(h^2 - \nu^2 - 2c(\nu - c))\rho\nabla\left(\frac{\Delta\sqrt{\rho}}{\sqrt{\rho}}\right).
\end{aligned}$$

Here, take $c = \nu, h_0^2 = h^2 - \nu^2 > 0$, thus, $(\rho, \boldsymbol{\omega})$ solves (2.2.5)–(2.2.7).

By virtue of Faedo–Galerkin method and compactness theory, there exists a global weak solution to (2.2.5)–(2.2.7). For the case of $h = \nu$ and $h < \nu$, one can refer to [33] and [49]. □

Theorem 2.2.1. *Let $T > 0$, $d \leq 3$, $h_0 > 0$, $\nu > 0$, $P(\rho) = \rho^\gamma, \gamma > 3(d = 3)$, $\gamma \geq 1(d = 2)$. (ρ_0, ω_0) is such that $\rho_0 \geq 0$ and $E(\rho_0, \omega_0)$ is finite. Then there exists a weak solution $(\rho, \boldsymbol{\omega})$ to (2.2.5)–(2.2.7) with the regularity*

$$\sqrt{\rho} \in L^\infty([0,T]; H^1(\mathbb{T}^d)) \cap L^2([0,T]; H^2(\mathbb{T}^d)), \rho \geq 0, \tag{2.2.8}$$

$$\rho \in H^1([0,T]; L^2(\mathbb{T}^d)) \cap L^\infty([0,T]; L^\gamma(\mathbb{T}^d)) \cap L^2([0,T]; W^{1,3}(\mathbb{T}^d)), \tag{2.2.9}$$

$$\sqrt{\rho}\boldsymbol{\omega} \in L^\infty([0,T]; L^2(\mathbb{T}^d)), \rho\boldsymbol{\omega} \in L^2([0,T]; W^{1,3/2}(\mathbb{T}^d)),$$

$$\sqrt{\rho}|\nabla\boldsymbol{\omega}| \in L^\infty([0,T]; L^2(\mathbb{T}^d)),$$

satisfying (2.2.5) pointwise, and for all smooth test functions η satisfying $\eta(\cdot, T) = 0$,

$$-\int_{\mathbb{T}^d} \rho_0^2\boldsymbol{\omega}_0 \cdot \boldsymbol{\eta}(\cdot, 0)\mathrm{d}x = \int_0^T \int_{\mathbb{T}^d} (\rho^2\boldsymbol{\omega} \cdot \partial_t\boldsymbol{\eta} - \rho^2\text{div}(\boldsymbol{\omega})\boldsymbol{\omega} \cdot \boldsymbol{\eta} - \nu(\rho\boldsymbol{\omega} \otimes \nabla_\rho) : \nabla\boldsymbol{\eta})$$

$$
\begin{aligned}
&+\rho\boldsymbol{\omega}\otimes\rho\boldsymbol{\omega}:\boldsymbol{\nabla\eta}+\frac{\gamma}{\gamma+1}\rho^{\gamma+1}\mathrm{div}\boldsymbol{\eta}\\
&-\nu\boldsymbol{\nabla}(\rho\boldsymbol{\omega}):(\rho\boldsymbol{\nabla\eta}+2\boldsymbol{\nabla\rho}\otimes\boldsymbol{\eta})\\
&-2h_0^2\Delta\sqrt{\rho}(\rho^{3/2}\mathrm{div}\boldsymbol{\eta}+2\sqrt{\rho}\boldsymbol{\nabla}\rho\cdot\boldsymbol{\eta}))\mathrm{d}x\mathrm{d}t.
\end{aligned}
\tag{2.2.10}
$$

Here

$$
\begin{aligned}
E(\rho(x,0),\boldsymbol{\omega}(x,0)) &= E(\rho_0,\boldsymbol{\omega}_0),\\
E(\rho(x,t),\boldsymbol{\omega}(x,t)) &= \int_{\mathbb{T}^d}\left(\frac{\rho}{2}|\boldsymbol{\omega}|^2+H(\rho)+2h^2|\boldsymbol{\nabla}\sqrt{\rho}|^2\right)\mathrm{d}x, H''(\rho)\\
&= \frac{P'(\rho)}{\rho}.
\end{aligned}
$$

Then the global existence of weak solutions to (2.2.1)–(2.2.3) is a consequence of Theorem 2.2.1.

Corollary 2.2.1. *Let $T>0$, $d\le 3$, $h>0$, $\nu>0$, $h>\nu$, $P(\rho)=\rho^\gamma, \gamma>3(d=3)$, $\gamma\ge 1(d=2)$. $(\rho_0,\boldsymbol{u}_0)$ is such that $\rho_0\ge 0$ and $E(\rho_0,\boldsymbol{u}_0+\nu\nabla\log\rho_0)$ is finite. Then there exists a weak solution $(\rho,\boldsymbol{u})$ to (2.2.1)–(2.2.3) with the regularity (2.2.8), (2.2.9) and*

$$
\begin{aligned}
&\sqrt{\rho\boldsymbol{u}}\in L^\infty([0,T];L^2(\mathbb{T}^d)), \rho\boldsymbol{u}\in L^2([0,T];W^{1,\frac{3}{2}}(\mathbb{T}^d)),\\
&\sqrt{\rho}|\boldsymbol{\nabla u}|\in L^\infty([0,T];L^2(\mathbb{T}^d)),
\end{aligned}
$$

satisfying (2.2.1) pointwise, and for all smooth test functions $\boldsymbol{\eta}$ satisfying $\boldsymbol{\eta}(\cdot,T)=0$,

$$
\begin{aligned}
-\int_{\mathbb{T}^d}\rho_0^2\boldsymbol{u}_0\cdot\boldsymbol{\eta}(\cdot,0)\mathrm{d}x = \int_0^T\int_{\mathbb{T}^d}&(\rho^2\boldsymbol{u}\cdot\partial_t\boldsymbol{\eta}-\rho^2\mathrm{div}(\boldsymbol{u}))\boldsymbol{u}\cdot\boldsymbol{\eta}\\
&+\rho\boldsymbol{u}\otimes\rho\boldsymbol{u}:\boldsymbol{\nabla\eta}\\
&-2\nu\rho D(\boldsymbol{u}):(\rho\boldsymbol{\nabla\eta}+\boldsymbol{\nabla}\rho\otimes\boldsymbol{\eta})\\
&+\frac{\gamma}{\gamma+1}\rho^{\gamma+1}\mathrm{div}\boldsymbol{\eta}\\
&-2h^2\Delta\sqrt{\rho}\rho^{3/2}\mathrm{div}\boldsymbol{\eta}+2\sqrt{\rho}\boldsymbol{\nabla\rho}\cdot\boldsymbol{\eta}))\mathrm{d}x\mathrm{d}t.
\end{aligned}
\tag{2.2.11}
$$

2.2.1 Faedo–Galerkin approximation

Let $T > 0$, and $(\boldsymbol{e}_n)$ be an orthonormal basis of $L^2(\mathbb{T}^d)$. Introduce the finite-dimensional space $X_n = \text{span } (\boldsymbol{e}_1, \dots, \boldsymbol{e}_n)$, $n \in \mathbb{N}$. Let $(\rho_0, \boldsymbol{u}_0) \in C^\infty(\mathbb{T}^d)^4$ be some initial data satisfying $\rho_0(x) \geq \delta > 0, x \in \mathbb{T}^d$, for given velocity $\boldsymbol{v} \in C^0(0, T; X_n)$, there exist some functions $\lambda_i(t)$, such that $\boldsymbol{v}$ can be written as

$$\boldsymbol{v}(x,t) = \sum_{i=1}^{n} \lambda_i(t)\boldsymbol{e}_i(x), t \in [0,T], x \in \mathbb{T}^d$$

and

$$\|\boldsymbol{v}\|_{C^0(0,T;X_n)} = \max_{t\in[0,T]} \sum_{i=1}^{n} |\lambda_i(t)|.$$

Thus, $\boldsymbol{v}$ is bounded in $C^0(0, T; C^k(\mathbb{T}^d))$, and there exists a positive constant C depending on k such that

$$\|\boldsymbol{v}\|_{C^0(0,T;C^k(\mathbb{T}^d))} \leq C\|\boldsymbol{v}\|_{C^0(0,T;L^2(\mathbb{T}^d))}.$$

Now, define the approximate system and let ρ be the classical solution to

$$\begin{aligned} \rho_t + \text{div}(\rho\boldsymbol{v}) &= \nu\Delta\rho, \quad x \in \mathbb{T}^d, t > 0, \\ \rho(x,0) &= \rho_0(x), x \in \mathbb{T}^d. \end{aligned} \tag{2.2.12}$$

For any $k \in \mathbb{N}$, $\rho \in C^0(0,T; C^k(\mathbb{T}^d))$, and $\int_{\mathbb{T}^d} \rho(x,t)\mathrm{d}x = \int_{\mathbb{T}^d} \rho_0(x)\mathrm{d}x$. Introduce $S : C^0(0,T;X_n) \to C^0(0,T;C^d(\mathbb{T}^d))$ by $S(\boldsymbol{v}) = \rho$. Since $\boldsymbol{v}$ is smooth, in view of the maximum principle, $\rho = S(\boldsymbol{v})$ has the lower and upper bounds. For $\|\boldsymbol{v}\|_{C^0(0,T;L^2(\mathbb{T}^d))} \leq c$, there exist positive constants $K_0(c)$ and $K_1(c)$ such that

$$0 < K_0(c) \leq (S(\boldsymbol{v}))(t,x) \leq K_1(c), t \in [0,T], x \in \mathbb{T}^d. \tag{2.2.13}$$

Moreover, for any $\boldsymbol{v}_1$, $\boldsymbol{v}_2 \in C^0(0,T;X_n)$, there exists a positive constant K_2 depending on k and n, such that

$$\|S(\boldsymbol{v}_1) - S(\boldsymbol{v}_2)\|_{C^0(0,T;C^k(\mathbb{T}^d))} \leq K_2\|\boldsymbol{v}_1 - \boldsymbol{v}_2\|_{C^0(0,T;L^2(\mathbb{T}^d))}. \tag{2.2.14}$$

Then for given $\rho = S(\boldsymbol{v})$, solve the following problem on X_n:

$$(\rho\boldsymbol{\omega}_n)_t + \operatorname{div}(\rho\boldsymbol{v}\otimes\boldsymbol{\omega}_n) + \boldsymbol{\nabla}P(\rho) - 2h_0^2\rho\boldsymbol{\nabla}\left(\frac{\Delta\sqrt{\rho}}{\sqrt{\rho}}\right)$$
$$= v\Delta(\rho\boldsymbol{\omega}_n) + \delta(\Delta\boldsymbol{\omega}_n - \boldsymbol{\omega}_n), \tag{2.2.15}$$

for all test functions $\boldsymbol{\eta} \in C^1(0,T;X_n), \boldsymbol{\eta}(\cdot,T) = 0$, looking for a function $\boldsymbol{\omega}_n \in C^1(0,T;X_n)$ such that

$$\begin{aligned} -\int_{\mathbb{T}^d}\rho_0\boldsymbol{\omega}_0\cdot\boldsymbol{\eta}(\cdot,0)\mathrm{d}x = \int_0^T\int_{\mathbb{T}^d} & [\rho\boldsymbol{\omega}_n\cdot\boldsymbol{\eta}_t + (\rho\boldsymbol{v}\otimes\boldsymbol{\omega}_n):\boldsymbol{\nabla}\boldsymbol{\eta} \\ & + P(\rho)\operatorname{div}\boldsymbol{\eta} \\ & - 2h_0^2\frac{\Delta\sqrt{\rho}}{\sqrt{\rho}}\operatorname{div}(\rho\boldsymbol{\eta}) - \nu\boldsymbol{\nabla}(\rho\boldsymbol{\omega}_n):\boldsymbol{\nabla}\boldsymbol{\eta} \\ & - \delta(\boldsymbol{\nabla}\boldsymbol{\omega_n}:\boldsymbol{\nabla}\boldsymbol{\eta} + \boldsymbol{\omega}_n\cdot\boldsymbol{\eta})]\mathrm{d}x\mathrm{d}t. \end{aligned} \tag{2.2.16}$$

For given $\rho \in K_{\underline{\rho}} = \{L^1(\mathbb{T}^d) : \inf_{x\in\mathbb{T}^d}\rho \geq \underline{\rho} > 0\}$, introduce the following family of operators:

$$M[\rho] : X_n \to X_n^*, (M[\rho]\boldsymbol{u},\boldsymbol{\omega}) = \int_{\mathbb{T}^d}\rho\boldsymbol{u}\boldsymbol{\omega}\mathrm{d}x, \boldsymbol{u},\boldsymbol{\omega} \in X_n.$$

These operators are symmetric and positive definite with the smallest eigenvalue

$$\inf_{\|\boldsymbol{\omega}\|_{L^2(\mathbb{T}^d)=1}}\langle M(\rho)\boldsymbol{\omega},\boldsymbol{\omega}\rangle = \inf_{\|\omega\|_{L^2(\mathbb{T}^d)=1}}\int_{\mathbb{T}^d}\rho\boldsymbol{\omega}^2\mathrm{d}x \geq \inf_{x\in\mathbb{T}^d}\rho(x) > \underline{\rho}.$$

Since X_n is finite-dimensional, the operators are invertible with

$$\|M^{-1}(\rho)\|_{L(X_n^*,X_n)} \leq \underline{\rho}^{-1},$$

where $L(X_n^*,X_n)$ is the set of bounded linear mappings from X_n^* to X_n. Moreover, M^{-1} is Lipschitz continuous in the sense

$$\|M^{-1}(\rho_1) - M^{-1}(\rho_2)\|_{L(X_n^*,X_n)} \leq K(n,\underline{\rho})\|\rho_1-\rho_2\|_{L^1(\mathbb{T}^d)} \tag{2.2.17}$$

for all $\rho_1,\rho_2 \in K_{\underline{\rho}}$.

Thus, the integral equation (2.2.15) can be rephrased as an ordinary differential equation on the finite-dimensional space X_n:

$$\frac{\mathrm{d}}{\mathrm{d}t}(M[\rho(t)]\boldsymbol{\omega}_n(t)) = N[\boldsymbol{v}, \boldsymbol{\omega}_n(t)], t > 0, M[\rho_0]\boldsymbol{\omega}_n(0) = M[\rho_0]\boldsymbol{\omega}_0, \tag{2.2.18}$$

where

$$(N[\boldsymbol{v}, \boldsymbol{\omega}_n(t)], \boldsymbol{\eta}) = \int_{\mathbb{T}^d} \Big((\rho \boldsymbol{v} \otimes \boldsymbol{\omega}_n) : \nabla \boldsymbol{\eta} + P(\rho)\mathrm{div}\boldsymbol{\eta} - 2h_0^2 \frac{\Delta\sqrt{\rho}}{\sqrt{\rho}} \mathrm{div}(\rho\boldsymbol{\eta})$$
$$- \nu \nabla(\rho\boldsymbol{\omega}_n) : \nabla\boldsymbol{\eta} - \delta(\nabla\boldsymbol{\omega}_n : \nabla\boldsymbol{\eta} + \boldsymbol{\omega}_n \cdot \boldsymbol{\eta}) \Big) \mathrm{d}x.$$

The above integral is well defined as $\rho = S(\boldsymbol{v})$ is bounded from below, and for any $t \in [0, T]$, the operator $N[\boldsymbol{v}, \cdot]$ from X_n to X_n^* is continuous in time. Therefore, standard theory for systems of ordinary differential equations provides the existence of a unique solution to (2.1.18). There exists a unique solution $\boldsymbol{\omega}_n \in C^1(0, T; X_n)$ to (2.1.15) for given $\boldsymbol{v} \in C^0(0, T; X_n)$.

Integrating (2.2.18) over $(0, t)$ yields

$$\boldsymbol{\omega}_n(t) = M^{-1}[S(\boldsymbol{\omega}_n)(t)](M[\rho_0]\boldsymbol{\omega}_0) + \int_0^t N[\boldsymbol{v}, \boldsymbol{\omega}_n(t)]\, \mathrm{d}s) \text{ in } X_n.$$

In view of the Lipschitz-type estimates (2.2.14), (2.2.17) for S and M^{-1}, this equation can be solved by virtue of the fixed-point theorem of Banach on a short time $[0, T']$, where $T' \leq T$, in the space $C^0([0, T']; X_n)$. In fact, $\boldsymbol{\omega}_n \in C^1([0, T']; X_n)$ can be obtained. Thus, there exists a unique local-in-time solution $(\rho_n, \boldsymbol{\omega}_n)$ to (2.2.12) and (2.2.16).

Next, in order to prove that the solution $(\rho_n, \boldsymbol{\omega}_n)$ constructed above exists on the whole time $[0, T]$, it's sufficient to show that $(\rho_n, \boldsymbol{\omega}_n)$ is bounded in X_n on $[0, T']$. This is achieved by using the energy estimate.

Lemma 2.2.2. *Let* $T' \leq T, \rho_n \in C^1([0, T']; C^3(\mathbb{T}^d)), \boldsymbol{\omega}_n \in C^1([0, T'];$ $X_n)$ *be a local-in-time solution to* (2.2.12), (2.2.16), *with* $\rho = \rho_n, \boldsymbol{v} =$

$\boldsymbol{\omega}_n$. *Then*

$$\frac{\mathrm{d}}{\mathrm{d}t}E(\rho_n,\boldsymbol{\omega}_n)$$
$$+\nu\int_{\mathbb{T}^d}(\rho_n|\boldsymbol{\nabla}\boldsymbol{\omega}_n|^2+H''(\rho_n)|\boldsymbol{\nabla}\rho_n|^2+h_0^2\rho_n|\boldsymbol{\nabla}^2\log\rho_n|^2)\mathrm{d}x$$
$$+\delta\int_{\mathbb{T}^d}(|\boldsymbol{\nabla}\boldsymbol{\omega}_n|^2+|\boldsymbol{\omega}_n|^2)\mathrm{d}x=0. \tag{2.2.19}$$

Proof. Using the test function $\boldsymbol{\omega}_n \in C^1([0,T];X_n)$ in (2.2.16), where $\boldsymbol{v}, = \boldsymbol{\omega}_n, \rho = \rho_n := S(\boldsymbol{v}) = S(\boldsymbol{\omega}_n)$, and integrating by parts yields

$$0=\int_{\mathbb{T}^d}(|\omega_n|^2\partial_t\rho_n+\frac{1}{2}\rho_n\partial_t|\boldsymbol{\omega}_n)^2$$
$$-\rho_n\boldsymbol{\omega}_n\otimes\boldsymbol{\omega}_n:\boldsymbol{\nabla}\boldsymbol{\omega}_n+P'(\rho_n)\boldsymbol{\nabla}\rho_n\cdot\boldsymbol{\omega}_n$$
$$+2h_0^2\frac{\Delta\sqrt{\rho_n}}{\sqrt{\rho_n}}\mathrm{div}(\rho_n\boldsymbol{\omega}_n)+\nu\boldsymbol{\nabla}\rho_n\cdot\boldsymbol{\nabla}\boldsymbol{\omega}_n\cdot\boldsymbol{\omega}_n+\nu\rho_n|\boldsymbol{\nabla}\boldsymbol{\omega}_n|^2$$
$$+\delta|\boldsymbol{\nabla}\rho_n|^2+\delta|\rho_n|^2)\mathrm{d}x. \tag{2.2.20}$$

Then multiplying (2.2.12) by $H'(\rho_n)-\frac{|\boldsymbol{\omega}_n|^2}{2}-2h_0^2\Delta\frac{\sqrt{\rho_n}}{\sqrt{\rho_n}}$, integrating over $\mathbb{T}^d$ and integrating by parts:

$$0=\int_{\mathbb{T}^d}(\partial_t H(\rho_n)-\frac{1}{2}|\boldsymbol{\omega}_n|^2\partial_t\rho_n+2h_0^2\partial_t|\boldsymbol{\nabla}\sqrt{\rho_n}|^2$$
$$-\rho_n H''(\rho_n)\boldsymbol{\nabla}\rho_n\cdot\boldsymbol{\omega}_n$$
$$+\rho_n\boldsymbol{\omega}_n\cdot\boldsymbol{\nabla}\boldsymbol{\omega}_n\cdot\boldsymbol{\omega}_n-2h_0^2\frac{\Delta\sqrt{\rho_n}}{\sqrt{\rho_n}}\mathrm{div}(\rho_n\boldsymbol{\omega}_n)+\nu H''(\rho_n)|\boldsymbol{\nabla}\rho_n|^2$$
$$-\nu\boldsymbol{\nabla}\rho_n\cdot\boldsymbol{\nabla}\boldsymbol{\omega}_n\cdot\boldsymbol{\omega}_n+2\nu h_0^2\frac{\Delta\sqrt{\rho_n}}{\sqrt{\rho_n}}\ \Delta\rho_n)\,\mathrm{d}x. \tag{2.2.21}$$

□

By virtue of the identity $2\rho_n\nabla(\Delta\frac{\sqrt{\rho_n}}{\sqrt{\rho_n}}) = \mathrm{div}(\rho_n\nabla^2\log\rho_n)$, we have

$$\begin{aligned}
2\int_{\mathbb{T}^d}\frac{\Delta\sqrt{\rho_n}}{\sqrt{\rho_n}}\Delta\rho_n\mathrm{d}x &= -2\int_{\mathbb{T}^d}\rho_n\nabla\log\rho_n\cdot\nabla\left(\frac{\Delta\sqrt{\rho_n}}{\sqrt{\rho_n}}\right)\mathrm{d}x \\
&= -\int_{\mathbb{T}^d}\nabla\log\rho_n\cdot\mathrm{div}(\rho_n\nabla^2\log\rho_n)\mathrm{d}x \\
&= \int_{\mathbb{T}^d}\rho_n|\nabla^2\log\rho_n|^2\mathrm{d}x, \qquad (2.2.22)
\end{aligned}$$

together with (2.2.20)–(2.2.22), we complete the proof of Lemma 2.2.2.

2.2.2 *A priori estimate*

Let $(\rho_n,\omega_n) \in C^1([0,T];C^3(\mathbb{T}^d)) \times C^1([0,T];X_n)$ be a solution to the approximate system (2.2.12) and (2.2.16). From Lemma 2.2.2 and Gronwall's lemma, we have the following uniform bounds

$$\|\sqrt{\rho_n}\|_{L^\infty([0,T];H^1(\mathbb{T}^d))} \le C, \qquad (2.2.23)$$

$$\|\rho_n\|_{L^\infty([0,T];L^\gamma(\mathbb{T}^d))} \le C, \qquad (2.2.24)$$

$$\|\sqrt{\rho_n}\omega_n\|_{L^\infty([0,T];L^2(\mathbb{T}^d))} + \|\sqrt{\rho_n}\nabla\omega_n\|_{L^2([0,T];L^2(\mathbb{T}^d))} \le C, \qquad (2.2.25)$$

$$\sqrt{\delta}\|\omega_n\|_{L^2([0,T];H^1(\mathbb{T}^d))} \le C, \qquad (2.2.26)$$

where $C > 0$ is a generic constant which is independent of n and δ. Since $H^1(\mathbb{T}^d) \hookrightarrow L^6(\mathbb{T}^d), d \le 3$. The $L^\infty([0,T];H^1(\mathbb{T}^d))$ estimate for $\sqrt{\rho_n}$ yields immediately $\rho_n \in L^\infty([0,T];L^3(\mathbb{T}^d))$. The estimate (2.2.24) improves this bound only if $\gamma > 3$. For $d = 2$, $\alpha < \infty$, $H^1(\mathbb{T}^d) \hookrightarrow L^\alpha(\mathbb{T}^d)$, thus, for $\gamma \ge 1$, $\rho_n \in L^\infty([0,T];L^\alpha(\mathbb{T}^d))$.

The above uniform bounds allow us to conclude some estimates.

Lemma 2.2.3. *For some constant $C > 0$ which is independent of n and δ, the following uniform estimate holds:*

$$\|\sqrt{\rho_n}\|_{L^2(0,T;H^2(\mathbb{T}^d))} + \|\sqrt[4]{\rho_n}\|_{L^4(0,T;W^{1,4}(\mathbb{T}^d))} \le C. \qquad (2.2.27)$$

Proof. From the energy estimates in Lemma 2.2.2 and the following inequalities

$$\int_{\mathbb{T}^d} \rho_n |\nabla^2 \log\rho_n|^2 dx \geq \kappa_d \int_{\mathbb{T}^d} |\nabla^2 \sqrt{\rho_n}|^2 dx,$$

$$\int_{\mathbb{T}^d} \rho_n |\nabla^2 \log\rho_n|^2 dx \geq \kappa \int_{\mathbb{T}^d} |\nabla \sqrt[4]{\rho_n}|^4 dx,$$

we can prove (2.2.27), where, $\kappa_{\mathrm{d}} = \frac{4d-1}{d(d+2)}, \kappa = \frac{16(4d-1)}{(d+2)^2}$. One can refer to [47], [50] for the detailed proof of these two inequalities. □

Lemma 2.2.4. *The following uniform estimates hold for some constant $C > 0$ not depending on n and δ:*

$$\|\rho_n \boldsymbol{\omega}_n\|_{L^2(0,T;W^{1,3/2}(\mathbb{T}^d))} \leq C, \tag{2.2.28}$$

$$\|\rho_n\|_{L^2(0,T;W^{2,p}(\mathbb{T}^d))} \leq C, \tag{2.2.29}$$

$$\|\rho_n\|_{L^{\frac{4\gamma+3}{3}}(0,T;L^{\frac{4\gamma+3}{3}}(\mathbb{T}^d))} \leq C, \tag{2.2.30}$$

where $p = 2\gamma/(\gamma+1)$ if $d = 3$ and $p < 2$ if $d = 2$.

Remark 2.2.1. For $\gamma > 3$, it holds that $p > 3/2$. Hence, the embedding $W^{2,p}(\mathbb{T}^d) \hookrightarrow W^{1,3}(\mathbb{T}^d)$ is compact.

Proof. Since $d \leq 3$, $H^2(\mathbb{T}^d)$ embeds continuously into $L^\infty(\mathbb{T}^d)$, using (2.2.27), $\{\sqrt{\rho_n}\} \in L^2(0,T;L^\infty(\mathbb{T}^d))$. Thus, (2.2.25) yields that $\rho_n\boldsymbol{\omega}_n = \sqrt{\rho_n}\sqrt{\rho_n}\boldsymbol{\omega}_n$ is uniformly bounded in $L^2(0,T;L^2(\mathbb{T}^d))$. By virtue of (2.2.23) and (2.2.27), $\{\nabla\sqrt{\rho_n}\} \in L^2(0,T;L^6(\mathbb{T}^d))$ and $\{\sqrt{\rho_n} \in L^\infty(0,T;L^6(\mathbb{T}^d))\}$. This together with (2.2.25), implies that

$$\nabla(\rho_n\boldsymbol{\omega}_n) = 2\nabla\sqrt{\rho_n} \otimes (\sqrt{\rho_n}\boldsymbol{\omega}_n) + \sqrt{\rho_n}\nabla\boldsymbol{\omega}_n\sqrt{\rho_n} \tag{2.2.31}$$

is uniformly bounded in $L^2(0,T;L^{3/2}(\mathbb{T}^d))$. This completes the proof of (2.2.28).

For (2.2.29), in view of Gagliardo–Nirenberg inequality, $p = 2\gamma/(\gamma+1)$, $\theta = 1/2$,

$$\begin{aligned}\|\boldsymbol{\nabla}\sqrt{\rho_n}\|^4_{L^4(0,T;L^{2p}(\mathbb{T}^d))} &\le C\int_0^T \|\sqrt{\rho_n}\|^{4\theta}_{H^2(\mathbb{T}^d)}\|\sqrt{\rho_n}\|^{4(1-\theta)}_{L^{2\gamma}(\mathbb{T}^d)}\mathrm{d}t\\ &\le C\|\sqrt{\rho_n}\|^2_{L^\infty(0,T;L^{2\gamma}(\mathbb{T}^d))}\int_0^T \|\sqrt{\rho_n}\|^2_{H^2(\mathbb{T}^d)}\mathrm{d}t\\ &\le C.\end{aligned}$$

Therefore, $\{\sqrt{\rho_n}\} \in L^4(0,T;W^{1,2p}(\mathbb{T}^d))$. Noting that $d = 3, p > 3/2$ implies that $\{\sqrt{\rho_n}\}$ is uniformly bounded in $L^4(0,T;L^\infty(\mathbb{T}^d))$. For $d = 2$, and any $\alpha < \infty, \{\sqrt{\rho_n}\} \in L^\infty(0,T;H^1(\mathbb{T}^d)) \hookrightarrow L^\infty(0,T;L^\alpha(\mathbb{T}^d))$. Hence, we may replace 2γ in the above estimate by α, obtaining an $L^4(0,T;W^{1,2p}(\mathbb{T}^d))$ bound for all $p < 2$. Then, in the two dimensional case, all $\gamma \ge 1$ are admissible. The estimate $\{\boldsymbol{\nabla}\sqrt{\rho_n}\} \in L^4(0,T;L^{2p}(\mathbb{T}^d))$ implies that

$$\boldsymbol{\nabla}^2\rho_n = 2(\sqrt{\rho_n}\boldsymbol{\nabla}^2\sqrt{\rho_n} + \nabla\sqrt{\rho_n}\otimes\boldsymbol{\nabla}\sqrt{\rho_n})$$

is bounded in $L^2(0,T;L^p(\mathbb{T}^d))$ which proves (2.2.29).

Finally, by the Gagliardo–Nirenberg inequality, with $\theta = 3/(4\gamma+3)$ and $q = 2(4\gamma+3)/3$

$$\begin{aligned}\|\sqrt{\rho_n}\|^q_{L^q(0,T;L^q(\mathbb{T}^d))} &\le C\int_0^T \|\sqrt{\rho_n}\|^{q\theta}_{H^2(\mathbb{T}^d)}\|\sqrt{\rho_n}\|^{q(1-\theta)}_{L^{2\gamma}(\mathbb{T}^d)}\mathrm{d}t\\ &\le C\|\rho_n\|^{q(1-\theta)}_{L^\infty(0,T;L^\gamma(\mathbb{T}^d))}\int_0^T \|\sqrt{\rho_n}\|^2_{H^2(\mathbb{T}^d)}\mathrm{d}t \le C.\end{aligned}$$

shows that $\{\rho_n\} \in L^{q/2}(0,T;L^{q/2}(\mathbb{T}^d))$. This completes the proof. □

Lemma 2.2.5. *For $s > d/2+1$ and some constant $C > 0$ which is independent of n and δ, the following uniform estimates hold:*

$$\|\partial_t\rho_n\|_{L^2(0,T;L^{\frac{3}{2}}(\mathbb{T}^d))} \le C, \tag{2.2.32}$$

$$\|\partial_t(\rho_n\boldsymbol{u}_n)\|_{L^{\frac{4}{3}}(0,T;(H^s(\mathbb{T}^d))^*)} \le C. \tag{2.2.33}$$

Proof. By (2.2.28) and (2.2.29), $\partial_t \rho_n = -\mathrm{div}(\rho_n \boldsymbol{\omega}_n) + \nu \Delta \rho_n$ is uniformly bounded in $L^2(0,T;L^{3/2}(\mathbb{T}^d))$ which finishes the proof of (2.2.32).

The sequence $\{\rho_n \boldsymbol{\omega}_n \otimes \boldsymbol{\omega}_n\} \in L^\infty(0,T;L^1(\mathbb{T}^d))$, hence, $\mathrm{div}(\rho_n \boldsymbol{\omega}_n \otimes \boldsymbol{\omega}_n) \in L^\infty(0,T;(W^{1,\infty}(\mathbb{T}^d))^*)$. For $s > d/2 + 1$, because of the continuous embedding of $H^s(\mathbb{T}^d)$ into $W^{1,\infty}(\mathbb{T}^d)$, we have $(\mathrm{div}(\rho_n \boldsymbol{\omega}_n \otimes \boldsymbol{\omega}_n)) \in L^\infty(0,T;(H^s(\mathbb{T}^d)^*)$. For all $\boldsymbol{\phi} \in L^4(0,T;W^{1,3}(\mathbb{T}^d))$, the estimate

$$
\begin{aligned}
&\int_0^T \int_{\mathbb{T}^d} \rho_n \boldsymbol{\nabla} \left(\frac{\Delta \sqrt{\rho_n}}{\sqrt{\rho_n}} \right) \cdot \boldsymbol{\phi} \mathrm{d}x \mathrm{d}t \\
&\quad = \int_0^T \int_{\mathbb{T}^d} \Delta \sqrt{\rho_n} (2 \boldsymbol{\nabla} \sqrt{\rho_n} \cdot \boldsymbol{\phi} + \sqrt{\rho_n} \mathrm{div} \boldsymbol{\phi}) \mathrm{d}x \mathrm{d}t \\
&\quad \leq \|\Delta \sqrt{\rho_n}\|_{L^2(0,T;L^2(\mathbb{T}^d))} \, (2 \, \|\sqrt{\rho_n}\|_{L^4(0,T;W^{1,3}(\mathbb{T}^d))} \|\boldsymbol{\phi}\|_{L^4(0,T;L^6(\mathbb{T}^d))} \\
&\qquad + \|\sqrt{\rho_n}\|_{L^\infty(0,T;L^6(\mathbb{T}^d))} \|\boldsymbol{\phi}\|_{L^2(0,T;W^{1,3}(\mathbb{T}^d))} \\
&\quad \leq C \|\boldsymbol{\phi}\|_{L^4(0,T;W^{1,3}(\mathbb{T}^d))}
\end{aligned}
$$

implies that $\rho_n \boldsymbol{\nabla}(\frac{\Delta \sqrt{\rho_n}}{\sqrt{\rho_n}})$ is uniformly bounded in $L^{4/3}(0,T; W^{1,3}(\mathbb{T}^d)^*) \hookrightarrow L^{4/3}(0,T;H^s(\mathbb{T}^d)^*)$. By (2.2.30), $\{\rho_n^\gamma\}$ is bounded in $L^{4/3}(0,T;L^{4/3}(\mathbb{T}^d)) \hookrightarrow L^{4/3}(0,T;H^s(\mathbb{T}^d)^*)$. Furthermore, by (2.2.28), $\Delta(\rho_n \boldsymbol{\omega}_n)$ is uniformly bounded in $L^2(0,T;W^{1,3}(\mathbb{T}^d)^*)$. Since $\{\delta \Delta \boldsymbol{\omega}_n\}$ is bounded in $L^2(0,T;H^1(\mathbb{T}^d)^*)$,

$$
\begin{aligned}
(\rho_n \boldsymbol{\omega}_n)_t &= -\mathrm{div}(\rho_n \boldsymbol{\omega}_n \otimes \boldsymbol{\omega}_n) - \boldsymbol{\nabla}(\rho_n^\gamma) + 2 h_0^2 \rho_n \boldsymbol{\nabla} \left(\frac{\Delta \sqrt{\rho_n}}{\sqrt{\rho_n}} \right) \\
&\quad + \nu \Delta(\rho_n \boldsymbol{\omega}_n) + \delta \Delta \boldsymbol{\omega}_n
\end{aligned}
$$

is uniformly bounded in $L^{4/3}(0,T;H^s(\mathbb{T}^d)^*)$.

The $L^4(0,T;W^{1,4}(\mathbb{T}^d))$ bound on $\sqrt[4]{\rho_n}$ provides a uniform estimate for $\partial t \sqrt{\rho_n}$. □

Lemma 2.2.6. *The following estimate holds*

$$
\|\partial t \sqrt{\rho_n}\|_{L^2(0,T;H^1(\mathbb{T}^d)^*)} \leq C. \tag{2.2.34}
$$

Proof. Dividing the mass equation by $\sqrt{\rho_n}$ yields

$$\begin{aligned}\partial_t\sqrt{\rho_n} &= -\nabla\sqrt{\rho_n}\cdot\boldsymbol{\omega}_n - \frac{1}{2}\sqrt{\rho_n}\mathrm{div}\boldsymbol{\omega}_n + \nu(\Delta\sqrt{\rho_n} + 4|\nabla\sqrt[4]{\rho_n}|^2)\\ &= -\mathrm{div}(\sqrt{\rho_n}\boldsymbol{\omega}_n) + \frac{1}{2}\sqrt{\rho_n}\mathrm{div}\boldsymbol{\omega}_n + \nu(\Delta\sqrt{\rho_n} + 4|\nabla\sqrt[4]{\rho_n}|^2).\end{aligned}$$

By (2.2.25) and (2.2.27), the first term on the right hand side is bounded in $L^2(0,T;H^1(\mathbb{T}^d)^*)$, and the remaining terms are uniformly bounded in $L^2(0,T;L^2(\mathbb{T}^d))$. □

2.2.3 *Limit* $n\to\infty$

First, for fixed $\delta>0$, perform the limit $n\to\infty$, the limit $\delta\to 0$ will be carried later. Here we consider both limits separately because the weak formulation (2.2.11) for the viscous quantum Euler model is different from its approximation (2.2.12) and (2.2.16).

By using Aubin's lemma, in view of the regularity (2.2.29) and (2.2.32) for ρ_n, the regularity (2.2.27) and (2.2.34) for $\sqrt{\rho_n}$, and the regularity (2.2.28) and (2.2.33) for $\rho_n\boldsymbol{\omega}_n$, there exist subsequences of $\{\rho_n\}$, $\{\sqrt{\rho_n}\}$, $\{\rho_n\boldsymbol{\omega}_n\}$, which are not relabeled, such that, for some functions $\rho, \boldsymbol{j}$, as $n\to\infty$,

$$\begin{aligned}\rho_n &\to \rho\,\text{strongly in}\,L^2(0,T;L^\infty(\mathbb{T}^d)),\\ \sqrt{\rho_n} &\rightharpoonup \sqrt{\rho}\,\text{weakly in}\,L^2(0,T;H^2(\mathbb{T}^d)),\\ \sqrt{\rho_n} &\to \sqrt{\rho}\,\text{strongly in}\,L^2(0,T;H^1(\mathbb{T}^d)),\\ \rho_n\boldsymbol{\omega}_n &\to \boldsymbol{j}\,\text{strongly in}\,L^2(0,T;L^2(\mathbb{T}^d)).\end{aligned}$$

We have used the fact that the embeddings $W^{2,p}(T^d)\hookrightarrow L^\infty(T^d), H^2(\mathbb{T}^d)\hookrightarrow H^1(\mathbb{T}^d)$, and $W^{1,3/2}(\mathbb{T}^d)\hookrightarrow L^2(\mathbb{T}^d)$ are compact. The estimate (2.2.26) on $\{\boldsymbol{\omega}_n\}$ provides the existence of a subsequence (not relabeled), such that, as $n\to\infty$,

$$\boldsymbol{\omega}_n \rightharpoonup \boldsymbol{\omega}\,\text{weakly in}\,L^2(0,T;H^1(\mathbb{T}^3)).$$

Since $\rho_n\boldsymbol{\omega}_n \rightharpoonup \rho\boldsymbol{\omega}$ weakly in $L^1(0,T;L^6(\mathbb{T}^3))$, we have $\boldsymbol{j}=\rho\boldsymbol{\omega}$.

Now we perform the limit $n\to\infty$ in the approximate system (2.2.12), (2.2.16), where $\rho=\rho_n, \boldsymbol{v}=\boldsymbol{\omega}_n$, it's easy to verify that as $n \to \infty$, ρ, $\boldsymbol{\omega}$ solves

$$\rho_t + \operatorname{div}(\rho \boldsymbol{u}) = \nu \Delta \rho, \text{ in } \mathbb{T}^3 \times (0, T).$$

Next, we consider the weak formulation (2.2.16) term by term, the strong convergence of $(\rho_n \boldsymbol{\omega}_n)$ in $L^2(0, T; L^2(\mathbb{T}^d))$ and the weak convergence of $\boldsymbol{\omega}_n$ in $L^2(0, T; L^6(\mathbb{T}^d))$ yield

$$\rho_n \boldsymbol{\omega}_n \otimes \boldsymbol{\omega}_n \rightharpoonup \rho \boldsymbol{\omega} \otimes \boldsymbol{\omega} \text{ weakly in } L^1(0, T; L^{3/2}(\mathbb{T}^d)). \tag{2.2.35}$$

By virtue of (2.2.28), furthermore

$$\boldsymbol{\nabla}(\rho_n \boldsymbol{\omega}_n) \rightharpoonup \boldsymbol{\nabla}(\rho \boldsymbol{\omega}) \text{ weakly in } L^2(0, T; L^{3/2}(\mathbb{T}^d)). \tag{2.2.36}$$

The $L^\infty(0, T; L^\gamma(\mathbb{T}^d))$ bound for (ρ_n) implies that for some function $z, \rho^\gamma \rightharpoonup z$ weakly * in $L^\infty(0, T; L^1(\mathbb{T}^d))$, since $\rho_n^\gamma \to \rho^\gamma$, a.e. $z = \rho^\gamma$. Finally, the above convergence shows that the limit $n \to \infty$ of

$$\int_{\mathbb{T}^d} \frac{\Delta \sqrt{\rho_n}}{\sqrt{\rho_n}} \operatorname{div}(\rho_n \boldsymbol{\phi}) \mathrm{d}x = \int_{\mathbb{T}^d} \Delta \sqrt{\rho_n} (2 \boldsymbol{\nabla} \sqrt{\rho_n} \cdot \boldsymbol{\phi} + \sqrt{\rho_n} \operatorname{div} \boldsymbol{\phi}) \mathrm{d}x \tag{2.2.37}$$

equals

$$\int_{\mathbb{T}^d} \Delta \sqrt{\rho} (2 \boldsymbol{\nabla} \sqrt{\rho} \cdot \boldsymbol{\phi} + \sqrt{\rho} \operatorname{div} \boldsymbol{\phi}) \mathrm{d}x, \tag{2.2.38}$$

for sufficiently smooth test functions.

We have proved that $(\rho, \rho\boldsymbol{\omega})$ solves $\rho_t + \operatorname{div}(\rho \boldsymbol{\omega}) = \nu \Delta \rho$ pointwise in $\mathbb{T}^d \times (0, T)$, and for any test function $\boldsymbol{\phi}$, such that below integrals are defined

$$\begin{aligned}
-\int_{\mathbb{T}^d} \rho_0 \boldsymbol{\omega}_0 \cdot \boldsymbol{\phi}(\cdot, 0) \mathrm{d}x = \int_0^T \int_{\mathbb{T}^d} & (\rho \boldsymbol{\omega} \cdot \boldsymbol{\phi}_t + (\rho \boldsymbol{\omega} \otimes \boldsymbol{\omega}) : \boldsymbol{\nabla} \boldsymbol{\phi} \\
& + P(\rho) \operatorname{div} \boldsymbol{\phi} \\
& - 2h_0^2 \Delta \sqrt{\rho} (\sqrt{\rho} \operatorname{div} \boldsymbol{\phi} + 2 \boldsymbol{\nabla} \sqrt{\rho} \cdot \boldsymbol{\phi}) \\
& - \nu \boldsymbol{\nabla}(\rho \boldsymbol{\omega}) : \boldsymbol{\nabla} \boldsymbol{\phi} \\
& - \delta(\boldsymbol{\nabla} \boldsymbol{\omega} : \boldsymbol{\nabla} \boldsymbol{\phi} + \boldsymbol{\omega} \cdot \boldsymbol{\phi})) \mathrm{d}x \mathrm{d}t.
\end{aligned} \tag{2.2.39}$$

2.2.4 *Limit* $\delta \to 0$

Let $(\rho_\delta, \boldsymbol{\omega}_\delta)$ be a solution to (2.2.12) and (2.2.39), employing the test function $\rho_\delta \boldsymbol{\phi}$ in (2.2.39) leads to

$$
\begin{aligned}
-\int_{\mathbb{T}^d} \rho_0^2 \boldsymbol{\omega}_0 \cdot \boldsymbol{\phi}(\cdot, 0)\mathrm{d}x = \int_0^T \int_{\mathbb{T}^d} & (\rho_\delta^2 \boldsymbol{\omega}_\delta) \cdot \partial_t \boldsymbol{\phi} - \rho_\delta^2 \mathrm{div}(\boldsymbol{\omega}_\delta) \boldsymbol{\omega}_\delta \cdot \boldsymbol{\phi} \\
& + \frac{\gamma}{\gamma+1} \rho_\delta^{\gamma+1} \mathrm{div} \boldsymbol{\phi} - \nu(\rho_\delta \boldsymbol{\omega}_\delta \otimes \boldsymbol{\nabla}_{\rho\delta}) : \boldsymbol{\nabla} \boldsymbol{\phi} \\
& - \nu \boldsymbol{\nabla}(\rho_\delta \boldsymbol{\omega}_\delta) : (\rho_\delta \boldsymbol{\nabla} \boldsymbol{\phi} + 2 \boldsymbol{\nabla} \rho_\delta \otimes \boldsymbol{\phi}) \\
& + \rho_\delta \boldsymbol{\omega}_\delta \otimes \rho_\delta \boldsymbol{\omega}_\delta : \boldsymbol{\nabla} \boldsymbol{\phi} - 2h_0^2 \Delta \sqrt{\rho_\delta} \\
& \times \Big(\rho_\delta^{\frac{3}{2}} \mathrm{div} \boldsymbol{\phi} + 2 \sqrt{\rho_\delta} \boldsymbol{\nabla} \rho_\delta \cdot \boldsymbol{\phi} \\
& - \delta \boldsymbol{\nabla} \boldsymbol{\omega}_\delta : (\rho_\delta \boldsymbol{\nabla} \boldsymbol{\phi} + \boldsymbol{\nabla} \rho_\delta \otimes \boldsymbol{\phi}) \\
& - \delta \rho_\delta \boldsymbol{\omega}_\delta \cdot \boldsymbol{\phi} \Big) \, \mathrm{d}x\mathrm{d}t. \qquad (2.2.40)
\end{aligned}
$$

By using Aubin's lemma and the obtained regularity results, we can deduce that for some functions ρ, j, there exist subsequences (not relabeled), such that as $\delta \to 0$,

$$\rho_\delta \to \rho \,\text{strongly in}\, L^2(0,T; W^{1,m}(\mathbb{T}^d)), 3 < m < 6\gamma/(\gamma+3), \quad (2.2.41)$$

$$\rho_\delta \boldsymbol{\omega}_\delta \to j \,\text{strongly in}\, L^2(0,T; L^q(\mathbb{T}^d)), 1 \le q < 3, \quad (2.2.42)$$

$$\sqrt{\rho_\delta} \to \sqrt{\rho} \,\text{strongly in}\, L^\infty(0,T; L^r(\mathbb{T}^d)), 1 \le r < 6. \quad (2.2.43)$$

Here we have used the embeddings $W^{2,p}(\mathbb{T}^d) \hookrightarrow W^{1,m}(\mathbb{T}^d)[3 < m < 6\gamma/(\gamma+3)]$, $W^{1,3/2}(\mathbb{T}^d) \hookrightarrow L^q(\mathbb{T}^d)(1 \le q < 3)$, and $H^1(\mathbb{T}^d) \hookrightarrow L^r(\mathbb{T}^d)(1 \le r < 6)$ to be compact. Estimate (2.2.25) and Fatou's lemma leads to

$$\int_{\mathbb{T}^d} \liminf_{\delta \to 0} \frac{|\rho_\delta \boldsymbol{\omega}_\delta|^2}{\rho_\delta} \mathrm{d}x < \infty.$$

This implies that in $\{\rho = 0\}$, $j = 0$, then we define the limit velocity $\boldsymbol{\omega} := j/\rho$ in $\{\rho \neq 0\}$ and $\boldsymbol{\omega} := 0$ in $\{\rho = 0\}$, then $j = \rho \boldsymbol{\omega}$. By (2.2.25), for some function g, there exists a subsequence (not relabeled) such that

$$\sqrt{\rho_\delta \boldsymbol{\omega}_\delta} \rightharpoonup g \,\text{weakly}^* \text{ in } L^\infty(0,T;L^2(\mathbb{T}^d)). \tag{2.2.44}$$

Hence, since $\sqrt{\rho_\delta}$ converges strongly in $L^2(0,\mathbb{T};L^\infty(\mathbb{T}^d))$ to $\sqrt{\rho}$, we infer $\rho_\delta\boldsymbol{\omega}_\delta = \sqrt{\rho_\delta}(\sqrt{\rho_\delta}\boldsymbol{\omega}_\delta) \rightharpoonup \sqrt{\rho}g$ weakly in $L^2(0,T;L^2(\mathbb{T}^d))$, and $\sqrt{\rho}g = \rho\boldsymbol{\omega} = j$. In particular, $g = j/\sqrt{\rho}$ in $\{\rho \neq 0\}$.

Now, we start to perform the limit $\delta \to 0$ in the weak formulation (2.2.40) term by term and the strong convergences of (2.2.41) and (2.2.42) lead to

$$\begin{aligned}
\rho_\delta^2\boldsymbol{\omega}_\delta &\to \rho^2\boldsymbol{\omega} \text{ strongly in } L^1(0,T;L^q(\mathbb{T}^d)), q<3,\\
\rho_\delta\boldsymbol{\omega}_\delta \otimes \boldsymbol{\nabla}\rho_\delta &\to \rho\boldsymbol{\omega}\otimes\boldsymbol{\nabla}\rho \text{ strongly in } L^1(0,T;L^{3/2}(\mathbb{T}^d)),\\
\rho_\delta\boldsymbol{\omega}_\delta \otimes \rho_\delta\boldsymbol{\omega}_\delta &\to \rho\boldsymbol{\omega}\otimes\rho\boldsymbol{\omega} \text{ strongly in } L^1(0,T;L^{q/2}(\mathbb{T}^d)), q<3.
\end{aligned}$$

Furthermore,

$$\begin{aligned}
\boldsymbol{\nabla}\rho_\delta &\to \boldsymbol{\nabla}\rho \,\text{strongly in } L^2(0,T;L^m(\mathbb{T}^d)), m>3,\\
\sqrt{\rho_\delta} &\to \sqrt{\rho}\,\text{strongly in } L^\infty(0,T;L^r(\mathbb{T}^d)), r<6,\\
\Delta\sqrt{\rho_\delta} &\to \Delta\sqrt{\rho}\,\text{strongly in } L^2(0,T;L^2(\mathbb{T}^d)).
\end{aligned}$$

This implies

$$\Delta\sqrt{\rho_\delta}\sqrt{\rho_\delta}\boldsymbol{\nabla}\rho_\delta \rightharpoonup \Delta\sqrt{\rho}\sqrt{\rho}\boldsymbol{\nabla}\rho \,\text{weakly in } L^1(0,T;L^1(\mathbb{T}^d)). \tag{2.2.45}$$

Here we need the assumption $\gamma > 3(d=3)$, which leads to the compactness of ρ_δ in $W^{1,m}(\mathbb{T}^d)(m>3)$. Since $\boldsymbol{\nabla}(\rho_\delta\boldsymbol{\omega}\delta)$ converges weakly in $L^2(0,T;L^{3/2}(\mathbb{T}^d))$, and $\boldsymbol{\nabla}\rho_\delta$ converges strongly in $L^2(0,T;L^3(\mathbb{T}^d))$, we have

$$\boldsymbol{\nabla}(\rho_\delta\boldsymbol{\omega}_\delta)\cdot\boldsymbol{\nabla}\rho_\delta \rightharpoonup \boldsymbol{\nabla}(\rho\boldsymbol{\omega})\cdot\boldsymbol{\nabla}\rho \,\text{weakly in } L^1(0,T;L^1(\mathbb{T}^d)). \tag{2.2.46}$$

Since ρ_δ is bounded in $L^{4\gamma/3+1}(0,T;L^{4\gamma/3+1}(\mathbb{T}^d))(4\gamma/3+1>\gamma+1)$, and ρ_δ converges almost everywhere to ρ, we have

$$\rho_\delta^{\gamma+1} \to \rho^{\gamma+1}\,\text{strongly in } L^1(0,T;L^1(\mathbb{T}^d)).$$

In view of the estimate (2.2.26) for $\sqrt{\delta}\boldsymbol{\omega}_\delta$, we have

$$\begin{aligned}
&\delta\int_{\mathbb{T}^d}\nabla\boldsymbol{\omega}_\delta:(\rho_\delta\nabla\boldsymbol{\phi}+\nabla\rho_\delta\otimes\boldsymbol{\phi})\mathrm{d}x\\
&\quad\le\sqrt{\delta}\|\sqrt{\delta}\nabla\boldsymbol{\omega}_\delta\|_{L^2(0,T;L^2(\mathbb{T}^d))}(\|\rho_\delta\|_{L^2(0,T;L^\infty(\mathbb{T}^d))}\|\boldsymbol{\phi}\|_{L^2(0,T;H^1(\mathbb{T}^d))}\\
&\qquad+\|\rho_\delta\|_{L^2(0,T;W^{1,3}(\mathbb{T}^d))}\|\boldsymbol{\phi}|_{L^\infty(0,T;L^6(\mathbb{T}^d)))}\to 0,\\
&\delta\int_{\mathbb{T}^d}\rho_\delta\boldsymbol{\omega}_\delta\cdot\boldsymbol{\phi}\mathrm{d}x\\
&\quad\le\delta|_{\rho_\delta\boldsymbol{\omega}_\delta}|_{L^2(0,T;L^3(\mathbb{T}^d))}\|\boldsymbol{\phi}\|_{L^2(0,T;L^{3/2}(\mathbb{T}^d))}\to 0\quad\text{as }\delta\to 0.
\end{aligned}$$

It remains to show the convergence of $\rho_\delta^2\mathrm{div}(\boldsymbol{\omega}_\delta)\boldsymbol{\omega}_\delta$. Similarly, as in [2], we introduce the function $G_\alpha\in C^\infty([0,\infty))$, $\alpha>0$, satisfying $G_\alpha(x)=1(x\ge 2\alpha)$, $G_\alpha(x)=0(x\le\alpha)$, and $0\le G_\alpha\le 1$. Then we estimate the low-density part of $\rho_\delta^2\mathrm{div}(\boldsymbol{\omega}_\delta)\,\boldsymbol{\omega}_\delta$ by

$$\begin{aligned}
&\|(1-G_\alpha(\rho_\delta))\rho_\delta^2\mathrm{div}(\boldsymbol{\omega}_\delta)\boldsymbol{\omega}_\delta\|_{L^1(0,T;L^1(\mathbb{T}^d))}\\
&\quad\le\|(1-G_\alpha(\rho_\delta))\sqrt{\rho_\delta}\|_{L^\infty(0,T;L^\infty(\mathbb{T}^d))}\|\sqrt{\rho_\delta}\mathrm{div}\boldsymbol{\omega}_\delta\|_{L^2(0,T;L^2(\mathbb{T}^d))}\\
&\qquad\times\|\rho_\delta\boldsymbol{\omega}_\delta\|_{L^2(0,T;L^2(\mathbb{T}^d))}\\
&\quad\le C\|(1-G_\alpha(\rho_\delta))\sqrt{\rho_\delta}\|_{L^\infty(0,T;L^\infty(\mathbb{T}^d))}\le C\sqrt{\alpha},
\end{aligned}$$

where $C>0$ is independent of δ and α. We write

$$\begin{aligned}
G_\alpha(\rho_\delta)\rho_\delta\mathrm{div}\boldsymbol{\omega}_\delta&=\mathrm{div}(G_\alpha(\rho_\delta)\rho_\delta\boldsymbol{\omega}_\delta)\\
&\quad-\rho_\delta\boldsymbol{\omega}_\delta\otimes\nabla\rho_\delta\left(G'_\alpha(\rho_\delta)+\frac{G_\alpha(\rho_\delta)}{\rho_\delta}\right),
\end{aligned}\tag{2.2.47}$$

as $\delta\to 0$, the first term on the right hand side converges strongly in $L^1(0,T;H^{-1}(\mathbb{T}^d))$ to $\mathrm{div}(G_\alpha(\rho)\rho\boldsymbol{\omega})$ since $G_\alpha(\rho_\delta)$ converges strongly in $L^p(0,T;L^p(\mathbb{T}^d))(p<\infty)$ to $G_\alpha(\rho)$, and $\rho_\delta\boldsymbol{\omega}_\delta$ converges strongly in $L^2(0,T;L^q(\mathbb{T}^d))(q<3)$ to $\rho\boldsymbol{\omega}$. By virtue of (2.2.43) and (2.2.44), we obtain that $\rho_\delta\boldsymbol{\omega}_\delta\rightharpoonup\rho\boldsymbol{\omega}$ converges weakly* in $L^\infty(0,T;L^{2r/(r+2)}(\mathbb{T}^d))(r<6)$. Therefore, by (2.2.41), we have

$$\rho_\delta\boldsymbol{\omega}_\delta\otimes\nabla\rho_\delta\rightharpoonup\rho\boldsymbol{\omega}\otimes\nabla\rho\,\text{weakly in }L^2(0,T;L^\theta(\mathbb{T}^d)),$$

where $\theta=2mr/(2m+2r+mr)$. Choose $3<m\le 6\gamma/(\gamma+3)$ and $r<6$ such that $\theta>1$. Thus, together with the strong convergence of $G'_\alpha(\rho_\delta) + \frac{G_\alpha(\rho_\delta)}{\rho_\delta}$ in $L^p(0,T;L^p(\mathbb{T}^d))(p<\infty)$, as $\delta \to 0$, the limit of (2.2.47) is

$$G_\alpha(\rho)\rho\text{div}(\boldsymbol{\omega}) = \text{div}(G_\alpha(\rho)\rho\boldsymbol{\omega}) - \rho\boldsymbol{\omega}\otimes\boldsymbol{\nabla}\rho\left(G'\alpha(\rho) + \frac{G\alpha(\rho)}{\rho}\right)$$

in $L^1(0,T;(H^2(\mathbb{T}^d))^*)$. Since $G_\alpha(\rho_\delta)\rho_\delta\,\text{div}\,\boldsymbol{\omega}_\delta$ is bounded in $L^2(0,T;L^2(\mathbb{T}^d))$, we have

$$G_\alpha(\rho_\delta)\rho_\delta\text{div}\,\boldsymbol{\omega}_\delta \rightharpoonup G_\alpha(\rho)\rho\,\text{div}\,\boldsymbol{\omega} \text{ weakly in } L^2(0,T;L^2(\mathbb{T}^d)).$$

Moreover, in view of the strong convergence of $\rho_\delta\boldsymbol{\omega}_\delta$ in $L^2(0,T;L^q(\mathbb{T}^d))(q<3)$ to $\rho\boldsymbol{\omega}$, we have

$$G_\alpha(\rho_\delta)\rho_\delta\,\text{div}\,(\boldsymbol{\omega}_\delta)\rho_\delta\boldsymbol{\omega}_\delta \rightharpoonup G_\alpha(\rho)\rho^2\,\text{div}\,(\boldsymbol{\omega})\boldsymbol{\omega} \text{ weakly in } L^1(0,T;L^{q/2}(\mathbb{T}^d)).$$

For $\boldsymbol{\phi} \in L^\infty(0,T;L^\infty(\mathbb{T}^d))$,

$$\begin{aligned}
&\int_{\mathbb{T}^d}(\rho_\delta^2\,\text{div}\,(\boldsymbol{\omega}_\delta)\boldsymbol{\omega}_\delta - \rho^2\,\text{div}\,(\boldsymbol{\omega})\boldsymbol{\omega})\cdot\boldsymbol{\phi}\text{d}x\\
&\quad= \int_{\mathbb{T}^d}(G_\alpha(\rho_\delta)\rho_\delta^2\,\text{div}\,(\boldsymbol{\omega}_\delta)\boldsymbol{\omega}_\delta) - G_\alpha(\rho)\rho^2\,\text{div}\,(\boldsymbol{\omega})\boldsymbol{\omega})\cdot\boldsymbol{\phi}\text{d}x\\
&\qquad+ \int_{\mathbb{T}^d}(G_\alpha(\rho) - G_\alpha(\rho_\delta))\rho^2\,\text{div}(\boldsymbol{\omega})\boldsymbol{\omega}\cdot\boldsymbol{\phi}\text{d}x\\
&\qquad+ \int_{\mathbb{T}^d}(1 - G_\alpha(\rho_\delta))(\rho_\delta^2\,\text{div}\,(\boldsymbol{\omega}_\delta)\boldsymbol{\omega}_\delta - \rho^2\,\text{div}\,(\boldsymbol{\omega})\boldsymbol{\omega})\cdot\boldsymbol{\phi}\text{d}x. \quad (2.2.48)
\end{aligned}$$

For fixed $\alpha > 0$, as $\delta \to 0$, the first integral converges to zero, the last integral can be estimated by $C\sqrt{\alpha}$ uniformly in δ. For the second integral, since $G_\alpha(\rho_\delta) \to G_\alpha(\rho)$ converges strongly in $L^p(0,T;L^p(\mathbb{T}^d))(p<\infty)$, by virtue of the interpolation inequality, the bounds of $\rho\boldsymbol{\omega}$ in $L^\infty(0,T;L^{3/2}(\mathbb{T}^d))$ and $L^2\left(0,T;W^{1,3/2}(\mathbb{T}^d)\right)$ imply that $\rho\boldsymbol{\omega} \in L^{5/2}(0,T;L^{5/2}(\mathbb{T}^d))$. Therefore, since $\sqrt{\rho}\,\text{div}\,\boldsymbol{\omega} \in$

$L^2(0,T;L^2(\mathbb{T}^d))$ and $\sqrt{\rho}\in L^q(0,T;L^q(\mathbb{T}^d))(q=8\gamma/3+2)$,

$$\rho^2\mathrm{div}(\boldsymbol{\omega})\boldsymbol{\omega}=\sqrt{\rho}(\sqrt{\rho}\,\mathrm{div}\boldsymbol{\omega})\rho\boldsymbol{\omega}\in L^r(0,T;L^r(\mathbb{T}^d)),r=\frac{20_\gamma+15}{18_\gamma+21}>1,$$

as a consequence, the second integral converges to zero as $\delta\to 0$. Thus, (2.2.48) can be made arbitrarily small, then

$$\rho_\delta^2\,\mathrm{div}\,(\boldsymbol{\omega}\delta)\boldsymbol{\omega}\delta\rightharpoonup\rho^2\,\mathrm{div}\,(\boldsymbol{\omega})\boldsymbol{\omega}\text{ weakly in }L^1(0,\mathbb{T};L^1(\mathbb{T}^d)).$$

Hence, for smooth initial data, we have proved that $(\rho,\boldsymbol{\omega})$ solves (2.2.5) and (2.2.11). Let $(\rho_0,\boldsymbol{\omega}_0)$ be some finite-energy initial data, i.e. $\rho_0>0$, $E(\rho_0,\boldsymbol{\omega}_0)<\infty$. Let $(\rho_0^\delta,\boldsymbol{\omega}_0^\delta)$ be smooth approximations in $\mathbb{T}^3$ satisfying $\rho_0^\delta\geq\delta>0$, as $\delta\to 0$, $\sqrt{\rho_0^\delta}\to\sqrt{\rho_0}$ converges strongly in $H^1(\mathbb{T}^3)$, and $\sqrt{\rho_0^\delta\boldsymbol{\omega}_0^\delta}\to\sqrt{\rho_0}\boldsymbol{\omega}_0$ converges strongly in $L^2(\mathbb{T}^3)$. Thus, for the initial data $(\rho_0,\boldsymbol{\omega}_0)$ satisfying all the bounds, there exists a weak solution to (2.2.5)–(2.2.7). Then $(\rho,\boldsymbol{u}=\boldsymbol{\omega}-\nu\boldsymbol{\nabla}\log\rho)$ is a weak solution to (2.2.1)–(2.2.3).

2.3 Global existence of weak solutions to the compressible quantum hydrodynamic equations with cold pressure

In this section, we consider the following QHD equations with cold pressure:

$$\begin{cases}\partial_t\rho+\mathrm{div}(\rho\boldsymbol{u})=0,\\ \partial_t(\rho\boldsymbol{u})+\mathrm{div}(\rho\boldsymbol{u}\otimes\boldsymbol{u})+\boldsymbol{\nabla}(p(\rho)+p_c(\rho))\\ \qquad-2h^2\rho\boldsymbol{\nabla}\left(\frac{\Delta\sqrt{\rho}}{\sqrt{\rho}}\right)=2\nu\mathrm{div}(\rho D(\boldsymbol{u})),\\ \rho|_{t=0}=\rho_0,(\rho\boldsymbol{u})|_{t=0}=\rho_0\boldsymbol{u}_0,\end{cases}\tag{2.3.1}$$

where $p_c(\rho)$ is a singular continuous function satisfying the following growth condition

$$\lim_{p\to 0}p_c(\rho)=+\infty,$$

and called cold pressure. More precisely, we assume

$$p_c'(\rho)\begin{cases} c\rho^{-4k-1}, \rho \leq 1, k > 1, \\ \rho^{\gamma-1}, \rho > 1, \gamma > 1 \end{cases} \tag{2.3.2}$$

for some constant $c > 0$. We write the quantum term in a different form to pass to the limit

$$<\rho\boldsymbol{\nabla}\left(\frac{\Delta\sqrt{\rho}}{\sqrt{\rho}}\right), \boldsymbol{\phi}> = <\sqrt{\rho}\boldsymbol{\nabla}\sqrt{\rho}, \boldsymbol{\nabla}\mathrm{div}\boldsymbol{\phi}> + 2<\boldsymbol{\nabla}\sqrt{\rho}\otimes\boldsymbol{\nabla}\sqrt{\rho}, \boldsymbol{\nabla}\boldsymbol{\phi}>.$$

Then we can define the weak formulation of momentum equation in (2.3.1)

$$\begin{aligned} &\int_\Omega \rho_0\boldsymbol{u}_0\cdot\boldsymbol{\phi}(\cdot,0)\mathrm{d}x + \int_0^T\int_\Omega (\rho\boldsymbol{u}\cdot\partial_t\boldsymbol{\phi} + \rho(\boldsymbol{u}\otimes\boldsymbol{u}) : \boldsymbol{\nabla}\boldsymbol{\phi})\mathrm{d}x\mathrm{d}t \\ &\quad + \int_0^T\int_\Omega (p(\rho) + p_c(\rho))\,\mathrm{div}\,\boldsymbol{\phi}\mathrm{d}x\mathrm{d}t \\ &= \int_0^T\int_\Omega (2\varepsilon^2\sqrt{\rho}\boldsymbol{\nabla}\sqrt{\rho}\cdot\boldsymbol{\nabla}\mathrm{div}\boldsymbol{\phi} + 4\varepsilon^2\boldsymbol{\nabla}\sqrt{\rho}\otimes\boldsymbol{\nabla}\sqrt{\rho} : \boldsymbol{\nabla}\boldsymbol{\phi} \\ &\quad + 2\nu\rho D(\boldsymbol{u})) : \boldsymbol{\nabla}\boldsymbol{\phi})\mathrm{d}x\mathrm{d}t. \end{aligned}$$

In this section, for fixed ε, we first prove that there exists a global weak solution $(\rho^\varepsilon, \boldsymbol{u}^\varepsilon)$ to (2.3.1). Secondly, as ε tends to zero, and $(\rho^\varepsilon, \boldsymbol{u}^\varepsilon)$ tends to $(\rho^0, \boldsymbol{u}^0)$, $(\rho^0, \boldsymbol{u}^0)$ is a global weak solution to

$$\begin{cases} \partial_t\rho^0 + \mathrm{div}(\rho^0\boldsymbol{u}^0) = 0, \\ \partial_t(\rho^0\boldsymbol{u}^0) + \mathrm{div}(\rho^0\boldsymbol{u}^0\otimes\boldsymbol{u}^0) + \boldsymbol{\nabla}(p(\rho^0) + p_c(\rho^0)) \\ \quad = 2\nu\mathrm{div}(\rho^0 D(\boldsymbol{u}^0)), \\ \rho^0|_{t=0} = \rho_0, (\rho^0\boldsymbol{u}^0)|_{t=0} = \rho_0\boldsymbol{u}_0. \end{cases} \tag{2.3.3}$$

Let us firstly give the definitions of the weak solution.

Definition 2.3.1. $(\rho, \boldsymbol{u})$ is a weak solution to (2.3.1), if the continuity equation

$$\begin{cases} \partial_t\rho + \mathrm{div}\,(\sqrt{\rho}\sqrt{\rho}\boldsymbol{u}) = 0, \\ \rho(0,x) = \rho_0(x), \end{cases} \tag{2.3.4}$$

is satisfied in the sense of distributions and the weak formulation of the momentum equation

$$\begin{aligned}
&\int_\Omega \rho_0 \boldsymbol{u}_0 \cdot \boldsymbol{\phi}(\cdot,0)\mathrm{d}x + \int_0^T \int_\Omega (\rho \boldsymbol{u} \cdot \partial_t \boldsymbol{\phi} + \rho(\boldsymbol{u} \otimes \boldsymbol{u}) : \boldsymbol{\nabla}\boldsymbol{\phi})\mathrm{d}x\mathrm{d}t \\
&\quad + \int_0^T \int_\Omega (p(\rho) + p_c(\rho)) \operatorname{div} \boldsymbol{\phi} \mathrm{d}x\mathrm{d}t \\
&= \int_0^T \int_\Omega (2\varepsilon^2 \sqrt{\rho}\boldsymbol{\nabla}\sqrt{\rho} \cdot \boldsymbol{\nabla} \operatorname{div} \boldsymbol{\phi} + 4\varepsilon^2 \boldsymbol{\nabla}\sqrt{\rho} \otimes \boldsymbol{\nabla}\sqrt{\rho} : \boldsymbol{\nabla}\boldsymbol{\phi} \\
&\quad + 2\nu\rho D(\boldsymbol{u}) : \boldsymbol{\nabla}\boldsymbol{\phi})\mathrm{d}x\mathrm{d}t \qquad (2.3.5)
\end{aligned}$$

holds for any smooth, compactly supported test function $\boldsymbol{\phi}$ such that $\boldsymbol{\phi}(T,\cdot) = 0$.

Definition 2.3.2. $(\rho^0, \boldsymbol{u}^0)$ is a weak solution to (2.3.3), if the continuity equation

$$\begin{cases} \partial_t \rho^0 + \operatorname{div}(\sqrt{\rho^0}\sqrt{\rho^0}\boldsymbol{u}^0) = 0, \\ \rho^0(0,x) = \rho_0(x), \end{cases} \qquad (2.3.6)$$

is satisfied in the sense of distributions and the weak formulation of the momentum equation

$$\begin{aligned}
&\int_\Omega \rho_0 \boldsymbol{u}_0 \cdot \boldsymbol{\phi}(\cdot,0)\mathrm{d}x + \int_0^T \int_\Omega (\rho^0 \boldsymbol{u}^0 \cdot \partial_t \boldsymbol{\phi} + \rho^0(\boldsymbol{u}^0 \otimes \boldsymbol{u}^0) : \boldsymbol{\nabla}\boldsymbol{\phi})\mathrm{d}x\mathrm{d}t \\
&\quad + \int_0^T \int_\Omega (p(\rho^0) + p_c(\rho^0))\operatorname{div}\boldsymbol{\phi}\mathrm{d}x\mathrm{d}t \\
&= \int_0^T \int_\Omega 2\nu\rho^0 D(\boldsymbol{u}^0) : \boldsymbol{\nabla}\boldsymbol{\phi}\mathrm{d}x\mathrm{d}t \qquad (2.3.7)
\end{aligned}$$

holds for any smooth, compactly supported test function $\boldsymbol{\phi}$ such that $\boldsymbol{\phi}(T,\cdot) = 0$.

One can refer to [59]–[64] for more background. Now we introduce the energy of the system, which is given by the sum of the kinetic, internal and quantum energies:

$$E_\varepsilon(\rho, \boldsymbol{u}) = \frac{\rho}{2}|\boldsymbol{u}|^2 + H(\rho) + H_c(\rho) + 2\varepsilon^2|\nabla\sqrt{\rho}|^2, \tag{2.3.8}$$

where $H(\rho)$ and $H_c(\rho)$ are given by

$$H''(\rho) = \frac{p'(\rho)}{\rho} \text{ and } H_c''(\rho) = \frac{p_c'(\rho)}{\rho}.$$

Remark 2.3.1. If $p(\rho) = \rho^\gamma$ and $\gamma > 1$, then

$$H(\rho) = \frac{\rho^\gamma}{\gamma - 1}.$$

The existence of global weak solution in the sense of Definition 2.3.1 is given by the following theorem.

Theorem 2.3.1. *Let* $\nu > 0$, $\varepsilon > 0$, $1 \leq d \leq 3$, $T > 0$, $\gamma \geq 1$. *Let* $(\rho_0, \boldsymbol{u}_0)$ *such that* $\rho_0 \geq 0$ *and* $E_\varepsilon(\rho_0, \boldsymbol{u}_0) \leq \infty$. *Then there exists a weak solution to* (2.3.1) *in the sense of Definition* 2.3.1 *such that*

$$\begin{aligned}
&\rho \geq 0, \sqrt{\rho} \in L^\infty(0, T; H^1(\Omega)) \cap L^2(0, T; H^2(\Omega)),\\
&\rho \in L^\infty(0, T; L^\gamma(\Omega)), \rho^\gamma \in L^{5/3}(0, T; L^{5/3}(\Omega)),\\
&\sqrt{\rho}\boldsymbol{u} \in L^\infty(0, T; L^2(\Omega)), \rho|\nabla \boldsymbol{u}| \in L^2(0, T; L^2(\Omega)),\\
&\sqrt{\rho}|\nabla \boldsymbol{u}| \in L^2(0, T; L^2(\Omega)),\\
&\nabla\left(\frac{1}{\sqrt{\rho}}\right) \in L^2(0, T; L^2(\Omega)).
\end{aligned}$$

The existence of global weak solution in the sense of Definition 2.3.2 is given by the following theorem.

Theorem 2.3.2. *Let* $1 \leq d \leq 3, T > 0, 0 < \varepsilon < \nu, \gamma \geq 1$. *Let* $(\rho_0, \boldsymbol{u}_0)$ *such that* $\rho_0 \geq 0$ and $E_\varepsilon(\rho_0, \boldsymbol{u}_0) \leq \infty$. *Then for the solution*

$(\rho^\varepsilon, \boldsymbol{u}^\varepsilon) \to$ (2.3.1), *as* $\varepsilon \to 0$:

$$\begin{aligned}
&\rho^\varepsilon \to \rho^0 \ \textit{strongly in}\ L^2(0,T;L^\infty(\Omega)),\\
&\sqrt{\rho^\varepsilon} \to \sqrt{\rho^0} \ \textit{strongly in}\ L^2(0,T;H^2(\Omega)),\\
&\sqrt{\rho^\varepsilon} \to \sqrt{\rho^0} \ \textit{strongly in}\ L^2(0,T;H^1(\Omega)),\\
&\frac{1}{\sqrt{\rho^\varepsilon}} \to \frac{1}{\sqrt{\rho^0}} \ \textit{almost everywhere},\\
&\sqrt{\rho^\varepsilon}\boldsymbol{u}^\varepsilon \to \sqrt{\rho^0}\boldsymbol{u}^0 \ \textit{strongly in}\ L^2(0,T;L^2(\Omega)),\\
&\boldsymbol{u}^\varepsilon \rightharpoonup \boldsymbol{u}^0 \ \textit{weakly in}\ L^p(0,T;L^{q^*}(\Omega)),
\end{aligned}$$

with $p = 8k/(4k+1), q^* = 24k/(12k+1)$, *and* $(\rho^0, \boldsymbol{u}^0)$ *is the solution to* (2.3.3).

Remark 2.3.2. Noting that the results are available for $1 \le d \le 3$, in the proofs we focus on the three-dimensional case. In fact, the most interesting case in terms of difficulty is the three-dimensional case.

2.3.1 *A priori estimate*

We now start to establish *a priori* estimate in order to prove the existence and convergence results. After formal calculation, it's easy to see that

$$\frac{\mathrm{d}}{\mathrm{d}t}\int_\Omega E_\varepsilon(\rho, \boldsymbol{u})\mathrm{d}x + \nu\int_\Omega \rho|D(\boldsymbol{u})|^2\mathrm{d}x = 0, \tag{2.3.9}$$

which leads to the following estimates.

Lemma 2.3.1. *Under the assumptions of Theorem* 2.3.1, *there exists a constant* C *independent of* ε, *such that*

$$\|\sqrt{\rho}\boldsymbol{u}\|_{L^\infty(0,T;L^2(\Omega))} \le C, \tag{2.3.10}$$

$$\|\rho^\gamma\|_{L^\infty(0,T;L^1(\Omega))} \le C, \tag{2.3.11}$$

$$\|\sqrt{\rho}D(\boldsymbol{u})\|_{L^2(0,T;L^2(\Omega))} \le C. \tag{2.3.12}$$

The following entropy inequality can be shown, which allows us to deduce high regularity for density in order to prove the convergence.

Proposition 2.3.1. *Under the assumptions of Theorem 2.3.1, there exists a constant C independent of ε, such that*

$$\frac{\mathrm{d}}{\mathrm{d}t}\int_{\Omega}\left(\frac{\rho}{2}|\boldsymbol{u}+\nu\boldsymbol{\nabla}\log\rho|^2+H(\rho)+H_c(\rho)+(2\varepsilon^2+4\nu^2)|\boldsymbol{\nabla}\sqrt{\rho}|^2\right)\mathrm{d}x$$
$$+\nu\int_{\Omega}(H''|\boldsymbol{\nabla}\rho|^2+H_c''|\boldsymbol{\nabla}\rho|^2+\varepsilon^2\rho|\boldsymbol{\nabla}^2\log\rho|^2+2\rho|\boldsymbol{\nabla}\boldsymbol{u}|^2)\mathrm{d}x=0. \tag{2.3.13}$$

Proof. Since

$$\partial_t(\rho\boldsymbol{\nabla}\log\rho)+\mathrm{div}(\rho(\boldsymbol{\nabla}\boldsymbol{u})^{\mathrm{T}})+\mathrm{div}(\rho\boldsymbol{u}\otimes\boldsymbol{\nabla}\log\rho)=0,$$

multiplying the above equation with ν, and adding below equation

$$\partial_t(\rho\boldsymbol{u})+\mathrm{div}(\rho\boldsymbol{u}\otimes\boldsymbol{u})+\boldsymbol{\nabla}(p(\rho)+pc(\rho))-2\varepsilon^2\rho\boldsymbol{\nabla}\left(\frac{\Delta\sqrt{\rho}}{\sqrt{\rho}}\right)$$
$$=2\nu\mathrm{div}(\rho D(\boldsymbol{u}))$$

imply that

$$\partial_t(\rho(\boldsymbol{u}+\nu\boldsymbol{\nabla}\log\rho))+\mathrm{div}(\rho\boldsymbol{u}\otimes(\boldsymbol{u}+\nu\boldsymbol{\nabla}\log\rho))$$
$$+\boldsymbol{\nabla}(p(\rho)+pc(\rho))-2\varepsilon^2\rho\boldsymbol{\nabla}\left(\frac{\Delta\sqrt{\rho}}{\sqrt{\rho}}\right)=\nu\mathrm{div}(\rho\boldsymbol{\nabla}\boldsymbol{u}).$$

Multiplying the above equation with $(\boldsymbol{u}+\nu\boldsymbol{\nabla}\log\rho)$, and integrating over Ω yield

$$\int_{\Omega}\partial_t(\rho(\boldsymbol{u}+\nu\boldsymbol{\nabla}\log\rho))\cdot(\boldsymbol{u}+\nu\boldsymbol{\nabla}\log\rho)\mathrm{d}x$$
$$+\int_{\Omega}\mathrm{div}(\rho\boldsymbol{u}\otimes(\boldsymbol{u}+\nu\boldsymbol{\nabla}\log\rho))$$
$$\cdot(\boldsymbol{u}+\nu\boldsymbol{\nabla}\log\rho)\mathrm{d}x+\int_{\Omega}\boldsymbol{\nabla}(p(\rho)+pc(\rho))$$

$$\cdot (\boldsymbol{u} + \nu \boldsymbol{\nabla} \log \rho) \mathrm{d}x - 2\varepsilon^2 \int_\Omega \rho \boldsymbol{\nabla} \left(\frac{\Delta \sqrt{\rho}}{\sqrt{\rho}} \right)$$

$$\cdot (\boldsymbol{u} + \nu \boldsymbol{\nabla} \mathrm{log} \rho) \mathrm{d}x = \nu \int_\Omega \mathrm{div}(\rho \boldsymbol{\nabla} \boldsymbol{u}) \cdot (\boldsymbol{u} + \nu \boldsymbol{\nabla} \log \rho) \mathrm{d}x. \tag{2.3.14}$$

Moreover,

$$\int_\Omega \partial_t(\rho(\boldsymbol{u} + \nu \boldsymbol{\nabla} \log \rho)) \cdot (\boldsymbol{u} + \nu \boldsymbol{\nabla} \mathrm{log} \rho) \mathrm{d}x$$

$$= \frac{\mathrm{d}}{\mathrm{d}t} \left(\int_\Omega \frac{\rho}{2} |\boldsymbol{u} + \nu \boldsymbol{\nabla}_{\mathrm{log}} \rho|^2 \right) \mathrm{d}x$$

$$- \frac{1}{2} \int_\Omega \mathrm{div}(\rho \boldsymbol{u}) |\boldsymbol{u} + \nu \boldsymbol{\nabla} \mathrm{log} \rho|^2 \mathrm{d}x.$$

$$\int_\Omega \boldsymbol{\nabla}(p(\rho) + pc(\rho)) \cdot (\boldsymbol{u} + \nu \boldsymbol{\nabla} \log \rho) dx = \int_\Omega \partial_t (H(\rho) + H_c(\rho)) \mathrm{d}x$$

$$+ \nu \int_\Omega H''(\rho) |\boldsymbol{\nabla} \rho|^2 \mathrm{d}x + \nu \int_\Omega H_c''(\rho) |\boldsymbol{\nabla} \rho|^2 \mathrm{d}x.$$

$$- 2\varepsilon^2 \int_\Omega \rho \boldsymbol{\nabla} \left(\frac{\Delta \sqrt{\rho}}{\sqrt{\rho}} \right) \cdot (\boldsymbol{u} + \nu \boldsymbol{\nabla} \log \rho) \mathrm{d}x$$

$$= 2\varepsilon^2 \int_\Omega \partial_t ((\boldsymbol{\nabla} \sqrt{\rho})^2) \mathrm{d}x$$

$$+ \nu \varepsilon^2 \int_\Omega \rho |\boldsymbol{\nabla}^2 \log \rho|^2 \mathrm{d}x.$$

$$\int_\Omega \mathrm{div}(\rho \boldsymbol{u} \otimes (\boldsymbol{u} + \nu \boldsymbol{\nabla} \log \rho)) \cdot (\boldsymbol{u} + \nu \boldsymbol{\nabla} \log \rho) \mathrm{d}x$$

$$= \int_\Omega \frac{1}{2} \mathrm{div}(\rho \boldsymbol{u}) |\boldsymbol{u} + \nu \boldsymbol{\nabla} \log \rho|^2 \mathrm{d}x.$$

Finally, using integration by parts,

$$\nu \int_\Omega \mathrm{div}(\rho \boldsymbol{\nabla} \boldsymbol{u}) \cdot (\boldsymbol{u} + \nu \boldsymbol{\nabla} \log \rho) \mathrm{d}x$$

$$= -\nu \int_\Omega \rho |\boldsymbol{\nabla} \boldsymbol{u}|^2 \mathrm{d}x + \nu^2 \int_\Omega \boldsymbol{u} \mathrm{div}(\rho \boldsymbol{\nabla}^2 \log \rho) \mathrm{d}x$$

$$= -\nu \int_\Omega \rho|\nabla \boldsymbol{u}|^2 \mathrm{d}x + 2\nu^2 \int_\Omega \rho \boldsymbol{u} \nabla \left(\frac{\Delta\sqrt{\rho}}{\sqrt{\rho}}\right) \mathrm{d}x$$

$$= -\nu \int_\Omega \rho|\nabla \boldsymbol{u}|^2 \mathrm{d}x - 2\nu^2 \int_\Omega \mathrm{div}(\rho \boldsymbol{u}) \left(\frac{\Delta\sqrt{\rho}}{\sqrt{\rho}}\right) \mathrm{d}x$$

$$= -\nu \int_\Omega \rho|\nabla \boldsymbol{u}|^2 \mathrm{d}x + 2\nu^2 \int_\Omega \partial_t \rho \left(\frac{\Delta\sqrt{\rho}}{\sqrt{\rho}}\right) \mathrm{d}x$$

$$= -\nu \int_\Omega \rho|\nabla \boldsymbol{u}|^2 \mathrm{d}x - 2\nu^2 \int_\Omega \partial_t(|\nabla\sqrt{\rho}|^2) \mathrm{d}x.$$

Using all the above inequalities in (2.3.14) leads to Proposition 2.3.1.

Directly from Proposition 2.3.1, we deduce the following estimates: □

Lemma 2.3.2. *Under the assumptions of Theorem* 2.3.1, *there exists a constant C independent of ε, such that*

$$\|\sqrt{\rho}(\boldsymbol{u} + \nu\nabla \log\rho)_{L^\infty(0,T;L^2(\Omega))} \le C, \tag{2.3.15}$$

$$\|\nabla\sqrt{\rho}\|_{L^\infty(0,T;L^2(\Omega))} \le C, \tag{2.3.16}$$

$$\|\nabla \rho^{\gamma/2}\|_{L2(0,T;L^2(\Omega))} \le C, \tag{2.3.17}$$

$$\|\sqrt{\rho}\nabla \boldsymbol{u}\|_{L2(0,T;L^2(\Omega))} \le C, \tag{2.3.18}$$

$$\varepsilon\|\sqrt{\rho}\nabla^2 \log\rho\|_{L^2(0,T;L^2(\Omega))} \le C, \tag{2.3.19}$$

$$\varepsilon\|\nabla^2\sqrt{\rho}\|_{L^2(0,T;L^2(\Omega))} \le C. \tag{2.3.20}$$

Remark 2.3.3. In contrast to (2.3.9), B-D entropy equality (2.3.13) allows us to obtain (2.3.16) with C independent of ε.

Proof. Estimates (2.3.15), (2.3.16), (2.3.18) and (2.3.19) are direct consequence from (2.3.13), and estimate (2.3.17) is derived from the following identity

$$H''(\rho)|\nabla\rho|^2 = \frac{p'(\rho)|\nabla\rho|^2}{\rho} = \gamma\rho^{\gamma-2}|\nabla\rho|^2, \nabla\rho^{\gamma/2} = \frac{\gamma}{2}\rho^{\gamma/2-1}\nabla\rho.$$

(2.3.20) is obtained by using the following identity

$$\int_0^T \int_\Omega \rho|\nabla^2 \log\rho|^2 \mathrm{d}x\mathrm{d}t \ge \int_0^T \int_\Omega |\nabla^2\sqrt{\rho}|^2 \mathrm{d}x\mathrm{d}t$$

and (2.3.19).

Next let us show some estimates concerning the two different pressure. □

Lemma 2.3.3. *Under the assumptions of Theorem* 2.3.1, *and in three-dimensional case, there exists a constant C independent of ε, such that*

$$\|\rho^\gamma\|_{L^{5/3}(0,T;L^{5/3}(\Omega))} \le C, \tag{2.3.21}$$

$$\|p_c(\rho)\|_{L^{5/3}(0,T;L^{5/3}(\Omega))} \le C. \tag{2.3.22}$$

Proof. Using (2.3.11), (2.3.17) and Sobolev embedding, we have

$$\|\rho^{\gamma/2}\|_{L^2(0,T;L^6(\Omega))} \le C,$$

which can also be written as

$$\|\rho^\gamma\|_{L^1(0,T;L^3(\Omega))} \le C.$$

Using interpolation, the last inequality and (2.3.11) imply that

$$\|\rho^\gamma\|_{L^{5/3}(0,T;L^{5/3}(\Omega))} \le C.$$

Recall that

$$p_c = \frac{c}{-4k}\rho^{-4k}, \rho \le 1, k > 1.$$

Let ζ be a smooth function such that

$$\zeta(y) = y, y \le 1/2; \zeta(y) = 0, y > 1.$$

By (2.3.9) and (2.3.13), we have

$$\int_0^T \int_\Omega |\nabla\zeta(\rho)^{-2k}|^2 \mathrm{d}x\mathrm{d}t \le C.$$

This means that $\nabla\zeta(\rho)^{-2k} \in L^2(0,T;L^2(\Omega))$ and close to vacuum,

$$\operatorname*{sub}_t \int_\Omega \rho^{-4k} \mathrm{d}x \le C. \tag{2.3.23}$$

This gives $\nabla\zeta(\rho)^{-2k} \in L^2(0,T;L^2(\Omega))$, thus $\zeta(\rho)^{-2k} \in L^2(0,T;H^1(\Omega))$. Using Sobolev embedding,

$$\|\zeta(\rho)^{-2k}\|_{L^2(0,T;L^6(\Omega))} \leq C. \tag{2.3.24}$$

Finally, interpolation and (2.3.23), (2.3.24) allow us to deduce (2.3.22). This finishes the proof. □

Lemma 2.3.4. *Under the assumptions of Theorem 2.3.1, and in three-dimensional case, there exists a constant C independent of ε, such that*

$$\|\nabla \boldsymbol{u}\|_{L^p(0,T;L^q(\Omega))} \leq C, \tag{2.3.25}$$

with $p = 8k/(4k+1), q = 24k/(12k+1)$.

Proof. Let us write

$$\nabla \boldsymbol{u} = \frac{1}{\sqrt{\rho}}\sqrt{\rho}\nabla \boldsymbol{u},$$

using (2.3.18), we have that $\sqrt{\rho}\nabla\ \boldsymbol{u}$ is uniformly bounded in $L^2(0,T;L^2(\Omega))$. Then it remains to obtain an estimate for $1/\sqrt{\rho}$. By (2.3.24), we have

$$\|\zeta(\rho)^{-2k}\|^2_{L^2(0,T;L^6(\Omega))} = \int_0^T \left(\int_\Omega \left(\frac{1}{\sqrt{\rho}}\right)^{24k} \mathrm{d}x\right)^{8k/24k} \mathrm{d}t,$$

then

$$\|1/\sqrt{\rho}\|_{L^{8k}(0,T;L^{24k}(\Omega))} \leq C. \tag{2.3.26}$$

Similarly, since $\nabla \boldsymbol{u} = (1/\sqrt{\rho})\sqrt{\rho}\nabla \boldsymbol{u}$, using (2.3.18) and (2.3.26), we finish the proof. □

Lemma 2.3.5. *Under the assumptions of Theorem 2.3.1, there exists a constant C independent of ε, such that*

$$\|\boldsymbol{u}\|_{L^p(0,T;L^{q^*}(\Omega))} \leq C, \tag{2.3.27}$$

with $p = 8k/(4k+1), q^* = 24k/(12k+1)$.

Lemma 2.3.6. *Under the assumptions of Theorem 2.3.1, there exists a constant C independent of ε, such that*

$$\|\nabla(1/\sqrt{\rho})\|_{L^2(0,T;L^2(\Omega))} \leq C. \tag{2.3.28}$$

Proof. From (2.3.16) and $\boldsymbol{\nabla}\sqrt{\rho} \in L^2(\Omega)$,

$$\boldsymbol{\nabla}(1/\sqrt{\rho}) = -\frac{1}{2}\frac{\boldsymbol{\nabla}\rho}{\rho^{3/2}} = -\frac{\boldsymbol{\nabla}\sqrt{\rho}}{\rho},$$

then, if $\rho > 1$, $\frac{1}{\rho} < 1$ and $\boldsymbol{\nabla}(1/\sqrt{\rho}) \in L^2(\Omega)$.

For the case of $\rho \leq 1$, from Proposition 2.3.1, we have $\int_0^T \int_\Omega H_c''(\rho)|\boldsymbol{\nabla}\rho|^2 \mathrm{d}x\mathrm{d}t < +\infty$, where $H_c''(\rho) = \frac{p_c'(\rho)}{\rho}$. For $\rho \leq 1, p_c'(\rho) = c\rho^{-4k-1}, H_c''(\rho) = c\rho^{-4k-2}$,

$$H_c''(\rho)|\boldsymbol{\nabla}\rho|^2 = c\left|\frac{\boldsymbol{\nabla}\rho}{\rho^{2k+1}}\right|^2 = \frac{c}{4k^2}\left|\boldsymbol{\nabla}\left(\frac{1}{\rho^{2k}}\right)\right|^2.$$

Making the connection between $\boldsymbol{\nabla}(\frac{1}{\rho^{2k}})$ *and* $\boldsymbol{\nabla}(\frac{1}{\sqrt{\rho}})$,

$$\begin{aligned}
\boldsymbol{\nabla}\left(\frac{1}{\sqrt{\rho}}\right) &= \boldsymbol{\nabla}\left(\frac{1}{\rho^{2k}}\rho^{2k-1/2}\right) \\
&= \rho^{2k-1/2}\boldsymbol{\nabla}\left(\frac{1}{\rho^{2k}}\right) + \frac{1}{\rho^{2k}}\boldsymbol{\nabla}(\rho^{2k-1/2}) \\
&= \rho^{2k-1/2}\boldsymbol{\nabla}\left(\frac{1}{\rho^{2k}}\right) + (2k-1/2)\rho^{-2k}\boldsymbol{\nabla}(\rho)\rho^{2k-3/2} \\
&= \rho^{2k-1/2}\boldsymbol{\nabla}\left(\frac{1}{\rho^{2k}}\right) - 2(2k-1/2)\boldsymbol{\nabla}\left(\frac{1}{\sqrt{\rho}}\right),
\end{aligned}$$

we have

$$4k\boldsymbol{\nabla}\left(\frac{1}{\sqrt{\rho}}\right) = \rho^{2k-1/2}\boldsymbol{\nabla}\left(\frac{1}{\rho^{2k}}\right),$$

and

$$\left|\boldsymbol{\nabla}\left(\frac{1}{\sqrt{\rho}}\right)\right|^2 = \frac{\rho^{4k-1}}{16k^2}\left|\boldsymbol{\nabla}\left(\frac{1}{\rho^{2k}}\right)\right|^2 = \frac{1}{4c}\rho^{4k-1}H_c''(\rho)|\boldsymbol{\nabla}_\rho|^2.$$

Since $\rho \leq 1$, *and* $\int_0^T \int_\Omega H_c''(\rho)|\boldsymbol{\nabla}_\rho|^2 \mathrm{d}x < +\infty$, we have $\boldsymbol{\nabla}(\frac{1}{\sqrt{\rho}}) \in L^2(0,T;L^2(\Omega))$.

Using the previous lemma, we establish the following proposition. □

Proposition 2.3.2. *Under the assumptions of Theorem 2.3.1, there exists a constant C independent of ε, such that*

$$\|\sqrt{\rho}\boldsymbol{u}\|_{L^{p'}(0,T;L^{q'}(\Omega))} \leq C \tag{2.3.29}$$

with $p', q' > 2$.

Proof. Let $r > 0$, which will be chosen later. We write

$$\sqrt{\rho}\boldsymbol{u} = (\sqrt{\rho}\boldsymbol{u})^{2r}\boldsymbol{u}^{1-2r}\rho^{1/2-r}.$$

Using (2.3.11),

$$\|\rho^{1/2-r}\|_{L^\infty(0,T;L^{\gamma/(1/2-r)}(\Omega))} \leq C$$

with C, a constant independent of ε. Using (2.3.10), (2.3.27) and (2.3.28), choosing

$$\frac{1}{p'} = \frac{1-2r}{p} \text{ and } \frac{1}{q'} = \frac{2r}{2} + \frac{1-2r}{q^*} + \frac{1/2-r}{\gamma}$$

with $r = 2/3$, this finishes the proof. □

2.3.2 *Global existence of weak solutions*

Now, we prove the global existence of the weak solution to (2.3.1), i.e. we prove Theorem 2.3.1. The proof can be split into two parts: in the first part, we construct an approximate solution; the second part is devoted to prove the stability of the approximate solution. This is the classical way to prove the existence of the solution for Navier–Stokes equations.

Construction of an approximate solution

By the use of the so-called effective velocity,

$$\boldsymbol{w} = \boldsymbol{u} + \nu \boldsymbol{\nabla} \log \rho.$$

We transform (2.3.1) into another form

$$\begin{cases} \partial_t \rho + \operatorname{div}(\rho \boldsymbol{w}) = \nu \Delta \rho, \\ \partial_t(\rho \boldsymbol{w}) + \operatorname{div}(\rho \boldsymbol{w} \otimes \boldsymbol{w}) + \boldsymbol{\nabla}(p(\rho) + pc(\rho)) \\ \qquad - 2\varepsilon_0 \rho \boldsymbol{\nabla} \left(\dfrac{\Delta \sqrt{\rho}}{\sqrt{\rho}} \right) = \nu \Delta(\rho \boldsymbol{w}), \\ \rho|_{t=0} = \rho_0, (\rho \boldsymbol{w})|_{t=0} = \rho_0 \boldsymbol{w}_0, \end{cases} \tag{2.3.30}$$

with $\boldsymbol{w}_0 = \boldsymbol{u}_0 + \nu \boldsymbol{\nabla} \log \rho_0, \varepsilon_0 = h^2 - \nu^2$. Next, we give the definition of the weak solution to (2.3.30).

Definition 2.3.3. We say that $(\rho, \boldsymbol{w})$ is a weak solution to (2.3.30), if the equation

$$\begin{cases} \partial_t \rho + \operatorname{div}(\sqrt{\rho}\sqrt{\rho} \boldsymbol{w}) = \nu \Delta \rho, \\ \rho(0, x) = \rho_0(x), \end{cases} \tag{2.3.31}$$

is satisfied in the sense of distributions, and the weak formulation of the momentum equation

$$\begin{aligned} &\int_\Omega \rho_0 \boldsymbol{w}_0 \cdot \boldsymbol{\phi}(\cdot, 0) \mathrm{d}x + \int_0^T \int_\Omega [\rho \boldsymbol{w} \cdot \partial_t \boldsymbol{\phi} + \rho(\boldsymbol{w} \otimes \boldsymbol{w}) : \boldsymbol{\nabla} \boldsymbol{\phi}] \mathrm{d}x \mathrm{d}t \\ &\quad + \int_0^T \int_\Omega (p(\rho) + pc(\rho)) \operatorname{div} \boldsymbol{\phi} \mathrm{d}x \mathrm{d}t \\ &= \int_0^T \int_\Omega (2\varepsilon_0 \frac{\Delta \sqrt{\rho}}{\sqrt{\rho}} \operatorname{div}(\rho \boldsymbol{\phi}) + \nu \boldsymbol{\nabla}(\rho \boldsymbol{w}) : \boldsymbol{\nabla} \boldsymbol{\phi}) \mathrm{d}x \mathrm{d}t \end{aligned} \tag{2.3.32}$$

holds for any smooth, compactly supported test function $\boldsymbol{\phi}$ such that $\boldsymbol{\phi}(T, \cdot) = 0$.

Now, the goal is to prove the existence of the solution to (2.3.30).

Global existence of approximate solutions

Let $T > 0$, and $(\boldsymbol{e}_p)$ is an orthonormal basis of $L^2(\Omega)$, noting that $(\boldsymbol{e}_p)$ is also an orthogonal basis of $H^1(\Omega)$. Introduce the finite space $X_N = \text{span}\{\boldsymbol{e}_1, \boldsymbol{e}_2, \ldots, \boldsymbol{e}_N\}$, $N \in \mathbb{N}$. Let $(\rho_0, \boldsymbol{w}_0) \in C^\infty(\Omega)^2$ be some initial data such that $\rho_0 \geq \delta > 0$, for all $x \in \Omega$ and for some $\delta > 0$. Let $\boldsymbol{v} \in C^0([0,T]; X_N)$ be a given velocity, and $\boldsymbol{v}$ can be written for $x \in \Omega$ and $t \in [0,T]$

$$\boldsymbol{v}(x,t) = \sum_{i=1}^{N} \lambda_i(t) \boldsymbol{e}_i(x),$$

for some functions λ_i. The norm of $\boldsymbol{v}$ in $C^0([0,T]; C^n(\Omega))$ reads as follows

$$\|\boldsymbol{v}\|_{C^0([0,T];C^n(\Omega))} = \max_{t\in[0,T]} \sum_{i=1}^{N} |\lambda_i(t)|.$$

Then $\boldsymbol{v}$ is bounded in $C^0([0,T]; C^n(\Omega))$, and there exists a constant depending on n such that

$$\|\boldsymbol{v}\|_{C^0([0,T];C^n(\Omega))} \leq C \|\boldsymbol{v}\|_{C^0([0,T];L^2(\Omega))}. \tag{2.3.33}$$

Let $\rho \in C^1([0,T]; C^3(\Omega))$ be a classical solution to the following equation

$$\begin{cases} \partial_t \rho + \text{div}(\rho \boldsymbol{w}) = \nu \Delta \rho, x \in \Omega, t \in [0,T], \\ \rho(0,x) = \rho_0(x), x \in \Omega. \end{cases} \tag{2.3.34}$$

By the maximum principle, the assumption $\rho_0 \geq \delta > 0$ and (2.3.33), for all $(x,t) \in \Omega \times [0,T]$, we have

$$0 < \underline{\rho}(c) \leq \rho(x,t) \leq \overline{\rho}(c).$$

Moreover, for given ρ_N, we look for a function $\boldsymbol{w}_N \in C^0([0,T]; X_N)$ such that

$$-\int_\Omega \rho_0 \boldsymbol{w}_0 \cdot \boldsymbol{\phi}(\cdot, 0) \mathrm{d}x = \int_0^T \int_\Omega (\rho_N \boldsymbol{w}_N \cdot \partial_t \boldsymbol{\phi} + \rho_N (\boldsymbol{v} \otimes \boldsymbol{w}_N) : \boldsymbol{\nabla} \boldsymbol{\phi}$$

$$+ (p(\rho_N) + p_c(\rho_N))\mathrm{div}\boldsymbol{\phi} - 2\varepsilon_0 \frac{\Delta\sqrt{\rho_N}}{\sqrt{\rho_N}}\mathrm{div}$$

$$\times (\rho_N \boldsymbol{\phi}) - \nu \nabla(\rho_N \boldsymbol{w}_N) : \nabla\boldsymbol{\phi}$$

$$- \delta\left(\nabla \boldsymbol{w}_N : \nabla\boldsymbol{\phi} + \boldsymbol{w}_N \cdot \boldsymbol{\phi})\right)\mathrm{d}x\mathrm{d}t. \quad (2.3.35)$$

Using a Banach fixed-point theorem, there exists a unique local-in-time solution $\rho_N, \boldsymbol{w}_N$ to (2.3.34) and (2.3.35) with $\rho_N \in C^1([0,T'];C^3(\Omega)), \boldsymbol{w}_N \in C^0([0,T];X_N)$, for $T' \leq T$. In order to prove the global existence of the approximate solution, we need the following energy estimate.

Lemma 2.3.7. *Let $T' \leq T, \rho_N \in C^1([0,T'];C^3(\Omega))$, and $\boldsymbol{w}_N \in C^0([0,T];X_N)$ be a local-in-time solution to (2.3.34) and (2.3.35), with $\rho = \rho_N, \boldsymbol{v} = \boldsymbol{w}_N$, then*

$$\frac{\mathrm{d}E_{\varepsilon 0}}{\mathrm{d}t}(\rho_N, \boldsymbol{w}_N) + \nu \int_\Omega (\rho_N |\nabla \boldsymbol{w}_N|^2 + H''|\nabla \rho_N|^2 + H_c''|\nabla\rho_N|^2)\mathrm{d}x$$

$$+ \varepsilon_0 \nu \int_\Omega \rho_N |\nabla^2 \log \rho_N|^2 \mathrm{d}x + \delta \int_\Omega (|\nabla \boldsymbol{w}_N)|^2 + |\boldsymbol{w}_N|^2)\, \mathrm{d}x = 0, \quad (2.3.36)$$

where

$$E_{\varepsilon_0}(\rho_N, \boldsymbol{w}_N) = \int_\Omega \left(\frac{\rho_N}{2}|\boldsymbol{w}_N|^2 + H(\rho_N) + H_c(\rho_N) + 2\varepsilon_0^2 |\nabla\sqrt{\rho}|^2\right)\mathrm{d}x.$$

Limits $N \to \infty$ and $\delta \to 0$.

First, for fixed $\delta > 0$, we perform the limit $N \to \infty$. This is achieved by the use of regularities of the approximate solution and Aubin–Simon's lemma. Here, the use of a cold pressure avoids the hypothesis $\gamma > 3$ in the previous section.

Now, we start to prove the following proposition.

Proposition 2.3.3. *Under the assumptions of Theorem* 2.3.1, *for fixed* ε, *there exists a subsequence (not relabeled), as* $N \to \infty$,

$$
\begin{aligned}
&\sqrt{\rho_N} \to \sqrt{\rho_\delta} \; strongly\, in\, L^2(0,T;H^1(\Omega)),\\
&p(\rho_N) \to p(\rho_\delta) \; strongly\, in\, L^1(0,T;L^1(\Omega)),\\
&p_c(\rho_N) \to p_c(\rho_\delta) \; strongly\, in\, L^1(0,T;L^1(\Omega)),\\
&\frac{1}{\sqrt{\rho_N}} \to \frac{1}{\sqrt{\rho_\delta}} \; almost\, everywhere,\\
&\sqrt{\rho_N}\boldsymbol{w}_N \to \sqrt{\rho_\delta}\boldsymbol{w}_\delta \text{ strongly in } L^2\left(0,T\,;L^2(\Omega)\right),\\
&\boldsymbol{\nabla}\boldsymbol{w}_N \rightharpoonup \boldsymbol{\nabla}\boldsymbol{w}_\delta \; weakly\, in\, L^p(0,T;L^q(\Omega)),\\
&\boldsymbol{w}_N \rightharpoonup \boldsymbol{w}_\delta \; weakly\, in\, L^p(0,T;L^{q*}(\Omega)),
\end{aligned}
$$

with p, q, q^* *given in Lemma* 2.3.4 *and Lemma* 2.3.5.

Proof. Firstly, using (2.3.36), we are able to prove estimates (2.3.10), (2.3.16)–(2.3.27) with $\rho = \rho_N$ and $\boldsymbol{u} = \boldsymbol{w}_N$, where constants C are all independent of N and δ, but can depend on ε.

Using (2.3.10), (2.3.16), (2.3.18), (2.3.20), (2.3.26), and rewriting equation (2.3.31) as

$$
\partial t(\sqrt{\rho_N}) + \frac{1}{2\sqrt{\rho_N}}\text{div}(\rho_N \boldsymbol{w}_N) = \nu\left(\Delta\sqrt{\rho_N} + \frac{|\boldsymbol{\nabla}\sqrt{\rho_N}|^2}{\sqrt{\rho_N}}\right),
$$

we have

$$
\|\partial_t(\sqrt{\rho_N})\|_{L^2(0,T;H^{-1}(\Omega))} \leq C, \tag{2.3.37}
$$

where C is independent of ε. Then using (2.3.20) and (2.3.37), we apply Aubin–Simon's lemma to deduce the strong convergence of $\sqrt{\rho_N}$ in $L^2(0,T;H^1(\Omega))$ to $\sqrt{\rho}$. Noting that estimate (2.3.20) depends on ε, this strong convergence is only true for a fixed ε.

Using (2.3.21), (2.3.22) and the almost everywhere convergence of ρ_N, we obtain the strong convergence of the two pressures in $L^1(0,T;L^1(\Omega))$.

Rewriting equation (2.3.31) as

$$\begin{aligned}\partial_t\left(\frac{1}{\sqrt{\rho_N}}\right)+\nabla\left(\frac{\boldsymbol{w}_N}{\sqrt{\rho_N}}\right)+\frac{3}{2\sqrt{\rho_N}}\operatorname{div}(\boldsymbol{w}_N)\\ =-\nu\left(\frac{\Delta\sqrt{\rho_N}}{\rho_N}+\frac{|\nabla\sqrt{\rho_N}|^2}{\rho_N^{3/2}}\right),\end{aligned}$$

using (2.3.10), (2.3.16), (2.3.18), (2.3.20), (2.3.23) and (2.3.26), we have

$$\left\|\partial_t\left(\frac{1}{\sqrt{\rho_N}}\right)\right\|_{L^2(0,T;W^{-1,1}(\Omega))}\leq C(\varepsilon),\tag{2.3.38}$$

where constant C is independent of N and δ, but can depend on ε. Then (2.3.38) and (2.3.26) allow us to employ the Aubin–Simon's lemma to deduce the almost everywhere convergence of $1/\sqrt{\rho_N}$.

Moreover, we have

$$\begin{aligned}\nabla(\rho_N\boldsymbol{w}_N)&=\rho_N\nabla\boldsymbol{w}_N+\boldsymbol{w}_N\nabla\rho_N\\&=\sqrt{\rho_N}\sqrt{\rho_N}\nabla\boldsymbol{w}_N\\&\quad+2\sqrt{\rho_N}\boldsymbol{w}_N\nabla\sqrt{\rho_N}\in L^2(0,T;L^1(\Omega)),\end{aligned}$$

thus $\rho_N\boldsymbol{w}_N\in L^2(0,T;W^{1,1}(\Omega))$. Using the second equation in (2.3.30), we obtain the estimate for $\partial_t(\rho_N\boldsymbol{w}_N)$. Thus by using Aubin–Simon's lemma, $\rho_N\boldsymbol{w}_N$ converges almost everywhere. Using (2.3.29) and the almost everywhere convergence of $\sqrt{\rho_N}\boldsymbol{w}_N, (\sqrt{\rho_N}\boldsymbol{w}_N)^2$ converges strongly in $L^1(0,T;L^1(\Omega))$. Thus $\sqrt{\rho_N}\boldsymbol{w}_N$ converges strongly in $L^2(0,T;L^2(\Omega))$.

Finally, the last two weak convergences come from estimates (2.3.25) and (2.3.27).

Using Proposition 2.3.3, performing the limit $N\to\infty$, then the solution $(\rho_\delta,\boldsymbol{w}_\delta)$ solves (2.3.36). Noting that estimates (2.3.10), (2.3.16)–(2.3.27) hold for $\rho=\rho_N$ and $\boldsymbol{u}=\boldsymbol{w}_N$, also for $\rho=\rho_\delta$ and $\boldsymbol{u}=\boldsymbol{w}_\delta$, where constants C are independent of N and δ, but can depend on ε. Similarly, we can deduce the following convergences. □

Proposition 2.3.4. *Under the assumptions of Theorem* 2.3.1, *for fixed* ε, *there exists a subsequence (not relabeled), as* $N \to \infty$,

$$\begin{aligned}
&\sqrt{\rho_N} \to \sqrt{\rho_\delta}\, strongly\, in\, L^2(0,T;H^1(\Omega)),\\
&p(\rho_N) \to p(\rho_\delta)\, strongly\, in\, L^1(0,T;L^1(\Omega)),\\
&p_c(\rho_N) \to p_c(\rho_\delta)\, strongly\, in\, L^1(0,T;L^1(\Omega)),\\
&\frac{1}{\sqrt{\rho_N}} \to \frac{1}{\sqrt{\rho_\delta}}\, almost\, everywhere,\\
&\sqrt{\rho_N}\boldsymbol{w}_N \rightharpoonup \sqrt{\rho_\delta}\boldsymbol{w}_\delta\, strongly\, in\, L^2(0,T;L^2(\Omega)),\\
&\boldsymbol{\nabla}\boldsymbol{w}_N \rightharpoonup \boldsymbol{\nabla}\boldsymbol{w}_\delta\, weakly\, in\, L^p(0,T;L^q(\Omega)),\\
&\boldsymbol{w}_N \rightharpoonup \boldsymbol{w}_\delta\, weakly\, in\, L^p(0,T;L^{q*}(\Omega)),
\end{aligned}$$

where p, q, q^* *are given in Lemma* 2.3.4 *and Lemma* 2.3.5.

The proof of Proposition 2.3.4 is similar to that of Proposition 2.3.3. By Proposition 2.3.4, there exists a weak solution to (2.3.30) in the sense of Definition 2.3.3 without the assumption $\gamma > 3$. Then we construct a global approximate solution.

2.3.3 *Planck limit*

Let us consider a sequence of solutions $(\rho^h, \boldsymbol{u}^h)_h$ to (2.3.1). Then we start to prove Theorem 2.3.2, namely, as $h \to 0$, there exists a subsequence $(\rho^h, \boldsymbol{u}^h)_h$, which converges to the solution to (2.3.3) in the sense of Definition 2.3.2.

Using the *a priori* estimate, obtained we can prove the stability results, the unique difference. Here is the strong convergence of $\sqrt{\rho^h}$ which is only in $L^2(0,T;L^2(\Omega))$, instead of $L^2(0,T;H^1(\Omega))$. Indeed, constant C in $L^2(0,T;H^1(\Omega))$ depends on ε.

Proposition 2.3.5. *Under the assumptions of Theorem* 2.3.2, *there exists a subsequence (not relabeled), as* $\varepsilon \to 0$,

$$\begin{aligned}
&\sqrt{\rho^h} \to \sqrt{\rho}\,\text{strongly in}\, L^2(0,T;L^2(\Omega)),\\
&p(\rho^h) \to p(\rho)\,\text{strongly in}\, L^1(0,T;L^1(\Omega)),
\end{aligned}$$

$$p_c(\rho^h) \to p_c(\rho) \text{ strongly in } L^1(0,T;L^1(\Omega)),$$
$$\frac{1}{\sqrt{\rho^h}} \to \frac{1}{\sqrt{\rho}} \text{ almost everywhere,}$$
$$\sqrt{\rho^h}\boldsymbol{u}^h \to \sqrt{\rho}\boldsymbol{u} \text{ strongly in } L^2(0,T;L^2(\Omega)).$$

Proof. In order to prove the first convergence, we use the estimate $\boldsymbol{\nabla}\sqrt{\rho} \in L^2(0,T;L^2(\Omega))$, together with estimate $\partial_t\sqrt{\rho} \in L^2(0,T;H^{-1}(\Omega))$, and Aubin–Simon's lemma. We finish the proof of the first convergence. One can refer to the previous subsection for the other convergences. □

Convergences of Proposition 2.3.5 allow us to pass to the limit $\varepsilon \to 0$ in the second and third integrals of the left hand side, and the last one of the right hand side of the weak formulation (2.3.5). Using $\sqrt{\rho} \in L^2(0,T;L^2(\Omega))$ and $\boldsymbol{\nabla}\sqrt{\rho} \in L^\infty(0,T;L^2(\Omega))$, we have

$$2^2_\varepsilon \int_0^T \int_\Omega \sqrt{\rho}\boldsymbol{\nabla}\sqrt{\rho} \cdot \boldsymbol{\nabla}\text{div}\boldsymbol{\phi}\text{d}x\text{d}t$$
$$+ 4\varepsilon^2 \int_0^T \int_\Omega \boldsymbol{\nabla}\sqrt{\rho} \otimes \boldsymbol{\nabla}\sqrt{\rho} : \boldsymbol{\nabla}\boldsymbol{\phi}\text{d}x\text{d}t \le C\varepsilon^2,$$

with constant C independent of ε. Therefore, as $\varepsilon \to 0$, these two integrals converge to zero. This the proof of Theorem 2.3.2.

Chapter 3

Existence of Finite Energy Weak Solutions of Inviscid Quantum Hydrodynamic Equations

3.1 Introduction and main result

In this chapter we study the Cauchy problem for the QHD system:

$$\begin{cases} \partial_t\rho + \boldsymbol{\nabla}\boldsymbol{J} = 0, \\ \partial_t J + \operatorname{div}\left(\dfrac{\boldsymbol{J}\otimes\boldsymbol{J}}{\rho}\right) + \nabla P(\rho) + \rho\boldsymbol{\nabla}V \\ \quad = \dfrac{h^2}{2}\rho\boldsymbol{\nabla}\left(\dfrac{\Delta\sqrt{\rho}}{\sqrt{\rho}}\right) \\ +f(\sqrt{\rho}, \boldsymbol{J}, \boldsymbol{\nabla}\sqrt{\rho}, \boldsymbol{\nabla}^2\sqrt{\rho}, \boldsymbol{\nabla}(\boldsymbol{J}/\sqrt{\rho})), \\ -\Delta V = \rho - C(x), \end{cases} \tag{3.1.1}$$

with initial data

$$\rho(0) = \rho_0, J(0) = J_0. \tag{3.1.2}$$

There is a considerable amount of literature giving some first results in the case in which $f = -\boldsymbol{J}$ however, we are able to treat a more general collision term, as it will be clarified in the following. Indeed, we show that it is possible to consider a general collision term of the form $f = -\alpha\boldsymbol{J} + \rho\boldsymbol{\nabla}g$, where $\alpha \geqslant 0$ and g is a nonlinear function of $\sqrt{\rho}, \boldsymbol{J}, \boldsymbol{\nabla}\sqrt{\rho}$ satisfying certain Caratheodory-type conditions. We are interested in studying the global existence in the class of finite energy initial data without higher regularity hypotheses or smallness assumptions. There is a formal analogy between (3.1.1)–(3.1.2) and the classical fluid mechanics system, in particular, when $h = 0$, the system (3.1.1)–(3.1.2) formally coincides with the compressible Euler–Poisson fluid system with damping, which is quite well understood in the literature.

The main tools in the proof of result of this chapter are energy estimates, entropy estimates, dispertive estimates, such as Strichartz estimate and Morawetz estimate; and harmonic analysis techniques, such as Paley–Littlewood frequency decomposition and double linear estimates.

There is a formal equivalence between the system (3.1.1)–(3.1.2) and the following non-linear Schrödinger–Poisson system through the Madelung transform $\psi = \sqrt{\rho}e^{iS/h}$

$$\begin{cases} ih\partial_t\psi + \frac{h^2}{2}\Delta\psi = |\psi|^{p-1}\psi + V_\psi + \tilde{V}_\psi, \\ -\Delta V = |\psi|^2, \end{cases} \tag{3.1.3}$$

where $\tilde{V} = \frac{1}{2i}\log(\frac{\psi}{\bar{\psi}})$. On the other hand, the QHD system can also be obtained by defining $\rho = |\psi|^2$, $\boldsymbol{J} = h\mathrm{Im}(\bar{\psi}\boldsymbol{\nabla}\psi)$ and by computing the related balance laws. But in the serious proof of this process, we encounter the possible existence of nodal regions, or vacuum in fluid terms, namely where $\rho = 0$. In dealing with this problem, we introduce the polarized factor decomposition method in which the wave function can be transformed to the density and velocity of the quantum system in the some meaning. Another question that we have to face is the ill-posed potential $\tilde{V}\psi$. We adopt the fractional steps to construct approximate solution of the system (3.1.1)–(3.1.2). Then we deduce the uniform estimates of the approximate solutions. These estimates have enough compactness to pass the limit passage, thus the limit functions satisfy the original equation (3.1.1). In this way we can prove the global large weak solution to the QHD system (3.1.1).

Before giving the main result, we introduce the definition of weak solutions:

Definition 3.1.1. We say the pair $(\rho, \boldsymbol{J})$ is a weak solution of the Cauchy problem (3.1.1), (3.1.2) in $[0,T) \times \mathbf{R}^n$ with Cauchy data $(\rho_0, \boldsymbol{J}_0) \in L^1_{loc}(\mathbf{R}^n)$, if there exist locally integrable functions $(\sqrt{\rho}, \boldsymbol{\Lambda})$

such that $\sqrt{\rho} \in L^2_{loc}([0,T); H^1_{loc}(\mathbf{R}^n)), \mathbf{\Lambda} \in L^2_{loc}([0,T); L^2_{loc}(\mathbf{R}^n))$ and by defining $\rho := (\sqrt{\rho})^2, \boldsymbol{J} := \sqrt{\rho}\mathbf{\Lambda}$, one has:

1. for any test function $\eta \in C_0^\infty([0,T) \otimes \mathbf{R}^n)$, we have

$$\int_\Omega \int_{\mathbf{R}^n} (\rho\partial_t\eta + \boldsymbol{J} \cdot \boldsymbol{\nabla}\eta)\mathrm{d}x\mathrm{d}t + \int_{\mathbf{R}^n} \rho_0\eta(0)dx = 0; \tag{3.1.4}$$

2. for any test function $\boldsymbol{\zeta} \in C_0^\infty([0,T) \otimes \mathbf{R}^n; \mathbf{R}^n)$, we have

$$\int_\Omega \int_{\mathbf{R}^n} (\boldsymbol{J} \cdot \partial_t\boldsymbol{\zeta} + \mathbf{\Lambda} \otimes \mathbf{\Lambda} : \boldsymbol{\nabla}\boldsymbol{\zeta} + \boldsymbol{\nabla}P\mathrm{div}\boldsymbol{\zeta} - \rho\boldsymbol{\nabla}\boldsymbol{V} \cdot \boldsymbol{\zeta} - \boldsymbol{J} \cdot \boldsymbol{\zeta}$$

$$+ h^2\boldsymbol{\nabla}\sqrt{\rho} \otimes \sqrt{\rho} : \boldsymbol{\nabla}\boldsymbol{\zeta} - \frac{h^2}{4}\rho\Delta\,\mathrm{div}\,\boldsymbol{\zeta})\mathrm{d}x\mathrm{d}t + \int_{\mathbf{R}^n} \boldsymbol{J}_0\boldsymbol{\zeta}(0)dx = 0; \tag{3.1.5}$$

3. (generalized irrotationality condition) for almost every $t \in (0,T)$, we have

$$\boldsymbol{\nabla} \wedge \boldsymbol{J} = 2\boldsymbol{\nabla}\sqrt{\rho} \wedge \mathbf{\Lambda} \tag{3.1.6}$$

which holds in the sense of distributions.

Definition 3.1.2. We say that the weak solution $(\rho, \boldsymbol{J})$ to the Cauchy problem (3.1.1), (3.1.2) is a finite energy weak solution (FEWS) in $t \in (0,T) \times \mathbf{R}^n$, if in addition for almost every $t \in [0,T)$, the energy (3.1.7) is finite.

$$E(t) := \int_{\mathbf{R}^3} \frac{h^2}{2}|\boldsymbol{\nabla}\sqrt{\rho(t)}|^2 + \frac{1}{2}|\mathbf{\Lambda}(t)|^2 + f(\rho(t)) + \frac{1}{2}|\boldsymbol{\nabla}V(t)|^2\mathrm{d}x. \tag{3.1.7}$$

where $\mathbf{\Lambda} := \boldsymbol{J}/\sqrt{\rho}$, and $f(\rho) = \frac{2}{\rho+1}\rho^{\frac{p+1}{2}}$. The function $f(\rho)$ denotes the internal energy which is related to the pressure through the identity $P(\rho) = \rho f'(\rho) - f(\rho)$. Therefore our initial data are required to satisfy

$$E_0 := \int_{\mathbf{R}^3} \frac{h^2}{2}|\boldsymbol{\nabla}\sqrt{\rho_0}|^2 + \frac{1}{2}|\mathbf{\Lambda}_0|^2 + f(\rho_0) + \frac{1}{2}|\boldsymbol{\nabla}V_0|^2\mathrm{d}x < \infty. \tag{3.1.8}$$

Or equivalently (if we have $1 \leqslant p \leqslant 5$),

$$\sqrt{\rho_0} \in H^1(\mathbf{R}^3) \text{ and } \mathbf{\Lambda}_0 := \boldsymbol{J}_0/\sqrt{\rho_0} \in L^2(\boldsymbol{R}^3). \tag{3.1.9}$$

In the following we give a main result, supposing the initial conditions $\rho_0, \boldsymbol{J}_0$ be as momentum of some wave function $\psi_0 \in H^1(\boldsymbol{R}^3)$. Then there exists a global finite energy weak solution for the problem (3.1.1), (3.1.2). The main reference can be seen in [1]–[13].

Theorem 3.1.1. *Let $\psi_0 \in H^1(\boldsymbol{R})^3$ and define*

$$\rho_0 := |\psi_0|^2, \boldsymbol{J}_0 := h \operatorname{Im}(\overline{\psi_0}\boldsymbol{\nabla}\psi_0). \tag{3.1.10}$$

Then, for each $0 < T < \infty$, there exists a finite energy weak solution to the QHD system (3.1.1), (3.1.2) in $[0,T) \times \boldsymbol{R}^n$, with initial data $(\rho_0, \boldsymbol{J}_0)$ defined as above.

3.2 Preliminaries and notations

3.2.1 *Notations*

For convenience of the reader we will set some notations which will be used in the sequel.

If X, Y are two quantities (typically non-negative), we use $X \preceq Y$ to denote $X \leqslant CY$, for some absolute constant $C > 0$. We will use the standard Lebesgue norms for complex-valued measurable functions $f : \mathbf{R}^d \to C$

$$\|f\|_{L^p(\boldsymbol{R}^d)} := \left(\int_{\mathbf{R}^d} |f(x)|^p \mathrm{d}x\right)^{\frac{1}{p}}.$$

If we replace C by a Banach space X, we will adopt the notation

$$\|f\|_{L^p(\mathbf{R}^d;X)} := \left(\int_{\mathbf{R}^d} \|f(x)\|X \mathrm{d}x\right)^{\frac{1}{p}},$$

to denote the norm of $f : \mathbf{R}^d \to X$. In particular, if X is a Lebesgue space $L^r(\mathbf{R}^n)$, and $d = 1$, we will make the notation shortened by writing

$$\|f\|_{L^q_t L^r_x(\mathbf{I}\times\mathbf{R}^n)} := \left(\int_{\mathbf{I}} \|f(x)\|^q_{L^r(\mathbf{R}^n)} \mathrm{d}x\right)^{\frac{1}{q}}$$
$$= \left(\int_{\mathbf{I}} \left(\int_{\mathbf{R}^n} \|f(t,x)\|^r \mathrm{d}x\right)^{\frac{q}{r}} \mathrm{d}t\right)^{1/q},$$

to denote the mixed Lebesgue norm $f : \boldsymbol{I} \to L^r(\mathbf{R}^n)$; moreover, we will write $L^q_t L^r_x(\mathbf{I} \times \mathbf{R}^n) := L^q(I; L^r(\mathbf{R}^n))$. If $q = r$, we will also write $L^q_{t,x}(\mathbf{I} \times \mathbf{R}^n)$ to mean $L^q(I; L^q(\mathbf{R}^n))$. For $s \in R$ we will define the Sobolev space $H^s(\mathbf{R}^n) := (1 - \Delta)^{-s/2} L^2(\mathbf{R}^n)$.

Definition 3.2.1. Let $n \geqslant 2$. We say that (q, r) is an admissible pair of Schrödinger exponents in $\boldsymbol{R}^n$. If $2 \leqslant q \leqslant \infty, 2 \leqslant r \leqslant \frac{2n}{n-2}$, $(q, r, n) \neq (2, \infty, 2)$ and

$$\frac{1}{q} = \frac{n}{2}\left(\frac{1}{2} - \frac{1}{r}\right). \tag{3.2.1}$$

Let $I \times \mathbf{R}^3$ be a space-time slab and we deduce the Strichartz norm $\dot{S}^0(\boldsymbol{I} \times \mathbf{R}^3)$,

$$\|u\|_{\dot{S}^0(I \times \mathbf{R}^3)} := \sup \left(\sum_N \|P_N u\|^2_{L^q_t L^r_x(\boldsymbol{I} \times \mathbf{R}^3)} \right)^{1/2}, \tag{3.2.2}$$

where the sup is taken over all the admissible pairs (q, r). Here P_N denotes the Paley–Littlewood projection operator, with the sum taken over dyadic numbers of the form $N = 2^j, j \in Z$. For any $k \geqslant 1$, we can define

$$\|u\|_{\dot{S}^k(\boldsymbol{I} \times \mathbf{R}^3)} := \|\boldsymbol{\nabla}^{k_u}\|_{\dot{S}^0(\mathbf{I} \times \mathbf{R}^3)}. \tag{3.2.3}$$

Note that, from Paley–Littlewood inequality we have

$$\begin{aligned} \|u\|_{L^q_t L^r_x(\boldsymbol{I} \times \mathbf{R}^3)} &\preceq \left\| \left(\sum_N |P_N u|^2 \right)^{1/2} \right\|_{L^q_t L^r_x(\boldsymbol{I} \times \mathbf{R}^3)} \\ &\preceq \left(\left\| \sum_N P_N u \right\|^2_{L^q_t L^r_x(\boldsymbol{I} \times \mathbf{R}^3)} \right)^{1/2}, \end{aligned} \tag{3.2.4}$$

and hence for each admissible pair of exponents, one has

$$\|u\|_{L^q_t L^r_x(\boldsymbol{I} \times \mathbf{R}^3)} \preceq \|u\|_{\dot{S}^0(\boldsymbol{I} \times \mathbf{R}^3)}. \tag{3.2.5}$$

Lemma 3.2.1.

$$\|u\|_{L^4_t L^\infty_x(\boldsymbol{I} \times \boldsymbol{R}^3)} \preceq \|u\|_{\dot{S}^1(\boldsymbol{I} \times \boldsymbol{R}^3)}. \tag{3.2.6}$$

3.2.2 Non-linear Schrödinger equation

In this section we briefly review some well-known results about the theory of non-linear Schrödinger equations: we will state some global well-posedness results and then we will show the main properties for the solutions to NLS equations, namely the Strichartz and the local smoothing estimates. The main results exposed in this section can be found, in a more detailed way, in the standard books by [10] and [63] and in the references therein. Let us consider the following non-linear Schrödinger–Poisson system

$$i h\partial_t \psi + \frac{h^2}{2}\Delta\psi = |\psi|^{p-1}\psi + V\psi + W\psi, \tag{3.2.7}$$

$$-\Delta V = |\psi|^2, \tag{3.2.8}$$

with the initial data

$$\psi(0) = \psi_0 \in H^1(\mathbf{R}^n), \tag{3.2.9}$$

where $1 \leqslant p < \frac{2n}{n-2}(1 \leqslant p < \infty, n = 2)$, and W is a real-valued potential such that $W \in L^p(\mathbf{R}^n) + L^\infty(\mathbf{R}^n), p > n/2$. Then it is well-known that (3.2.1)–(3.2.2) are globally well-posed, i.e. there exists a unique strong solution $\psi \in C(\mathbf{R}; H^1(\mathbf{R}^n)) \cap C^1(\mathbf{R}; H^{-1}(\mathbf{R}^n))$ which continuously depends on the initial data. Furthermore, the total energy

$$E(t) := \int_{\mathbf{R}^n} \left(\frac{h^2}{2}|\nabla\psi(t,x)|^2 + \frac{2}{p+1}|\psi(t,x)|^{p+1} + V(x,t)|\psi(t,x)|^2 \right.$$
$$\left. + W(t,x)|\psi(t,x)|^2\right) \mathrm{d}x \tag{3.2.10}$$

is conserved, i.e.

$$E(t) := E_0, \quad \text{for all} \quad t \in \mathbf{R}. \tag{3.2.11}$$

Moreover, we can exploit the dispersive nature of the Schrödinger equation to get some further estimates, which show the integrability and regularity properties for the solutions.

Let us consider the free Schrödinger equation

$$i\partial_t u + \Delta u = 0,$$
$$u(0) = u_0.$$

We shall denote $U(\cdot)$ as the free Schrödinger group, defined by the relation $U(t)u_0 = u(t)$, where $u(t)$ is the solution of the free Schrödinger equation above, with initial data u_0.

Theorem 3.2.1. *Let $(q, r), (\tilde{q}, \tilde{r})$ be two arbitrary admissible pairs of exponents and let $U(\cdot)$ be the free Schrödinger group. Then we have,*

$$\|U(t)f\|L_t^q L_x^r \lesssim \|f\|_{L^2(\boldsymbol{R}^n)}, \tag{3.2.12}$$

$$\left\| \int_{s<t} U(t)F(s)\mathrm{d}s \right\|_{L_t^q L_x^r} \lesssim \|F\|_{L_t^{\tilde{q}'} L_x^{\tilde{r}'}}, \tag{3.2.13}$$

$$\left\| \int U(t)F(s)\mathrm{d}s \right\|_{L_t^q L_x^r} \lesssim \|F\|_{L_t^{\tilde{q}'} L_x^{\tilde{r}'}}. \tag{3.2.14}$$

Then by using the Theorem above, we can state the following.

Theorem 3.2.2. *Let $\boldsymbol{I}$ be a compact interval, and let $u : \boldsymbol{I}\times \boldsymbol{R}^3 \to C$ be a Schwartz solution to the Schrödinger equation*

$$i\partial_t u + \Delta u = F_1 + \cdots + F_M,$$

for some Schwartz functions $F_1, \ldots, F_M$. Then we have

$$\|u\|_{\dot{S}^0(\boldsymbol{I}\times\boldsymbol{R}^n)} \lesssim \|u_0\|_{L^2(\boldsymbol{R}^n)} + \|F_1\|_{L_t^{\tilde{q}_1'} L_x^{\tilde{r}_1'}} + \cdots + \|F_M\|_{L_t^{\tilde{q}_1'} L_x^{\tilde{r}_1'}},$$

where $(q_1, r_1), \ldots, (q_M, r_M)$ are arbitrary admissible pairs of exponents.

Moreover, the dispersion allows us to gain also some further regularity properties for the solutions to the Schrödinger equation. Especially, what they could prove is that actually the solution to a dispersive equation is locally smoother than the initial data.

Theorem 3.2.3. *Let u solve the free Schrödinger equation*

$$i\partial_t u + \Delta u = 0, u(0) = u_0.$$

Let $\chi \in C_0^\infty(\boldsymbol{R}^{1+n})$ of the form

$$\chi(t,x) = \chi_0(t)\chi_1(x_1)\dots\chi_n(x_n),$$

with $\chi_j \in C_0^\infty(\boldsymbol{R}), j = 1, 2, \dots, n$. Then we have

$$\int_{\boldsymbol{R}^1+n}\int_0^T \chi^2(t,x)|(\boldsymbol{I} - \nabla)^{1/4}u(x,t)|^2 \mathrm{d}x\mathrm{d}t \leqslant C^2\|u_0\|_{L^2(\boldsymbol{R}^n)}.$$

In particular, if $u_0 \in L^2(\boldsymbol{R}^n)$, one has for all $T > 0$,

$$u \in L^2([0,T]; H^{1/2}_{loc}(\boldsymbol{R}^n)).$$

We also have a similar result for the non-homogeneous case:

Theorem 3.2.4. *Let u be the solution of*

$$i\partial_t u + \Delta u = F, u(0) = u_0 \in L^2(\boldsymbol{R}^n),$$

where $F \in L^1([0,T]; L^2(\boldsymbol{R}^n))$. Then it follows

$$u \in L^2([0,T]; H^{1/2}_{loc}(\boldsymbol{R}^n)).$$

Moreover, let $\chi \in C_0^\infty(\boldsymbol{R}^{1+n})$ be of the form

$$\chi(t,x) = \chi_0(t)\chi_1(x_1)\dots\chi_n(x_n),$$

with $\chi_j \in C_0^\infty(\boldsymbol{R}), j = 1, 2, \dots, n$, supp $\chi_0 \subset [0,T]$. Then the following local smoothing estimate holds

$$\left(\int_{\boldsymbol{R}^{1+n}}\int_0^T \chi^2(t,x)|(\boldsymbol{I} - \Delta)^{1/4}u(x,t)|^2 \mathrm{d}x\mathrm{d}t\right)^{\frac{1}{2}}$$
$$\leqslant C(\|u_0\|_{L^2(\boldsymbol{R}^n)} + \|F\|_{L^1_tL^2_x([0,T]\times\boldsymbol{R}^n)}).$$

This results imply that, the free Schrödinger group $U(\cdot)$ fulfills the following inequalities

$$\|U(\cdot)u_0\|L^2([0,T]; H^{1/2}_{loc}(\boldsymbol{R}^n)) \lesssim \|u_0\|_{L^2(\boldsymbol{R}^n)}$$

$$\|\int_0^t U(t-s)F(s)\mathrm{d}s\|_{L^2([0,T];H^{1/2}_{loc}(\boldsymbol{R}^n))} \lesssim \|F\|_{L^1(0,T];L^2(\boldsymbol{R}^n))}.$$

3.2.3 *Compactness tools*

In this subsection we will give a compactness result.

Theorem 3.2.5. *Let $(V, \|\cdot\|_V), (H, \|\cdot\|_H)$ be two separable Hilbert spaces. Assume that $V \subset H$ with a compact and dense embedding. Consider a sequence $\{u^\varepsilon\}$ converging weakly to a function u in $L^2([0,T];V), T < \infty$. Then u^ε converges strongly to u in $L^2([0,T];V)$, if and only if*

1. *u^ε converges to $u(t)$ weakly in H for a.e.t;*
2. $\lim\limits_{|E|\to 0, E\subset[0,T]} \sup\limits_{\varepsilon>0} \int_E \|U^\varepsilon\|^2 H^{\mathrm{dt}} = 0.$

3.2.4 *Tools in two-dimension*

Finally we recall some theorems which will be useful at studying the two-dimensional case of the Cauchy problem for the QHD system. First of all we recall the Gagliardo–Nirenberg–Sobolev inequality, which is the following

$$\|u\|_{L^p} \leqslant \|u\|_{L^2}^{\frac{2}{p}} \|\nabla u\|_{L^2}^{1-\frac{2}{p}}, \tag{3.2.15}$$

for $2 < p < \infty$. Then we want to recall the logarithmic Sobolev inequality.

Theorem 3.2.6. *Let f be a non-negative function in $L^1(\boldsymbol{R}^2)$ such that $f \log f, f \log(1 + |x|^2) \in L^1(\boldsymbol{R}^2)$. If $\int f \mathrm{d}x = M$, then*

$$\begin{aligned} &\frac{M}{2} \int_{\boldsymbol{R}^2} f \log f \mathrm{d}x + \int_{\boldsymbol{R}^2\times\boldsymbol{R}^2} f(x) \log|x-y| f(y) \mathrm{d}x\mathrm{d}y \\ &\geqslant C(M) := \frac{M^2}{2}(1 + \log \pi + \log M). \end{aligned}$$

3.3 Polar decomposition

In this section we will explain how to decompose an arbitrary wave function ψ into its amplitude $\sqrt{\rho} = |\psi|$ and its unitary factor φ, namely a function taking its values into the unitary circle of the complex plane, such that $\psi = \sqrt{\rho}\varphi$. The idea is similar in the spirit to the one used in [6], to find the measure preserving maps needed to write vector valued functions as compositions of gradients of convex functions and measure preserving maps. Our case is much simpler than [6] and it can be studied directly in a simpler setting. Brenier's idea looks for projections of L^2 functions u onto the set of measure preserving maps S, contained in a given sphere of L^2, i.e. one has to find $s \in S$, which minimizes the distance $\|u - s\|_{L^2}$, or equivalently, which maximizes $(u, s)_{L^2}$. In our case we should maximize Re $(u, s)_{L^2}$, within complex-valued functions with the constraint to take values in the unit ball of L^∞.

Let us consider a wave function $\psi \in L^2(\mathbf{R}^n)$, and define the set

$$P(\psi) := \{\varphi \mid \|\varphi\|_{L^\infty(\mathbf{R}^n)} \leqslant 1, \sqrt{\rho}\varphi = \psi \text{ a.e.in } \mathbf{R}^n\}, \tag{3.3.1}$$

where $\sqrt{\rho} = |\psi|$. Of course, if we consider $\varphi \in P(\psi)$, then by the definition of the set $P(\psi)$ it is immediate that $|\varphi| = 1$ a.e. $\sqrt{\rho} - dx$ in $\mathbf{R}^n$ and $\varphi \in P(\psi)$ is uniquely determined a.e. $\sqrt{\rho} - dx$ in $\mathbf{R}^n$.

Lemma 3.3.1. *Let $\psi \in H^1(\boldsymbol{R}^n), \sqrt{\rho} := |\psi|$, then there exists $\phi \in L^\infty(\boldsymbol{R}^n)$ such that $\psi = \sqrt{\rho}\phi$ a.e. in $\boldsymbol{R}^n, \sqrt{\rho} \in H^1(\boldsymbol{R}^n), \nabla\sqrt{\rho} = \mathrm{Re}(\bar{\phi}\boldsymbol{\nabla}\psi)$. If we set $\Lambda := h\mathrm{Im}(\bar{\phi}\boldsymbol{\nabla}\psi)$, one has $\boldsymbol{\Lambda} \in L^2(\boldsymbol{R}^n)$ and the following identity holds*

$$h^2\mathrm{Re}(\partial_j\overline{\psi}\partial_k\psi) = h^2\partial_j\sqrt{\rho}\partial_k\sqrt{\rho} + \Lambda^j\Lambda^k. \tag{3.3.2}$$

Furthermore, let $\{\psi_n\} \to \psi$ strongly in $H^1(\boldsymbol{R}^n)$, then it follows

$$\boldsymbol{\nabla}\sqrt{\rho_n} \to \boldsymbol{\nabla}\sqrt{\rho}, \boldsymbol{\Lambda}_n \to \boldsymbol{\Lambda} \; inL^2(\boldsymbol{R}^n), \tag{3.3.3}$$

where $\boldsymbol{\Lambda}_n := h\mathrm{Im}(\overline{\varphi_n}\nabla\psi_n)$.

Proof. Let us consider a sequence $\{\psi_n\} \subset C\infty(\boldsymbol{R}^n), \psi_n \to \psi$ in $H^1(\boldsymbol{R}^n)$. Thanks to the regularity of the sequence ψ_n, we can define

$$\phi_n(x) = \begin{cases} \dfrac{\psi_n(x)}{|\psi_n(x)|} & if\ \psi_n(x) \neq 0, \\ 0 & if\ \psi_n(x) = 0. \end{cases}$$

Then, there exists $\phi \in L^\infty(\boldsymbol{R}^n)$ such that $\phi_n \overset{*}{\rightharpoonup} \phi$ in $L^\infty(\mathbf{R}^n)$ and $\boldsymbol{\nabla}\psi_n \to \boldsymbol{\nabla}\psi$ in $L^2(\mathbf{R}^n)$, hence

$$\mathrm{Re}(\bar{\phi}_n\boldsymbol{\nabla}\psi_n) \rightharpoonup \mathrm{Re}(\overline{\varphi}\boldsymbol{\nabla}\psi) \; in \; L^2(\mathbf{R}^n). \tag{3.3.4}$$

with the definition of φ_n, *one has*

$$Re(\overline{\phi_n}\boldsymbol{\nabla}\psi_n) = \boldsymbol{\nabla}|\psi_n| \text{ a.e. on } \mathbf{R}^n. \tag{3.3.5}$$

It follows

$$\boldsymbol{\nabla}\sqrt{\rho_n} \rightharpoonup \mathrm{Re}(\overline{\phi}\boldsymbol{\nabla}\psi) \text{ in } L^2(\mathbf{R}^n). \tag{3.3.6}$$

Moreover, one has $\boldsymbol{\nabla}\sqrt{\rho_n} \rightharpoonup \boldsymbol{\nabla}\sqrt{\rho}$ in $L^2(\mathbf{R}^n)$, therefore

$$\boldsymbol{\nabla}\sqrt{\rho} = \mathrm{Re}(\overline{\phi}\boldsymbol{\nabla}\psi), \tag{3.3.7}$$

where ϕ is a unitary factor of ψ. The identity (3.3.2) follows immediately from the following

$$\begin{aligned} h^2\mathrm{Re}(\partial_j\overline{\psi}\partial_k\psi) &= h^2\mathrm{Re}((\phi\partial_j\overline{\psi})(\overline{\phi}\partial_k\psi)) \\ &= h^2\mathrm{Re}(\overline{\phi}\partial_j\overline{\psi})\mathrm{Re}(\overline{\phi}\partial_k\psi) - h^2\mathrm{Im}(\phi\partial_j\overline{\psi})\mathrm{Im}(\overline{\phi}\partial_k\psi) \\ &= h^2\partial_j\sqrt{\rho}\partial_k\sqrt{\rho} + \Lambda^j\Lambda^k. \end{aligned} \tag{3.3.8}$$

Now we prove (3.3.3). Let us take $\psi_n \to \psi$ strongly in $H^1(\mathbf{R}^n)$, and consider $\boldsymbol{\nabla}\sqrt{\rho_n} = \mathrm{Re}(\overline{\phi_n}\boldsymbol{\nabla}\psi_n)$, $\boldsymbol{\Lambda}_n := h\mathrm{Im}(\overline{\phi_n}\boldsymbol{\nabla}\psi_n)$. As before, $\phi_n \overset{*}{\rightharpoonup} \phi$ in $L^\infty(\mathbf{R}^n), \phi$ is a polar factor of ψ; indeed, $\sqrt{\rho_n}\phi_n \rightharpoonup \sqrt{\rho}\phi$ and $\sqrt{\rho_n}\phi_n = \psi_n \to \psi$. Then $\boldsymbol{\nabla}\sqrt{\rho_n} \rightharpoonup \boldsymbol{\nabla}\sqrt{\rho}$, $\mathrm{Re}(\overline{\phi n}\boldsymbol{\nabla}\psi_n) \rightharpoonup \mathrm{Re}(\overline{\phi}\boldsymbol{\nabla}\psi)$ and $\boldsymbol{\nabla}\sqrt{\rho} = \mathrm{Re}(\overline{\phi}\boldsymbol{\nabla}\psi)$. Moreover, $\boldsymbol{\Lambda}_n := h\mathrm{Im}(\overline{\phi_n}\boldsymbol{\nabla}\psi_n) \rightharpoonup h\mathrm{Im}(\overline{\phi}\boldsymbol{\nabla}\psi) := \boldsymbol{\Lambda}$. To upgrade the weak convergence into the strong one, simply notice that one has

$$\begin{aligned} h^2\|\boldsymbol{\nabla}\psi\|_{L^2}^2 &= h^2\|\boldsymbol{\nabla}\sqrt{\rho}\|_{L^2}^2 + \|\boldsymbol{\nabla}\|_{L^2}^2 \\ &\leqslant \liminf_{n\to\infty}(h^2\|\boldsymbol{\nabla}\sqrt{\rho_n}\|_{L^2}^2 + \|\boldsymbol{\Lambda}_n\|_{L^2}^2) \\ &= h^2\|\boldsymbol{\nabla}\psi_n\|_{L^2}^2 = h^2\|\boldsymbol{\nabla}\psi\|_{L^2}^2. \end{aligned} \tag{3.3.9}$$

Corollary 3.3.1. *Let* $\psi \in H^1(\boldsymbol{R}^n)$, *then*

$$h\boldsymbol{\nabla}\overline{\psi} \wedge \boldsymbol{\nabla}\psi = 2i\boldsymbol{\nabla}\sqrt{\rho} \wedge \boldsymbol{\Lambda}. \tag{3.3.10}$$

Proof. It suffices to note that we can write

$$h\nabla\overline{\psi}\wedge\nabla\psi = h(\phi\nabla\psi)\wedge(\overline{\phi}\nabla\psi),$$

where $\phi \in L^\infty(\mathbf{R}^n)$ is the polar factor of ψ such that $\nabla\sqrt{\rho} = \mathrm{Re}(\overline{\phi}\nabla\psi), \mathbf{\Lambda} = h\ \mathrm{Im}\ (\overline{\phi}\nabla\psi)$. By splitting $\overline{\phi}\nabla\psi$ into its real and imaginary part, we get the identity (3.3.10).

Now we state a technical lemma which will be used in the next sections. It summarizes the results of this section we are going to use later on and will be handy for the application to the fractional step method.

Lemma 3.3.2. *Let $\psi \in H^1(\boldsymbol{R}^n)$, and let $\tau, \varepsilon > 0$ be two arbitrary (small) real numbers. Then there exists $\psi \in H^1(\boldsymbol{R}^n)$ such that*

$$\begin{aligned}\tilde{\rho} &= \rho,\\ \tilde{\mathbf{\Lambda}} &= (1-\tau)\mathbf{\Lambda} + r_\varepsilon,\end{aligned}$$

$\sqrt{\rho} := |\psi|, \sqrt{\tilde{\rho}} := |\tilde{\psi}|, \mathbf{\Lambda} := h\ \mathrm{Im}(\overline{\varphi}\nabla\psi), \tilde{\mathbf{\Lambda}} := h\,\mathrm{Im}\ (\overline{\tilde{\varphi}}\nabla\tilde{\psi}), \phi, \overline{\phi}$ *are polar factors of $\psi, \tilde{\psi}$, respectively, and*

$$\|r_\varepsilon\|_{L^2(\boldsymbol{R}^n)} \leqslant \varepsilon. \tag{3.3.11}$$

Furthermore, we have

$$\nabla\tilde{\psi} = \nabla\psi - i\frac{\tau}{h}\phi^*\mathbf{\Lambda} + r_{\varepsilon,\tau}, \tag{3.3.12}$$

where $\|\varphi^*\|_{L^\infty(\boldsymbol{R}^n)} \leqslant 1$ and

$$\|r_{\varepsilon,\tau}\|_{L^2(\boldsymbol{R}^n)} \leqslant C(\tau\|\nabla\psi\|_{L^2(\boldsymbol{R}^n)} + \varepsilon). \tag{3.3.13}$$

Proof. Consider a sequence $\psi_n \subset C^\infty(\mathbf{R}^n)$ converging to ψ in $H^1(\mathbf{R}^n)$, in $\psi_n \to \psi$, and define

$$\phi_n(x) = \begin{cases}\dfrac{\psi_n(x)}{|\psi_n(x)|} & \textit{if}\ \psi_n(x) = 0\\ 0 & \textit{if}\ \psi_n(x) = 0\end{cases}$$

as polar factors of the wave functions ψ_n. Since $\psi_n \subset C^\infty(\boldsymbol{R}^n)$, then ψ_n is piecewise smooth, and $\Omega_n := \{x \in \mathbf{R}^n : |\psi n(x)| > 0\}$ is an open set, with smooth boundary. Therefore we can say there exists a function $\theta_n : \Omega_n \to [0, 2\pi)$, piecewise smooth in Ω_n and

$$\psi_n(x) - e^{i\theta_n(x)}, \quad for\ all\ x \in \Omega_n \tag{3.3.14}$$

Moreover, by the previous lemma, we have $\phi_n \overset{*}{\rightharpoonup} \phi$ in $L^\infty(\mathbf{R}^n)$, where φ is a polar factor of ψ, and $\mathbf{\Lambda}_n := h\,\mathrm{Im}(\overline{\phi_n}\nabla\psi_n) \to \mathbf{\Lambda} := h\,\mathrm{Im}\,(\overline{\phi}\nabla\psi)$ in $L^2(\mathbf{R}^n)$. Thus there exists $n \in \mathbb{N}$ such that

$$\|\psi - \psi_n\|_{H^1(\mathbf{R}^n)} + \|\mathbf{\Lambda} - \mathbf{\Lambda}_n\|_{L^2(\mathbf{R}^n)} \leqslant \varepsilon. \tag{3.3.15}$$

Now we can define

$$\tilde{\psi} := e^{i(1-\tau)\theta_n}\sqrt{\rho_n}. \tag{3.3.16}$$

Furthermore,

$$\begin{aligned}
\nabla\tilde{\psi} &= e^{-i\tau\theta_n}\nabla\psi_n - i\frac{\tau}{h}e^{i(1-\tau)^{\theta_n}}\mathbf{\Lambda}_n \\
&= \nabla\psi - i\frac{\tau}{h} - e^{i(1-\tau)^{\theta_n}}\mathbf{\Lambda}_n + \tau\left(\sum_{j=1}^{\infty}\frac{(-i\theta_n)^j\tau^{j-1}}{j!}\right)\nabla\psi_n \\
&\quad + (\nabla\psi_n - \nabla\psi) - i\frac{\tau}{h}e^{i(1-\tau)^{\theta_n}}(\mathbf{\Lambda}_n - \mathbf{\Lambda})
\end{aligned} \tag{3.3.17}$$

and $r_{\varepsilon,\tau}$ is given as

$$r_{\varepsilon,\tau} = \tau\left(\sum_{j=1}^{\infty}\frac{(-i\theta_n)^j\tau^{j-1}}{j!}\right)\nabla\psi_n + (\nabla\psi_n - \nabla\psi) - i\frac{\tau}{h}e^{i(1-\tau)^{\theta_n}}(\mathbf{\Lambda}_n - \mathbf{\Lambda}). \tag{3.3.18}$$

Moreover,

$$\tilde{\mathbf{\Lambda}} = h\,\mathrm{Im}\,(e^{i(1-\tau)^{\theta_n}}\nabla\tilde{\psi} = (1-\tau)\mathbf{\Lambda} + (1-\tau)(\mathbf{\Lambda}_n - \mathbf{\Lambda}) \tag{3.3.19}$$

and $r_\varepsilon := (1-\tau)(\mathbf{\Lambda}_n - \mathbf{\Lambda})$ has L^2 norm less than ε by assumption. □

3.4 Quantum hydrodynamic equations without collision term

In this section we will focus on the three-dimensional case, which is physically most relevant. Now let us summarize some key points regarding the existence of weak solutions to the quantum hydrodynamic system, in the collisionless case.

The balance equations can be written in the following way:

$$\partial_t \rho + \nabla \boldsymbol{J} = 0, \tag{3.4.1}$$

$$\partial_t \boldsymbol{J} + \operatorname{div}\left(\frac{\boldsymbol{J} \otimes \boldsymbol{J}}{\rho}\right) + \nabla P(\rho) + \rho \nabla V = \frac{h^2}{2} \rho \nabla \left(\frac{\Delta\sqrt{\rho}}{\sqrt{\rho}}\right), \tag{3.4.2}$$

$$-\Delta V = \rho, \tag{3.4.3}$$

where $P(\rho) = \frac{p-1}{p+1}\rho^{(p+1)/2}, 1 \leqslant p < 5$.

Definition 3.4.1. We say the pair $(\rho, \boldsymbol{J})$ is a weak solution in $[0,T) \times \mathbf{R}^n$ to the system (3.4.1)–(3.4.3), with Cauchy data $(\rho_0, \boldsymbol{J}_0) \in L^1_{loc}(\mathbf{R}^3)$, if and only if there exist locally integrable functions $(\sqrt{\rho}, \boldsymbol{\Lambda})$ such that $\rho \in L^2_{loc}([0,T); H^1_{loc}(\mathbf{R}^n))$, $\boldsymbol{\Lambda} \in L^2_{loc}([0,T); L^2_{loc}(\mathbf{R}^n))$.

If we define $\rho := (\sqrt{\rho})^2$, $\boldsymbol{J} := \sqrt{\rho}\boldsymbol{\Lambda}$, then

1. for any test function $\eta \in C_0^\infty([0,T) \otimes \mathbf{R}^n)$, we have

$$\int_\Omega \int_{\mathbf{R}^n} (\rho \partial_t \eta + \boldsymbol{J} \cdot \nabla_\eta) \mathrm{d}x\mathrm{d}t + \int_{\mathbf{R}^n} \rho_0 \eta(0) \mathrm{d}x = 0; \tag{3.4.4}$$

2. for any test function $\zeta \in C_0^\infty([0,T) \otimes \mathbf{R}^n; \mathbf{R}^n)$, we have

$$\int_\Omega \int_{\mathbf{R}^n} g(\boldsymbol{J} \cdot \partial_t \zeta + \boldsymbol{\Lambda} \otimes \boldsymbol{\Lambda} : \nabla \zeta + \nabla P \operatorname{div} \zeta - \rho \nabla V \cdot \zeta$$
$$+ h^2 \nabla \sqrt{\rho} \otimes \sqrt{\rho} : \nabla \zeta - \frac{h^2}{4} \rho \Delta \div \zeta g) \mathrm{d}x\mathrm{d}t + \int_{\mathbf{R}^n} \boldsymbol{J}_0 \zeta(0) \mathrm{d}x = 0; \tag{3.4.5}$$

3. (generalized irrotationality condition) for almost every $t \in (0,T)$, we have

$$\nabla \wedge \boldsymbol{J} = 2\nabla\sqrt{\rho} \wedge \boldsymbol{\Lambda} \tag{3.4.6}$$

holds in the sense of distributions. We say that $(\rho, \boldsymbol{J})$ is a finite energy weak solution if and only if it is a weak solution and the energy is finite a.e. in $[0, T)$.

The next existence result, roughly speaking, shows how to get a weak solution to the system (3.4.1)–(3.4.3) out of a strong solution to the Schrödinger–Poisson system

$$i h \partial_t \psi + \frac{h^2}{2} \Delta \psi = |\psi|^{p-1} \psi + V \psi, \tag{3.4.7}$$

$$-\Delta V = |\psi|^2. \tag{3.4.8}$$

The quadratic non-linearity in (3.4.2) is originated by a term of the form Re $(\nabla \overline{\psi} \otimes \nabla \psi)$ since formally

$$h^2 \mathrm{Re}(\nabla \overline{\psi} \otimes \nabla \psi) = h^2 \mathrm{Re}(\nabla \sqrt{\rho} \otimes \nabla \sqrt{\rho}) + \frac{\boldsymbol{J} \otimes \boldsymbol{J}}{\rho}. \tag{3.4.9}$$

However, this identity can be justified in the nodal region $\{\rho = 0\}$ only by means of the polar factorization discussed in the previous section.

Indeed, we stress that, in the previous identity the right hand side is written in terms of ρ and $\boldsymbol{J}$ which exist in the whole space $\mathbf{R}^3$, via the Madelung transform. However, the term $\frac{\boldsymbol{J} \otimes \boldsymbol{J}}{\rho}$ should be interpreted as $\boldsymbol{\Lambda} \otimes \boldsymbol{\Lambda}$, where $\boldsymbol{\Lambda}$ is the Radon–Nikodym derivative of $\boldsymbol{J} \mathrm{d}x$ with respect to $\sqrt{\rho} \mathrm{d}x$. Unfortunately the Madelung transformations are unable to define $\boldsymbol{\Lambda}$ on the whole $\mathbf{R}^3$, hence one need to use the polar factorization discussed in the previous section to define $\boldsymbol{\Lambda}$ in the whole $\mathbf{R}^3$.

Furthermore, the study of the existence of weak solutions of (3.4.1)–(3.4.3) is done with Cauchy data of the form $(\rho_0, \boldsymbol{J}_0) = (|\psi|^2, h \mathrm{Im}(\overline{\psi_0} \nabla \psi_0))$, for some $\psi_0 \in H^1(\mathbf{R}^3)$. These special initial data yield to consider the Cauchy problem for the QHD system (3.4.7)–(3.4.8) only for solutions compatible with a wave mechanics point of view.

Proposition 3.4.1. *Let* $0 < T < \infty, \psi_0 \in H^1(\boldsymbol{R}^3)$ *and define the initial data for* (3.4.1) – (3.4.3), $(\rho_0, \boldsymbol{J}_0) = (|\psi|^2, h \text{ Im } (\overline{\psi_0} \nabla \psi_0))$. *Then there exists a finite energy weak solution* $(\rho, \boldsymbol{J})$ *to the Cauchy* *problem (3.4.1) – (3.4.3) in the space-time slab* $[0,T) \times \boldsymbol{R}^3$. *Furthermore, the energy* $E(t)$ *is conserved for all times* $t \in [0,T)$.

The idea behind the proof of Proposition 3.4.1 is as follows. Let us consider the Schrödinger–Poisson system (3.4.7)–(3.4.8), with initial data $\psi(0) = \psi_0$. It is well known that it is globally well-posed for initial data in $H^1(\mathbf{R}^3)$, and the solution satisfies $\psi \in C^0(\mathbf{R}; H^1(\mathbf{R}^3))$.

Then it makes sense to define for each time $t \in [0,T)$ the quantities $(\rho, \boldsymbol{J}) = (|\psi|^2, h\mathrm{Im}(\overline{\psi}\boldsymbol{\nabla}\psi))$ and we can see that $(\rho, \boldsymbol{J})$ is a finite energy weak solution of (3.4.1)–(3.4.3): indeed, for the solution ψ of (3.4.7)–(3.4.8), we have

$$\partial_t \boldsymbol{\nabla}\psi = \frac{ih}{2}\Delta\boldsymbol{\nabla}\psi - \frac{i}{h}\boldsymbol{\nabla}((|\psi|^{p-1} + V)\psi)$$

in the sense of distributions. Thus formally we get the following identities for $(\rho, \boldsymbol{J})$

$$\begin{aligned}
\partial_t \rho &= \frac{ih}{2}\Delta\psi - \frac{\mathrm{i}}{h}\boldsymbol{\nabla}((|\psi|^{p-1} + V)\psi),\\
\partial_t \boldsymbol{J} &= h\,\mathrm{Im}(\boldsymbol{\nabla}\psi(-\frac{ih}{2}\Delta\overline{\psi} + \frac{\mathrm{i}}{h}(|\psi|^{p-1} + V)\overline{\psi}))\\
&\quad + h\,\mathrm{Im}(\overline{\psi}(-\frac{ih}{2}\Delta\overline{\psi} - \frac{\mathrm{i}}{h}\boldsymbol{\nabla}(|\psi|^{p-1} + V)\psi - \frac{i}{h}(|\psi|^{p-1} + V)\boldsymbol{\nabla}\psi))\\
&= \frac{h^2}{4}\boldsymbol{\nabla}\Delta|\psi|^2 - h^2\,\mathrm{div}\,(\mathrm{Re}(\boldsymbol{\nabla}\overline{\psi} \otimes \boldsymbol{\nabla}\psi))\\
&\quad - \frac{p-1}{p+1}\boldsymbol{\nabla}(|\psi|^{p+1}) - |\psi|^2\boldsymbol{\nabla}V.
\end{aligned}$$

Thanks to Lemma 3.3.1, we can write

$$h^2\mathrm{Re}(\boldsymbol{\nabla}\overline{\psi} \otimes \boldsymbol{\nabla}\psi) = h^2\mathrm{Re}(\boldsymbol{\nabla}\sqrt{\rho} \otimes \boldsymbol{\nabla}\sqrt{\rho}) + \boldsymbol{\Lambda} \otimes \boldsymbol{\Lambda}$$

and they satisfy

$$\boldsymbol{J} = \sqrt{\rho}\boldsymbol{\Lambda}.$$

Hence formally the following identity holds:

$$\begin{aligned}\partial_t \boldsymbol{J} + \operatorname{div}(\boldsymbol{\Lambda}\otimes\boldsymbol{\Lambda}) + \boldsymbol{\nabla} P(\rho) + \rho\boldsymbol{\nabla} V \\ = \frac{h^2}{4}\boldsymbol{\nabla}\Delta\rho - h^2\operatorname{div}(\boldsymbol{\nabla}\sqrt{\rho}\otimes\boldsymbol{\nabla}\sqrt{\rho}). \qquad (3.4.10)\end{aligned}$$

Of course these calculations are just formal, since ψ does not have the necessary regularity to implement them.

Furthermore, it is well known that the energy

$$\begin{aligned}E(t) := \int_{\mathbf{R}^3}\left(\frac{h^2}{2}|\boldsymbol{\nabla}\psi|^2 + \frac{2}{p+1}|\psi|^{p+1}\right)\mathrm{d}x \\ + \frac{1}{2}\int_{\mathbf{R}^3_x\times\mathbf{R}^3_y}|\psi(t,y)|^2\frac{1}{|x-y|}|\psi(t,x)|^2\mathrm{d}x\mathrm{d}y \quad (3.4.11)\end{aligned}$$

is conserved for the Schrödinger–Poisson. Thus, it only remains to note that by Lemma 3.3.1 the energy (3.4.11) and (3.2.10) is equal.

Proof. Let us consider the Schrödinger–Poisson system (3.4.7)–(3.4.8). From standard theory about non-linear Schrödinger equations it is well known that (3.4.7)–(3.4.8) is globally well-posed for initial data in the space of energy: $\psi(0) = \psi_0 \in H^1(\boldsymbol{R}^3)$. Now, let us take a sequence of mollifiers χ_ε converging to the Dirac mass and define $\psi^\varepsilon := \chi_\varepsilon * \psi$, where $\psi \in C(R; H^1(\mathbf{R}^3))$ is the solution to (3.4.7)–(3.4.8). Then $\psi^\varepsilon \in C^\infty(\mathbf{R}^{1+3})$ and, moreover,

$$\mathrm{i}h\partial_t\psi^\varepsilon + \frac{h^2}{2}\Delta\psi^\varepsilon = \chi_\varepsilon * (|\psi|^{p-1}\psi + V\psi), \qquad (3.4.12)$$

where V is the classical Hartree potential. Therefore, differentiating $|\psi^\varepsilon|^2$ with respect to time, we get

$$\begin{aligned}&\partial_t|\psi^\varepsilon|^2 \\ &= 2\mathrm{Re}(\overline{\psi^\varepsilon}\partial_t\psi^\varepsilon) = 2\mathrm{Re}\left(\overline{\psi^\varepsilon}\left(\frac{\mathrm{i}h^2}{2}\Delta\psi^\varepsilon - \frac{\mathrm{i}}{h}\chi_\varepsilon * (|\psi|^{p-1}\psi + V\psi)\right)\right) \\ &= -h\operatorname{div}(\mathrm{Im}(\overline{\psi^\varepsilon}\boldsymbol{\nabla}\psi^\varepsilon)) + \frac{2}{h}\mathrm{Im}(\overline{\psi^\varepsilon}\chi_\varepsilon * (|\psi|^{p-1}\psi + V\psi)).\end{aligned}$$

If we differentiate with respect to time h Im $(\overline{\psi^\varepsilon}\nabla\psi^\varepsilon)$, we get

$$\begin{aligned}
&\partial_t(h\,\mathrm{Im}(\overline{\psi^\varepsilon}\nabla\psi^\varepsilon))\\
&\quad= h\,\mathrm{Im}\left(-\frac{\mathrm{i}h^2}{2}\Delta\overline{\psi^\varepsilon}+\frac{\mathrm{i}}{h}\chi_\varepsilon*(|\psi|^{p-1}\psi)+\frac{\mathrm{i}}{h}\chi_\varepsilon*(V\psi)\nabla\psi^\varepsilon\right)\\
&\qquad+h\mathrm{Im}\left[\overline{\psi^\varepsilon}(\frac{\mathrm{i}h^2}{2}\Delta\nabla\psi^\varepsilon-\frac{\mathrm{i}}{h}\chi_\varepsilon*(|\psi|^{p-1}\psi)-\frac{\mathrm{i}}{h}\chi_\varepsilon*(V\psi)\right]\\
&\quad=\frac{h^2}{2}\mathrm{Re}(-\nabla\psi^\varepsilon\Delta\overline{\psi^\varepsilon}+\overline{\psi^\varepsilon}\Delta\nabla\psi^\varepsilon)\\
&\qquad+\mathrm{Re}(\chi_\varepsilon*(|\psi|^{p-1}\overline{\psi})\nabla\psi^\varepsilon-\overline{\psi^\varepsilon}\chi_\varepsilon*(|\psi|^{p-1}))\\
&\qquad+\mathrm{Re}(\chi_\varepsilon(V\overline{\psi})\nabla\psi^\varepsilon-\chi_\varepsilon*(V\psi)\overline{\psi^\varepsilon})=:A+B+C.
\end{aligned}$$

Let us discuss these three terms separately. For the first term A, it is immediate that

$$\begin{aligned}
&\frac{h^2}{2}\mathrm{Re}(-\nabla\psi^\varepsilon\Delta\overline{\psi^\varepsilon}+\overline{\psi^\varepsilon}\Delta\nabla\psi^\varepsilon)\\
&\qquad= h^2\mathrm{div}(\mathrm{Re}(\nabla\psi^\varepsilon\otimes\nabla\overline{\psi^\varepsilon}))+\frac{h^2}{4}\Delta\nabla|\psi|^2.
\end{aligned}\tag{3.4.13}$$

Let us deal with the second and the third terms, B and C. We can recast B in the following way

$$\begin{aligned}
&\mathrm{Re}(\chi_\varepsilon*(|\psi|^{p-1}\overline{\psi})\nabla\psi^\varepsilon-\overline{\psi^\varepsilon}\chi_\varepsilon*(|\psi|^{p-1}))\\
&\quad=\mathrm{Re}(|\psi|^{p-1}\overline{\psi^\varepsilon}\nabla\psi^\varepsilon-\overline{\psi^\varepsilon}\chi_\varepsilon*(\psi\nabla|\psi|^{p-1}+\nabla\psi|\psi|^{p-1}))+R_1^\varepsilon\\
&\quad=-\mathrm{Re}(\overline{\psi^\varepsilon}\chi_\varepsilon*(\psi\nabla|\psi|^{p-1})+R_1^\varepsilon+R_2^\varepsilon\\
&\quad=-|\psi^\varepsilon|^2\nabla|\psi^\varepsilon|^{p-1}+R_1^\varepsilon+R_2^\varepsilon+R_3^\varepsilon,
\end{aligned}$$

where

$$\begin{aligned}
R_1^\varepsilon&:=\mathrm{Re}((\chi_\varepsilon*(|\psi|^{p-1}\overline{\psi})-|\psi|^{p-1}\overline{\psi^\varepsilon})\nabla\psi^\varepsilon),\\
R_2^\varepsilon&:=\mathrm{Re}(\overline{\chi_\varepsilon}(|\psi^\varepsilon|^{p-1}\nabla\psi^\varepsilon-\chi_\varepsilon*(|\psi|^{p-1}\nabla\psi))),\\
R_3^\varepsilon&:=\mathrm{Re}(\overline{\psi^\varepsilon}(\psi|^\varepsilon\nabla|\psi^\varepsilon|^{p-1}-\chi_\varepsilon*(\psi|\nabla|\psi|^{p-1}))).
\end{aligned}$$

Now it only remains to analyze how the remainder terms $R_j^\varepsilon(j=1,2,3)$ go to zero. We will prove it in the next lemma. Furthermore, the following identity holds

$$|\psi^\varepsilon|^2 \nabla |\psi^\varepsilon|^{p-1} = \frac{p-1}{p+1} \nabla |\psi^\varepsilon|^{p+1}.$$

For the third term C, after similar computations we get

$$C = |\psi^\varepsilon|^2 \nabla (\chi * V) + R_4^\varepsilon,$$

where the remainder term R_4^ε will be analyzed in the next lemma. Hence we can conclude that $h\,\mathrm{Im}(\overline{\psi^\varepsilon} \nabla \psi^\varepsilon)$ satisfies

$$\begin{aligned}\partial_t(h\,\mathrm{Im}(\overline{\psi^\varepsilon} \nabla \psi^\varepsilon)) &= -h^2 \mathrm{div}(\mathrm{Re}(\nabla \psi^\varepsilon \otimes \nabla \overline{\psi^\varepsilon})) + \frac{h^2}{4} \Delta \nabla |\psi^\varepsilon|^2 \\ &\quad + \frac{p-1}{p+1} \nabla |\psi^\varepsilon|^{p+1} + |\psi^\varepsilon|^2 \nabla(\chi * V) + R^\varepsilon,\end{aligned}$$

where $R^\varepsilon := R_1^\varepsilon + R_2^\varepsilon + R_3^\varepsilon + R_4^\varepsilon$. Now, as $\varepsilon \to 0$, we get

$$\begin{aligned}\partial_t(h\,\mathrm{Im}(\overline{\psi^\varepsilon} \nabla \psi^\varepsilon)) &= h^2 \mathrm{div}(\mathrm{Re}(\nabla \psi^\varepsilon \otimes \nabla \overline{\psi \varepsilon})) + \frac{h^2}{4} \Delta \nabla |\psi^\varepsilon|^2 \\ &\quad + \frac{p-1}{p+1} \nabla |\psi^\varepsilon|^{p+1} + |\psi^\varepsilon|^2 \nabla V.\end{aligned}$$

This identity is equivalent to (3.4.10) since, by Lemma 3.3.1, we have

$$h^2 \mathrm{Re}(\nabla \overline{\psi} \otimes \nabla \psi) = h^2 \mathrm{Re}(\nabla \sqrt{\rho} \otimes \nabla \sqrt{\rho}) + \boldsymbol{\Lambda} \otimes \boldsymbol{\Lambda},$$

where $\sqrt{\rho}\boldsymbol{\Lambda} = \boldsymbol{J}, \boldsymbol{J} := h\,\mathrm{Im}(\overline{\psi} \nabla \psi)$.

Finally, let us note that, by the definition of $\boldsymbol{J}$, we have

$$\nabla \wedge \boldsymbol{J} = h(\nabla \overline{\psi} \wedge \nabla \psi),$$

then by Corollary 3.3.1, we get $\nabla \wedge \boldsymbol{J} = 2\nabla\sqrt{\rho} \wedge \boldsymbol{\Lambda}$.

Lemma 3.4.1. *Let* $0 < T < \infty$, *then*

$$\|R_j^\varepsilon\|_{L^1_{t,x}([0,T)\times \boldsymbol{R}^3)} \to 0, as\ \varepsilon \to 0.$$

Proof. It is a direct consequence of Strichartz estimates for the solution of the Schrödinger–Poisson system

$$\|\psi\|_{L_t^q W^{1,r}([0,T]\times\mathbf{R}^3)} \leqslant C(E_0, \|\psi_0\|_{L^2(\mathbf{R}^3)}, T),$$

where (q,r) is a pair of admissible exponents and E_0 is the initial energy. The first error can be controlled in the following way

$$\begin{aligned}
&\|R_1^\varepsilon\|_{L^1_{t,x}([0,T]\times\mathbf{R}^3)}\\
&\leqslant \|\chi_\varepsilon * (|\psi|^{p-1}\overline{\psi}) - |\psi^\varepsilon|^{p-1}\overline{\psi^\varepsilon}\|_{L_t^{\frac{4(p+1)}{p+7}} L_x^{\frac{p+1}{p}}([0,T]\times\mathbf{R}^3)}\\
&\quad\times\|\nabla\psi^\varepsilon\|_{L_t^{\frac{4(p+1)}{3(p-1)}} L_x^p([0,T]\times\mathbf{R}^3)}\\
&\leqslant T^{\frac{4(p+1)}{p+7}}(\|\chi_\varepsilon * (|\psi|^{p-1}\overline{\psi}) - |\psi|^{p-1}\overline{\psi}\|_{L_t^\infty L_x^{p+1}([0,T]\times\mathbf{R}^3)}\\
&\quad+\||\psi|^{p-1}\overline{\psi} - |\psi^\varepsilon|^{p-1}\overline{\psi^\varepsilon}\|_{L_t^\infty L_x^{p+1}([0,T]\times\mathbf{R}^3)})\\
&\quad\times\|\nabla\psi^\varepsilon\|_{L_t^{\frac{4(p+1)}{3(p-1)}} L_x^p([0,T]\times\mathbf{R}^3)}.
\end{aligned}$$

The remaining error terms can be computed in the same way. We remark that by using Strichartz estimates it follows $\psi\nabla|\psi|^{p-1}$ that lies in $L_t^{\frac{4(p+1)}{3(p-1)}} L_x^p([0,T]\times\mathbf{R}^3)$. □

3.5 Fractional step method: Definition and uniformity

In this section we make use of the results of the previous section to construct a sequence of approximate solutions of the QHD system

$$\partial_t\rho + \nabla\boldsymbol{J} = 0, \tag{3.5.1}$$

$$\partial_t\boldsymbol{J} + \operatorname{div}\left(\frac{\boldsymbol{J}\otimes\boldsymbol{J}}{\rho}\right) + \nabla P(\rho) + \rho\nabla V + \boldsymbol{J} = \frac{h^2}{2}\rho\nabla\left(\frac{\Delta\sqrt{\rho}}{\rho}\right), \tag{3.5.2}$$

$$-\Delta V = \rho, \tag{3.5.3}$$

with Cauchy data

$$\rho(0) = \rho_0, \boldsymbol{J}(0) = \boldsymbol{J}_0. \tag{3.5.4}$$

Definition 3.5.1. We say $\{(\rho^\tau, \boldsymbol{J}^\tau)\}$ is a sequence of approximate solutions to the system (3.5.1)–(3.5.4) in $[0,T) \times \mathbf{R}^n$, with initial data $(\rho_0, \boldsymbol{J}_0) \in L^1_{loc}(\mathbf{R}^3)$. If there exist locally integrable functions $(\sqrt{\rho^\tau}, \boldsymbol{\Lambda}^\tau)$, such that $\rho^\tau \in L^2_{loc}([0,T); H^1_{loc}(\mathbf{R}^n))$, $\boldsymbol{\Lambda}^\tau \in L^2_{loc}([0,T); L^2_{loc}(\boldsymbol{R}^n))$ and if we define $\rho^\tau := (\sqrt{\rho^\tau})^2, \boldsymbol{J} := \sqrt{\rho^\tau}\boldsymbol{\Lambda}^\tau$, then

1. for any test function $\eta \in C_0^\infty([0,T) \otimes \mathbf{R}^n)$, we have

$$\int_\Omega \int_{\mathbf{R}^n} (\rho^\tau \partial_t \eta + \boldsymbol{J}^\tau \cdot \boldsymbol{\nabla}_\eta) \mathrm{d}x\mathrm{d}t + \int_{\mathbf{R}^n} \rho_0 \eta(0) \mathrm{d}x = 0; \qquad (3.5.5)$$

   as $\tau \to 0$,
2. for any test function $\zeta \in C_0^\infty([0,T) \otimes \mathbf{R}^n; \mathbf{R}^n)$, we have

$$\int_\Omega \int_{\mathbf{R}^n} \bigg(\boldsymbol{J} \cdot \partial_t \zeta + \boldsymbol{\Lambda} \otimes \boldsymbol{\Lambda} : \boldsymbol{\nabla}\zeta + \boldsymbol{\nabla} P \mathrm{div}\zeta - \rho \boldsymbol{\nabla} V \cdot \zeta$$
$$+ h^2 \boldsymbol{\nabla}\sqrt{\rho} \otimes \sqrt{\rho} : \boldsymbol{\nabla}\zeta - \boldsymbol{J}\zeta + \frac{h^2}{4}\rho\Delta\mathrm{div}\zeta \bigg) \mathrm{d}x\mathrm{d}t$$
$$+ \int_{\mathbf{R}^n} \boldsymbol{J}_0 \zeta(0) \mathrm{d}x = 0; \qquad (3.5.6)$$

3. (generalized irrotationality condition) for almost every $t \in (0,T)$, we have

$$\boldsymbol{\nabla} \wedge \boldsymbol{J} = 2\boldsymbol{\nabla}\sqrt{\rho} \wedge \boldsymbol{\Lambda}. \qquad (3.5.7)$$

Our fractional step method is based on the following simple idea. We split the evolution of our problem into two separate steps. Let us fix a (small) parameter $\tau > 0$, then in the former step we solve a non-collisional QHD problem, while in the latter one we solve the collisional problem without QHD, and at this point we can start again with the non-collisional QHD problem. The main difficulty is the updating of the initial data at each time step. Indeed, as remarked in the previous chapter we are able to solve the non-collisional QHD only in the case of Cauchy data compatible with the Schrödinger picture. This restriction is imposed to reconstruct a wave function at each time step.

Now we explain how to set up the fractional step procedure which generates the approximate solutions. We first remark that, as in the previous section, this method requires special type of initial data $(\rho_0, \boldsymbol{J}_0)$. Namely, we assume that there exists $(\rho_0, \boldsymbol{J}_0)$ and that there exists $\psi_0 \in H^1(\boldsymbol{R}^3)$, such that the hydrodynamic initial data are given via the Madelung transforms

$$\rho_0 = |\psi_0|^2, \boldsymbol{J}_0 = h \operatorname{Im}(\overline{\psi_0} \boldsymbol{\nabla} \psi_0). \tag{3.5.8}$$

This assumption is physically relevant since it implies the compatibility of our solutions to the QHD problem with the wave mechanics approach. The iteration procedure can be defined in this following way. First of all, we take $\tau > 0$, which will be the time mesh unit. Therefore we define the approximate solutions in each time interval $[k\tau, (k+1)\tau)$, for any integer $k \geqslant 0$.

At the first step, $k = 0$, we solve the Cauchy problem for the Schrödinger–Poisson system

$$ih\partial_t \psi^\tau + \frac{h^2}{2} \Delta \psi^\tau = |\psi^\tau|^{p-1} \psi^\tau + V^\tau \psi^\tau, \tag{3.5.9}$$

$$-\Delta V^\tau = |\psi^\tau|^2, \tag{3.5.10}$$

$$\psi^\tau(0) = \psi_0, \tag{3.5.11}$$

by looking for the restriction of the unique strong solution $\psi \in C(\mathbf{R}; H^1(\mathbf{R}^3))$ in $[0, \tau)$.

Let us define $\rho^\tau := |\psi^\tau|^2$, $\boldsymbol{J}^\tau := h \operatorname{Im}(\psi^\tau \boldsymbol{\nabla} \psi^\tau)$. Then, as shown in the previous section, $(\rho^\tau, \boldsymbol{J}^\tau)$ is a weak solution to the non-collisional QHD system. Let us assume that we know ψ^τ in the space-time slab $[(k-1)\tau, k\tau) \times \mathbf{R}^3$ and we want to set up a recursive method. Hence we have to show how to define $\psi^\tau, \rho^\tau, \boldsymbol{J}^\tau$ in the strip $[k\tau, (k+1)\tau)$.

In order to take into account the presence of the collisional term $f = -\boldsymbol{J}$, we update ψ in $t = k\tau$. Namely, we define $\psi(k\tau+)$. The construction of $\psi(k\tau+)$ will be done by means of the polar decomposition described in the previous section.

Let us apply Lemma 3.3.2, with $\psi = \psi^\tau(k\tau-), \varepsilon = \tau 2^{-k} \|\psi_0\|_{H^1(\boldsymbol{R}^3)}$, then we can define

$$\psi^\tau(k\tau+) = \tilde{\psi}, \tag{3.5.12}$$

by using the wave function $\tilde{\psi}$. Therefore we have

$$\rho^\tau(k\tau+) = \rho^\tau(k\tau-), \tag{3.5.13}$$

$$\boldsymbol{\Lambda}^\tau(k\tau+) = (1-\tau)\boldsymbol{\Lambda}^\tau(k\tau-) + R_k, \tag{3.5.14}$$

where $\|R_k\|_{L^2(\boldsymbol{R}^3)} \leqslant \tau 2^{-k}\|\psi_0\|_{H^1(\boldsymbol{R}^3)}$ and

$$\boldsymbol{\nabla}\psi^\tau(k\tau+) = \boldsymbol{\nabla}\psi^\tau(k\tau-) - i\frac{\tau}{h}\psi^*\boldsymbol{\Lambda}^\tau(k\tau-) + r_{k,\tau}, \tag{3.5.15}$$

$\|\psi^*\|_{L^\infty} \leqslant 1$ and

$$\|r_{k,\tau}\|_{L^2} \leqslant C(\tau\|\boldsymbol{\nabla}\psi^\tau(k\tau-)\| + \tau 2^{-k}\|\psi_0\|_{H^1(\mathbf{R}^3)}) \leqslant \tau E_0^{\frac{1}{2}}. \tag{3.5.16}$$

We then solve the Schrödinger–Poisson system with initial data $\psi(0) = \psi^\tau(k\tau+)$. We define ψ^τ in the time strip $[k\tau, (k+1)\tau)$ as the restriction of the Schrödinger–Poisson solution just found in $[0,\tau)$. Furthermore, we define $\rho^\tau := |\psi^\tau|^2, \boldsymbol{J}^\tau := h\,\mathrm{Im}\,(\overline{\psi^\tau}\boldsymbol{\nabla}\psi^\tau)$ as the solution of the non-collisional QHD system.

With this procedure we can go on every time strip and then construct an approximate solution $(\rho^\tau, \boldsymbol{J}^\tau, V^\tau)$ of the QHD system.

Theorem 3.5.1. *Let us consider a sequence of approximate solutions $\{(\rho^{\tau k}, \boldsymbol{J}^{\tau k})\}_{k\geqslant 0}$ constructed via the fractional step method, and assume there exists $0 < T < \infty, \sqrt{\rho} \in L^2_{loc}([0,T]; H^1_{loc}(\boldsymbol{R}^n))$ and $\boldsymbol{\Lambda} \in L^2_{loc}([0,T]; L^2_{loc}(\boldsymbol{R}^n))$ such that*

$$\sqrt{\rho^{\tau k}} \to \sqrt{\rho}, in L^2_{loc}([0,T]; H^1_{loc}(\boldsymbol{R}^n)), \tag{3.5.17}$$

$$\boldsymbol{\Lambda}^{\tau k} \to \boldsymbol{\Lambda}, in L^2_{loc}([0,T]; L^2_{loc}(\boldsymbol{R}^n)). \tag{3.5.18}$$

Then the limit function $(\rho, \boldsymbol{J})$ is a weak solution of the QHD system in $[0,T) \times \boldsymbol{R}^n$, with Cauchy data $(\rho_0, \mathbf{J}_0)$.

Proof. Along this proof we omit the index k. Let us plug the approximate solutions $(\rho^\tau, \boldsymbol{J}^\tau)$ in the weak formulation and let

$\zeta \in C_0^\infty([0,T) \times \mathbf{R}^n)$, then we have

$$\int_0^\infty \int_{\mathbf{R}^n} [(\boldsymbol{J}^\tau \cdot \partial_t \boldsymbol{\zeta} + \boldsymbol{\Lambda}^\tau \otimes \boldsymbol{\Lambda}^\tau : \boldsymbol{\nabla}\boldsymbol{\zeta} + \boldsymbol{\nabla} P(\rho^\tau)\mathrm{div}\boldsymbol{\zeta} - \rho^\tau \boldsymbol{\nabla} V^\tau \cdot \boldsymbol{\zeta}$$

$$+ h^2 \boldsymbol{\nabla}\sqrt{\rho^\tau} \otimes \sqrt{\rho^\tau} : \boldsymbol{\nabla}\zeta - \frac{h^2}{4}\rho^\tau \Delta \mathrm{div}\zeta)]\mathrm{d}x\mathrm{d}t + \int_{\mathbf{R}^n} \boldsymbol{J}_0 \zeta(0)\mathrm{d}x$$

$$= \sum_{k=0}^\infty \int_{k\tau}^{(k+1)\tau} \int_{\mathbf{R}^n} [(\boldsymbol{J}^\tau \cdot \partial_t \zeta + \boldsymbol{\Lambda}^\tau \otimes \boldsymbol{\Lambda}^\tau : \boldsymbol{\nabla}\zeta$$

$$+ \boldsymbol{\nabla} P(\rho^\tau)\mathrm{div}\boldsymbol{\zeta} - \rho^\tau \boldsymbol{\nabla} V^\tau \cdot \zeta$$

$$+ h^2 \boldsymbol{\nabla}\sqrt{\rho^\tau} \otimes \sqrt{\rho^\tau} : \boldsymbol{\nabla}\zeta - \frac{h^2}{4}\rho^\tau \Delta \mathrm{div}\zeta)]\mathrm{d}x\mathrm{d}t + \int_{\mathbf{R}^n} \boldsymbol{J}_0 \zeta(0)\mathrm{d}x$$

$$= \sum_{k=0}^\infty \int_{k\tau}^{(k+1)\tau} \int_{\mathbf{R}^n} -\boldsymbol{J}^\tau \cdot \zeta \mathrm{d}x\mathrm{d}t$$

$$+ \int_{\mathbf{R}^n} \{\boldsymbol{J}^\tau[(k+1)\tau-] \cdot \zeta[(k+1)\tau]$$

$$- \boldsymbol{J}^\tau(k\tau) \cdot \zeta(k\tau)\}\mathrm{d}x + \int_{\mathbf{R}^n} J_0 \zeta(0)\mathrm{d}x$$

$$= \sum_{k=0}^\infty \int_{k\tau}^{(k+1)\tau} \int_{\mathbf{R}^n} -\boldsymbol{J}^\tau \cdot \zeta \mathrm{d}x\mathrm{d}t + \sum_{k=0}^\infty \int_{\mathbf{R}^n} [\boldsymbol{J}^\tau(k\tau-) - \boldsymbol{J}^\tau(k\tau)]$$

$$\cdot \zeta(k\tau)\mathrm{d}x$$

$$= \sum_{k=0}^\infty \int_{k\tau}^{(k+1)\tau} \int_{\mathbf{R}^n} -\boldsymbol{J}^\tau \cdot \zeta \mathrm{d}x\mathrm{d}t + \sum_{k=0}^\infty \tau \int_{\mathbf{R}^n} \boldsymbol{J}^\tau(k\tau-) \cdot \zeta(k\tau)\mathrm{d}x$$

$$+ \sum_{k=0}^\infty \int_{\mathbf{R}^n} (1-\tau)\sqrt{\rho^\tau(k\tau)} R_k \cdot \zeta(k\tau)\mathrm{d}x$$

$$= \sum_{k=0}^\infty \int_{k\tau}^{(k+1)\tau} \int_{\mathbf{R}^n} \{\boldsymbol{J}^\tau[(k+1)\tau-] \cdot \zeta[(k+1)\tau] - \boldsymbol{J}^\tau \cdot \zeta(t)\}\mathrm{d}x$$

$$+ (1-\tau)\sum_{k=0}^\infty \int_{\mathbf{R}^n} \sqrt{\rho^\tau(k\tau)} R_k \cdot \zeta(k\tau)\mathrm{d}x$$

$$= o(1) + O(\tau) as\ \tau \to 0.$$

□

3.6 A priori estimate and convergence

In this section we obtain various priori estimates necessary to show the compactness of the sequence of approximate solutions $(\rho^\tau, \boldsymbol{J}^\tau)$ in some appropriate function spaces. As we stated in Theorem 3.5.1, we wish to prove the strong convergence of $\{\sqrt{\rho^\tau}\}$ in $L^2_{loc}[0,T); H^1_{loc}(\mathbf{R}^3))$ and of $\{\boldsymbol{\Lambda}^\tau\}$ in $L^2_{loc}([0,T); L^2_{loc}(\boldsymbol{R}^3))$. To achieve this goal we use a compactness result in the class of the Aubin–Lions's type lemma. The plan of this section is as follows. First of all, we get a discrete version of the (dissipative) energy inequality for the system (3.1.1). Later we use the Strichartz estimates for $\{\boldsymbol{\nabla}\psi^\tau\}$. Consequently, via the Strichartz estimates and by using the local smoothing results of Theorems 3.2.3, 3.2.4, we deduce some further regularity properties of the sequence $\{\boldsymbol{\nabla}\psi^\tau\}$. In this way it is possible to get the regularity properties of the sequence $\{\sqrt{\rho^\tau}\}$ and hence to get the convergence of $\{\sqrt{\rho^\tau}\}, \{\boldsymbol{\Lambda}^\tau\}$.

Let us begin with the energy inequality. First of all, note that if we have a sufficiently regular solution of the QHD system in $[0,T)\times\mathbf{R}^n$, one has

$$E(t) = -\int_0^t \int_{\mathbf{R}^n} |\boldsymbol{\Lambda}|^2 \mathrm{d}x\mathrm{d}t + E_0, t \in [0,T). \tag{3.6.1}$$

Now we would like to find a discrete version of the energy dissipation for the approximate solutions.

Lemma 3.6.1 (Discrete Energy Inequality). *Let $(\rho^\tau, \boldsymbol{J}^\tau)$ be an approximate solution of the QHD system, with $0<\tau<1$. Then, for $t \in [N\tau, (N+1)\tau)$, we have*

$$E^\tau(t) \leqslant -\frac{\tau}{2}\sum_{k=1}^{N} \|\boldsymbol{\Lambda}(k\tau-)\|_{L^2(\boldsymbol{R}^n)} + (1+\tau)E_0. \tag{3.6.2}$$

Proof. For all $k \geqslant 1$, we have

$$\begin{aligned} E^\tau(k\tau+) - E^\tau(k\tau-) &= \int \left(\frac{1}{2}|\boldsymbol{\Lambda}^\tau(k\tau+)|^2 - \frac{1}{2}|\boldsymbol{\Lambda}^\tau(k\tau-)|^2\right)\mathrm{d}x \\ &= \frac{1}{2}\int [(-2\tau+\tau^2)\boldsymbol{\Lambda}^\tau(k\tau-)^2 - 2(1-\tau) \end{aligned}$$

$$
\begin{aligned}
&\times\mathbf{\Lambda}^\tau(k\tau-)\cdot R_k+|R_k|^2]\mathrm{d}x\\
&\leqslant\frac{1}{2}\int[(-2\tau+\tau^2)|\mathbf{\Lambda}^\tau(k\tau-)|^2-(1-\tau)\alpha|\\
&\times\mathbf{\Lambda}^\tau(k\tau-)|^2+\frac{1-\tau}{\alpha}|R_k|^2+|R_k|^2]\mathrm{d}x\\
&=\frac{1}{2}\int[(-2\tau+\tau^2+\alpha-\alpha\tau)|\\
&\times\mathbf{\Lambda}^\tau(k\tau-)|^2+\left(\frac{1-\tau+\alpha}{\alpha}\right)|R_k|^2]\mathrm{d}x.
\end{aligned}
$$

Here R_k denotes the error term as in expression (3.5.14). If we choose $\alpha=\tau$, it follows

$$
\begin{aligned}
E^\tau(k\tau+)-E^\tau(k\tau-)&\leqslant-\frac{\tau}{2}|\mathbf{\Lambda}^\tau(k\tau-)|_{L^2}^2+\frac{1}{2\tau}\|R_k\|_{L^2}\\
&\leqslant-\frac{\tau}{2}|\mathbf{\Lambda}^\tau(k\tau-)|_{L^2}^2+\tau^{2-k-1}\|\psi_0\|_{H^1}.
\end{aligned}
$$

The inequality (3.6.2) follows by summing up all the terms and by the energy conservation in each time strip $[k\tau,(k+1)\tau)$. □

Unfortunately the energy estimates are not sufficient to get enough compactness to show the convergence of the sequence of the approximate solutions. Indeed, from the discrete energy inequality, we get only the weak convergence of $\nabla\psi^\tau$ in $L^\infty([0,\infty);H^1(\mathbf{R}^3))$ and therefore the quadratic terms in (3.5.4) could exhibit some concentration phenomena in the limit.

More precisely, from energy inequality we get the sequence $\{\psi^\tau\}$ which is uniformly bounded in $L^\infty([0,\infty);H^1(\mathbf{R}^3))$. Hence there exists $\psi\in L^\infty([0,\infty);H^1(\mathbf{R}^3))$, such that, up to subsequences

$$
\psi^\tau\rightharpoonup\psi,\ \text{in}\ L^\infty([0,\infty);H^1(\mathbf{R}^3)).
$$

Therefore we get

$$
\begin{aligned}
\sqrt{\rho^\tau}&\rightharpoonup\sqrt{\rho},\text{in}\,L^\infty([0,\infty);H^1(\mathbf{R}^3));\\
\mathbf{\Lambda}^\tau&\rightharpoonup\mathbf{\Lambda},\text{in}\,L^\infty([0,\infty);L^2(\mathbf{R}^3)).
\end{aligned}
$$

The need to pass into the limit of the quadratic expressions leads us to look for priori estimates in stronger norms. The relationships with the Schrödinger equation bring naturally into this search of the Strichartz-type estimates. The following results are concerned with these estimates. However, they are not an immediate consequence of the Strichartz estimates for the Schrödinger equation since we have to take into account the effects of the updating procedure which we implement at each step.

Lemma 3.6.2. *Let ψ^τ be the wave function defined by the fractional step method, and let $t \in [\tau, (N+1)\tau)$. Then we have*

$$\nabla\psi^\tau(t) = U(t) - i\frac{\tau}{h}\sum_{k=1}^{N} U(t-k\tau)(\psi_k^\tau \Lambda^\tau(k\tau-))$$

$$-\mathrm{i}\int_0^t U(t-s)F(s)\mathrm{d}s + \sum_{k=1}^{N} U(t-k\tau)r_k^\tau \quad (3.6.3)$$

where as before $U(t)$ is the free Schrödinger group,

$$\|\psi_k^\tau\|_{L^\infty(\boldsymbol{R}^3)} \leqslant 1, \quad \|r_k^\tau\|_{L^2(\boldsymbol{R}^3)} \leqslant \tau\|\psi_0\|_{H^1(\boldsymbol{R}^3)} \quad (3.6.4)$$

and $F = \nabla(|\nabla\psi^\tau|^{p-1}\psi^\tau + V^\tau\psi^\tau)$.

Proof. Since ψ^τ is the solution of the Schrödinger–Poisson in the spacetime slab $[N_\tau, (N+1)\tau) \times \mathbf{R}^3$, then we can write

$$\psi^\tau(t) = U(t-N_\tau)\nabla\psi^\tau(N\tau+) - i\int_{N\tau}^t U(t-s)F(s)\mathrm{d}s, \quad (3.6.5)$$

where F is defined in the statement of Lemma 3.6.2. Now there exists a piecewise smooth function θ_N, such that

$$\psi(N\tau+) = e^{i(1-\tau)\theta_N}\sqrt{\rho n}, \quad (3.6.6)$$

with $\psi = \psi^\tau(N\tau-), \tilde{\psi} = \psi^\tau(N\tau+)$ and $\varepsilon = 2^{-N}\tau\|\psi\|_{H^1(\mathbf{R}^3)}$. Therefore, we have

$$\nabla\psi^\tau(N\tau+) = \nabla\psi^\tau(N\tau-) - \mathrm{i}\frac{\tau}{h}e^{i(1-\theta)\theta_N}\Lambda^\tau(N\tau-) + r_N^\tau, \quad (3.6.7)$$

where $\|r_N^\tau\|_{L^2} \leqslant \tau\|\psi_0\|_{H^1(\mathbf{R}^3)}$. By plugging (3.6.7) into (3.6.5) we deduce

$$\nabla\psi^\tau(t) = U(t-N\tau)\nabla\psi^\tau(N\tau-) - i\frac{\tau}{h}U(t-N\tau)(e^{i(1-\theta)\theta_N}\mathbf{\Lambda}^\tau(N\tau-)) + U(t-N\tau)r_N^\tau - i\int_{N\tau}^t U(t-s)F(s)\mathrm{d}s. \tag{3.6.8}$$

Let us iterate this formula, repeating the same procedure for $\nabla\psi^\tau(N\tau-)$, then (3.6.3) holds. □

Proposition 3.6.1 (Strichartz estimates for $\{\nabla\psi^\tau\}$). *Let $0 < T < \infty, \psi^\tau$ be as in the previous subsection, then one has*

$$\|\nabla\psi^\tau\|_{L_t^q L_r^x([0,T]\times\boldsymbol{R}^3)} \leqslant C(E_0^{\frac{1}{2}}, \|\rho_0\|_{L^1(\boldsymbol{R}^3)}, T) \tag{3.6.9}$$

for each admissible pair of exponents (q,r).

Proof. First of all, let us prove, for a small time $0 < T_1 \leqslant T$ and (q,r) be an admissible pair of exponents. We choose $T_1 > 0$ later. Let N be a positive integer such that $T_1 \leqslant N\tau$, we get

$$\begin{aligned}
&\|\nabla\psi^\tau\|_{L_t^q L_r^x([0,T_1]\times\mathbf{R}^3)} \\
&\leqslant \|U(t)\nabla\psi_0\|_{L_t^q L_r^x([0,T_1]\times\mathbf{R}^3)} \\
&\quad + \frac{\mathrm{i}}{h}\sum_{k=1}^{\tau}\|U(t-kr)(e^{\mathrm{i}(1-\theta)\theta_k}\mathbf{\Lambda}^\tau(N\tau-))\|_{L_t^q L_r^x([0,T_1]\times\mathbf{R}^3)} \\
&\quad + \sum_{k=1}^{\tau}\|U(t-k\tau)r_k^\tau\|_{L_t^q L_r^x([0,T_1]\times\mathbf{R}^3)} \\
&\quad + \|\int_0^t U(t-s)F(s)\mathrm{d}s\|_{L_t^q L_r^x([0,T_1]\times\mathbf{R}^3)} \\
&=: A+B+C+D.
\end{aligned}$$

Now we estimate term by term the expression above. The estimate of A is straightforward, since

$$\|U(t)\nabla\psi_0\|_{L_t^q L_r^x([0,T_1]\times\mathbf{R}^3)} \leqslant \|\nabla\psi_0\|_{L^2(\mathbf{R}^3)}. \tag{3.6.10}$$

The estimate of B follows from

$$\frac{\mathrm{i}}{h}\sum_{k=1}^{\tau}\|U(t-k\tau)(e^{\mathrm{i}(1-\theta)\theta_k}\mathbf{\Lambda}^{\tau}(N\tau-))\|_{L_t^q L_r^x([0,T_1]\times\mathbf{R}^3)}$$

$$\leqslant \sum_{k=1}^{N}\|\mathbf{\Lambda}^{\tau}(N\tau-)\|_{L^2(\mathbf{R}^3)} \leqslant T_1 E_0^{\frac{1}{2}}.$$

The term C can be estimated in a similar way, namely,

$$\sum_{k=1}^{\tau}\|U(t-k\tau)r_k^{\tau}\|_{L_t^q L_r^x([0,T_1]\times\mathbf{R}^3)} \lesssim \sum_{k=1}^{\tau}\|r_k^{\tau}\|_{L^2(\mathbf{R}^3)} \lesssim T_1\|\psi_0\|_{H^1(\mathbf{R}^3)}.$$

The last term is a little bit more tricky to estimate. First of all, we decompose F into three terms, $F =: F_1 + F_2 + F_3$, where $F_1 = \mathbf{\nabla}(|\psi^{\tau}|^{p-1}\psi^{\tau}), F_2 = \mathbf{\nabla}V^{\tau}\psi^{\tau}, F_3 = V^{\tau}\mathbf{\nabla}\psi^{\tau}$. By the Strichartz estimates, we have

$$\begin{aligned}\|\int_0^t U(t-s)F(s)\mathrm{d}s\|_{L_t^q L_x^r([0,T_1]\times\mathbf{R}^3)} \preceq \|F_1\|&_{L_t^{q_1'} L_x^{r_1'}([0,T_1]\times\mathbf{R}^3)}\\ +\|F_2\|&_{L_t^{q_2'} L_x^{r_2'}([0,T_1]\times\mathbf{R}^3)}\\ +\|F_3\|&_{L_t^{q_3'} L_x^{r_3'}([0,T_1]\times\mathbf{R}^3)} \quad \square\end{aligned}$$

where (q_i', r_i') are pairs of admissible exponents. Let us start with the first term.

Lemma 3.6.3. *There exists $\alpha > 0$, depending on p, such that*

$$\||\psi^{\tau}|^{p-1}\mathbf{\nabla}\psi^{\tau}\|_{L_t^{\bar{q}'} L_x^{\bar{r}'}([0,T]\times\boldsymbol{R}^3)} \preceq T_1^{\alpha}\|psi_{\tau}\|_{\dot{S}([0,T_1]\times\boldsymbol{R}^3)}. \quad (3.6.11)$$

Proof. First of all, let us apply Hölder inequality, we have

$$\begin{aligned}&\||\psi^{\tau}|^{p-1}\mathbf{\nabla}\psi_{\tau}\|_{L_t^{\bar{q}'} L_x^{\bar{r}'}([0,T]\times\mathbf{R}^3)}\\ &\preceq T_1^{\alpha}\||\psi^{\tau}|^{p-1}_{L_t^{q1} L_x^{r1}([0,T_1]\times\mathbf{R}^3)}\|\mathbf{\nabla}\psi_{\tau}\|_{L_t^{q1} L_x^{r1}([0,T_1]\times\mathbf{R}^3)}\\ &= T_1^{\alpha}\|\psi^{\tau}\|^{p-1}_{L_t^{q1(p-1)} L_x^{r1(p-1)}([0,T_1]\times\mathbf{R}^3)}\|\mathbf{\nabla}\psi_{\tau}\|_{L_t^{q1} L_x^{r1}([0,T_1]\times\mathbf{R}^3)}.\end{aligned} \quad (3.6.12)$$

Now we want $\frac{1}{q1(p-1)} = \frac{3}{2}\left(\frac{1}{6} - \frac{1}{r_1(p-1)}\right)$ and $\frac{1}{q2} = \frac{3}{2}(\frac{1}{2} - \frac{1}{r_2})$, in such a way that $\|f\|^{p-1}_{L_t^{q1(p-1)}L_x^{r1(p-1)}}, \|\nabla f\|_{L_t^{q1}L_x^{r1}} \leqslant \|f\|_{\dot{S}^1} = \|\nabla f\|_{\dot{S}^0}$. We already know $\frac{1}{\tilde{q}'} = \frac{3}{2}(\frac{1}{2} - \frac{1}{\tilde{r}'})$, then putting together the conditions on $(\tilde{q}, \tilde{r}), (q_j, r_j)$, it follows

$$\begin{aligned}\frac{1}{\tilde{q}'} = \frac{1}{\alpha} + \frac{1}{q1} + \frac{1}{q2} &= \frac{1}{\alpha} + (p-1)\frac{3}{2}\left(\frac{1}{6} - \frac{1}{r_1(p-1)}\right) + \frac{3}{2}\left(\frac{1}{2} - \frac{1}{r_2}\right) \\ &= 1 + \frac{3}{2}\left(\frac{1}{2} - \frac{1}{\tilde{r}'}\right)\end{aligned}$$

and when $1 \leqslant p < 5$,

$$\alpha = \frac{5-p}{4} > 0.$$

This means that we can always choose pairs $(\tilde{q}', \tilde{r}'), (q_1, r_1), (q_2, r_2)$ satisfying the previous conditions so that the inequality (3.6.12) holds, with $\alpha > 0$. For instance, if $1 \leqslant p \leqslant 3$, we can choose $\frac{1}{r1} = \frac{p-1}{6}, \frac{1}{r2} = \frac{1}{2}$ then $q1 = q2 = \infty$, hence we have $\frac{1}{\tilde{r}'} = \frac{2+p}{6}, \frac{1}{\tilde{q}'} = \frac{5-p}{4}$. In the case $3 \geqslant p < 5$, we take $\frac{1}{r1} = \frac{p-1}{6}, \frac{1}{r2} = \frac{1}{6}$, then $q1 = \infty, q2 = 2$, hence we have $\frac{1}{\tilde{r}'} = \frac{p}{6}, \frac{1}{\tilde{q}'} = \frac{7-p}{4}$. □

Now let us consider the second term: $V^\tau \nabla \psi^\tau$, here we choose $(q_2', r_2') = (1, 2)$, so by the Hardy–Littlewood–Sobolev and the Hölder inequalities, one has

$$\begin{aligned}\|V^\tau \nabla \psi^\tau\|_{L_t^1 L_x^2([0,T]\times\mathbf{R}^3)} &\leqslant T_1^{\frac{1}{2}} \|V^\tau\|_{L_t^\infty L_x^2} \|\nabla \psi^\tau\|_{L_t^2 L_x^6} \\ &\preceq T_1^{\frac{1}{2}} \|\psi^\tau\|^2_{L_t^\infty L_x^2} \|\nabla \psi^\tau\|_{L_t^2 L_x^6}.\end{aligned}$$

For the third term, we choose $(q_2', r_2') = (\frac{2}{2-3\varepsilon}, \frac{2}{1+2\varepsilon})$, and again by using the Hardy–Littlewood–Sobolev and the Hölder inequalities, we have

$$\|\nabla V^\tau \psi^\tau\|_{L_t^{\frac{2}{2-3\varepsilon}} L_x^{\frac{2}{1+2\varepsilon}}([0,T]\times\mathbf{R}^3)} \leqslant T_1^{\frac{1}{2}} \|\nabla V^\tau\|_{L_t^{\frac{2}{1-3\varepsilon}} L_x^{\frac{1}{\varepsilon}}} \|\psi^\tau\|_{L_t^\infty L_x^2}$$

$$\preceq T_1^{\frac{1}{2}} \|\nabla |\psi^\tau|^2\|_{L_t^{\frac{2}{1-3\varepsilon}} L_x^{\frac{3}{2+3\varepsilon}}} \|\tilde{\psi}^\tau\|_{L_t^\infty L_x^2}$$

$$\preceq T_1^{\frac{1}{2}} \|\nabla |\psi^\tau|^2\|_{L_t^{\frac{2}{1-3\varepsilon}} L_x^{\frac{6}{1+6\varepsilon}}} \|\psi^\tau\|^2_{L_t^\infty L_x^2}.$$

Now, we summarize the previous estimates in the following way

$$\begin{aligned} \|\nabla\psi^\tau\psi^\tau\|_{\dot{S}^0([0,T_1]\times\mathbf{R}^3)} &\preceq \|\nabla\psi_0\|_{L^2(\mathbf{R}^3)} + T_1 E_0^{\frac{1}{2}} \\ &\quad + T_1^\alpha \|\nabla\psi^\tau\psi^\tau\|^p_{\dot{S}^1([0,T]\times\mathbf{R}^3)} \\ &\leqslant T_1^{\frac{1}{2}} \|\psi_0\|_{L^2(\mathbf{R}^3)} \|\nabla\psi^\tau\psi^\tau\|_{\dot{S}^1([0,T]\times\mathbf{R}^3)} \\ &\preceq (1+T)E_0^{\frac{1}{2}} + T_1^\alpha \|\nabla\psi^\tau\psi^\tau\|^p_{\dot{S}^1([0,T]\times\mathbf{R}^3)} \\ &\leqslant T_1^{\frac{1}{2}} \|\psi_0\|_{L^2(\mathbf{R}^3)} \|\nabla\psi^\tau\psi^\tau\|_{\dot{S}^1([0,T]\times\mathbf{R}^3)}. \end{aligned}$$

Lemma 3.6.4. *There exists* $T_1(E_0, \|\psi_0\|_{L^2(\boldsymbol{R}^3)}, T) > 0$ *and* $C_1(E_0, \|\psi_0\|_{L^2(\boldsymbol{R}^3)}, T) > 0$, *independent on* τ, *such that*

$$\|\nabla\psi^\tau\|_{\dot{S}^0([0,\tilde{T}]\times\boldsymbol{R}^3)} \leqslant C_1(E_0^{\frac{1}{2}}, \|\rho_0\|_{L^1(\boldsymbol{R}^3}, T), \tag{3.6.13}$$

for all $0 < \tilde{T} \leqslant T_1(E_0, \|\rho_0\|_{L^1(\boldsymbol{R}^3)})$.

Proof. Let us consider the non-trivial case $\|\psi_0\|_{L^2} > 0$. Assume that $X \in (0,\infty)$ satisfies

$$X \leqslant A + \mu X + \lambda X^p = \psi(X), \tag{3.6.14}$$

with $p > 1, A > 0$ and for all $0 < \mu < 1, \lambda > 0$. Let X_* be such that $\psi'(X_*) = 1$, namely $X_* = (\frac{1-mu}{p\lambda})^{\frac{1}{p-1}}$, hence one has $\psi(X_*) < X_*$ and each time the following inequality is satisfied

$$\left(\frac{1}{\frac{1}{p^{p-1}}} - \frac{1}{\frac{1}{\frac{p}{p^{p-1}}}}\right) \frac{(1-\mu)^{\frac{p}{p-1}}}{\lambda^{\frac{1}{p-1}}} > A. \tag{3.6.15}$$

Therefore the convexity of ψ implies that, if the condition (3.6.15) holds, there exist two roots $X_\pm, X_+(\mu,\lambda,A) > X_-(\mu,\lambda,A)$, to the equation $\phi(X) = X$. Then it follows either $0 \leqslant X \leqslant X_-$, or $X \geqslant X_+$.

In the our case $\mu = T_1^{1/2}\|\psi_0\|_{L^2}^2$, $\lambda = T_1^\alpha$, $A = (1+T)E_0^{1/2}$, hence we assume

$$\mu = T_1^{1/2}\|\psi_0\|_{L^2}^2 < \frac{1}{2},$$

$$\lambda = T_1^\alpha = T_1^{\frac{5-p}{4}} < \left[\frac{p^{-\frac{1}{p-1}} - p^{-\frac{p}{p-1}}}{2^{\frac{p}{p-1}}(1+T)E_0^{1/2}}\right]^{p-1}.$$

Therefore we choose

$$T_1 := \min\left[(2\|\psi_0\|_{L^2}^2)^{-2}, \left[\frac{p^{-\frac{1}{p-1}} - p^{-\frac{p}{p-1}}}{2^{\frac{p}{p-1}}(1+T)E_0^{1/2}}\right]^{\frac{4(p-1)}{5-p}}\right]. \tag{3.6.16}$$

Clearly we cannot have

$$X_* = \left(\frac{1 - T_1\|\psi_0\|_{L^2}}{pT_1^\alpha}\right)^{\frac{1}{p-1}} \leqslant X_+ \leqslant \|\nabla\psi^\tau\|_{\dot{S}^0([0,\tilde{T}]\times\mathbf{R}^3)},$$

since we get a contradiction as $T_1 \to 0$,

$$\|\nabla\psi^\tau\|_{\dot{S}^0([0,\tilde{T}]\times\mathbf{R}^3)} \leqslant X_-.$$

□

Corollary 3.6.1. *Let* $0 < T < \infty$, *and* $\sqrt{\rho^\tau}, \mathbf{\Lambda}^\tau$ *be as in previous subsection, then*

$$\|\nabla\sqrt{\rho^\tau}\|_{L_t^q L_r^x([0,T]\times\boldsymbol{R}^3)} + \|\nabla\sqrt{\mathbf{\Lambda}^\tau}\|_{L_t^q L_r^x([0,T]\times\boldsymbol{R}^3)}$$
$$\leqslant C(E_0^{\frac{1}{2}}, \|\rho_0\|_{L^1(\boldsymbol{R}^3)}, T),$$

for each admissible pair of exponents (q, r).

Unfortunately, this is not enough to achieve the convergence of the quadratic terms. We need some additional compactness estimates on the sequence $\{\nabla\psi^\tau\}$ in order to apply Theorem 3.2.5. In particular, we need some tightness and regularity properties on the sequence $\{\nabla\psi^\tau\}$. Therefore we apply some results concerning local smoothing.

Proposition 3.6.2 (Local smoothing for $\{\nabla\psi^\tau\}$). *Let $0 < T < \infty$, then one has $C_1(E_0, \|\psi_0\|_{L^2(\mathbf{R}^3)}, T) > 0$, such that*

$$\|\nabla\psi^\tau\|_{L^2([0,T];H^{1/2}_{loc}(\mathbf{R}^3))} \leqslant C(E_0, \|\rho_0\|_{L^1(\mathbf{R}^3)}, T) \tag{3.6.17}$$

Proof. Using the Strichartz estimates obtained above, we can apply Theorems 3.2.3, 3.2.4, and we deduce

$$\begin{aligned}
\|\nabla\psi^\tau\|_{L^2([0,T];H^{1/2}(\mathbf{R}^3))} \preceq \|\nabla\psi_0\|_{L^2(\mathbf{R}^3)} &+ \frac{\mathrm{i}}{h}\sum_{k=1}^{N}\|\mathbf{\Lambda}^\tau(k\tau-)\|_{L^2(\mathbf{R}^3)} \\
&+ \tau\sum_{k=1}^{N}\|\nabla\psi_{nk}\|_{L^2(\mathbf{R}^3)} \\
&+ \sum_{k=1}^{N}\|\nabla\psi_{nk}\nabla\psi^\tau(k\tau-) \\
&+ \frac{\tau}{h}(\mathbf{\Lambda}_{nk} - \mathbf{\Lambda}^\tau(k\tau-))\|_{L^2(\mathbf{R}^3)} \\
&+ \|F\|_{L^1([0,T];L^2(\mathbf{R}^3))}.
\end{aligned}$$

The first three terms are clearly estimated by a constant $C(E_0, T)$ depending on the initial energy and time. The fourth term is $O(\tau)$. The last term will be estimated using the previous Strichartz estimates. As before, we split F into three parts $F =: F_1 + F_2 + F_3$, then one has

$$\begin{aligned}
&\||\psi^\tau|^{p-1}\nabla\psi^\tau\|_{L^1_tL^2_x([0,T]\times\mathbf{R}^3)} \\
&\leqslant T^{\frac{4}{5-p}}\||\psi^\tau|^{p-1}\|_{L_t^{\frac{4}{p-1}}L_x^\infty([0,T]\times\mathbf{R}^3)}\|\nabla\psi^\tau\|_{L_t^\infty L_x^2([0,T]\times\mathbf{R}^3)} \\
&\leqslant T^{\frac{4}{5-p}}\|\psi^\tau\|^{(p-1)}_{L_t^4L_x^\infty([0,T]\times\mathbf{R}^3)}\|\nabla\psi^\tau\|_{L_t^\infty L_x^2([0,T]\times\mathbf{R}^3)},
\end{aligned}$$

while

$$\begin{aligned}
&\|\nabla V^\tau\psi^\tau\|_{L^1_tL^2_x([0,T]\times\mathbf{R}^3)} \leqslant T^{\frac{1}{2}}\|\nabla V^\tau\|_{L_t^{\frac{2}{1-2\varepsilon}}L_x^{\frac{1}{\varepsilon}}([0,T]\times\mathbf{R}^3)} \\
&\times\|\psi^\tau\|_{L_t^{\frac{2}{3\varepsilon}}L_x^{\frac{2}{1-2\varepsilon}}([0,T]\times\mathbf{R}^3)},
\end{aligned}$$

and now the remaining calculations are similar to those already done for the Strichartz estimates. Regarding the term $V^\tau \nabla \psi^\tau$ we already estimated its $L_t^1 L_x^2$ norm. □

Since $H_{loc}^{1/2}$ is compactly embedded in L_{loc}^2, we can apply Theorem 3.2.5.

Proposition 3.6.3. *The sequence $\{\nabla \psi^\tau\}$ is, up to subsequences, relatively compact in $L^2([0,T]; L_{loc}^2(\boldsymbol{R}^3))$, namely,*

$$\|\nabla \psi\| := s - \lim_{k\to\infty} \nabla \psi^{\tau_k} in\, L^2([0,T]; L_{loc}^2(\boldsymbol{R}^3)). \tag{3.6.18}$$

In particular, one has $\nabla\sqrt{\rho^\tau} \to \nabla\sqrt{\rho}$ and $\boldsymbol{\Lambda}^\tau \to \boldsymbol{\Lambda}$ in $L^2([0,T]; L_{loc}^2(\boldsymbol{R}^3))$.

Proof. Proposition 3.6.2 implies that the sequence $\{\nabla\psi^\tau\}_{\tau>0}$ is uniformly bounded in $L^2([0,T]; H_{loc}^{1/2}(\mathbf{R}^3))$ and then, up to subsequences, $\nabla\psi^\tau \rightharpoonup \nabla\psi$ in $L^2([0,T]; H_{loc}^{1/2}(\mathbf{R}^3))$. Now $H_{loc}^{1/2}$ is compactly embedded in L_{loc}^2, since $\nabla\psi^\tau(t) \rightharpoonup \nabla\psi(t)$ for all most every $t \geqslant 0$ and

$$\lim_{|E|\to, E\subset[0,T]} \sup_{\tau>0} \int_E \|\nabla\psi^\tau(t)\|_{L_{loc}^2}^2 \,\mathrm{d}t = 0,$$

as $\nabla\psi^\tau \in L^\infty([0,T]; L^2(\boldsymbol{R}^3))$. Hence we can apply Theorem 3.2.5 and we get (3.6.18). □

Proposition 3.6.4. *$(\rho, \boldsymbol{J})$ is a weak solution to Cauchy problems* (3.1.1) *and* (3.1.2).

Proof. It follows directly by combining Theorem 3.5.1 and Proposition 3.6.2 □

3.7 Further generalization

3.7.1 *Case with impurity distribution*

In this section, we are going to study the QHD system in which we have a non-zero doping profile. That is, in the Poisson equation for the electrostatic potential there is a given non- zero source. The QHD system indeed reads:

$$\begin{cases} \partial_t \rho + \operatorname{div} \boldsymbol{J} = 0, \\ \partial_t \boldsymbol{J} + \operatorname{div}\left(\dfrac{\boldsymbol{J} \otimes \boldsymbol{J}}{\rho}\right) + \boldsymbol{\nabla} P(\rho) + \rho \boldsymbol{\nabla} V \\ \quad = \dfrac{h^2}{2} \rho \boldsymbol{\nabla}\left(\dfrac{\Delta\sqrt{\rho}}{\sqrt{\rho}}\right) \\ + f(\sqrt{\rho}, \boldsymbol{J}, \boldsymbol{\nabla}\sqrt{\rho} \boldsymbol{\nabla}^2 \sqrt{\rho} \boldsymbol{\nabla}(\boldsymbol{J}/\sqrt{\rho})), \\ -\Delta V = \rho - C(x), \end{cases} \tag{3.7.1}$$

where $C(x)$ is the doping profile, on which we make the following assumption:

$$C \in L^{p1}(\mathbf{R}^3) + L^{p2}(\mathbf{R}^3), \quad \text{where } 6/5 < p_1 < p_2 < 3/2. \tag{3.7.2}$$

Actually, this is the important case in the semiconductor device modeling, but we postponed the discussion to this chapter from a mathematical point of view. This is completely similar to the problem treated in the previous chapters. Only there are some further calculations to do, due to the presence of a source in the Poisson equation. Hence what we are going to do in this section would be to follow the steps that we already studied in the previous chapters, writing down the missing calculations involving the doping profile. Hence, also for this slightly modified system of PDEs we can state a result of the global existence of the finite energy weak solutions, similar to the main Theorem 3.1.1.

Theorem 3.7.1. *Let $\psi \in H^1(\boldsymbol{R}^3)$ and let us define*

$$\rho_0 := |\psi_0|^2, \boldsymbol{J}_0 := h \operatorname{Im}(\overline{\psi_0} \boldsymbol{\nabla} \psi_0).$$

Then for each $0 < T < \infty$, there exists a finite energy weak solution to the QHD system (3.7.1), (3.7.2) in $[0, T) \times \boldsymbol{R}^n$, with initial data $(\rho_0, \boldsymbol{J}_0)$.

Remark 3.7.1 Let us remark that each function C satisfying the assumption in (3.7.2) can be written as the sum of two functions $C = C_1 + C_2$, where $C_1 \in L^{r1}, C_2 \in L^{r2}, r_1 > 6/5$ is close to $6/5$

and $r_2 < 3/2$ is close to $3/2$. So, from now on we will assume that $p_1 > 6/5$ is close to $6/5$, and $p_2 < 3/2$ is close to $3/2$, and we will write $p_1 = 6/5 + \varepsilon_1, p_2 = 3/2 - \varepsilon_2$.

In the case of a non-zero doping profile, we split V into two parts $V := V_1 + V_2$, where

$$-\Delta V_1 = \rho,\ -\Delta V_2 = -C(x).$$

In this way we can use the same fractional step method as in the zero doping profile case, but now on each space-time slab we solve the following non-linear Schrödinger–Poisson system

$$\begin{cases} i h\partial_t \psi + \dfrac{h^2}{2}\Delta\psi = |\psi|^{p-1}\psi + V_1\psi + V_2\psi, \\ -\Delta V_1 = |\psi|^2, \end{cases} \tag{3.7.3}$$

where V_2 is the solution of the linear elliptic equation $\Delta V_2 = C$ in $\boldsymbol{R}^3$. That is

$$V_2 = -\int_{\mathbf{R}^3} \frac{1}{|x-y|} C(y)\mathrm{d}y. \tag{3.7.4}$$

By applying the Hardy's inequality, it is easy to see that

$$V_2 \in L^{6+\varepsilon_1}(\mathbf{R}^3) + L^{\infty-\varepsilon_2}(\mathbf{R}^3), \tag{3.7.5}$$

$$V_2 \in L^{2+\varepsilon_1}(\mathbf{R}^3) + L^{3-\varepsilon_2}(\mathbf{R}^3). \tag{3.7.6}$$

Remark 3.7.2 Of course let us note that $\varepsilon_1, \varepsilon_2$ in the definition of p_1, p_2 and in the Lebesgue spaces above are not the same, but clearly they are related each other.

By standard theory on non-linear Schrödinger equation, the system (3.7.3) is globally well-posed. Thus, what it only remains to do is to give some Strichartz estimates as in previous section, in order to apply the same compactness tools as we did previously.

Proposition 3.7.1. *Let $\psi \in C([0,\infty); H^1(\boldsymbol{R}^3))$ be a solution to the Cauchy problem associated to equation (3.7.3), and let $0 < T < \infty$. Then we have*

$$\|\nabla\psi\|_{L_t^q L_x^r([0,T]\times\boldsymbol{R}^3)} \leqslant C(T, \|\psi_0\|_{L^2(\boldsymbol{R}^3)}, E_0), \tag{3.7.7}$$

for each admissible pair (q,r) of Schrödinger exponents in $\boldsymbol{R}^3$.

Proof. By applying the Strichartz estimates of Theorem 3.2.2 to the solution of equation (3.2.7), we obtain

$$\|\nabla\psi\|_{L_t^q L_x^r} \lesssim \||\psi|^{p-1}\nabla\psi_0\|_{L^2} + \|\nabla\psi\|_{L_t^{q_1'} L_x^{r_1'}} + \|\nabla V_1\psi\|_{L_t^{q_2'} L_x^{r_2'}}$$
$$+ \|V_1\nabla\psi\|_{L_t^{q_3'} L_x^{r_3'}} + \|\nabla V_2\psi\|_{L_t^{q_4'} L_x^{r_4'}} + \|V_2\nabla\psi\|_{L_t^{q_5'} L_x^{r_5'}},$$

clearly all the terms except $\nabla V_2\psi$ and $V_2\nabla\psi$ can be estimated as before. Thus let us estimate the $\nabla V_2\psi$ and $V_2\nabla\psi$. First consider $\nabla V_2\psi$ and write it as

$$(\nabla V_2)^1\psi + (\nabla V_2)^2\psi, \ \textit{where}\ (\nabla V_2)^1 \in L^{2+\varepsilon 1}(\mathbf{R}^3), (\nabla V_2)^2$$
$$\times \in L^{3-\varepsilon_2}(\mathbf{R}^3). \tag{3.7.8}$$

By the Hölder inequality, for each one of these terms we have

$$\|(\nabla V_2)^j\psi\|_{L_t^{\overline{q}'} L_x^{\overline{r}'}} \leqslant T^\alpha \|(\nabla V_2)^j\|_{L_x^{r_j}} \|\psi\|_{L_t^q L_x^r}, j = 1, 2, \tag{3.7.9}$$

where the Lebesgue exponents satisfy the following algebraic conditions:

$$\begin{cases} \dfrac{1}{\overline{q}'} = 1 + \dfrac{3}{2}\left(\dfrac{1}{2} - \dfrac{1}{\overline{r}'}\right), \\ \dfrac{1}{q} = \frac{3}{2}\left(\dfrac{1}{6} - \dfrac{1}{r}\right), \alpha = \dfrac{3}{2}\left(1 - \dfrac{1}{q_j}\right), \end{cases}$$

where the last condition is deduced from the first two. Note that as above the conditions on the exponents ensure us that $\alpha \in (0, 1)$. This means that by applying a standard bootstrap argument, the proposition is proved.

Now we can come back to the proof of Proposition 3.7.1. By following the same steps as in the previous chapters we can apply the same fractional step method, with the same updating arguments, and then, we can recover a formula.

$$\begin{aligned}\nabla\psi^\tau(t) = U(t) - \mathrm{i}\frac{\tau}{h}\sum_{k=1}^{N} U(t-k\tau)(\psi_k^\tau \boldsymbol{\Lambda}^\tau(k\tau-))\\ - \mathrm{i}\int_0^t U(t-s)F(s)\mathrm{d}s + \sum_{k=1}^{N} U(t-k\tau)r_k^\tau,\end{aligned} \tag{3.7.10}$$

but where now $F = \nabla(|\psi^\tau|^{p-1}\psi^\tau + (V_1^\tau + V_2^\tau)\psi^\tau)$. Thus we can use the Strichartz estimates to get a bound like the following one

$$\|\nabla\psi^\tau\|_{L_t^q L_x^r([0,T]\times\mathbf{R}^3)} \leqslant C(T, \|\psi_0\|_{L^2(\mathbf{R}^3)}, E_0). \tag{3.7.11}$$

Now, as in Section 3.2 we need the local smoothing estimates of Theorems 3.2.3 and 3.2.4 in order to have the sufficient compactness on the sequence of the approximate solutions and in order to pass to the limit the quadratic terms in (3.3.5). We proceed as in Section 3.2. Clearly the only new term is

$$\|\nabla(V_2\psi^\tau)\|_{L^1([0,T];L^2(\boldsymbol{R}^3))}. \tag{3.7.12}$$

As before, we decompose it into four parts $\nabla(V_2\psi^\tau) = (\nabla V_2)^1\psi^\tau + (\nabla V_2)^2\psi^\tau + V_2^1\nabla\psi^\tau + V_2^2\nabla\psi^\tau$, where

$$\begin{cases}(\nabla V_2)^1 \in L^{2+\varepsilon_1}(\mathbf{R}^3)\\ (\nabla V_2)^2 \in L^{3+\varepsilon_1}(\mathbf{R}^3)\\ V_2^1 \in L^{6+\varepsilon_1}(\mathbf{R}^3)\\ V_2^2 \in L^{\infty-\varepsilon_2}(\mathbf{R}^3).\end{cases}$$

Now we are going to estimate $\|\nabla(V_2\psi^\tau)\|_{L_t^1 L_x^2}$ term by term: for the first one we have

$$\begin{aligned}\|(\nabla V_2)^1\psi^\tau\|_{L^1([0,T];L^2)} \leqslant T^{\frac{1}{q'}}\|(\nabla V_2)^1\|_{L^{2+\varepsilon_1}}\|\psi^\tau\|_{L_t^q L_x^r} \lesssim T^{\frac{1}{q'}}\\ \times\|(\nabla V_2)^1\|_{L^{2+\varepsilon_1}}\|\psi^\tau\|_{S^0},\end{aligned} \tag{3.7.13}$$

where $1/r$ is small and $\frac{1}{q} = \frac{3}{2}(\frac{1}{6} - \frac{1}{r})$. For the other terms the process is similar. Thus resuming, we get the following bound

$$\|\nabla\psi^\tau\|_{L^2([0,T];H_{loc}^{1/2}(\mathbf{R}^3))} \leqslant C(T, \|\psi_0\|_{L^2(\mathbf{R}^3)}, E_0), \tag{3.7.14}$$

which allows us to apply the theorem by Rakotoson, Temam. We can finally say there exists $\psi \in L^\infty([0,\infty); H^1(\mathbf{R}^3))$ such that $(\rho, \boldsymbol{J})$ defined by $(\rho, \boldsymbol{J}) := (|\psi|^2, h\,\mathrm{Im}(\overline{\psi}\boldsymbol{\nabla}\psi))$ is a finite energy weak solution of (3.7.1) and (3.7.2). Thus Proposition 3.7.1 is proved.

3.7.2 *Two dimensional case*

In this subsection we treat the 2D analogue for the Cauchy problem of the QHD system.

$$\begin{cases} \partial_t \rho + \mathrm{div}\boldsymbol{J} = 0, \\ \partial_t \boldsymbol{J} + \mathrm{div}\left(\dfrac{\boldsymbol{J}\otimes\boldsymbol{J}}{\rho}\right) + \boldsymbol{\nabla}P(\rho) + \rho\boldsymbol{\nabla}V = \dfrac{h^2}{2}\rho\boldsymbol{\nabla}\left(\dfrac{\Delta\sqrt{\rho}}{\sqrt{\rho}}\right), \\ -\Delta V = \rho, \end{cases} \tag{3.7.15}$$

with initial condition

$$\rho(0) = \rho_0, \boldsymbol{J}(0) = \boldsymbol{J}_0, \tag{3.7.16}$$

such that there exists a wave function $\psi_0 \in H^1(\mathbf{R}^2)$ whose $\rho_0, \boldsymbol{J}_0$ are its first two momentum:

$$\rho_0 := |\psi_0|^2, \boldsymbol{J}_0 := h\,\mathrm{Im}(\overline{\psi_0}\boldsymbol{\nabla}\psi_0).$$

Clearly, in order to prove the global existence result for a weak solution to (3.7.15) and (3.7.16), we are going to use the same tools that we developed for the three-dimensional case. However, there are some minor changes. First of all, in the two-dimensional case we can treat a more general pressure term, allowing a power-like behavior for $P(\rho) = \frac{p-1}{p+1}\rho^{\frac{P+1}{2}}$, for each $1 \leqslant p < \infty$. Another important change is that now the electrostatic potential is more singular, in the sense that the Green's function for the Poisson equation in $\mathbf{R}^2$ is

$$\Phi(x) := -\frac{1}{2\pi}\log|x|, x \in \mathbf{R}^2,$$

and consequently the electrostatic potential which is given by

$$V(t,x) := -\frac{1}{2\pi}\int_{\mathbf{R}^2} \log|x-y|\rho(t,y)\mathrm{d}y \tag{3.7.17}$$

could be unbounded at infinity, even if we assume ρ is sufficiently rapidly decaying at infinity. Thus some hypotheses on the electrostatic potential at the initial time are required, in order to assure it is bounded and lies in some Lebesgue spaces. We can now state the main theorem of the subsection.

Theorem 3.7.2. *Let $\rho_0, \boldsymbol{J}_0$ be such that there exists $\psi \in H^1(\boldsymbol{R}^3)$ satisfying*

$$\rho_0 := |\psi_0|^2, \boldsymbol{J}_0 := h \operatorname{Im}(\overline{\psi_0} \boldsymbol{\nabla} \psi_0).$$

Furthermore, let us assume that

$$\int_{\boldsymbol{R}^2} \rho_0 \log \rho_0 \mathrm{d}x < \infty, \tag{3.7.18}$$

and that

$$V(0,x) := -\frac{1}{2\pi} \int_{\boldsymbol{R}^2} \log|x-t| \rho_0(y) \mathrm{d}y \tag{3.7.19}$$

satisfies $V(0,\cdot) \in L^r(\boldsymbol{R}^2)$, for some $2 < r < \infty$. Then, for $0 < T < \infty$, there exists a finite energy weak solution of (3.7.15) *and* (3.7.16), *in $[0,T) \times \boldsymbol{R}^n$.*

As we already saw for the three-dimensional case, if we drop out the collision term $\boldsymbol{J}$ in the equation for the current in (3.7.15), then the QHD system (3.7.15) is closely related to the following non-linear Schrödinger equation

$$\begin{cases} ih\partial_t \psi + \dfrac{h^2}{2} \Delta \psi = |\psi|^{p-1}\psi + V_\psi, \\ -\Delta V = |\psi|^2, \end{cases} \tag{3.7.20}$$

with initial data

$$\psi(0) = \psi_0. \tag{3.7.21}$$

Since we can recover a finite energy weak solution $(\rho, \boldsymbol{J})$ to (3.7.15) by defining $(\rho, \boldsymbol{J}) := (|\psi|^2, h \operatorname{Im}(\overline{\psi} \boldsymbol{\nabla} \psi))$, where ψ is a solution to (3.7.15).

As in the three-dimensional case we want to use the fractional step method in order to construct a sequence of approximate solutions for the system (3.7.15), thus we need a global well-posedness result for the Cauchy problem associated to the equation (3.7.20), and also we need the solution of (3.7.20) which satisfies some Strichartz estimates, in order to obtain the sufficient compactness properties for the sequence of approximate solutions and to pass it to the limit. Thus we now state a global well-posedness result for (3.7.20).

Theorem 3.7.3. *Let us consider the Cauchy problem* (3.7.15) *with initial data* $\psi(0) = \psi_0$ *such that*

$$\psi_0 \in H^1(\boldsymbol{R}^3), \int_{\boldsymbol{R}^2} |\psi_0|^2 \log |\psi_0|^2 \mathrm{d}x < \infty \tag{3.7.22}$$

and such that

$$V(0,\cdot) := -\frac{1}{2\pi}\int_{\boldsymbol{R}^2} \log|\cdot - y|\rho_0(y)\mathrm{d}y \in L^r(\boldsymbol{R}^2), \tag{3.7.23}$$

for some $2 < r < \infty$. *Then, there exists a unique solution* $\psi \in C([0,\infty); H^1(\boldsymbol{R}^3))$ *to* (3.7.20) – (3.7.21). *Furthermore, for every* $T > 0$ *and for every* (q, r) *admissible pair of Schrödinger exponents in* $\boldsymbol{R}^2$, *we have*

$$\|\nabla\psi\|_{L_t^q L_x^r([0,T]\times\boldsymbol{R}^3)} \leqslant C(T, \|\psi_0\|_{L^2(\boldsymbol{R}^3)}, E_0). \tag{3.7.24}$$

Remark 3.7.3. As we will see, hypothesis (3.7.22) is needed to ensure the energy which is bounded, where the energy for (3.7.20) is defined by

$$\begin{aligned} E(t) &:= \int_{\mathbf{R}^2} \left(\frac{h^2}{2}|\nabla\psi|^2 + \frac{2}{p+1}|\psi|^{p+1} + \frac{1}{2}V|\psi|^2\right)\mathrm{d}x \\ &= \int_{\mathbf{R}^2} \left(\frac{h^2}{2}|\nabla\psi|^2 + \frac{2}{p+1}|\psi|^{p+1}\right)\mathrm{d}x \\ &\quad -\frac{1}{4\pi}\int\int_{\mathbf{R}^2\times\mathbf{R}^2} \rho(t,x)\log|x-y|\rho(t,y)\mathrm{d}x\mathrm{d}y, \end{aligned} \tag{3.7.25}$$

while hypothesis (3.7.22) is needed in order to ensure the electrostatic potential which lies in some Lebesgue spaces for all times.

Proof. Let us remark that now we are going to deal with the case in which the power-like non-linearity $|\psi|^{p-1}\psi$ satisfies $1 < p < \infty$; the case $p = 1$ easily follows with slighter changes. Firstly, by the logarithmic Sobolev inequality we can say that the initial energy is bounded, and furthermore, since it is conserved for all times, we have

$$E(t) = E_0 \lesssim \|\psi_0\|^2_{L^2(\mathbf{R}^2)} + \|\nabla\psi_0\|^2_{L^2(\mathbf{R}^2)} + \int_{\mathbf{R}^2} \rho_0 \log\rho_0 \mathrm{d}x < \infty. \tag{3.7.26}$$

Secondly, since we have $\psi \in H^1(\boldsymbol{R}^2)$, from the Gagliardo–Nirenberg–Sobolev inequality we can say that $\psi \in L^p(\mathbf{R}^2), 2 \leqslant p < \infty$, and consequently $\rho \in L^p(\mathbf{R}^2)$, for each $1 \leqslant p < \infty$.

Now, we want to give a local well-posedness result for the solutions to (3.7.20). By standard arguments, it is sufficient to give a priori estimate for a solution to (3.7.20) in the Strichartz spaces. But to apply Strichartz estimates, we should first see in which spaces do the electrostatic potential to live. □

Lemma 3.7.1. *Let V satisfy the Poisson equation of* (3.7.20) *in* $[0,T) \times \boldsymbol{R}^2$ *for some* $T > 0$, *and be such that* $V(0) \in L^r(\boldsymbol{R}^2)$, *for some* $2 \leqslant p < \infty$. *Then*

$$\begin{cases} \nabla V \in L_t^\infty L_x^p([0,T) \times \boldsymbol{R}^2) p \in [2,\infty), \\ V \in L_t^\infty L_x^r([0,T) \times \boldsymbol{R}^2). \end{cases} \tag{3.7.27}$$

Proof. Clearly by the elliptic regularity theory and the Sobolev embedding theorem we have

$$\|\nabla V\|_{L^\infty L^2} \leqslant \||\nabla|^{-1}\rho\|_{L^\infty L^2} \lesssim \|\rho_0\|_{L^1}. \tag{3.7.28}$$

Furthermore, by taking the gradient of (3.7.4), we get

$$\nabla V(t,x) = -\frac{1}{2\pi}\int_{\mathbf{R}^3} \frac{|x-y|}{|x-y|^2}\rho(t,y)\mathrm{d}y, \tag{3.7.29}$$

and by using the generalized Hardy–Littlewood–Sobolev inequality we obtain that $\nabla V \in L^\infty([0,T); L^p(\mathbf{R}^2))$, for $2 < p < \infty$. Resuming, we can say $\nabla V \in L^\infty([0,T); L^p(\mathbf{R}^2)), 2 \leqslant p < \infty$.

Now let us differentiate with respect to time the formula (3.7.4), thus we obtain, by using the continuity equation,

$$\partial_t V(t,x) = -\frac{1}{2\pi}\int_{\mathbf{R}^3} \frac{|x-y|}{|x-y|^2}\,\boldsymbol{J}(t,y)\mathrm{d}y, \tag{3.7.30}$$

and consequently

$$V(t,x) = V(0,x) - \frac{1}{2\pi}\int_0^t \int_{\mathbf{R}^3} \frac{|x-y|}{|x-y|^2}\cdot \boldsymbol{J}(s,y)\mathrm{d}y\mathrm{d}s. \tag{3.7.31}$$

Now, from the bounds on the energy and by the Gagliardo–Nirenberg–Sobolev inequality we can see that $\boldsymbol{J} \in L^\infty([0,T); L^p(\mathbf{R}^2))$, for $1 \leqslant p < 2$. Thus by using the generalized Hardy–Littlewood–Sobolev inequality we can say that $\partial_t V \in L^\infty([0,T); L^r(\mathbf{R}^2)), 1 \leqslant p < 2$, with $2 < r < \infty$. Hence,

$$\|V(t)\|_{L^r} \leqslant \|V(0)\|_{L^r} + t\|\partial_t V\|_{L_t^\infty L_x^r([0,\infty)\times\mathbf{R}^2)}, 2 < r < \infty, \tag{3.7.32}$$

and consequently $V \in L_t^\infty L_x^r([0,\infty)\times\mathbf{R}^2)$.

Now let us come back to the proof of Theorem 3.7.3. By Strichartz estimates we have

$$\begin{aligned}\|\boldsymbol{\nabla}\psi\|_{L_t^q L_x^r([0,T)\times\mathbf{R}^2)} \lesssim \|\boldsymbol{\nabla}\psi_0\|_{L^2(\mathbf{R}^2)} + \||\psi|^{p-1}\boldsymbol{\nabla}\psi\|_{L_t^{q_1'} L_x^{r_1'}([0,T)\times\mathbf{R}^2)}\\ + \|V\boldsymbol{\nabla}\psi\|_{L_t^{q_2'} L_x^{r_2'}} + \|\boldsymbol{\nabla}V\psi\|_{L_t^{q_3'} L_x^{r_3'}},\end{aligned}$$

where (q_i', q_i') are dual exponents of an admissible pair for Schrödinger in $\mathbf{R}^2$. Let us consider the second term on the right hand side, $\||\psi|^{p-1}\boldsymbol{\nabla}\psi\|_{L_t^{q_1'} L_x^{r_1'}([0,T)\times\mathbf{R}^2)}$. By Hölder inequality we have

$$\||\psi|^{p-1}\boldsymbol{\nabla}\psi\|_{L_t^{q_1'} L_x^{r_1'}([0,T)\times\mathbf{R}^2)} \leqslant T^{1/q_1'}\|\psi\|^{p-1}_{L_t^\infty L_x^{r(p-1)}}\|\boldsymbol{\nabla}\psi\|_{L_t^\infty L_x^2},$$

where we have the following algebraic conditions on the exponents r_1', r,

$$\begin{cases} 0 < \dfrac{1}{r} \leqslant \dfrac{p-1}{2}, \\ \dfrac{1}{2} \leqslant \dfrac{1}{r_1'} = \dfrac{1}{r} + \dfrac{1}{2}. \end{cases}$$

Then, by applying the Gagliardo–Nirenberg–Sobolev inequality we get

$$\begin{aligned}\||\psi|^{p-1}\nabla\psi\|_{L_t^{q_1'}L_x^{r_1'}([0,T)\times\mathbf{R}^2)} &\leqslant T^{1/q_1'}\|\psi\|^{p-1}_{L_t^\infty L_x^{r(p-1)}}\|\nabla\psi\|_{L_t^\infty L_x^2}\\ &\lesssim T^{1/q_1'}\|\psi\|^{\frac{2}{r}}_{L_t^\infty L_x^2}\|\nabla\psi\|^{\frac{r(p-1)-2}{r(p-2)}(p-1)}_{L_t^\infty L_x^2}\\ &\quad\times\|\nabla\psi\|_{L_t^\infty L_x^2}=T^{1/q_1'}\|\psi\|^{\frac{2}{r}}_{L_t^\infty L_x^2}\\ &\quad\times\|\nabla\psi\|^{\frac{rp-2}{r}}_{L_t^\infty L_x^2}.\end{aligned}$$

For the term $\|V\nabla\psi\|_{L_t^{q_2'}L_x^{r_2'}}$ we choose $(q_2', r_2')=(r', \frac{2r}{r+2})$, where r is the exponent such that $V(0)\in L^r$, and hence

$$\begin{aligned}\|V\nabla\psi\|_{L_t^{r'}L_x^{\frac{2r}{r+2}}} &\leqslant T^{\frac{1}{r'}}\|V\|_{L_t^\infty L_x^r}\|\nabla\psi\|_{L_t^\infty L_x^2}\\ &\lesssim T^{\frac{1}{r'}}(\|V(0)\|_{L_x^r}+T\|\partial_t V\|_{L_t^\infty L_x^r})\|\nabla\psi_{L_t^\infty L_x^2}.\end{aligned}$$

Finally, for the last term $\|\nabla V\psi\|_{L_t^{q_3'}L_x^{r_3'}}$, we can easily choose three exponents $r_3', p, p1$, such that

$$\begin{aligned}\|\nabla V\psi\|_{L_t^{q_3'}L_x^{r_3'}} &\leqslant T^{\frac{1}{q_3'}}\|\nabla V\|_{L_t^\infty L_x^p}\|\psi\|_{L_t^\infty L_x^{p1}}\\ &\lesssim T^{\frac{1}{q_3'}}\|\nabla V\|_{L_t^\infty L_x^p}\|\psi\|^{\frac{2}{p1}}_{L_t^\infty L_x^{p1}}\|\nabla\psi\|^{1-\frac{2}{p1}}_{L_t^\infty L_x^{p1}},\end{aligned}$$

and

$$\begin{cases}0<\dfrac{1}{p}\leqslant\dfrac{1}{2}, 0<\dfrac{1}{p1}\leqslant\dfrac{1}{2},\\ \dfrac{1}{2}\leqslant\dfrac{1}{r_3'}=\dfrac{1}{p}+\dfrac{1}{p1}<1.\end{cases}$$

Hence we obtain the following estimate

$$\|\nabla\psi\|_{L_t^qL_x^r([0,T)\times\mathbf{R}^2)}\lesssim\|\nabla\psi_0\|_{L^2(\mathbf{R}^2)}+T^\alpha C(\|\psi_0\|_{L^2}, E_0), \quad (3.7.33)$$

for some $\alpha < 0$. Thus by a standard argument we can state that there exists a $T^* = T^*(\|\psi_0\|_{L^2}, \|\nabla\psi_0\|_{L^2})$, such that there exists a unique solution

$$\psi \in C([0,T^*); H^1(\mathbf{R}^2)) \cap L^q([0,T^*); L^r(\mathbf{R}^2)),$$
$$\text{for every admissible pair } (q,r).$$

Furthermore, again by standard arguments, namely as the energy is conserved for every time, we can repeat the same argument for each space-time slab $[kT^*, (k+1)T^*]$. Since T^* depends only on the initial data, we can extend this solution globally.

Hence Theorem 3.7.3 is proved.

Now we use Theorem 3.7.3 to construct, as in the three-dimensional case, a sequence of approximate solutions, just by repeating the same updating procedure.

Further, since again we have the formula (3.6.3), for $t \in [N_\tau, (N+1)\tau)$,

$$\begin{aligned} \nabla\psi^\tau(t) &= U(t)\nabla\psi_0 - \mathrm{i}\int_0^t U(t-s)F(s)\mathrm{d}s \\ &\quad - \frac{\tau}{h}\sum_{k=1}^N U(t-k\tau)[\phi_k^\tau \mathbf{\Lambda}^\tau(k\tau-)+] + \sum_{k=1}^N U(t-k\tau))r_k^\tau \end{aligned} \tag{3.7.34}$$

holds, we can exploit the Strichartz estimates to obtain a bound like the one in (3.7.24),

$$\|\nabla\psi^\tau\|_{L_t^q L_x^r([0,T]\times\mathbf{R}^3)} \leqslant C(T, \|\psi\|_{L^2}, E_0), \tag{3.7.35}$$

for every admissible pair (q,r) of Schrödinger exponents in $\mathbf{R}^2$. Hence, in order to obtain the sufficient compactness properties on the sequence of approximate solutions $\{\nabla\psi^\tau\}$, it only remains to recover the smoothing estimates for $\{\nabla\psi^\tau\}$. But this is straightforward since, by Theorems 3.2.3 and 3.2.4, we have

$$\begin{aligned}
\|\nabla\psi^\tau\|_{L^2([0,T];H^{1/2}_{loc}(\mathbf{R}^2))} &\preceq \|\nabla\psi_0\|_{L^2} + \||\psi^\tau|^{p-1}\nabla\psi^\tau\|_{L^1_tL^2_x} \\
&\quad + \|V\nabla\psi\|_{L^1_tL^2_x} + \|\nabla V\psi\|_{L^1_tL^2_x} \\
&\preceq \|\nabla\psi_0\|_{L^2} + T^{\frac{1}{q_1'}}\|\psi\|^{\alpha}_{L^\infty_t L_x^{r_2(p-1)}}\|\nabla\psi\|_{L^{q1}_tL^{r1}_x} \\
&\quad + T_1^{\frac{1}{r'}}\|V\|_{L^\infty_tL^r_x}\|\nabla\psi\|_{L^r_tL_x^{\frac{2r}{r-2}}} \\
&\quad + T\|\nabla V\|_{L^\infty_tL^p_x}\|\psi\|_{L^\infty_tL_x^{\frac{2p}{p-2}}} \\
&\preceq \|\nabla\psi_0\|_{L^2} + T^{\frac{1}{q_1'}}\|\psi\|^{\frac{2}{r2}}_{L^\infty_tL^2_x}\|\nabla\psi\|^{\frac{r2(p-1)-2}{r2}}_{L^\infty_tL^2_x} \\
&\quad \times T^{\frac{1}{r_1'}}\|V\|_{L^\infty_tL^2_x}\|\nabla\psi\|^{\frac{2r}{r-2}}_{L^\infty_tL^2_x} \\
&\quad + T\|\nabla V\|_{L^\infty_tL^p_x}\|\psi\|^{1-\frac{p}{2}}_{L^\infty_tL^2_x}\|\nabla\psi\|^{\frac{p}{2}}_{L^\infty_tL^2_x},
\end{aligned} \tag{3.7.36}$$

where (q_1, r_1) is a Schrödinger admissible pair in $\mathbf{R}^2, r_2$ is such that $2 \leqslant r_2(p-1) < \infty, r$ is such that $V(0) \in L^r$ and $2 < p < \infty$. Note that in the second inequality we used Hölder inequality, while in the third one we used the Gagliardo–Nirenberg–Sobolev inequality.

Hence, by the estimate (3.7.33) we can say

$$\|\nabla\psi^\tau\|_{L^2([0,T];H^{1/2}_{loc}(\mathbf{R}^2))} \leqslant C(T, \|\psi_0\|_{L^2}, E_0). \tag{3.7.37}$$

Therefore, we can apply the Theorem 3.2.5 by Rakotoson, Temam. This means there exists $\psi \in L^\infty([0,T]; H^1(\mathbf{R}^2)) \cap L^2([0,T]; H^{3/2}_{loc}(\mathbf{R}^2))$ such that

$$\psi^\tau \to \psi \text{ strongly in } L^\infty([0,T]; H^1(\mathbf{R}^2)). \tag{3.7.38}$$

Finally, by applying the consistency Theorem 3.5.1 we can state that $(\rho, \boldsymbol{J}) := (|\psi|^2, h\ \mathrm{Im}(\overline{\psi}\nabla\psi))$ is a finite energy weak solution to (3.7.15) and (3.7.16). Thus Theorem 3.7.2 is proved.

Chapter 4

Non-isentropic Quantum Navier–Stokes Equations with Cold Pressure

In this chapter we mainly introduce a non-isothermal quantum Navier–Stokes model with cold pressure effect. We consider the following quantum hydrodynamic model for Cauchy problem

$$\partial_t \rho + \operatorname{div}(\rho \boldsymbol{u}) = 0, \tag{4.0.1}$$

$$\begin{aligned}&\partial_t(\rho \boldsymbol{u}) + \operatorname{div}(\rho \boldsymbol{u} \otimes \boldsymbol{u}) + \boldsymbol{\nabla} P - 2h^2 \operatorname{div}(\rho(\boldsymbol{\nabla} \otimes \boldsymbol{\nabla}) \log \rho) \\ &\quad = \nu \operatorname{div}(\rho D(\boldsymbol{u})),\end{aligned} \tag{4.0.2}$$

$$\begin{aligned}&\partial t(\rho E) + \operatorname{div}(\rho E \boldsymbol{u}) + \operatorname{div}(P \boldsymbol{u}) - 2h^2 \operatorname{div}(\rho \boldsymbol{u}(\boldsymbol{\nabla} \otimes \boldsymbol{\nabla}) \log \rho) \\ &\quad - h^2 \operatorname{div}(\rho \Delta \boldsymbol{u}) = \operatorname{div} q + \nu \operatorname{div}(\rho D(\boldsymbol{u}) \boldsymbol{u}),\end{aligned} \tag{4.0.3}$$

with the total energy E, the thermal diffusion flux q and symmetric part of the velocity gradient $D\ (\boldsymbol{u})$ respectively.

$$\begin{aligned}&\rho E = \rho e + \frac{1}{2}\rho|\boldsymbol{u}|^2 - h^2 \rho \Delta \log \rho, q = \kappa(\rho, \theta)\boldsymbol{\nabla}\theta, D(\boldsymbol{u}) \\ &\quad = \frac{\boldsymbol{\nabla}\boldsymbol{u} + (\boldsymbol{\nabla}\boldsymbol{u})^{\mathrm{T}}}{2},\end{aligned}$$

where the unknown functions $\rho, \boldsymbol{u}, \theta, e$ denote the density, velocity, temperature of the fluid and internal energy, P is the pressure field, κ is the thermal conductivity coefficient, h is Plank constant, ν is the viscosity coefficient.

System (4.0.1)–(4.0.3) supplemented with initial conditions:

$$\rho|_{t=0} = \rho_0, \rho\boldsymbol{u}|_{t=0} = m_0, \ \rho E|_{t=0} = (\rho E)_0.$$

Here we assume the functions ρ_0, m_0 satisfy:

$$m_0 = 0 \ a.e. \text{ on } \{x \in R^n : \ \rho_0 = 0\}. \tag{4.0.4}$$

Interestingly, if we know that the solutions are smooth, quantum terms can be cancelled in the energy equation. In fact, by substituting the above expression for the energy density in the equation in (4.0.3), a computation yields

$$\partial_t(\rho e) + \text{div}\,(\rho e\boldsymbol{u}) + P\boldsymbol{u} = \text{div}(\kappa(\rho,\theta)\boldsymbol{\nabla}\theta) + \nu\rho|D(\boldsymbol{u})|^2. \tag{4.0.5}$$

In the following we will prove the global existence of weak solutions of the system (4.0.1)–(4.0.3) for large initial values. The main idea is divided into several steps. Firstly, using the Galerkin method we can project the system into finite-dimensional space. Next, by adding to the system the artificial high regularizes terms, we can get more regularities of the solution which can insure the existence of the solutions. Therefore we construct the approximate solutions of the hydrodynamic quantum system. In the next step, we can establish the local existence for these approximate solutions through the fixed-point theorem. Also, we can deduce the uniform estimates which are independent of dimension. Thus the global existence of approximate solutions is proved. Finally, according to the physical energy estimate, entropy estimate and B-D effective energy estimate which are independent of the regularized parameters, we can pass the limit with these parameters that the limit solutions satisfy the definitions of the origin system. Thus the main theorem is proved.

4.1 Preliminaries and main result

4.1.1 *Preliminaries*

This subsection deals with assumptions regarding physical coefficients, such as initial values, the thermal conductivity coefficient κ and the equation of state.

1. Initial values: the initial state of the system is determined by ρ_0, m_0, $(\rho E)_0$. The initial density ρ_0 is a non-negative measure function satisfying:

$$\rho_0 \in L^{5\gamma^+/3}(\Omega), \quad 1/\rho_0 \in L^{(5\gamma^+-1)/3}(\Omega), \quad \int_\Omega \rho_0 \mathrm{d}x = M_0 > 0, \tag{4.1.1}$$

the initial momentum $\boldsymbol{m}_0$ satisfying a compatibility condition:

$$\boldsymbol{m}_0 = 0 \ a.e. \text{ on } \{x \in \mathbf{R}^n : \rho_0 = 0\}, \tag{4.1.2}$$

and the knetic energy satisfying

$$\int_\Omega \frac{|\boldsymbol{m}_0|^2}{\rho_0} dx \leqslant \infty. \tag{4.1.3}$$

Finally we assume that the initial energy is finite, more specifically

$$E_0 = \int_\Omega \left(\frac{|\boldsymbol{m}_0|^2}{2\rho_0} + \rho_0 e(\rho_0, \theta_0) + h^2 |\boldsymbol{\nabla} \sqrt{\rho_0}|^2 \right) \mathrm{d}x \leqslant \infty. \tag{4.1.4}$$

2. The thermal conductivity coefficient κ: the assumption of the growth conditions satisfies:

$$\kappa(\rho, \theta) = \kappa_0 + \rho + \rho\theta^2 + \beta\theta^B, \tag{4.1.5}$$

where $\kappa_0 > 0$, $B \geqslant 8$, $\beta > 0$ are constants.

3. The equation of state: we assume that the state P and e satisfy the following growth condition:

$$\begin{aligned} P &= P_m + P_r = R\rho\theta + P_c + \frac{\beta}{3}\theta^4, \\ e &= e_m + e_r = C_\mu \theta + e_c + \beta\frac{\theta^4}{\rho}, \end{aligned} \tag{4.1.6}$$

where R and C_μ are two positive constants, $P_m = R\rho\theta + P_c$ denotes the internal pressure, $P_r = \frac{\beta}{3}\theta^4$ is the radiative pressure. Moreover, the additional pressure and the internal energy P_c and e_c are associated with the "zero Kelvin isothermal". We require that e_c

is a C^2 non-negative function on R_+ and the following constraint is satisfied in order to satisfy assumption

$$P_c(\rho) = \rho^2 \frac{de_c}{d\rho}(\rho). \tag{4.1.7}$$

We also require that P_c is a continuous function satisfying the following growth condition

$$P_c'(\rho) = \begin{cases} c_2\rho^{-\gamma^- -1} & \text{when} \quad \rho \leqslant 1, \\ c_3\rho^{\gamma^+ -1} & \text{when} \quad \rho > 1, \end{cases} \tag{4.1.8}$$

for positive constants c_2, c_3 and $\gamma^-, \gamma^+ > 1$. It was proposed in [1] to encompass plasticity and elasticity effect of solid materials, for which low densities may lead to negative pressures. By this modification, the compactness of velocity can be obtained. In later section we will use the notation: $P_\beta = R\rho\theta + \frac{\beta}{3}\theta^4$.

The second law of thermodynamics asserts the irreversible transfer of the mechanical energy into heat valid for all physical systems. Therefore there exists a state function and we call it entropy function, satisfying Gibbs relation:

$$\theta Ds(\rho,\theta) = De(\rho,\theta) + P(\rho,\theta)D\left(\frac{1}{\rho}\right), \quad \text{for } \rho,\theta > 0. \tag{4.1.9}$$

By this relation, if we know that the solution is smooth, we can deduce the evolution of the entropy which can be described by the following equation:

$$\partial_t(\rho s) + \operatorname{div}(\rho s\boldsymbol{u}) - \operatorname{div}\left(\frac{\kappa(\rho,\theta)\boldsymbol{\nabla}\theta}{\theta}\right) = \frac{\kappa(\rho,\theta)|\boldsymbol{\nabla}\theta|^2}{\theta^2} + \frac{\upsilon\rho|D(\boldsymbol{u})|^2}{\theta}. \tag{4.1.10}$$

In addition, the form of entropy is defined as follows

$$s = s_m + s_r, \quad \frac{\partial s_m}{\partial\theta} = \frac{1}{\theta}\frac{\partial e_m}{\partial\theta}, \quad \rho s_r = \frac{4}{3}\beta\theta^3, \tag{4.1.11}$$

where

$$|s_m(\rho,\theta)| \leqslant C(1 + |\log\rho| + |\log\theta|), \text{ for all } \rho,\theta > 0. \tag{4.1.12}$$

In order to utilize this Gibbs relation for obtaining *a priori* bound, we introduce a remarkable quantity:

$$H_{\bar{\theta}}(\rho,\theta) = \rho(e(\rho,\theta) - \bar{\theta}s(\rho,\theta)), \tag{4.1.13}$$

where $\bar{\theta}$ is a positive constant.

4.1.2 *Main result*

Before we state the main result,we need to specify the definition of weak solutions which we will address. It is necessary to require that the weak solutions should satisfy the natural energy estimates and from the viewpoint of physics, the conservation laws on mass, momentum and energy should also be satisfied at least in the sense of distributions. Based on those considerations, the definition of reasonable global weak-in-time weak solutions goes as follows. A couple $(\rho, \boldsymbol{u}, \theta)$ is called a weak solution of the system (4.1.1)–(4.1.3) if and only if for any positive number T, the following conditions are satisfied:

1. $\rho, \boldsymbol{u}, \theta$ belong to the classes:

$$\begin{aligned}
&\rho \in L^{\infty}([0,T]; L^{\gamma+}(\Omega)),\ \rho^{-1} \in L^{\infty}([0,T];\ L^{\gamma-}(\Omega)),\\
&\sqrt{\rho}\boldsymbol{u} \in L^{\infty}([0,T]; L^2(\Omega)), \sqrt{\rho}\boldsymbol{\nabla}\boldsymbol{u} \in L^2([0,T];\ L^2(\Omega)),\\
&\theta \in L^{\infty}([0,T]; L^4(\Omega)) \cap L^2([0,T]; W^{1,2}(\Omega));
\end{aligned} \tag{4.1.14}$$

2. The following identities are fulfilled:
   (i) the continuity equation

$$\partial_t \rho + \operatorname{div}(\rho \boldsymbol{u}) = 0$$

   is satisfied pointwisely on $[0,T] \times \Omega$;
   (ii) the momentum equation

$$\begin{aligned}
&\int_0^T \int_\Omega \rho\boldsymbol{u} \cdot \partial_t \boldsymbol{\phi} \mathrm{d}x\mathrm{d}t + \int_0^T \int_\Omega (\rho\boldsymbol{u} \otimes \boldsymbol{u}) \boldsymbol{\nabla}\boldsymbol{\phi} \mathrm{d}x\mathrm{d}t\\
&- \int_0^T \int_\Omega \upsilon\rho D\boldsymbol{u} : D\boldsymbol{\phi} \mathrm{d}x\mathrm{d}t + \int_0^T \int_\Omega P \operatorname{div} \boldsymbol{\phi} \mathrm{d}x\mathrm{d}t
\end{aligned}$$

$$-2h^2 \int_0^T \int_\Omega \rho \boldsymbol{\nabla}^2 \log \rho \otimes \boldsymbol{\nabla}\boldsymbol{\phi} \mathrm{d}x\mathrm{d}t = -\int_\Omega \boldsymbol{m}_0 \cdot \boldsymbol{\phi}(0) dx, \tag{4.1.15}$$

holds for any test function $\boldsymbol{\phi}$ of smooth function such that $\boldsymbol{\phi}(\cdot, T) = 0$, where

$$\begin{aligned} -2h^2 \int_0^T \int_\Omega \rho \boldsymbol{\nabla}^2 \log \rho \otimes \boldsymbol{\nabla}\boldsymbol{\phi} \mathrm{d}x\mathrm{d}t & \\ = -4h^2 \int_0^T \int_\Omega \sqrt{\rho} \nabla \sqrt{\rho} \cdot \boldsymbol{\nabla} \mathrm{div} \boldsymbol{\phi} \mathrm{d}x\mathrm{d}t & \\ - 8h^2 \int_0^T \int_\Omega \boldsymbol{\nabla}\sqrt{\rho} \otimes \boldsymbol{\nabla}\sqrt{\rho} \ \nabla^2 \boldsymbol{\phi} \mathrm{d}x\mathrm{d}t; & \end{aligned} \tag{4.1.16}$$

(iii) the total energy equation

$$\begin{aligned} & \int_0^T \int_\Omega \left(\rho \frac{|\boldsymbol{u}|^2}{2} + \rho e - h^2 \rho \Delta \log \rho \right) \partial_t \boldsymbol{\phi} \mathrm{d}x\mathrm{d}t \\ & + \int_0^T \int_\Omega \left(\rho \frac{|\boldsymbol{u}|^2}{2} \boldsymbol{u} - h^2 \rho \Delta \log \rho \boldsymbol{u} \right) \cdot \boldsymbol{\nabla}\boldsymbol{\phi} \mathrm{d}x\mathrm{d}t \\ & + \int_0^T \int_\Omega P \boldsymbol{u} \cdot \boldsymbol{\nabla}\boldsymbol{\phi} \mathrm{d}x\mathrm{d}t - 2h^2 \int_0^T \int_\Omega \rho \boldsymbol{\nabla}^2 \log \rho \boldsymbol{u} \mathrm{d}x\mathrm{d}t \\ & - h^2 \int_0^T \int_\Omega \rho \Delta \boldsymbol{u} \cdot \boldsymbol{\nabla}\boldsymbol{\phi} \mathrm{d}x\mathrm{d}t - \int_0^T \int_\Omega \kappa \boldsymbol{\nabla}\theta \cdot \boldsymbol{\nabla}\boldsymbol{\phi} \mathrm{d}x\mathrm{d}t \\ & - \int_0^T \int_\Omega \upsilon \rho D(\boldsymbol{u}) \boldsymbol{u} \cdot \boldsymbol{\nabla}\boldsymbol{\phi} \mathrm{d}x\mathrm{d}t \\ & + \int_0^T \int_\Omega \left(\rho \frac{|\boldsymbol{u}|^2}{2} + \rho e + \frac{\lambda}{2} |\boldsymbol{\nabla}\Delta^s \rho|^2 - h^2 \rho \Delta \log \rho \right) \\ & \times (0) \boldsymbol{\phi}(0) \mathrm{d}x\mathrm{d}t = 0 \end{aligned} \tag{4.1.17}$$

holds for any smooth test function ϕ such that $\phi(\cdot, T) = 0$, where

$$\begin{aligned}
&-h^2 \int_0^T \int_\Omega \rho\Delta \log \rho \boldsymbol{u} \cdot \boldsymbol{\nabla}\phi \mathrm{d}x\mathrm{d}t \\
&= -h^2 \int_0^T \int_\Omega \nabla \rho \boldsymbol{u} \cdot \boldsymbol{\nabla}\phi \mathrm{d}x\mathrm{d}t \\
&+ h^2 \int_0^T \int_\Omega \frac{|\boldsymbol{\nabla}\rho|^2}{\rho} \boldsymbol{u} \cdot \boldsymbol{\nabla}\phi \mathrm{d}x\mathrm{d}t, \qquad (4.1.18)
\end{aligned}$$

and

$$\begin{aligned}
&-2h^2 \int_0^T \int_\Omega \rho \boldsymbol{\nabla}^2 \log \rho \boldsymbol{u} \mathrm{d}x\mathrm{d}t \\
&= -2h^2 \int_0^T \int_\Omega \boldsymbol{\nabla}^2 \rho \boldsymbol{u} \otimes \boldsymbol{\nabla}\phi \mathrm{d}x\mathrm{d}t \\
&+ 2h^2 \int_0^T \int_\Omega \frac{\nabla\rho \otimes \nabla\rho}{\rho} u \otimes \nabla\phi \mathrm{d}x\mathrm{d}t, \quad (4.1.19)
\end{aligned}$$

and

$$\begin{aligned}
-h^2 \int_0^T \int_\Omega \rho\Delta \boldsymbol{u} \cdot \boldsymbol{\nabla}\phi \mathrm{d}x\mathrm{d}t = -h^2 \int_0^T \int_\Omega \boldsymbol{u}(\Delta\rho\boldsymbol{\nabla}\phi \\
+ 2\boldsymbol{\nabla}\rho\Delta\phi + \rho\boldsymbol{\nabla}\delta\phi)\mathrm{d}x\mathrm{d}t.
\end{aligned} \qquad (4.1.20)$$

Now our main result of this chapter can be read as follows, the reference of this chapter can be seen [1]–[13].

Theorem 4.1.1. *Let Ω be the three-dimensional periodic domains. Assume that κ, e_c, P_c satisfy the hypotheses (4.1.6)–(4.1.12). Let the initial data $\rho_0 \in L^{5_\gamma{}^+/3}(\Omega)$, $1/\rho_0 \in L^{(5_\gamma{}^+-1)/3}(\Omega)$, $\boldsymbol{m}_0 \in L^1(\Omega)$, $\theta_0 \in L^4(\Omega)$ such that $\frac{(m_0)^2}{\rho_0} \in L^1(\Omega)$. Assume the parameter $\gamma^+ > 3$, $\gamma^- > \frac{5\gamma^+-3}{\gamma^+-3}$, $B \geqslant 8$. Let $T > 0$ be arbitrary. Then there exists a weak solution to (4.1.1)–(4.1.3) in the distribution sense. Moreover, the density $\rho > 0$ and the temperature $\theta > 0$ a.e. in $(0, T) \times \Omega$.*

4.2 Approximation

The aim of this section is to present two levels of approximation. First, we take $\varepsilon, \lambda > 0$ and fix s being a sufficiently large positive integer. Our aim is to consider the regularized problem given below, in which ε is the rate of dissipation in the continuity equation. By λ we insert the momentum equation to the artificial smoothing operator $\lambda \nabla \Delta^{2s+1}\rho$ with s sufficiently large, inspired by the works of Bresh and Desjardins. We introduce another regularization of the momentum $\lambda \nabla \Delta^{2s+1}(\rho \boldsymbol{u})$. Note that at the end, after passing subsequently with $\varepsilon, \lambda \to 0^+$, we recover our original problem.

We look for space periodic functions $(\rho, \rho\boldsymbol{u}, \theta)$ such that

$$\begin{aligned} &\rho \in L^2(0,T;\ W^{2s+2}(\Omega)), \partial t\rho \in L^2(0,T; L^2(\Omega)),\\ &\boldsymbol{u} \in L^2(0,T; W^{2s+1}(\Omega)), \\ &\theta \in L^2(0,T; W^{1,2}(\Omega)) \cap L^B(0,T; L^{3B}(\Omega)), \end{aligned} \tag{4.2.1}$$

solving the following problem:

(i) the approximate continuity equation

$$\partial_t \rho + \operatorname{div}(\rho \boldsymbol{u}) - \varepsilon \Delta \rho = 0, \rho(0,x) = \rho_\lambda^0(x) \tag{4.2.2}$$

is satisfied pointwisely on $[0,T] \times \Omega$ and initial condition holds in the strong L^2 sense; here $\rho_\lambda^0 \in C^\infty(\Omega)$ is a regularized initial condition such that $\rho_\lambda^0 \to \rho^0$ in $L^{\gamma^+}(\Omega)$ for $\lambda \to 0^+$ such that $\lambda||\nabla^{2s+1}\rho_\lambda^0|| \to 0$ for $\lambda \to 0^+$, and

$$\inf_{x\in\omega} \rho_\lambda^0 > 0; \tag{4.2.3}$$

(ii) the weak formulation of the approximate momentum equation

$$\begin{aligned} &\int_0^T \int_\Omega \rho\boldsymbol{u} \cdot \partial_t \boldsymbol{\phi} \mathrm{d}x\mathrm{d}t - \int_0^T \int_\Omega \lambda \Delta^s \nabla(\rho\boldsymbol{u}) : \Delta^s \nabla(\rho\boldsymbol{\phi}) \mathrm{d}x\mathrm{d}t \\ &\quad + \int_0^T \int_\Omega (\rho\boldsymbol{u} \otimes \boldsymbol{u}) \nabla \boldsymbol{\phi} \mathrm{d}x\mathrm{d}t - \int_0^T \int_\Omega \upsilon\rho D\boldsymbol{u} : D\boldsymbol{\phi} \mathrm{d}x\mathrm{d}t \end{aligned}$$

$$
\begin{aligned}
&+\int_0^T\int_\Omega P\,\mathrm{div}\boldsymbol{\phi}\mathrm{d}x\mathrm{d}t-\int_0^T\int_\Omega \lambda\Delta^s\mathrm{div}(\rho\boldsymbol{\phi}) : \Delta^{s+1}(\rho)\mathrm{d}x\mathrm{d}t\\
&-2h^2\int_0^T\int_\Omega \rho\boldsymbol{\nabla}^2\log\rho\cdot\boldsymbol{\nabla}\boldsymbol{\phi}\mathrm{d}x\mathrm{d}t-\varepsilon\int_0^T\int_\Omega(\boldsymbol{\nabla}_\rho\cdot\boldsymbol{\nabla})\boldsymbol{u}\cdot\boldsymbol{\phi}\mathrm{d}x\mathrm{d}t\\
&=-\int_\Omega \boldsymbol{m}_0\cdot\boldsymbol{\phi}(0)\mathrm{d}x, \qquad (4.2.4)
\end{aligned}
$$

holds for any test function $\boldsymbol{\phi}\in L^2(0,T;W^{2s+1}(\Omega))\cap W^{1,2}(0,T;W^{1,2}(\Omega))$ such that $\boldsymbol{\phi}(\cdot,T)=0$;

(iii) the weak formulation of the energy equality

$$
\begin{aligned}
&\int_0^T\int_\Omega\left(\rho\frac{|\boldsymbol{u}|^2}{2}+\rho e+\frac{\lambda}{2}|\boldsymbol{\nabla}\Delta^s\rho|^2-h^2\rho\Delta\log\rho\right)\partial_t\boldsymbol{\phi}\mathrm{d}x\mathrm{d}t\\
&-\int_0^T\int_\Omega \kappa\boldsymbol{\nabla}\theta\cdot\boldsymbol{\nabla}\boldsymbol{\phi}\mathrm{d}x\mathrm{d}t\\
&+\int_0^T\int_\Omega\left(\rho e+\rho\frac{|\boldsymbol{u}|^2}{2}\boldsymbol{u}-h^2\rho\Delta\log\rho\boldsymbol{u}\right)\cdot\boldsymbol{\nabla}\boldsymbol{\phi}\mathrm{d}x\mathrm{d}t\\
&+\int_0^T\int_\Omega P\boldsymbol{u}\cdot\boldsymbol{\nabla}\boldsymbol{\phi}\mathrm{d}x\mathrm{d}t\\
&-2h^2\int_0^T\int_\Omega\rho\boldsymbol{u}\boldsymbol{\nabla}^2\log\rho\boldsymbol{\nabla}\boldsymbol{\phi}\mathrm{d}x\mathrm{d}t-h^2\int_0^T\int_\Omega\rho\Delta\boldsymbol{u}\boldsymbol{\nabla}\boldsymbol{\phi}\mathrm{d}x\mathrm{d}t\\
&-\int_0^T\int_\Omega\nu\rho D(\boldsymbol{u})\boldsymbol{u}\cdot\boldsymbol{\nabla}\boldsymbol{\phi}\mathrm{d}x\mathrm{d}t=-\int_0^T\int_\Omega\left(\frac{\varepsilon}{\theta^2}-\varepsilon\theta^5\right)\boldsymbol{\phi}\mathrm{d}x\\
&+\int_0^T\int_\Omega R_{\varepsilon,\lambda}\mathrm{d}x\mathrm{d}t-\int_0^T\int_\Omega\\
&\times\left(\rho\frac{|\boldsymbol{u}|^2}{2}+\rho e+\frac{\lambda}{2}|\boldsymbol{\nabla}\Delta^s\rho|^2-h^2\rho\Delta\log\rho\right)(0)\boldsymbol{\phi}(0)\mathrm{d}x\mathrm{d}t, \qquad (4.2.5)
\end{aligned}
$$

and

$$
\begin{aligned}
R_{\varepsilon,\lambda}(\rho,\theta,\boldsymbol{u},\boldsymbol{\phi})=&\lambda[\Delta^s(\mathrm{div}(\rho\boldsymbol{u}\boldsymbol{\phi}))\Delta^{s+1}\rho-\Delta^s\mathrm{div}(\rho\boldsymbol{u})\Delta^{s+1}\rho\boldsymbol{\phi}]\\
&-\lambda\Delta^s\mathrm{div}(\rho\boldsymbol{u})\boldsymbol{\nabla}\Delta^{s+1}\rho\cdot\boldsymbol{\nabla}\boldsymbol{\phi}
\end{aligned}
$$

$$
\begin{aligned}
&-\lambda[|\Delta^s\nabla(\rho\boldsymbol{u})|^2\phi-\Delta^s\nabla(\rho\boldsymbol{u}):\Delta^s\nabla(\rho\boldsymbol{u}\phi)]\\
&+\lambda\varepsilon\Delta^{s+1}\rho\nabla\Delta^s\rho\cdot\nabla\phi+\frac{\varepsilon}{2}|\boldsymbol{u}|^2\nabla\rho\cdot\nabla\phi\\
&+\varepsilon\nabla\rho\cdot\nabla\phi\left(e_c(\rho)+\frac{P_c(\rho)}{\rho}\right), \qquad (4.2.6)
\end{aligned}
$$

is satisfied for any $\phi\in C^\infty([0,T]\times\Omega)$ with $\phi(T,\cdot)=0$; here $\boldsymbol{u}_\lambda^0=\frac{m_0}{\rho_\lambda^0}$ and $\theta_\lambda^0\in C^\infty(\Omega)$, $\theta_\lambda^0\to\theta^0$ for $\lambda\to 0^+$ in $L^4(\Omega)$, and

$$
0<\inf_{x\in\Omega}\theta_\lambda^0(x)=\underline{\theta^0}\leqslant\theta^0(x)\leqslant\sup_{x\in\Omega}\theta_\lambda^0(x)=\overline{\theta^0}<\infty. \qquad (4.2.7)
$$

We prove the following result.

Theorem 4.2.1. *Under the assumptions of Theorem* 4.1.1 *and the assumptions specified in this section, for any* $T>0$, ε, $\lambda>0$, *there exists a solution to problem* (4.2.1)–(4.2.7) *in the sense defined above.*

Indeed, the proof of this result is far from being obvious. To prove Theorem 4.2.1, we have to introduce another level of approximation, based on regularization of certain quantities and finite-dimensional projection (Faedo–Galerkin approximation) of the momentum equation. More precisely, we look for functions $(\rho,\boldsymbol{u},\theta)$ such that

$$
\begin{aligned}
&\rho\in L^2(0,T;W^{2s+2}(\Omega)),\ \partial_t\rho\in L^2(0,T;L^2(\Omega)),\\
&\boldsymbol{u}\in C(0,T;X_N), \qquad (4.2.8)\\
&\theta\in L^2(0,T;W^{1,2}(\Omega))\cap L^\infty((0,T)\times\Omega),
\end{aligned}
$$

solving the following problem:

(i) the approximate continuity equation

$$
\partial_t\rho+\operatorname{div}(\rho\boldsymbol{u})-\varepsilon\Delta\rho=0, \rho(0,x)=\rho_\lambda^0(x) \qquad (4.2.9)
$$

is satisfied pointwisely on $[0,T]\times\Omega$ and initial condition holds in the strong L^2 sense; here ρ_λ^0 is as above;

(ii) the Faedo–Galerkin approximationfor the weak formulation of the approximate momentum balance: we look for $\boldsymbol{u} \in C([0,T]; X_N)$ such that

$$\int_0^T \int_\Omega \rho \boldsymbol{u} \cdot \partial_t \boldsymbol{\phi} \mathrm{d}x\mathrm{d}t - \int_0^T \int_\Omega \lambda \Delta^s \boldsymbol{\nabla}(\rho \boldsymbol{u}) : \Delta^s \boldsymbol{\nabla}(\rho \boldsymbol{\phi}) \mathrm{d}x\mathrm{d}t$$
$$+ \int_0^T \int_\Omega (\rho \boldsymbol{u} \otimes \boldsymbol{u}) \boldsymbol{\nabla} \boldsymbol{\phi} \mathrm{d}x\mathrm{d}t - \int_0^T \int_\Omega \nu \rho D(\boldsymbol{u}) : D(\boldsymbol{\phi}) \mathrm{d}x\mathrm{d}t$$
$$+ \int_0^T \int_\Omega P \,\mathrm{div} \boldsymbol{\phi} \mathrm{d}x\mathrm{d}t$$
$$- \int_0^T \int_\Omega \lambda \Delta^s \mathrm{div}(\rho \boldsymbol{\phi}) : \Delta^{s+1}(\rho) \mathrm{d}x\mathrm{d}t$$
$$- 2h^2 \int_0^T \int_\Omega \rho \boldsymbol{\nabla}^2 \log \rho \cdot \boldsymbol{\nabla} \boldsymbol{\phi} \mathrm{d}x\mathrm{d}t$$
$$-\varepsilon \int_0^T \int_\Omega (\boldsymbol{\nabla} \rho \cdot \boldsymbol{\nabla}) \boldsymbol{u} \cdot \boldsymbol{\phi} \mathrm{d}x\mathrm{d}t = - \int_\Omega m_0 \cdot \boldsymbol{\phi}(0) \mathrm{d}x \quad (4.2.10)$$

holds for any test function $\boldsymbol{\phi} \in X_N$, and $X_N = \mathrm{span}\{\boldsymbol{\phi}_i\}_{i=1}^N$, where $\{\boldsymbol{\phi}_i\}_{i=1}^N$ is an orthonormal basis in $L^2(\Omega)$, such that $\boldsymbol{\phi}_i \in C^\infty(\Omega)$ for all $i \in N$;

(iii) the approximate thermal energy equation

$$\partial_t(\rho\theta + \beta\theta^4) + \mathrm{div}\,(\boldsymbol{u}\rho\theta + \beta \boldsymbol{u}\theta^4) - \mathrm{div}(\kappa \boldsymbol{\nabla}\theta)$$
$$+ \left(P_m + \frac{\beta}{3}\right) \mathrm{div}\, \boldsymbol{u}$$
$$= \frac{\varepsilon}{\theta^2} - \varepsilon\theta^5 + \frac{4\varepsilon}{\gamma} |\boldsymbol{\nabla}\rho^{\frac{\lambda}{2}}|^2 + \nu\rho|D(\boldsymbol{u})|^2 + \lambda|\Delta^s \boldsymbol{\nabla}(\rho\boldsymbol{u})|^2$$
$$+ \lambda|\Delta^s \boldsymbol{\nabla}(\rho\boldsymbol{u})|^2 + \lambda\varepsilon|\Delta^{s+1}\rho|^2 + \varepsilon \frac{1}{\rho}\frac{\partial P_c(\rho)}{\partial\rho}|\boldsymbol{\nabla}\rho|^2$$
$$+ 2h^2\rho|\boldsymbol{\nabla}^2 \log\rho|^2 \quad (4.2.11)$$

is satisfied pointwisely on $[0,T] \times \Omega$ and initial condition θ_λ^0 is as above.

Theorem 4.2.2. *Let $N \in \mathcal{N}, \varepsilon, h$ and $\lambda > 0$. Let $\boldsymbol{m}_0, \rho_\lambda^0, \theta_\lambda^0$ be as above. Under the assumptions of* Theorem 4.2.1 *and the assumptions* *specified in this section, for any* $T > 0, \varepsilon, \lambda > 0$, *there exists a solution to problem* (4.2.1)–(4.2.7) *in the sense defined above.*

4.3 Proof of Theorem 4.2.2

This section is dedicated to the proof of Theorem 4.2.2. The strategy of the proof can be summarized as follows:

(i) We fix $\boldsymbol{u}(t,x)$ in the space $C(0,T;\ X_N)$ and use it to find a unique smooth solution to (4.2.9) $\rho = \rho(\boldsymbol{u})$ and a unique strong solution to (4.2.11) $\theta = \theta(\rho, \boldsymbol{u})$.
(ii) We find the local-in-time solution to the momentum equation (4.2.10) by a fixed-point argument.
(iii) We extend the local-in-time solution for the whole time interval using uniform estimates.

4.3.1 *Continuity equation*

We first prove the existence of a smooth, unique solution to the approximate continuity equation in the situation when the vector field $\boldsymbol{u}(x,t)$ is given and belongs to $C([0,T];X_N)$.

The following result can be proven by the Galerkin approximation and the well known statements about the regularity of the linear parabolic systems.

Lemma 4.3.1. *Let* $\boldsymbol{u}\,(x,t) \in C([0,T];X_N)$ *for* N *fixed and let* ρ_λ^0 *be as above. Then there exists the unique classical solution to* (4.2.9), *i.e.* $\rho \in V^\rho_{[0,T]}$, *where*

$$V^\rho_{[0,T]} = \begin{cases} \rho \in C([0,T];\ C^{2+\nu}(\Omega)), \\ \rho_{t\rho} \in C([0,T]; C^{0,\nu}(\Omega)), \end{cases} . \tag{4.3.1}$$

Moreover, the mapping $\boldsymbol{u} \mapsto \rho(\boldsymbol{u})$ *maps bounded sets in* $C([0,T];X_N)$ *into bounded sets in* $V^\rho_{[0,T]}$ *and is continuous with values in* $C([0,T];$ $C^{2+\nu'}(\Omega))$, $0 < \nu' < \nu < 1$,

$$\underline{\rho}^0 e^{-\int_0^T \|\mathrm{div}\,\boldsymbol{u}\|_\infty dt} \leqslant \rho(\tau, x) \leq \overline{\rho}^0 e^{-\int_0^T \|\mathrm{div}\,\boldsymbol{u}\|_\infty dt}. \tag{4.3.2}$$

Finally, for fixed N, *the function* ρ *is smooth in the space variable.*

For the proof, please see [75].

4.3.2 *Internal energy equation*

The existence of unique solution to (4.2.11) can be proven as in [5]. The rough idea of the proof is to transform and regularize equation (4.2.11) in such a way that the classical theory for quasilinear parabolic equations could be applied. We have the following lemma.

Lemma 4.3.2. *Let* $\boldsymbol{u}(x,t) \in C([0,T];\, X_N)$ *be a given vector field and let* $\rho = \rho_u$ *be the unique solution of the approximate problem constructed in Lemma* 4.3.1. *Then there exists the unique strong solution to* (4.2.11) *which belongs to*

$$V^{\theta}_{[0,T]} = \begin{cases} \partial_t\theta \in L^2((0,T)\times\Omega), \Delta\theta \in L^2((0,T)\times\Omega), \\ \theta \in L^{\infty}(0,T;W^{1,2}(\Omega)), \theta^{-1} \in L^{\infty}((0,T)\times\Omega). \end{cases} \tag{4.3.3}$$

Moreover, the mapping $\boldsymbol{u} \mapsto \theta(\boldsymbol{u})$ *maps bounded sets in* $C([0,T];\, X_N)$ *into bounded sets in* $V^{\theta}_{[0,T]}$ *and is continuous with values in* $L^2([0,T];\, W^{1,2}(\Omega))$.

For the proof, see [75].

4.3.3 *Fixed-point method*

At this stage, we are ready to show the existence of approximate solutions on a possibly short time interval $(0,\tau)$. We use the Schauder fixed-point theorem to find a solution of the momentum equations.

More precisely, we prove that there exists $\tau = \tau(N)$ such that $\boldsymbol{u}$ solves the approximate momentum equation (4.2.10). To this purpose we consider the following mapping

$$\begin{aligned} &\mathcal{T} : C([0,\tau];\, X_N) \to C([0,\tau]; X_N), \\ &\mathcal{T}(\boldsymbol{v}) = \boldsymbol{u}, \end{aligned} \tag{4.3.4}$$

which attain a solution to the following problem

$$\boldsymbol{v} = \mathcal{M}_{\rho(v)}(\boldsymbol{m}_0 + \int_0^t P_{X_N}\mathcal{N}(\boldsymbol{v})(\boldsymbol{s})d\boldsymbol{s}), \tag{4.3.5}$$

where

$$
\begin{aligned}
\langle \mathcal{N}(v), \phi\rangle = & \int_{\Omega} (\rho v \otimes v : \boldsymbol{\nabla}\phi \mathrm{d}x\mathrm{d}t - \int_{\Omega} v\rho D v : D\phi \mathrm{d}x\mathrm{d}t \\
& + \int_{\Omega} P \operatorname{div}\phi \mathrm{d}x\mathrm{d}t + \lambda \int_{\Omega} \rho \nabla \Delta^{2s+1}\rho \cdot \phi \mathrm{d}x\mathrm{d}t \\
& + \lambda \int_{\Omega} \rho \Delta^{s} \operatorname{div}(\Delta^{s} \boldsymbol{\nabla}(v\rho)) \cdot \phi \mathrm{d}x\mathrm{d}t \\
& - 2\delta^2 \int_{\Omega} \rho \boldsymbol{\nabla}^2 \log\rho \cdot \boldsymbol{\nabla}\phi \mathrm{d}x\mathrm{d}t \\
& - \varepsilon \int_{\Omega} (\boldsymbol{\nabla}_{\rho} \cdot \boldsymbol{\nabla}) v \cdot \phi \mathrm{d}x\mathrm{d}t
\end{aligned} \tag{4.3.6}
$$

and

$$
\mathcal{M}_{\rho} : X_N \to X_N, \int_{\Omega} \rho \mathcal{M}_{\rho}[w]\phi \mathrm{d}x =< w, \phi > \cdot \quad w, \phi \in X_N. \tag{4.3.7}
$$

Next, we consider a ball $\mathcal{B}_R$ in the space $C([0, T]; X_N)$:

$$
\mathcal{B}_R = \{v \in C([0, \tau]; X_N) : ||\boldsymbol{v}||_{C([0,\tau];X_N)} \leqslant R\}.
$$

We need to show that the operator $\mathcal{T}$ is continuous and maps $\mathcal{B}_R$ into itself, provided τ is sufficiently small. First observe that we have

$$
\begin{aligned}
||\mathcal{N}(v)||_{X_N} \leqslant & C[||\rho||_{L^\infty(\Omega)}(||v||_{X_N} + ||v||_{X_N}^2) \\
& + ||\rho||_{L^\infty(\Omega)}^{\gamma} + ||\theta||_{L^\infty(\Omega)}^4 \\
& + ||\rho||_{L^\infty(\Omega)}||\theta||_{L^\infty(\Omega)} + ||\rho||_{L^\infty(\Omega)}(||\rho||_{W^{4s+3,\infty}(\Omega)} \\
& + ||\rho||_{W^{4s+2,\infty}(\Omega)}||v||X_N)].
\end{aligned} \tag{4.3.8}
$$

From estimates (4.3.8) and the estimates established in Lemma 4.3.1 and Lemma 4.3.2, it follows that for sufficiently small τ, operator $\mathcal{T}$ maps the ball $\mathcal{B}_R$ into itself. Moreover, $\mathcal{T}$ is a continuous mapping and its image is Lipschitz functions, thus it is compact in $\mathcal{B}_R$. It allows us to apply Schauder fixed-point theorem to infer that there exists at least one fixed-point v solving (4.2.10) on $[0, \tau]$.

4.3.4 *A uniform priori estimate and global existence of approximate equations*

In order to extend this solution for the whole time interval $[0, \mathcal{T}]$, we need a uniform bound of the solution. It follows from (4.3.5) that $\boldsymbol{u}$ is continuously differentiable function, thus, system (4.2.10) may be transformed to

$$\begin{aligned}
&\int_\Omega \partial_t(\rho\boldsymbol{u})\cdot\boldsymbol{\phi}\mathrm{d}x\mathrm{d}t+\int_0^T\int_\Omega \lambda\Delta^s\boldsymbol{\nabla}(\rho\boldsymbol{u}) : \Delta^s\boldsymbol{\nabla}(\rho\boldsymbol{\phi})\mathrm{d}x\mathrm{d}t\\
&\quad-\int_\Omega(\rho\boldsymbol{u}\otimes\boldsymbol{u}) : \boldsymbol{\nabla}\boldsymbol{\phi}\mathrm{d}x\mathrm{d}t+\int_\Omega \nu\rho D\boldsymbol{u} : D\boldsymbol{\phi}\mathrm{d}x\mathrm{d}t\\
&\quad-\int_\Omega P\,\mathrm{div}\,\boldsymbol{\phi}\mathrm{d}x\mathrm{d}t-\int_0^T\int_\Omega\lambda_\rho\boldsymbol{\nabla}\Delta^{2s+1}(\rho\boldsymbol{u})\cdot\boldsymbol{\phi}\mathrm{d}x\mathrm{d}t\\
&\quad+2h^2\int_\Omega\rho\boldsymbol{\nabla}^2\log\rho\otimes\boldsymbol{\nabla}\boldsymbol{\phi}\mathrm{d}x\mathrm{d}t+\varepsilon\int_\Omega(\boldsymbol{\nabla}\rho\cdot\boldsymbol{\nabla})\boldsymbol{u}\cdot\boldsymbol{\phi}\mathrm{d}x\mathrm{d}t=0,
\end{aligned}\tag{4.3.9}$$

satisfied for any $\boldsymbol{\phi}\in X_N$. Therefore we can test (4.3.9) by $\boldsymbol{u}$. For the approximate momentum equation, using continuity equation, we obtain the kinetic energy balance

$$\begin{aligned}
&\frac{\mathrm{d}}{\mathrm{d}t}\int_\Omega\left(\frac{1}{2}\rho|\boldsymbol{u}|^2+\frac{\lambda}{2}|\boldsymbol{\nabla}^{2s+1}\rho|^2+\rho e(\rho,\theta)+4h^2|\boldsymbol{\nabla}\sqrt{\rho}|^2\right)\mathrm{d}x\\
&\quad+\nu\int_\Omega\rho|D\boldsymbol{u}|^2\mathrm{d}x+\lambda\int_\Omega|\Delta^s\boldsymbol{\nabla}(\rho\boldsymbol{u})|^2\mathrm{d}x+\varepsilon\lambda\int_\Omega|\Delta^{s+1}\rho|^2\mathrm{d}x\\
&\quad+2\varepsilon h^2\int_\Omega\rho|\boldsymbol{\nabla}^2\log\rho|^2\mathrm{d}x=\int_\Omega P\,\mathrm{div}\,\boldsymbol{u}\mathrm{d}x.
\end{aligned}\tag{4.3.10}$$

Adding to this equality (4.2.11) integrating with respect to space and integrating the sum with respect to time we obtain

$$\begin{aligned}
&\int_\Omega\left(\frac{1}{2}\rho|u|^2(t)+\frac{\lambda}{2}|\boldsymbol{\nabla}^{2s+1}\rho|^2(t)+\rho e(\rho,\theta)(t)+4h^2|\boldsymbol{\nabla}\sqrt{\rho}|^2(t)\right)\mathrm{d}x\\
&\quad+\varepsilon\int_0^t\int_\Omega\theta^5\mathrm{d}x\mathrm{d}t=\varepsilon\int_0^t\int_\Omega\frac{1}{\theta^2}\mathrm{d}x\mathrm{d}t+\int_\Omega\left(\frac{1}{2}\rho|\boldsymbol{u}|^2(0)\right.\\
&\quad\left.+\frac{\lambda}{2}|\boldsymbol{\nabla}^{2s+1}\rho|^2(0)+\rho e(\rho,\theta)(0)+4h^2|\boldsymbol{\nabla}\sqrt{\rho}|^2(0)\right)\mathrm{d}x.
\end{aligned}\tag{4.3.11}$$

4.3.5 Entropy estimate

Our aim now is to derive a fundamental estimate for our system. It can be viewed as a total global entropy balance.

From Lemma 4.3.2, it follows in particular that θ is bounded from below by a constant. Therefore, dividing internal energy equation by θ is possible and

$$\begin{aligned}&\partial_t(\rho s(\rho,\theta)) + \operatorname{div}(\rho s(\rho,\theta)\boldsymbol{u}) - \operatorname{div}\left(\frac{\kappa\nabla\theta}{\theta}\right)\\&= \frac{\kappa|\nabla\theta|^2}{\theta^2} + \varepsilon\frac{\Delta\rho}{\theta}\left(\theta s(\rho,\theta) - e(\rho,\theta) - \frac{P(\rho,\theta)}{\rho}\right) + \frac{\varepsilon}{\theta^3} - \varepsilon\theta^4\\&\quad + \frac{\nu\rho|D(\boldsymbol{u})|^2 + \lambda|\Delta^s\nabla(\rho\boldsymbol{u})|^2 + \lambda\varepsilon|\Delta^{s+1}\rho|^2 + 2\varepsilon h^2\rho|\nabla^2\log\rho|^2}{\theta},\end{aligned} \tag{4.3.12}$$

where

$$\theta s(\rho,\theta) - e(\rho,\theta) - \frac{P(\rho,\theta)}{\rho} = \theta s_m(\rho,\theta) - e_m(\rho,\theta) - \frac{p_m(\rho,\theta)}{\rho}, \tag{4.3.13}$$

integrated (4.3.12) by $\bar{\theta}$ and here $\bar{\theta}$ is a arbitrary constant. In the next step integrating over Ω and subtracting from (4.3.11), we get

$$\begin{aligned}&\int_\Omega\left(\frac{1}{2}\rho|\boldsymbol{u}|^2 + \frac{\lambda}{2}|\nabla^{2s+1}\rho|^2 + H_{\bar{\theta}}(\rho,\theta) + 4h^2|\nabla\sqrt{\rho}|^2\right)(t)dx\\&\quad + \bar{\theta}\int_0^t\int_\Omega\frac{\kappa|\nabla\theta|^2}{\theta^2}\mathrm{d}x\mathrm{d}t\\&\quad + \bar{\theta}\int_0^t\int_\Omega\frac{\nu\rho|D(\boldsymbol{u})|^2 + \lambda|\Delta^s\nabla(\rho\boldsymbol{u})|^2 + \lambda\varepsilon|\Delta^{s+1}\rho|^2 + 2\varepsilon h^2\rho|\nabla^2\log\rho|^2}{\theta}\mathrm{d}x\mathrm{d}t\\&\quad + \varepsilon\int_0^t\int_\Omega\theta^5\mathrm{d}x\mathrm{d}t + \bar{\theta}\int_0^t\int_\Omega\frac{\varepsilon}{\theta^3}\mathrm{d}x\mathrm{d}t = \bar{\theta}\epsilon\int_0^t\int_\Omega\theta^4\mathrm{d}x\mathrm{d}t + \varepsilon\int_0^t\int_\Omega\frac{1}{\theta^2}\mathrm{d}x\mathrm{d}t\\&\quad + \int_\Omega\left(\frac{1}{2}\rho_0|u_0|^2 + \frac{\lambda}{2}|\nabla^{2s+1}\rho_0|^2 + H_{\bar{\theta}}(\rho_0,\theta_0) + 4h^2|\nabla\sqrt{\rho 0}|^2\right)\mathrm{d}x\\&\quad - \bar{\theta}\epsilon\int_0^t\int_\Omega\frac{\Delta\rho}{\theta}\left(\theta s(\rho,\theta) - e(\rho,\theta) - \frac{P(\rho,\theta)}{\rho}\right)\mathrm{d}x\mathrm{d}t,\end{aligned} \tag{4.3.14}$$

where

$$
\begin{aligned}
&-\bar{\theta}\epsilon \int_0^t \int_\Omega \frac{\Delta\rho}{\theta}\left(\theta s(\rho,\theta) - e(\rho,\theta) - \frac{P(\rho,\theta)}{\rho}\right) \mathrm{d}x\mathrm{d}t \\
&\quad = \varepsilon \int_0^T \int_\Omega \frac{\bar{\theta}}{\theta^2}\left(e_m + \rho\frac{\partial e_m}{\partial \rho}\right) \nabla\rho\nabla\theta \mathrm{d}x\mathrm{d}t, \qquad (4.3.15)
\end{aligned}
$$

here $H_{\bar{\theta}} = \rho e(\rho,\theta) - \bar{\theta}\rho s(\rho,\theta)$ is similar to previous definition. To control the right hand side of above identities, $\frac{\varepsilon}{\theta^2}$ can be absorbed by the corresponding left hand term $\frac{\varepsilon}{\theta^3}$. In the same way, the term $\varepsilon\theta^4$ can be absorbed by the corresponding left hand term $\varepsilon\theta^5$. Therefore there remains only the last term, to deal with this term. We take advantage of the fact of the heat conductivity coefficient. We write

$$
\left|\frac{1}{\theta^2}\left(e_m + \rho\frac{\partial e_m}{\partial \rho}\right) \nabla_\rho\nabla_\theta\right| \leqslant C\left(\frac{\frac{P_c}{\rho}+\theta}{\theta^2}\right)|\nabla_\rho||\nabla\theta|\Bigg|. \qquad (4.3.16)
$$

Furthermore, we have

$$
\frac{|\nabla_\rho||\nabla\theta|}{\theta} \leqslant \varepsilon\frac{|\nabla\rho|^2}{\theta} + C(\varepsilon)\frac{|\nabla\theta|^2}{\theta}, \qquad (4.3.17)
$$

similarly,

$$
\frac{\frac{P_c}{\rho}|\nabla\rho||\nabla\theta|}{\theta^2} \leqslant \varepsilon\frac{\left(\frac{P_c}{\rho}\right)^2|\nabla\rho|^2}{\theta} + C(\varepsilon)\frac{|\nabla\theta|^2}{\theta^3}. \qquad (4.3.18)
$$

Therefore choosing $\varepsilon > 0$ sufficiently small enough, we have

$$
\begin{aligned}
&\varepsilon \int_0^t \int_\Omega \frac{\bar{\theta}}{\theta^2}\left(e_m + \rho\frac{\partial e_m}{\partial \rho}\right) \nabla\rho\nabla\theta \mathrm{d}x\mathrm{d}t \\
&\quad \leqslant \frac{1}{2}\bar{\theta}\int_0^t \int_\Omega \frac{\kappa|\nabla\theta|^2}{\theta^2}\mathrm{d}x\mathrm{d}t + \frac{1}{2}\bar{\theta}\varepsilon\int_0^t \int_\Omega \frac{h^2\rho|\nabla^2\log\rho|^2}{\theta}\mathrm{d}x\mathrm{d}t.
\end{aligned}
$$
(4.3.19)

Summarizing, we have shown the following estimate

$$\sup_{\tau\in[0,t]}\int_\Omega\left(\frac{1}{2}\rho|\boldsymbol{u}|^2+\frac{\lambda}{2}|\boldsymbol{\nabla}^{2s+1}\rho|^2+H_{\bar\theta}(\rho,\theta)+4h^2|\boldsymbol{\nabla}\sqrt{\rho}|^2\right)(\tau)\mathrm{d}x$$
$$+\int_0^t\int_\Omega\frac{\nu\rho|D(\boldsymbol{u})|^2+\lambda|\Delta^s\boldsymbol{\nabla}(\rho\boldsymbol{u})|^2+\lambda\varepsilon|\Delta^{s+1}\rho|^2+2h^2\rho|\boldsymbol{\nabla}^2\log\rho|^2}{\theta}dxdt$$
$$+\int_0^t\int_\Omega\frac{\kappa(\rho,\theta)|\boldsymbol{\nabla}\theta|^2}{\theta^2}\mathrm{d}x\mathrm{d}t+\varepsilon\int_0^t\int_\Omega\theta^5\mathrm{d}x\mathrm{d}t+\int_0^t\int_\Omega\frac{\varepsilon}{\theta^3}\mathrm{d}x\mathrm{d}t\leqslant C\quad(4.3.20)$$

Taking s from the density-regularizing term sufficiently large, one can show that the density is separated from 0 uniformly with respect to all approximation parameter except for λ. This property was observed in [1] where the case of single-component heat-conducting fluid was discussed. Recalling their reasoning we may use the Sobolev embedding $||\rho-1||_{L^\infty(\Omega)}\leqslant C||\rho^{-1}||_{W^{3,2}(\Omega)}$ and

$$||\boldsymbol{\nabla}^3\rho^{-1}||_{L^2(\Omega)}\leqslant C(1+||\boldsymbol{\nabla}^3\rho|_{L^2(\Omega)})^3(1+||\rho^{-1}||_{L^4(\Omega)})^4 \tag{4.3.21}$$

where the last term is bounded on account of (4.3.20) and the assumption that $\gamma^-\geqslant 4$. So, provided that $(2s+1)\geqslant 3$ we have

$$||\rho^{-1}||_{L^\infty((0,\tau)\times\Omega)}\leqslant C(\lambda)\quad\text{a.e. in }(0,\tau)\times\Omega \tag{4.3.22}$$

4.3.6 *Global existence of first level approximate equations*

The uniform estimates for $\boldsymbol{u}$ can be summarized as follows

$$||\sqrt{\rho}\boldsymbol{u}||L^\infty(0,\tau;L^2(\Omega))+\sqrt{\lambda}||\Delta^s\boldsymbol{\nabla}(\rho\boldsymbol{u})||_{L^2(0,\tau;L^2(\Omega))}\leqslant C. \tag{4.3.23}$$

Moreover, the density ρ is bounded from below by a positive constant on account of (4.3.20). By the equivalence of norms on the finite-dimensional spaces X_N we can thus deduce the uniform bounds for $\boldsymbol{u}$ in $C([0,\tau);X_N)$. Therefore we get a solution defined on $[0,T]$ for arbitrary but finite $T>0$.

4.4 Faedo–Galerkin limit

The purpose of this section is to let $N \to \infty$ in the equations of approximate system in Section 4.3. We start with summarizing all the estimates that are uniform with respect to N derived mostly from (4.3.20) and its consequences. This will be done in Subsection 4.4.1, then in Subsection 4.4.2 we use these estimates to extract the weekly convergent subsequences and to prove that the limit $N \to \infty$ can be performed.

4.4.1 *A uniform priori estimate with respect to N*

Note that the above estimates are not only uniform with respect to time but also with respect to N. From (4.3.20) and the property of Helmholtz function, we get $\rho_N s_N \in L^\infty(0,T;L^1(\Omega))$, more specifically we have

$$\begin{aligned}&||\rho_N e_c(\rho_N)||_{L^\infty(0,T;L^1(\Omega))} + ||\rho_N\theta_N||_{L^\infty(0,T;L^1(\Omega))}\\&\quad + ||a\rho_N^4||_{L^\infty(0,T;L^1(\Omega))} \leqslant C,\end{aligned} \tag{4.4.1}$$

also, from (4.3.20), we get that

(i) the density estimates:

$$\begin{aligned}&4h^2\|\nabla\sqrt{\rho_N}\|_{L^\infty(0,T;L^2(\Omega))} + \left\|\sqrt{\lambda\varepsilon}\frac{\Delta^{s+1}\rho_N}{\sqrt{\theta_N}}\right\|_{L^2(0,T;L^2(\Omega))}\\&\quad + \left\|\sqrt{2\varepsilon h^2}\frac{\rho|\nabla^2\log\rho|}{\sqrt{\theta_N}}\right\|_{L^2(0,T;L^2(\Omega))} \leqslant C;\end{aligned} \tag{4.4.2}$$

(ii) the velocity estimates:

$$\begin{aligned}&\left\|\sqrt{\frac{\nu\rho_N}{\theta_N}}D(\boldsymbol{u}_N)\right\|_{L^2(0,T;L^2(\Omega))}\\&\quad + \left\|\sqrt{\frac{\lambda}{\theta_N}}\Delta^s\nabla(\rho_N\boldsymbol{u}_N)\right\|_{L^2(0,T;L^2(\Omega))} \leqslant C;\end{aligned} \tag{4.4.3}$$

(iii) the temperature estimates:

$$\left\|\frac{\sqrt{\kappa(\rho,\theta)}\boldsymbol{\nabla}\theta_N}{\theta_N}\right\|_{L^2(0,T;L^2(\Omega))}+\left\|\frac{\varepsilon}{\theta_N^3}\right\|_{L^1(0,T;L^1(\Omega))}$$
$$+\left\|\varepsilon\theta_N^5\right\|_{L^1(0,T;L^1(\Omega))}\leqslant C. \tag{4.4.4}$$

(iv) further temperature estimates:

$$||(1+\sqrt{\rho_N})\boldsymbol{\nabla}\log\theta_N||_{L^2(0,T;L^2(\Omega))}+||\sqrt{\rho_N}\boldsymbol{\nabla}\theta_N||_{L^2(0,T;L^2(\Omega))}$$
$$+||\sqrt{\beta}\boldsymbol{\nabla}\theta_N^a||_{L^2(0,T;L^2(\Omega))}\leqslant C, \tag{4.4.5}$$

where $a\in[0,\frac{B}{2}]$ and $B\geqslant 8$. To control the full norm of θ_N^a in $L^2(0,T;W^{1,2}(\Omega))$ we combine the above estimates (4.4.5) with (4.4.2). Therefore, the Sobolev embedding gives

$$||\sqrt{\beta}\boldsymbol{\nabla}\theta_N||_{L^B(0,T;L^{3B}(\Omega))}\leqslant C; \tag{4.4.6}$$

(v) kinetic energy estimates:
We now integrate (4.3.10) with respect to time to get

$$\int_\Omega\left(\frac{1}{2}\rho|\boldsymbol{u}|^2+\frac{\lambda}{2}|\boldsymbol{\nabla}^{2s+1}\rho|^2+\rho e(\rho,\theta)+4h^2|\boldsymbol{\nabla}\sqrt{\rho}|^2\right)(T)\mathrm{d}x$$
$$+\int_0^T\int_\Omega(\nu\rho|D\boldsymbol{u}|^2+\lambda|\Delta^s\boldsymbol{\nabla}(\rho\boldsymbol{u})|^2$$
$$+\lambda\varepsilon|\Delta^{s+1}(\rho)|^2+2\varepsilon h^2\rho|\boldsymbol{\nabla}^2\log\rho|^2)\mathrm{d}x\mathrm{d}t=\int_0^T\int_\Omega P\,\mathrm{div}\,\boldsymbol{u}\mathrm{d}x\mathrm{d}t$$
$$+\int_\Omega\left(\frac{1}{2}\rho|\boldsymbol{u}|^2+\frac{\lambda}{2}|\boldsymbol{\nabla}^{2s+1}\rho|^2+\rho e(\rho,\theta)+4h^2|\boldsymbol{\nabla}\sqrt{\rho}|^2\right)(0)\mathrm{d}x. \tag{4.4.7}$$

In the following we show the right hand side of the equality (4.4.7) is bounded.

$$\int_0^T\int_\Omega R\rho_N\theta_N\mathrm{div}\,\boldsymbol{u}_N\mathrm{d}x$$

$$\leq C||\sqrt{\rho_N}D(\boldsymbol{u}_N)||_{L^2(0,T;L^2(\Omega))}||\sqrt{\rho_N}||_{L^\infty(0,T;L^6(\Omega))}$$
$$||\theta_N||_{L^2(0,T;L^3(\Omega))} \leqslant C, \tag{4.4.8}$$

and

$$\int_0^T \int_\Omega \frac{\beta}{3}\theta_N^4 \operatorname{div} \boldsymbol{u}_N \mathrm{d}x$$
$$\leq C||\theta_N||^4_{L^\infty(0,T;L^4(\Omega))}||\boldsymbol{\nabla}\boldsymbol{u}_N||_{L^2(0,T;L^\infty(\Omega))}$$
$$\times||\boldsymbol{\nabla}\boldsymbol{u}_N||_{L^2(0,T;L^\infty(\Omega))} \tag{4.4.9}$$
$$\leqslant ||\boldsymbol{\nabla}\boldsymbol{u}_N||_{L^2(0,T;W^{3,2}(\Omega))} \tag{4.4.10}$$

and the right hand side is bounded provided $(2s+1) \geqslant 3$. To see it, we write

$$\boldsymbol{\nabla}^3\boldsymbol{u}_N = \boldsymbol{\nabla}^3(\rho_N^{-1}\rho_N\boldsymbol{u}_N)$$
$$= \left(\frac{\boldsymbol{\nabla}^3\rho_N}{\rho_N^2} + \frac{(\boldsymbol{\nabla}\rho_N)^3}{\rho_N^4}\right)\rho_N^{\boldsymbol{u}_N}x + \rho_N^{-1}\boldsymbol{\nabla}^3(\rho_N\boldsymbol{u}_N) \tag{4.4.11}$$

and the boundedness of the right hand side follows from (4.3.20), (4.3.18) and the Cauchy inequality.

4.4.2 *Limit $N \to \infty$*

This subsection is devoted to the limit passage $N \to \infty$. Using estimates from the previous subsection we can extract weakly subsequences, whose limits satisfy the approximate system. It should be, however, emphasized that this level we replace weak formulation of the thermal energy by the weak formulation of the total energy.

4.4.3 *Strong convergence of the density and passage to the limit in the continuity equation*

From (4.4.8)–(4.4.12) we deduce that

$$\boldsymbol{u}_N \rightharpoonup \boldsymbol{u} \text{ weakly in } L^2(0,T;\ W^{2s+1,2}(\Omega)) \tag{4.4.12}$$

and

$$\rho_N \rightharpoonup \rho \text{ weakly in } L^2(0,T;\ W^{2s+2,2}(\Omega)) \tag{4.4.13}$$

at least for a suitable subsequence. In addition the right hand side of the linear parabolic problem,

$$\begin{aligned} &\partial_{t\rho} + \operatorname{div}(\rho \boldsymbol{u}) - \varepsilon\Delta\rho = 0, \\ &\rho(0,x) = \rho_\lambda^0(x), \end{aligned} \tag{4.4.14}$$

is uniformly bounded in $L^2(0,\,T; W^{2s,2}(\Omega))$ and the initial condition is sufficiently smooth. Thus, applying the $L^p - L^q$ theory to this problem we conclude that $\{\partial_{t\rho_N}\}_{n=1}^{\infty}$ is uniformly bounded in $L^2(0,T; W^{2s,2}(\Omega))$. Hence, the standard compact embedding implies $\rho_N \rightharpoonup \rho$ a.e. in $(0,T) \times \Omega$, and therefore passage to the limit in the approximate continuity equation is straightforward.

4.4.4 *Strong convergence of the temperature*

For the temperature uniform estimates we have

$$\theta_N \rightharpoonup \theta \text{ weakly in } L^2(0,T;\ W^{1,2}(\Omega)), \tag{4.4.15}$$

note that at this level, the time-compactness can be proved directly from the internal energy equation (4.2.11). Indeed, due to the continuity equation (4.2.11), we have

$$\begin{aligned} &\partial t(\rho_N\theta_N + a\theta_N^4) = -\operatorname{div}(\boldsymbol{u}_{N\rho_N}\theta_N + \beta\boldsymbol{u}_N\theta_N^4) + \operatorname{div}(\kappa\boldsymbol{\nabla}\theta_N) \\ &+\frac{\varepsilon}{\theta^2} - \varepsilon\theta^5 - \left(P_m + \frac{a}{3}\right)\operatorname{div}\boldsymbol{u}_N + \frac{1}{\rho_N}\frac{\partial P_c}{\partial_{\rho_N}}|\boldsymbol{\nabla}_{\rho_N}|^2 + \frac{4\varepsilon}{\gamma}|\boldsymbol{\nabla}\rho^{\frac{\gamma}{2}}|^2 \\ &+\nu\rho_N|D(u_N)|^2 + \lambda|\Delta^s\boldsymbol{\nabla}(_{\rho_N}\,\boldsymbol{u}_N)|^2 + \lambda\varepsilon|\Delta_{\rho_N}^{s+1}|^2 \\ &+2h_{\rho_N}^2|\boldsymbol{\nabla}^2\log_{\rho_N}|^2. \end{aligned} \tag{4.4.16}$$

On the account of (4.4.5) and (4.4.7), the last 9 terms of (4.4.16) are bounded in $L^1((0,T)\times\Omega)$. Then it follows from (4.4.2), (4.4.7) and

(4.4.9) that I_1 can be estimated as

$$\begin{aligned}&||\boldsymbol{u}_{N\rho_N}\theta_N||_{L^2((0,T)\times\Omega)}\\&\quad\leqslant C||\sqrt{\rho_N}\boldsymbol{u}_N||_{L^\infty(0,T;L^2\Omega)}||\sqrt{\rho_N}||_{L^\infty(0,T;L^6(\Omega))}\\&||\theta_N||_{L^\infty(0,T;L^4(\Omega))}\leqslant C\end{aligned}\tag{4.4.17}$$

and

$$||\boldsymbol{u}_N\theta_N^4||_{L^2((0,T)\times\Omega)}\leqslant C||\boldsymbol{u}_N||_{L^2(0,T;L^\infty(\Omega)}||\theta_N^4||_{L^{\frac{8}{3}}((0,TT\times)\Omega)}\tag{4.4.18}$$

where we used the interpolation

$$||\theta_N||_{L^{\frac{32}{3}}((0,T)\times\Omega)}\leqslant||\theta_N||^{\frac{1}{4}}_{L^\infty(0,T;L^{24}(\Omega))}||\theta_N||^{\frac{3}{4}}_{L^8(0,T;L^{24}(\Omega))'}\tag{4.4.19}$$

hence the last term of (4.4.19) is bounded provided $B\geqslant 8$.

Recall that we have $\kappa_\varepsilon\boldsymbol{\nabla}\theta_N=(\kappa_0+\rho_N+\rho_N\theta_N^2+\beta\theta_N^B)\boldsymbol{\nabla}\theta_N$, therefore using estimate (4.4.5) and (4.4.2) we verify that the most restrictive terms are bounded. Indeed,

$$\begin{aligned}&||\rho_N\boldsymbol{\nabla}\theta_N||_{L^p((0,T)\times\Omega)}\\&\quad\leqslant||\sqrt{\rho_N}\boldsymbol{\nabla}\log\theta_N||_{L^2(0,T;L^2(\Omega))}||\sqrt{\rho_N}||_{L^\infty((0,T)\times(\Omega)}\\&\qquad\times||\theta_N||_{L^{\frac{32}{3}}((0,T)\times\Omega)}\end{aligned}\tag{4.4.20}$$

with $p>1$, furthermore,

$$\begin{aligned}&||\rho_N\theta_N^2\boldsymbol{\nabla}\theta_N||_{L^{\frac{2B}{B+4}}(0,T;L^{\frac{3B}{B+2}}(\Omega))}\\&\quad\leqslant||\sqrt{\rho_N}||_{L^\infty((0,T)\times\Omega)}||\sqrt{\rho_N}\boldsymbol{\nabla}\theta_N||_{L^2(0,T;L^{\frac{3}{2}}(\Omega))}||\theta_N||^2_{L^B(0,T;L^{3B}(\Omega))}.\end{aligned}\tag{4.4.21}$$

Finally, since $B\geqslant 8$, θ^{B+1} can be bounded using (4.4.19).

As a conclusion, we have that

$$\partial_t(\rho_N\theta_N + \beta\theta_N^4) \in L^1(0,T;W^{-1,p}(\Omega)) \cup L^p(0,T;\ W^{-2,q}(\Omega)) \tag{4.4.22}$$

for some $p, q > 1$. On the other hand, since $\partial_t\rho$ is uniformly bounded in $L^2(0,T;W^{2s,2}(\Omega))$, $\rho > C(\lambda)$ and $\theta > 0$, we have

$$\begin{aligned}&||\partial t\theta_N||_{L^1(0,T;W^{-1,p}(\Omega))\cup L^p(0,T;W^{-2,q}(\Omega))}\\ &\quad\leqslant C\partial_t(\rho_N\theta_N + a\theta_N^4) \in L^1(0,T;W^{-1,p}(\Omega)) \cup L^p(0,T;W^{-2,q}(\Omega)),\end{aligned} \tag{4.4.23}$$

thus an application of the Aubin–Lions lemma gives precompactness of the sequence approximating the temperature

$$\theta_N \to \theta \text{ strongly in} L^{p'}((0,T)\times\Omega) \tag{4.4.24}$$

for any $1 \leqslant p' < \frac{32}{3}$.

Passage to the limit in the momentum equation

Having the strong convergence of the density, we start to identify the limit for $N \to \infty$ in the non-linear terms of the momentum equation.

(i) The convective term.
First, according to the estimate (4.4.13) and the strong convergence of the density, one observes that

$$\rho_N\boldsymbol{u}_N \rightharpoonup \rho\boldsymbol{u} \text{ weakly}^* \text{ in } L^\infty(0,T;L^2(\Omega)).$$

Next, using the momentum equation and the estimates (4.4.13)–(4.4.14), one can show that, for any $\boldsymbol{\phi} \in \cap_{n=1}^\infty X_N$, the family of functions $\int_\Omega \rho_N\boldsymbol{u}_N\phi dx$ is bounded and equicontinuous in $C(0,T)$, thus, via the Arzela–Ascoli theorem and density of smooth functions in $L^2(\Omega)$, we get that

$$\rho_N\boldsymbol{u}_N \rightharpoonup \rho\boldsymbol{u} \text{ weakly in } C([0,T];L^2(\Omega)). \tag{4.4.25}$$

Finally, by the compact embedding $L^2(\Omega) \subset W^{-1,2}(\Omega)$ and the weak convergence of $\boldsymbol{u}_N$ we verify that

$$\rho_N\boldsymbol{u}_N \otimes \boldsymbol{u}_N \rightharpoonup \rho\boldsymbol{u}\otimes\boldsymbol{u} \text{ weakly in } L^2((0,T)\times\Omega). \tag{4.4.26}$$

(ii) The capillarity term.
We write it in the form

$$\int_0^T \int_\Omega \rho_N \nabla \Delta^{2s+1} \rho_N \cdot \boldsymbol{\phi} \, \mathrm{d}x\mathrm{d}t$$
$$= \int_0^T \int_\Omega \Delta^s \operatorname{div}(\rho_N) \boldsymbol{\phi} \Delta^{s+1} \rho_N \, \mathrm{d}x\mathrm{d}t.$$

Due to (4.4.14) and the boundedness of the time derivative of ρ_N, we infer that

$$\rho_N \to \rho \text{ strongly in } L^2(0,T;W^{2s+1,2}(\Omega)). \tag{4.4.27}$$

Thus,

$$\int_0^T \int_\Omega \Delta^s \operatorname{div}(\rho_N \boldsymbol{\phi}) \Delta^{s+1} \rho_N \, \mathrm{d}x\mathrm{d}t$$
$$\to \int_0^T \int_\Omega \Delta^s \operatorname{div}(\rho \boldsymbol{\phi}) \Delta^{s+1} \rho \, \mathrm{d}x\mathrm{d}t$$

for any $\boldsymbol{\phi} \in C^\infty((0,T) \times \bar{\Omega})$.

(iii) The momentum term.
We write it in the form

$$-\lambda \int_0^T \int_\Omega \rho_N \Delta^{2s+1} (\rho_N \boldsymbol{u}_N) \cdot \boldsymbol{\phi} \, \mathrm{d}x\mathrm{d}t$$
$$= -\lambda \int_0^T \int_\Omega \Delta^s \nabla(\rho_N \boldsymbol{u}_N) : \Delta^s \nabla(\rho_N \boldsymbol{\phi}) \mathrm{d}x\mathrm{d}t,$$

so the convergence established in (4.4.13) and (4.4.28) is sufficient to pass to the limit here.

Strong convergence of the density and temperature enables us to perform in the momentum equation (4.2.10) satisfied for any function $\boldsymbol{\phi} \in C^1([0,T];(X_N))$ such that $\boldsymbol{\phi}(T) = 0$, and by the density argument we can take all such test functions from $C^1([0,T];W^{2s+1}(\Omega))$.

4.4.5 *Passage to the limit in the internal energy balance equation*

We are going to pass the limit in the terms $\nu\rho|D(u)|^2$, $\lambda|\Delta^s\nabla(\rho\boldsymbol{u})|^2$, $\lambda\varepsilon|\Delta^{s+1}\rho|^2$ and $2h^2\rho|\nabla^2\log\rho|^2$ which require a sort of strong convergence of these quantities. This will be deduced from the kinetic energy balance. For this purpose we need to show that $\boldsymbol{u}$ can be a test function in the limit momentum equation. Indeed, in (4.2.4) all terms are bounded due to the estimate (4.4.8) above. Moreover, thanks to the lower bound of ρ, we can verify that $\boldsymbol{u}$ is actually a continuous function with respect to time and that it is continuously differentiable. To see this it is enough to differentiate (4.2.4) with respect to time and use the kinetic energy balance.

Now, using $\boldsymbol{u}$ as a test function and taking advantage of the fact that the limit continuity equation is satisfied pointwisely, we obtain

$$\begin{aligned}&\int_0^T\int_\Omega(\nu\rho|D(\boldsymbol{u})|^2+\lambda|\Delta^s\nabla(\rho\boldsymbol{u})|^2+\lambda\varepsilon|\Delta^{s+1}(\rho)|^2\\&\quad+2h^2\rho|\nabla^2\log\rho|^2)\mathrm{d}x\mathrm{d}t+\int_\Omega\left(\frac{1}{2}\rho|\boldsymbol{u}|^2+\frac{\lambda}{2}|\nabla^{2s+1}\rho|^2\right.\\&\quad\left.+4h^2|\nabla\sqrt{\rho}|^2\right)(t)\mathrm{d}x=\int_0^T\int_\Omega P\,\mathrm{div}\,\boldsymbol{u}\,\mathrm{d}x\\&\quad+\int_\Omega\left(\frac{1}{2}\rho|\boldsymbol{u}|^2+\frac{\lambda}{2}|\nabla^{2s+1}\rho|^2+4h^2|\nabla\sqrt{\rho}|^2\right)(0)\mathrm{d}x\end{aligned}\tag{4.4.28}$$

for any $t\in[0,T]$. On the other hand, due to (4.3.10), we have

$$\begin{aligned}&\lim_{N\to\infty}\int_0^T\int_\Omega(\nu\rho_N|D(\boldsymbol{u}_N)|^2+\lambda|\Delta^s\nabla(\rho_N\boldsymbol{u}_N)|^2\\&\quad+\lambda\varepsilon|\Delta^{s+1}(\rho_N)|^2+2h^2\rho_N|\nabla^2\log\rho_N|^2)\mathrm{d}x\mathrm{d}t\\&\quad+\int_\Omega\left(\frac{1}{2}\rho_N|\boldsymbol{u}_N|^2+\frac{\lambda}{2}|\nabla^{2s+1}\rho_N|^2+4h^2|\nabla\sqrt{\rho_N}|^2\right)(t)\mathrm{d}x\\&=\int_0^T\int_\Omega P\,\mathrm{div}\,\boldsymbol{u}\,\mathrm{d}x\mathrm{d}t+\int_\Omega\left(\frac{1}{2}\rho|\boldsymbol{u}|^2+\frac{\lambda}{2}|\nabla^{2s+1}\rho|^2\right.\\&\quad\left.+4h^2|\nabla\sqrt{\rho}|^2\right)(0)\mathrm{d}x.\end{aligned}\tag{4.4.29}$$

The comparison of these two expressions (4.4.28) and (4.4.29) yields

$$\begin{aligned}
&\nu\|\sqrt{\rho_N}D(\boldsymbol{u}_N)\|^2_{L^2((0,T)\times\Omega)} \to \nu\|\sqrt{\rho}D(\boldsymbol{u})\|^2_{L^2((0,T)\times\Omega)},\\
&\lambda\|\Delta^s\boldsymbol{\nabla}(\rho_N\boldsymbol{u}_N)\|^2_{L^2((0,T)\times\Omega)} \to \lambda\|\Delta^s\boldsymbol{\nabla}(p\boldsymbol{u})\|^2_{L^2((0,T)\times\Omega)},\\
&\lambda\varepsilon\|\Delta^{s+1}\rho_N\|^2_{L^2((0,T)\times\Omega)} \to \lambda\varepsilon\|\Delta^{s+1}\rho\|^2_{L^2((0,T)\times\Omega)},\\
&2h^2\|\sqrt{\rho_N}|\boldsymbol{\nabla}^2\log\rho_N|\|^2_{L^2((0,T)\times\Omega)} \to 2h^2\|\sqrt{\rho}|\boldsymbol{\nabla}^2\log\rho|\|^2_{L^2((0,T)\times\Omega)},
\end{aligned} \tag{4.4.30}$$

and for all $t \in [0,T]$, we have

$$\begin{aligned}
\|\rho_N|\boldsymbol{u}_N|^2(t)\|_{L^1(\Omega)} &\to \|\rho|\boldsymbol{u}|^2(t)\|_{L^1(\Omega)},\\
\lambda\|\boldsymbol{\nabla}^{2s+1}\rho_N(t)\|^2_{L^2(\Omega)} &\to \lambda\|\boldsymbol{\nabla}^{2s+1}\rho(t)\|^2_{L^2(\Omega)},\\
4h^2\|\boldsymbol{\nabla}\sqrt{\rho_N}\|^2_{L^2(\Omega)} &\to 4h^2\|\boldsymbol{\nabla}\|^2_{L^2(\Omega)}.
\end{aligned} \tag{4.4.31}$$

Having convergence of these norms and relevant weakly convergent sequences we deduce the strong convergence. On account of that we are able to perform the limit passage in the internal energy equation (4.2.11),

$$\begin{aligned}
&\int_0^T\int_\Omega(\rho\theta+\beta\theta^4)\partial_t\phi\mathrm{d}x\mathrm{d}t + \int_0^T\int_\Omega \boldsymbol{u}(\rho\theta+\beta\theta^4)\cdot\boldsymbol{\nabla}\phi\mathrm{d}x\mathrm{d}t\\
&\quad - \int_0^T\int_\Omega \kappa\boldsymbol{\nabla}\theta\cdot\boldsymbol{\nabla}\phi\mathrm{d}x\mathrm{d}t = \int_0^T\int_\Omega\left(\frac{\varepsilon}{\theta^2}-\varepsilon\theta^5\right)\phi\mathrm{d}x\mathrm{d}t\\
&\quad + \int_0^T\int_\Omega\left(P_m+\frac{\beta}{3}\theta^4\right)\operatorname{div}\boldsymbol{u}\phi\mathrm{d}x\mathrm{d}t - \int_\Omega(\rho\theta+\beta\theta^4)(0)\phi(0)\mathrm{d}x\\
&\quad - \int_0^T\int_\Omega\left(\nu_\rho|D\boldsymbol{u}|^2+\lambda|\Delta^s\boldsymbol{\nabla}(\rho\boldsymbol{u})|^2+\lambda\varepsilon|\Delta^{s+1}(\rho)|^2\right.\\
&\quad \left.+2h^2\rho|\boldsymbol{\nabla}^2\log\rho|^2+\varepsilon\frac{1}{\rho}\frac{\partial P_c}{\partial\rho}|\boldsymbol{\nabla}_\rho|^2\right)\mathrm{d}x\mathrm{d}t,
\end{aligned} \tag{4.4.32}$$

for any smooth ϕ vanishing at $t = \mathbb{T}$.

4.4.6 Passage to the limit in the total energy balance equation

Now we use $\boldsymbol{u}\phi$ as a test function in the limit momentum equation (4.2.4), using again the limit continuity equation and after integrating by parts, we get

$$
\int_0^T \int_\Omega \left(\rho \frac{|\boldsymbol{u}|^2}{2} - h^2 \rho \Delta \log \rho \right) \partial_t \phi \, dxdt
$$

$$
+ \int_0^T \int_\Omega \left(\rho \frac{|\boldsymbol{u}|^2}{2} \boldsymbol{u} - h^2 \rho \Delta \log \rho \boldsymbol{u} \right) \cdot \boldsymbol{\nabla} \phi \, dxdt
$$

$$
- \int_0^T \int_\Omega (\nu \rho D(\boldsymbol{u}) \boldsymbol{u} - P \boldsymbol{u}) \cdot \boldsymbol{\nabla} \phi dxdt
$$

$$
+ 2h^2 \int_0^T \int_\Omega \rho \boldsymbol{\nabla}^2 \rho : \boldsymbol{\nabla} \boldsymbol{u} \phi \, dxdt
$$

$$
= \int_0^T \int_\Omega \nu_\rho |D\boldsymbol{u}|^2 \phi \, dxdt + \int_0^T \int_\Omega \lambda \Delta^s \boldsymbol{\nabla}(\rho \boldsymbol{u}) : \Delta^s \boldsymbol{\nabla}(\rho \boldsymbol{u} \phi) dxdt
$$

$$
- h^2 \int_0^T \int_\Omega \operatorname{div}(\rho \Delta \boldsymbol{u}) \phi dxdt + \frac{\varepsilon}{2} \int_0^T \int_\Omega |\boldsymbol{u}|^2 \boldsymbol{\nabla} \rho : \boldsymbol{\nabla} \phi \, dxdt
$$

$$
- \int_0^T \int_\Omega P \operatorname{div} \boldsymbol{u} \phi \, dxdt + \int_0^T \int_\Omega \lambda \Delta^s \operatorname{div}(\rho \boldsymbol{u} \phi) : \Delta^{s+1} \rho \, dxdt
$$

$$
- h^2 \varepsilon \int_0^T \int_\Omega \rho \Delta \left(\frac{\Delta \rho}{\rho} \right) dxdt - \int_0^T \int_\Omega \rho \frac{|\boldsymbol{u}|^2}{2}(0)\phi(0)dxdt
$$

$$
- \int_0^T \int_\Omega h^2 \rho \Delta \log \rho(0)\phi(0)dxdt. \tag{4.4.33}
$$

We apply to the approximate continuity equation. The operator Δ^s and then we test it by $\lambda \operatorname{div}(\boldsymbol{\nabla}\Delta^s \rho \phi)$ in order to obtain

$$
\int_0^T \int_\Omega \frac{\lambda}{2} |\boldsymbol{\nabla}\Delta^s \rho|^2 \partial_t \phi dxdt + \lambda \int_0^T \int_\Omega \Delta^s \operatorname{div}(\rho \boldsymbol{u}) \Delta^{s+1} \rho \phi \, dddt
$$

$$
+ \lambda \int_0^T \int_\Omega \Delta^s \operatorname{div}(\rho \boldsymbol{u}) \boldsymbol{\nabla}\Delta^s \rho \cdot \boldsymbol{\nabla} \phi dxdt
$$

$$
\begin{aligned}
&- \lambda\varepsilon \int_0^T \int_\Omega |\Delta^{s+1}\rho|^2 \phi \,\mathrm{d}x\mathrm{d}t \\
&- \lambda\varepsilon \int_0^T \int_\Omega \Delta^{s+1}\rho \nabla \Delta^s \rho \cdot \nabla \phi \mathrm{d}x\mathrm{d}t \\
&+ \frac{\lambda}{2} \int_0^T \int_\Omega |\nabla \Delta^s \rho|^2(0)\phi(0) \mathrm{d}x\mathrm{d}t = 0. \qquad (4.4.34)
\end{aligned}
$$

Now summing (4.4.32) with (4.4.33), (4.4.34) and using the limit continuity equation to rewrite the term $\int_0^T \int_\Omega P_c \operatorname{div} \boldsymbol{u} \mathrm{d}x\mathrm{d}t$, we get the weak formulation of the total energy plus some terms which will appear in the subsequent limit passages

$$
\begin{aligned}
&\int_0^T \int_\Omega \left(\rho \frac{|\boldsymbol{u}|^2}{2} + \rho e + \frac{\lambda}{2} |\nabla \Delta^s \rho|^2 - h^2 \rho \Delta \log \rho \right) \partial_t \phi \,\mathrm{d}x\mathrm{d}t \\
&+ 2h^2 \int_0^T \int_\Omega \rho \nabla^2 \rho : \nabla \boldsymbol{u} \phi \,\mathrm{d}x\mathrm{d}t \\
&+ \int_0^T \int_\Omega \left(\rho e + \rho \frac{|u|^2}{2} \boldsymbol{u} - h^2 \rho \Delta \log \rho \boldsymbol{u} \right) \cdot \nabla \phi \,\mathrm{d}x\mathrm{d}t \\
&- \int_0^T \int_\Omega \kappa \nabla \theta \cdot \nabla \phi \,\mathrm{d}x\mathrm{d}t + \int_0^T \int_\Omega P \boldsymbol{u} \cdot \nabla \phi \,\mathrm{d}x\mathrm{d}t \\
&- \int_0^T \int_\Omega \nu_\rho D(\boldsymbol{u}) \boldsymbol{u} \cdot \nabla \phi \,\mathrm{d}x\mathrm{d}t = - \int_0^T \int_\Omega \left(\frac{\varepsilon}{\theta^2} - \varepsilon \theta^5 \right) \phi \,\mathrm{d}x\mathrm{d}t \\
&- h^2 \int_0^T \int_\Omega \operatorname{div}(\rho \Delta \boldsymbol{u}) \phi \,\mathrm{d}x\mathrm{d}t - h^2 \varepsilon \int_0^T \int_\Omega \rho \Delta \left(\frac{\Delta \rho}{\rho} \right) \mathrm{d}x\mathrm{d}t \\
&+ \int_0^T \int_\Omega R_{\varepsilon,\lambda} \,\mathrm{d}x\mathrm{d}t - \int_0^T \int_\Omega \left(\rho \frac{|\boldsymbol{u}|^2}{2} + \rho e + \frac{\lambda}{2} |\nabla \Delta^s \rho|^2 \right) \\
&\times (0) \phi(0) \,\mathrm{d}x\mathrm{d}t, \qquad (4.4.35)
\end{aligned}
$$

and

$$
\begin{aligned}
&R_{\varepsilon,\lambda}(\rho, \theta, \boldsymbol{u}, \phi) \\
&\quad = \lambda [\Delta(\operatorname{div}(\rho \boldsymbol{u} \phi)) \Delta^{s+1} \rho - \Delta \operatorname{div}(\rho \boldsymbol{u}) \Delta^{s+1} \rho \phi]
\end{aligned}
$$

$$
\begin{aligned}
&- \lambda\Delta^s \operatorname{div}(\rho\boldsymbol{u})\boldsymbol{\nabla}\Delta^{s+1}\rho\cdot\boldsymbol{\nabla}\boldsymbol{\phi} - \lambda[|\Delta^s\boldsymbol{\nabla}(\rho\boldsymbol{u})|^2\boldsymbol{\phi}\\
&- \Delta^s\boldsymbol{\nabla}(\rho\boldsymbol{u}) : \Delta^s\boldsymbol{\nabla}(\rho\boldsymbol{u}\boldsymbol{\phi})] + \lambda\varepsilon\Delta^{s+1}\rho\boldsymbol{\nabla}\Delta^{s+1}\rho\cdot\boldsymbol{\nabla}\boldsymbol{\phi}\\
&+ \frac{\varepsilon}{2}|\boldsymbol{u}|^2\boldsymbol{\nabla}\rho\cdot\boldsymbol{\nabla}\boldsymbol{\phi} + \varepsilon\boldsymbol{\nabla}\rho\cdot\boldsymbol{\nabla}\boldsymbol{\phi}\left(e_c(\rho) + \frac{P_c(\rho)}{\rho}\right). \qquad (4.4.36)
\end{aligned}
$$

4.5 B-D entropy inequality

At this level we are only left with two parameters of approximation: ε and λ. From the so-far obtained priori estimates, only the ones following from (4.3.11) and (4.3.20) were independent of these parameters. However, having the ε-dependent estimate for $\Delta^{s+1}\rho$ allows us to derive a type of B-D estimate, from which it follows that this estimate depends only on λ. As a product, we will derive the energy estimate independent of λ. Note that, so far in (4.4.8), we were only able to estimate the right hand side. using the λ-dependent bounds for $\boldsymbol{u}$. We will prove the following lemma. The main advantage of the B-D entropy inequality gives the space compactness of the density.

Lemma 4.5.1. *For any positive constant $r > 1$, we have*

$$
\begin{aligned}
&\frac{\mathrm{d}}{\mathrm{d}t}\int_\Omega\left(\frac{1}{2}\rho|\boldsymbol{u} + \boldsymbol{\nabla}\boldsymbol{\phi}(\rho)|^2 + \frac{r-1}{2}\rho|\boldsymbol{u}|^2 + \frac{r\lambda}{2}|\boldsymbol{\nabla}\Delta^s\rho|^2 + r\rho e_c(\rho)\right)\mathrm{d}x\\
&+ \int_\Omega \boldsymbol{\nabla}\boldsymbol{\phi}(\rho)\cdot\boldsymbol{\nabla}P\,\mathrm{d}x + \frac{1}{2}\int_\Omega \rho|\boldsymbol{\nabla}\boldsymbol{u} - \boldsymbol{\nabla}^{\mathrm{T}}\boldsymbol{u}|^2\,\mathrm{d}x\\
&+ 2\lambda\int_\Omega |\Delta^{s+1}\rho|^2\,\mathrm{d}x + 2(r-1)\int_\Omega \rho|D(u)|^2\,\mathrm{d}x\\
&+ r\int_\Omega(\lambda\varepsilon|\Delta^{s+1}\rho|^2 + \lambda|\Delta^s\boldsymbol{\nabla}(p\boldsymbol{u})|^2)\mathrm{d}x + \int_\Omega 4h^2\rho|\boldsymbol{\nabla}^2\log\rho|^2\,\mathrm{d}x\\
&+ \frac{4\varepsilon}{\gamma}\int_0^t\int_\Omega |\boldsymbol{\nabla}\rho^{\frac{\gamma}{2}}|^2\,\mathrm{d}x\mathrm{d}t = -\varepsilon\int_\Omega(\boldsymbol{\nabla}\rho\cdot\boldsymbol{\nabla})\boldsymbol{u}\,\mathrm{d}x
\end{aligned}
$$

$$
\begin{aligned}
&+\varepsilon\int_{\Omega}\rho\boldsymbol{\nabla}\boldsymbol{\phi}(\rho)\cdot\boldsymbol{\nabla}(\boldsymbol{\phi}'(\rho)\Delta\rho)\mathrm{d}x-\varepsilon\int_{\Omega}\mathrm{div}(\rho\boldsymbol{u})\boldsymbol{\phi}'(\rho)\Delta\rho\,\mathrm{d}x\\
&+\varepsilon\int_{\Omega}\left(R\rho\theta+\frac{\beta}{3}\theta^{4}\right)\mathrm{div}\,\boldsymbol{u}\,\mathrm{d}x-2\lambda\int_{\Omega}\Delta^{s}\boldsymbol{\nabla}(\rho\boldsymbol{u})\cdot\Delta^{s}\boldsymbol{\nabla}^{2}\rho\,\mathrm{d}x,
\end{aligned}
\tag{4.5.1}
$$

$\mathcal{D}'(0,T)$, *where* $\boldsymbol{\nabla}\boldsymbol{\phi}(\rho)=2\boldsymbol{\nabla}\log\rho$, $e_c(\rho)=\int_0^{\rho}y^{-2}P_c(\rho)\mathrm{d}y\geq 0$.

Proof. The basic idea of the proof is to find the explicit form of the terms

$$
\frac{\mathrm{d}}{\mathrm{d}t}\int_{\Omega}\left(\frac{1}{2}\rho|\boldsymbol{u}|^{2}+\rho\boldsymbol{u}\cdot\boldsymbol{\nabla}\boldsymbol{\phi}(\rho)+\frac{1}{2}\rho|\boldsymbol{\nabla}(\rho)|^{2}\right)\mathrm{d}x. \tag{4.5.2}
$$

The first term can be evaluated by means of the main energy inequality, i.e.

$$
\begin{aligned}
&\int_{\Omega}\left(\frac{1}{2}\rho|\boldsymbol{u}|^{2}(t)+\frac{\rho^{\gamma}}{\gamma-1}(t)+\frac{\lambda}{2}|\boldsymbol{\nabla}^{2s+1}\rho|^{2}(t)\right.\\
&\qquad\left.+\rho e(\rho)(t)+4h^{2}|\boldsymbol{\nabla}\sqrt{\rho}|^{2}(t)\right)\mathrm{d}x+\varepsilon\int_{0}^{t}\int_{\Omega}\theta^{5}\,\mathrm{d}x\mathrm{d}t\\
&=\varepsilon\int_{0}^{t}\int_{\Omega}\frac{1}{\theta^{2}}\,\mathrm{d}x\mathrm{d}t+\int_{\Omega}\left(\frac{1}{2}\rho|\boldsymbol{u}|^{2}(0)+\frac{\lambda}{2}|\boldsymbol{\nabla}^{2s+1}\rho|^{2}(0)\right.\\
&\qquad\left.+\rho e(\rho)(0)+\frac{\rho^{\gamma}}{\gamma-1}(0)+4h^{2}|\boldsymbol{\nabla}\sqrt{\rho}|^{2}(0)\right)\mathrm{d}x.
\end{aligned}
\tag{4.5.3}
$$

To get a relevant expression for the third term in (4.5.2), we multiply the approximate continuity equation by $\frac{|\boldsymbol{\nabla}\boldsymbol{\phi}(\rho)|^{2}}{2}$ and we obtain the following sequence of equalities

$$\begin{aligned}
&\frac{\mathrm{d}}{\mathrm{d}t}\int_\Omega \frac{1}{2}\rho|\nabla\phi(\rho)|^2\,\mathrm{d}x\\
&\quad=\int_\Omega\left(\rho\partial_t\frac{|\nabla\phi(\rho)|^2}{2}-\frac{|\nabla\phi(\rho)|^2}{2}\mathrm{div}(\rho\boldsymbol{u})+\varepsilon\frac{|\nabla\phi(\rho)|^2}{2}\Delta_\rho\right)\mathrm{d}x\\
&\quad=\int_\Omega\left(\rho\nabla\phi(\rho)\cdot\nabla(\phi'(\rho)\partial_t\rho)-\frac{|\nabla\phi(\rho)|^2}{2}\mathrm{div}(\rho\boldsymbol{u})\right.\\
&\qquad\left.+\varepsilon\frac{|\nabla\phi(\rho)|^2}{2}\Delta_\rho\right)\mathrm{d}x. \qquad (4.5.4)
\end{aligned}$$

Using the approximate continuity equation, we get

$$\begin{aligned}
&\int_\Omega\rho\nabla\phi(\rho)\cdot\nabla(\phi'(\rho)\partial_{t\rho})\mathrm{d}x\\
&\quad=\int_\Omega\varepsilon\rho\nabla\phi(\rho)\cdot\nabla(\phi'(\rho)\Delta\rho)\mathrm{d}x-\int_\Omega\rho\nabla\boldsymbol{u}:\nabla\phi(\rho)\otimes\nabla\phi(\rho)\mathrm{d}x\\
&\qquad-\int_\Omega\rho\nabla\phi(\rho)\cdot\nabla(\phi'(\rho)\rho\,\mathrm{div}\,\boldsymbol{u})\mathrm{d}x\\
&\qquad-\int_\Omega\rho\boldsymbol{u}\otimes\nabla\phi(\rho):\nabla^2\phi(\rho)\mathrm{d}x. \qquad (4.5.5)
\end{aligned}$$

Integrating by parts the two last terms from the right hand side of (4.5.5), we have

$$\begin{aligned}
&\int_\Omega\rho\nabla\phi(\rho)\cdot\nabla(\phi'(\rho)\partial_{t\rho})\mathrm{d}x\\
&\quad=\int_\Omega\varepsilon\rho\nabla\phi(\rho)\cdot\nabla(\phi'(\rho)\Delta\rho)\mathrm{d}x-\int_\Omega\rho\nabla\boldsymbol{u}:\nabla\phi(\rho)\otimes\nabla\phi(\rho)\mathrm{d}x\\
&\qquad+\int_\Omega\rho|\nabla\phi(\rho)|^2\,\mathrm{div}\,\boldsymbol{u})\mathrm{d}x+\int_\Omega\rho^2\phi'(\rho)\Delta\phi(\rho)\mathrm{div}\,\boldsymbol{u}\mathrm{d}x\\
&\qquad+\int_\Omega|\nabla\phi(\rho)|^2\,\mathrm{div}(\rho\boldsymbol{u})\mathrm{d}x+\int_\Omega\rho\boldsymbol{u}\cdot\nabla(\nabla\phi(\rho))\cdot\nabla(\phi(\rho))\mathrm{d}x. \\
&\qquad\qquad (4.5.6)
\end{aligned}$$

Combining the three previous equalities, we finally obtain

$$\begin{aligned}
&\frac{\mathrm{d}}{\mathrm{d}t}\int_\Omega \frac{1}{2}\rho|\boldsymbol{\nabla}\phi(\rho)|^2\mathrm{d}x \\
&= \int_\Omega \varepsilon\rho\boldsymbol{\nabla}\phi(\rho)\cdot\boldsymbol{\nabla}(\phi'(\rho)\Delta\rho)\mathrm{d}x - \int_\Omega \rho\boldsymbol{\nabla}\boldsymbol{u} : \boldsymbol{\nabla}\phi(\rho)\otimes\boldsymbol{\nabla}\phi(\rho)\mathrm{d}x \\
&\quad + \int_\Omega \rho|\boldsymbol{\nabla}\phi(\rho)|^2 \operatorname{div}\boldsymbol{u}\mathrm{d}x + \int_\Omega \rho^2\phi'(\rho)\Delta\phi(\rho)\operatorname{div}\boldsymbol{u}\mathrm{d}x \\
&\quad + \int_\Omega \varepsilon\frac{|\boldsymbol{\nabla}\phi(\rho)|^2}{2}\Delta\rho\mathrm{d}x. \qquad (4.5.7)
\end{aligned}$$

In the above series of equalities, each one holds ponitwisely with respect to time due to the regularity of ρ and $\boldsymbol{\nabla}\phi$. This is not the case of the middle integrant of (4.5.2), for which one should really think of weak-in-time formulation. Denote

$$V = W^{2s+1,2}(\Omega), \quad \text{and} v = \rho\boldsymbol{u}, \quad h = \boldsymbol{\nabla}\phi. \qquad (4.5.8)$$

We know that $v \in L^2(0,T;V)$ and its weak derivative with respect to time variable $v' \in L^2(0,T;V^*)$, where V^* denotes the dual space to V. Moreover, $h \in L^2(0,T;V)$, $h' \in L^2(0,T;W^{2s-1,2}(\Omega))$. Now, let v_m, h_m denote the standard mollifications in time of v, h, respectively. By the properties of mollifiers we know that

$$v_m, v_m' \in C^\infty(0,T;V), \quad h_m, h_m' \in C^\infty(0,T;V), \qquad (4.5.9)$$

and

$$\begin{aligned}
&v_m \to v \quad L^2(0,T;V), \quad h_m \to h \quad L^2(0,T;V), \\
&v_m' \to v' \quad L^2(0,T;V^*), \quad h_m' \to h' \quad L^2(0,T;V^*).
\end{aligned} \qquad (4.5.10)$$

For these regularized sequences we may write

$$\begin{aligned}
\frac{\mathrm{d}}{\mathrm{d}t}\int_\Omega v_m\cdot h_m\mathrm{d}x = \frac{\mathrm{d}}{\mathrm{d}t}(v_m, h_m)_V = (v_m', h_m)_V \\
+ (v_m, h_m')_V, \quad \forall\psi \in \mathcal{D}(0,T). \qquad (4.5.11)
\end{aligned}$$

Using the Riesz representation theorem we verify that $v_m' \in C^\infty(0,T;V)$ uniquely determines the functional $\Phi_{v_m'} \in V^*$ such that $(v_m', \psi)_V = (\Phi_{v_m'}, \psi)_V^*, V = \int_\Omega v_m'\cdot\psi\mathrm{d}x$, $\forall\psi \in V$; for the second term from the right hand side of (4.5.11) we can simply replace $V = L^2(\Omega)$, and thus we obtain

$$-\int_0^T (v_m, h_m)_V \psi' \mathrm{d}t = \int_0^T (v_m', h_m)_{V^*,V} \psi \mathrm{d}t + \int_0^T (v_m', h_m)_{L^2(\Omega)} \psi \mathrm{d}t \quad \forall \psi \in \mathcal{D}(0,T). \tag{4.5.12}$$

Observe that both integrands from the right hand side are uniformly bounded in $L^1(0,T)$. Thus, using (4.5.10), we let $m \to \infty$ in order to obtain

$$\frac{\mathrm{d}}{\mathrm{d}t} \int_\Omega v \cdot h \mathrm{d}x = (v'h)_V + (v, h')_V, \quad \forall \psi \in \mathcal{D}(0,T). \tag{4.5.13}$$

Coming back to our original notation, this means that the operation

$$\frac{\mathrm{d}}{\mathrm{d}t} \int_\Omega \rho \boldsymbol{u} \cdot \boldsymbol{\nabla} \phi(\rho) \mathrm{d}x = \langle \partial_t(\rho \boldsymbol{u}), \boldsymbol{\nabla} \phi(\rho) \rangle_{V^*,V} + \int_\Omega \rho \boldsymbol{u} \cdot \partial_t \boldsymbol{\nabla} \phi(\rho) \mathrm{d}x \tag{4.5.14}$$

is well defined and is nothing but equality between two scalar distributions. By the fact that $\partial_t \boldsymbol{\nabla} \phi(\rho)$ exists a.e. in $(0,T) \times \Omega$, we may use approximation to write

$$\int_\Omega \rho \boldsymbol{u} \cdot \boldsymbol{\nabla} \phi(\rho) \mathrm{d}x = \int_\Omega (\mathrm{div}(\rho \boldsymbol{u}))^2 \phi'(\rho) \mathrm{d}x - \varepsilon \int_\Omega \mathrm{div}(\rho \boldsymbol{u}) \phi'(\rho) \Delta \rho \mathrm{d}x \tag{4.5.15}$$

whence the first term on the right hand side of (4.5.14) may be evaluated by testing the approximate momentum equation

$$\begin{aligned}
&\langle\partial_t(\rho\boldsymbol{u}),\boldsymbol{\nabla}\phi(\rho)\rangle_{V^*,V}\\
&\quad=-\int_\Omega 2\rho\Delta\phi(\rho)\,\mathrm{div}\boldsymbol{u}\mathrm{d}x+2\int_\Omega\boldsymbol{\nabla}\boldsymbol{u}:\boldsymbol{\nabla}\phi(\rho)\otimes\boldsymbol{\nabla}\phi(\rho)\mathrm{d}x\\
&\qquad-2\int_\Omega\Delta\phi(\rho)\cdot\boldsymbol{\nabla}_\rho\,\mathrm{div}\boldsymbol{u}\mathrm{d}x-\int_\Omega\boldsymbol{\nabla}\phi(\rho)\cdot\boldsymbol{\nabla}P\mathrm{d}x\\
&\qquad-\int_\Omega\Delta^{s+1}\rho\Delta^s\,\mathrm{div}(\rho\boldsymbol{\nabla}\phi(\rho))\mathrm{d}x\\
&\qquad-\lambda\int_\Omega\Delta^s\boldsymbol{\nabla}(\rho\boldsymbol{u}):\Delta^s\boldsymbol{\nabla}(\rho\boldsymbol{\nabla}\phi(\rho))\mathrm{d}x\\
&\qquad-\int_\Omega\boldsymbol{\nabla}\phi(\rho)\cdot\mathrm{div}(\rho\boldsymbol{u}\otimes\boldsymbol{u})\mathrm{d}x-\varepsilon\int_\Omega(\boldsymbol{\nabla}_\rho\cdot\boldsymbol{\nabla})\boldsymbol{u}\cdot\boldsymbol{\nabla}\phi(\rho)\mathrm{d}x\\
&\qquad+\int_\Omega 2h^2\,\mathrm{div}(\rho\boldsymbol{\nabla}^2\log\rho)\cdot\boldsymbol{\nabla}\phi(\rho)\mathrm{d}x,
\end{aligned}\tag{4.5.16}$$

recalling the form of $\phi(\rho)$ it can be deduced as follows

$$\begin{aligned}
&\frac{\mathrm{d}}{\mathrm{d}t}\int_\Omega(\rho\boldsymbol{u}\cdot\boldsymbol{\nabla}\phi(\rho)+\frac{1}{2}\rho|\boldsymbol{\nabla}\phi(\rho)|^2)\mathrm{d}x\\
&\qquad+\int_\Omega\boldsymbol{\nabla}\phi(\rho)\cdot\boldsymbol{\nabla}P\mathrm{d}x+2\lambda\int_\Omega|\Delta^{s+1}\rho|^2\mathrm{d}x\\
&\quad=-\int_\Omega\boldsymbol{\nabla}\phi(\rho)\cdot\mathrm{div}(\rho\boldsymbol{u}\otimes\boldsymbol{u})\mathrm{d}x+\int_\Omega(\mathrm{div}(\rho\boldsymbol{u}))^2\phi'(\rho)\mathrm{d}x\\
&\qquad-2\lambda\int_\Omega\Delta^s\boldsymbol{\nabla}(\rho\boldsymbol{u}):\Delta^s\boldsymbol{\nabla}^2\rho\mathrm{d}x-\varepsilon\int_\Omega\mathrm{div}(\rho\boldsymbol{u})\phi'(\rho)\Delta\rho\mathrm{d}x\\
&\qquad+\int_\Omega\varepsilon\frac{|\boldsymbol{\nabla}\phi(\rho)|^2}{2}\Delta\rho\mathrm{d}x-\varepsilon\int_\Omega(\boldsymbol{\nabla}_\rho\cdot\boldsymbol{\nabla})\boldsymbol{u}\cdot\boldsymbol{\nabla}\phi(\rho)\mathrm{d}x\\
&\qquad+\int_\Omega\varepsilon_\rho\boldsymbol{\nabla}\phi(\rho)\cdot\boldsymbol{\nabla}(\phi'(\rho)\Delta\rho)\mathrm{d}x-\int_\Omega 4h^2\rho|\boldsymbol{\nabla}^2\log\boldsymbol{u}|^2\mathrm{d}x.
\end{aligned}\tag{4.5.17}$$

The first two terms from the right hand side of (4.5.17) can be transformed

$$\begin{aligned}&\int_\Omega[(\operatorname{div}(\rho\boldsymbol{u}))^2\phi'(\rho)-\boldsymbol{\nabla}\phi(\rho)\cdot\operatorname{div}(\rho\boldsymbol{u}\otimes\boldsymbol{u})]\mathrm{d}x\\&=\int_\Omega[\rho^2\phi'(\rho)(\operatorname{div}\boldsymbol{u})^2+\rho\phi'\cdot\boldsymbol{\nabla}_\rho\operatorname{div}\boldsymbol{u}-\rho\phi'\boldsymbol{\nabla}_\rho(\boldsymbol{u}\cdot\boldsymbol{\nabla}\boldsymbol{u})]\mathrm{d}x\\&=2\int_\Omega\rho\partial_i\boldsymbol{u}_j\partial_j\boldsymbol{u}_i\mathrm{d}x=2\int_\Omega\rho|D(\boldsymbol{u})|^2\mathrm{d}x-2\int_\Omega\\&\quad\times\rho\left(\frac{\partial_i\boldsymbol{u}_j-\partial_j\boldsymbol{u}_i}{2}\right)^2\mathrm{d}x,\end{aligned}\tag{4.5.18}$$

thus, the assertion of Lemma 4.5.1 follows by adding (4.5.3) multiplied by r to (4.5.17).

In order to deduce the uniform estimates from (4.5.1), we need to control all the non-positive contribution to the left hand side as well as the terms from the right hand side. The ε-dependent terms can be bounded similarly as in [6], so we focus on the new aspect.

Estimate of $\boldsymbol{\nabla}P\cdot\boldsymbol{\nabla}\phi$. Using the assumption of P and $\boldsymbol{\nabla}\phi=2\boldsymbol{\nabla}\log\rho$, we obtain

$$\boldsymbol{\nabla}P\boldsymbol{\nabla}\log\rho=P_c'(\rho)\frac{|\boldsymbol{\nabla}_\rho|^2}{\rho}+\frac{\boldsymbol{\nabla}(R\rho\theta)\boldsymbol{\nabla}_\rho}{\rho}+\frac{\beta}{3}\frac{\boldsymbol{\nabla}\theta^4\cdot\boldsymbol{\nabla}_\rho}{\rho},\tag{4.5.19}$$

so the integral can be written as

$$\begin{aligned}\int_\Omega\boldsymbol{\nabla}P\boldsymbol{\nabla}\log\rho\mathrm{d}x=&\int_\Omega P_c'(\rho)\frac{|\boldsymbol{\nabla}_\rho|^2}{\rho}\mathrm{d}x+R\int_\Omega\frac{\theta|\boldsymbol{\nabla}_\rho|^2}{\rho}\mathrm{d}x\\&+R\int_\Omega\boldsymbol{\nabla}_\rho\boldsymbol{\nabla}\theta\mathrm{d}x+\int_\Omega\frac{\beta}{3}\frac{\boldsymbol{\nabla}\theta^4\cdot\boldsymbol{\nabla}_\rho}{\rho}\mathrm{d}x.\end{aligned}\tag{4.5.20}$$

The first and second terms are non-negative in view of definition of P_c, so it can be considered on the left hand side of (4.5.1) and we only need to estimate the third one and the fourth one as follows:

$$R\int_\Omega \nabla_\rho \nabla\theta \mathrm{d}x \leq \frac{1}{2}R^2\int_\Omega \kappa(\rho,\theta)\frac{|\nabla\theta|^2}{\theta^2}\mathrm{d}x + \frac{1}{2}\int_\Omega \frac{\theta^2}{\kappa(\rho,\theta)}|\nabla_\rho|^2\mathrm{d}x$$

$$\leq C\int_\Omega \kappa(\rho,\theta)\frac{|\nabla\theta|^2}{\theta^2}\mathrm{d}x + C \leq C, \tag{4.5.21}$$

the fourth one can be estimated as follows:

$$\frac{\beta}{3}\frac{\nabla\theta^4\cdot\nabla_\rho}{\rho} \leq C(\varepsilon)\beta\|\nabla\theta^4\|^2_{L^2((0,T)\times\Omega)} + \varepsilon\|\nabla\log\rho\|^2_{L^2((0,T)\times\Omega)}. \tag{4.5.22}$$

The first term of (4.5.22) is bounded for $B \geq 8$ while the second term of (4.5.22) can be estimated differently in two cases:

(i) $\rho \geq 1$, then $\rho^{-1} \leq 1$ and $\rho^{-2}|\nabla_\rho|^2 \leq \rho^{-1}|\nabla_\rho|^2$ which is then bounded by the Gronwall inequality applied to (4.5.1);
(ii) $\rho < 1$, then $\rho^{-\gamma} \geq 1$ and $\varepsilon\rho^{-2}|\nabla_\rho|^2 \leq \rho^{-2-\gamma}|\nabla_\rho|^2 \leq \varepsilon P_c'(\rho)\rho^{-1}|\nabla_\rho|^2$ which is absorbed by the analogous term from the left hand side of (4.5.1) [the first term of (4.5.20)].

Estimate of $\left(P_m + \frac{\beta}{3}\theta^4\right)\operatorname{div}\boldsymbol{u}$. By the assumption of P_m, we have

$$\int_\Omega \left(P_m + \frac{\beta}{3}\theta^4\right)\operatorname{div}\boldsymbol{u}\,\mathrm{d}x = \int_\Omega R_\rho\theta\operatorname{div}\boldsymbol{u}\,\mathrm{d}x + \int_\Omega \frac{\beta}{3}\theta^4\operatorname{div}\boldsymbol{u}\,\mathrm{d}x.$$

Furthermore, by the Young' inequality,

$$\left|\int_\Omega R_\rho\theta\operatorname{div}\boldsymbol{u}\,\mathrm{d}x\right| \leq \varepsilon\|\sqrt{\rho}\operatorname{div}\boldsymbol{u}\|^2_{L^2(\Omega)} + C(\varepsilon)\|\sqrt{\rho}\theta\|^2_{L^2(\Omega)}, \tag{4.5.23}$$

the last term on the right hand side of above inequality can be written as

$$\|\sqrt{\rho}\theta\|_{L^2(\Omega)} \leq \|\rho\theta^2\|^{\frac{1}{2}}_{L^1(\Omega)} \leq C\|\rho\|^{\frac{1}{2}}_{L^{\frac{3}{2}}(\Omega)}\|\theta\|_{L^6(\Omega)} \leq C.$$

On account of (4.4.6). $\theta \in L^2(0,T;L^6(\Omega))$. Moreover, the Sobolev embedding theorem implies that $\|\rho\|_{L^{\frac{p}{2}}(\Omega)} \leq c\|\nabla\sqrt{\rho}\|_{L^2(\Omega)}$ for $1 \leq p \leq 6$, hence the last term on the right hand side of (4.5.23) is bounded whereas the first term can be absorbed by the left hand side of (4.5.1).

The radiative term is slightly more difficult, however, we still can write

$$\begin{aligned}
&\int_0^T\int_\Omega \theta^4|\mathrm{div}\,\boldsymbol{u}|\mathrm{d}x\,\mathrm{d}t \\
&\quad = \int_0^T\int_\Omega \theta^4\rho^4\rho^{-1/2}|\sqrt{\rho}\,\mathrm{div}\,\boldsymbol{u}|\mathrm{d}x\,\mathrm{d}t \\
&\quad \le \|\theta\|^4_{L^p(0,T;L^q(\Omega))}\|\rho^{-1/2}\|_{L^{2\gamma^-}(0,T;L^{6\gamma^+}(\Omega))}\|\sqrt{\rho}\,\mathrm{div}\,\boldsymbol{u}\|_{L^2((0,T)\times\Omega)},
\end{aligned} \tag{4.5.24}$$

where $p = \frac{8\gamma^-}{\gamma^- - 1}$, $q = \frac{24\gamma^-}{3\gamma^- - 1}$.

By the interpolation, $\|\theta\|_{L^p(0,T;L^q(\Omega))} \le \|\theta\|^{1-a}_{L^\infty(0,T;L^4(\Omega))}\|\theta\|^a_{L^G(0,T;L^{3G}(\Omega))}$ for $a = \frac{2}{3}$ and $G = \frac{16\gamma^-}{3(\gamma^- - 1)}$, where $G \le B$ is provided $\gamma^- \ge 3$. Thus, we can estimate

$$\begin{aligned}
&\int_0^T\int_\Omega \theta^4|\mathrm{div}\,\boldsymbol{u}|\mathrm{d}x\,\mathrm{d}t \\
&\quad \le C(\varepsilon)\left(\|\theta\|^{\frac{1}{3}}_{L^\infty(0,T;L^4(\Omega))}\|\theta\|^{\frac{2}{3}}_{L^B(0,T;L^{3B}(\Omega))}\right)^{\frac{2(\gamma^- - 1)}{\gamma^-}} \\
&\qquad + \varepsilon\|\rho^{-\gamma^-/2}\|_{L^2((0,T)\times\Omega)} + \varepsilon\|\sqrt{\rho}\,\mathrm{div}\,\boldsymbol{u}\|^2_{L^2((0,T)\times\Omega)},
\end{aligned} \tag{4.5.25}$$

and the last two terms are estimated by the right hand side of (4.5.1) and (4.5.3), while the boundedness of the first ones follows from (4.4.1) and (4.4.8).

Estimate of $\lambda\Delta^s\nabla(\rho\boldsymbol{u}) : \Delta^s\nabla^2\rho$. We have

$$\begin{aligned}
&2\lambda\int_\Omega |\Delta^s\nabla(\rho\boldsymbol{u}) : \Delta^s\nabla^2\rho|\mathrm{d}x \\
&\quad \le C\lambda\|\Delta^s\nabla(\rho\boldsymbol{u})\|^2_{L^2((0,T)\times\Omega)} + \lambda\|\Delta^{s+1}\rho\|^2_{L^2((0,T)\times\Omega)},
\end{aligned} \tag{4.5.26}$$

therefore, for r sufficiently large, such that $r\lambda^{-1} > c$, both terms of (4.5.26) are bounded by the left hand side of (4.5.1).

4.6 Artificial viscosity limit $\varepsilon \to 0$, $\lambda \to 0$

In this section we first present the new uniform bounds arising from the estimate of B-D entropy, performed in Section 4.5, and then we let the last two approximation parameters tend to 0. Note that the limit passage $\lambda \to 0$, $\varepsilon \to 0$ could be done in a single step, however, for the sake of transparency of this proof we do it separately.

We complete the set uniform bounds by the following ones

$$\begin{aligned}&\|\sqrt{\lambda}\Delta^{s+1}\rho\|_{L^2((0,T)\times\Omega)} + \|\sqrt{\theta_\rho - 1}\boldsymbol{\nabla}\rho\|_{L^2((0,T)\times\Omega)}\\ &\quad + \|\sqrt{P_c'(\rho)\rho^{-1}}\boldsymbol{\nabla}\rho\|_{L^2((0,T)\times\Omega)} \le C,\end{aligned} \tag{4.6.1}$$

moreover,

$$\begin{aligned}&\|\sqrt{\lambda}\boldsymbol{\nabla}^{2s+1}\rho\|_{L^\infty(0,T;L^2\Omega))} + \|\boldsymbol{\nabla}\sqrt{\rho}\|_{L^\infty(0,T;L^2(\Omega))}\\ &\quad + 4\delta^2\|\sqrt{\rho}\boldsymbol{\nabla}^2\log\rho\|_{L^2((0,T)\times\Omega)} \le C,\end{aligned} \tag{4.6.2}$$

The uniform estimates for the velocity vector field are as follows:

$$\begin{aligned}&\|\sqrt{\lambda}\Delta^s\boldsymbol{\nabla}(\rho\boldsymbol{u})\|_{L^2((0,T)\times\Omega)} + \|\sqrt{\rho}\boldsymbol{\nabla}\boldsymbol{u}\|_{L^2((0,T)\times\Omega)}\\ &\quad + \|\sqrt{\theta^{-1}}\rho\boldsymbol{\nabla}\boldsymbol{u}\|_{L^2((0,T)\times\Omega)} \le C,\end{aligned} \tag{4.6.3}$$

and the constants on the right hand side are independent of ε and λ.

We now present several additional estimates of ρ and $\boldsymbol{u}$ based on embedding of Sobolev spaces and simple interpolation inequalities. Once the B-D estimates can be proven exactly as in the paper of Bresch and Desjardins, which is devoted to the Navier–Stokes–Fourier system. However, for the sake of completeness, we recall them below.

Further estimates of ρ. From (4.5.1) we deduce that there exist functions $\xi_1(\rho) = \rho$ for $\rho < (1-\delta)$, $\xi_1(\rho) = 0$ for $\rho > 1$ and $\xi_2(\rho) = 0$ for $\rho < 1$, $\xi_2(\rho) = \rho$ for $\rho > (1+\delta)$, $\delta > 0$, such that

$$\|\boldsymbol{\nabla}\xi_1^{-\frac{\gamma^-}{2}}\|_{L^2((0,T)\times\Omega)}, \|\boldsymbol{\nabla}\xi_2^{\frac{\gamma^+}{2}}\|_{L^2((0,T)\times\Omega)} \le C. \tag{4.6.4}$$

Additionally, in accordance to (4.5.3), we are allowed to use the Sobolev embedding theorem, thus,

$$\|\xi_1^{-\frac{\gamma^-}{2}}\|_{L^2(0,T;L^6(\Omega))}, \|\xi_2^{\frac{\gamma^+}{2}}\|_{L^2(0,T;L^6(\Omega))} \le C. \tag{4.6.5}$$

From (4.5.1) we can also derive the following estimates,

$$\|\xi_1\|_{L^\infty(0,T;L^{\gamma^-}(\Omega))}, \|\xi_2\|_{L^\infty(0,T;L^{\gamma^+}(\Omega))} \le C, \tag{4.6.6}$$

and

$$\|\sqrt{\rho}\|_{L^2(0,T;H^2(\Omega))} \le C. \tag{4.6.7}$$

From (4.6.7), we further deduce

$$\|\rho\|_{L^2(0,T;W^{2,\frac{2\gamma^+}{\gamma^++1}}(\Omega))} \le C. \tag{4.6.8}$$

Due to (4.5.1) and Sobolev embedding theorem, we have

$$\rho \in L^\infty(0,T;L^3(\Omega)). \tag{4.6.9}$$

Remark 4.6.1. Note in particular that the first of the estimate (4.6.4) implies that

$$\rho > 0 \quad \text{a.e. on } (0,T)\times\Omega.$$

Estimate of the velocity vector field. We use the Hölder inequality to write

$$\|\nabla \boldsymbol{u}\|_{L^p(0,T;L^q(\Omega))} \le c\left(1 + \|\xi_1(\rho)^{-1/2}\|_{L^{2\gamma^-}(0,T;L^{6\gamma^-}(\Omega))}\right)$$
$$\times \|\sqrt{\rho}\nabla \boldsymbol{u}\|_{L^2((0,T)\times\Omega)}, \tag{4.6.10}$$

where $p = \frac{2\gamma^-}{\gamma^-+1}$, $q = \frac{6\gamma^-}{3\gamma^-+1}$. Therefore, the Korn inequality together with the Sobolev embedding theorem implies

$$\boldsymbol{u} \in L^{\frac{2\gamma^-}{\gamma^-+1}}(0,T;L^{\frac{6\gamma^-}{\gamma^-+1}}(\Omega)). \tag{4.6.11}$$

Next, by a similar argument,

$$\|\boldsymbol{u}\|_{L^{p'}(0,T;L^{q'}(\Omega))} \le c\left(1 + \|\xi_1(\rho)^{-1/2}\|_{L^{2\gamma^-}(0,T;L^{6\gamma}-(\Omega))}\right)$$
$$\times \|\sqrt{\rho}\boldsymbol{u}\|_{L^2((0,T)\times\Omega)} \tag{4.6.12}$$

with $p' = 2\gamma^-$, $q' = \frac{6\gamma^-}{3\gamma^-+1}$. By a simple interpolation between (4.6.11) and (4.6.12), we obtain

$$\boldsymbol{u} \in L^{\frac{10\gamma^-}{3\gamma^-+10}}(0,T,L^{\frac{10\gamma^-}{3\gamma^-+10}}(\Omega)) \tag{4.6.13}$$

and since $\gamma^- > 3$, we can see in particular that $\boldsymbol{u} \in L^{\frac{5}{2}}(0,T;L^{\frac{5}{2}}(\Omega))$ uniformly with respect to ε and λ.

Strict positivity of the absolute temperature. We now give the proof of uniform with respect to ε, positivity of θ_ε. Note that so far this was clear on account of the bound for $\varepsilon\theta^{-3}$ in $L^1((0,T)\times\Omega)$ following from (4.3.18).

Lemma 4.6.1. *We have, uniformly with respect to* ε *and* λ,

$$\theta_\varepsilon > 0 \text{ a.e. on } (0,T)\times\Omega. \tag{4.6.14}$$

Proof. The above statement is a consequence of the following estimate,

$$\int_0^T\int_\Omega (|\log\theta_\varepsilon|^2 + |\boldsymbol{\nabla}\log\theta_\varepsilon|^2)\mathrm{d}x\mathrm{d}t \le C, \tag{4.6.15}$$

which can be obtained by the application of generalized Korn inequality provided that we control the $L^1(\Omega)$ norm of $\rho|\log\theta|$. By virtue of (4.3.18) we have

$$\int_\Omega (\rho_\varepsilon s_\varepsilon)^0\,\mathrm{d}x \le \int_\Omega \rho_\varepsilon s_\varepsilon(T)\mathrm{d}x, \tag{4.6.16}$$

thus, substituting the form of ρs we obtain

$$-\int_\Omega \rho_\varepsilon\log\theta_\varepsilon(T)\mathrm{d}x \le \int_\Omega \rho_\varepsilon\log\rho_\varepsilon(T)\mathrm{d}x - \int_\Omega (\rho_\varepsilon s_\varepsilon)^0\mathrm{d}x, \tag{4.6.17}$$

and the right hand side is bounded on account of (4.5.1) and the initial condition. On the other hand, the positive part of the integrand $\rho_\varepsilon\log\theta_\varepsilon$ is bounded from above by $\rho_\varepsilon\theta_\varepsilon$, which belongs to $L^\infty(0,T;L^1(\Omega))$ due to (4.2.12), so we end up with

$$ess \sup_{t\in(0,T)}\int_\Omega |\rho_\varepsilon\log\theta_\varepsilon(t)|\mathrm{d}x \le C, \tag{4.6.18}$$

which completes the proof of (4.6.13). □

4.6.1 *Limit* $\varepsilon \to 0$

With the B-D estimate at hand, especially with the bound on $\Delta^{s+1}\rho_\varepsilon$ in $L^2((0,T)\times\Omega)$, which is now uniform with respect to ε, we may perform the limit passage similarly as in the previous step. Indeed, the uniform estimates allow us to extract subsequences, such that

$$\varepsilon\Delta^s\nabla\boldsymbol{u}_\varepsilon, \varepsilon\nabla\rho_\varepsilon, \varepsilon\Delta^{s+1}\rho_\varepsilon \to 0 \quad \text{strongly in } L^2((0,T)\times\Omega), \tag{4.6.19}$$

therefore,

$$\varepsilon\nabla\rho_\varepsilon\nabla\boldsymbol{u}_\varepsilon \to 0 \quad \text{strongly in } L^1((0,T)\times\Omega). \tag{4.6.20}$$

The strong convergence of the density as well as the velocity can be obtained identically as in the previous step. Therefore we focus only on the strong convergence of the temperature and the limit passage in the total energy balance.

Recall that from (4.3.18) and (4.4.7) it follows that

$$\theta_\varepsilon \rightharpoonup \theta \text{ weakly in } L^2(0,T;W^{1,2}(\Omega)) \tag{4.6.21}$$

and

$$\varepsilon\theta_\varepsilon^{-2}, \varepsilon\theta_\varepsilon^5 \to 0 \quad \text{strongly in } L^1((0,T)\times\Omega). \tag{4.6.22}$$

The pointwise convergence of the temperature is to be deduced from the version of the Aubin–Lions lemma, see [5].

Lemma 4.6.2. *Let $\{\nu_\varepsilon\}$ be sequence of functions bounded in $L^2(0,T;L^q(\Omega))$ and in $L^\infty(0,T;L^1(\Omega))$, where $q > \frac{6}{5}$. Furthermore, assume that*

$$\partial_t\nu_\varepsilon \geq g_\varepsilon \ \textit{in } \mathcal{D}'((0,T)\times\Omega), \tag{4.6.23}$$

where g_ε is bounded in $L^1(0,T;W^{-m,r}(\Omega))$ for some $m \geq 0$, $r > 1$ independently of ε. Then there exists a subsequence $\{\nu_\varepsilon\}$ which converges to ν strongly in $L^2(0,T;W^{-1,2}(\Omega))$.

We will apply this lemma to $\nu_\varepsilon = \rho_\varepsilon\theta_\varepsilon + \beta\theta_\varepsilon^4$.

$$-\nabla\varepsilon - |\nabla\rho_\varepsilon|^2\nu\rho_\varepsilon|(\boldsymbol{u}_\varepsilon)|\nabla(\rho_\varepsilon\boldsymbol{u}_\varepsilon)|\varepsilon|\rho_\varepsilon|2h^2\rho_\varepsilon|\nabla\rho_\varepsilon|^2. \tag{4.6.24}$$

Moreover, the right hand side is bounded in $L^1(0,T;W^{-1,p}(\Omega))\cup L^1(0,T;W^{-2,q}(\Omega))$ for some $p,q > 1$. Therefore, the above lemma and the strong convergence of ρ_ε imply in particular that

$$\theta_\varepsilon^4 \to \overline{\theta_\varepsilon^4} \quad \text{strongly in } L^2(0,T;W^{-1,2}(\Omega)).$$

On the other hand, we also know that $\theta_\varepsilon \to \theta$ weakly in $L^2(0,T;W^{1,2}(\Omega))$, therefore a simple argument based on the monotonicity of $f(x) = x^4$ implies strong convergence of θ_ε in $L^q(0,T;L^{3q}(\Omega))$ for any $q < B$.

Let us finish this subsection with the list of the limit equations.

(i) The continuity equation

$$\partial_t \rho + \operatorname{div}(\rho \boldsymbol{u}) = 0$$

is satisfied pointwisely on $[0,T] \times \Omega$;

(ii) The momentum equation

$$\begin{aligned}
&-\int_0^T \int_\Omega \rho \boldsymbol{u} \cdot \partial_t \boldsymbol{\phi} \mathrm{d}x\mathrm{d}t - \int_\Omega m^0 \cdot \boldsymbol{\phi}(0)\mathrm{d}x \\
&+ \int_0^T \int_\Omega \lambda \Delta^s \boldsymbol{\nabla}(\rho \boldsymbol{u}) : \Delta^s \boldsymbol{\nabla}(\rho \boldsymbol{\phi})\mathrm{d}x\mathrm{d}t \\
&- \int_0^T \int_\Omega (\rho \boldsymbol{u} \otimes \boldsymbol{u}) \boldsymbol{\nabla} \boldsymbol{\phi} \,\mathrm{d}x\mathrm{d}t + \int_0^T \int_\Omega \nu_\rho D(\boldsymbol{u}) : D(\boldsymbol{\phi})\mathrm{d}x\mathrm{d}t \\
&- \int_0^T \int_\Omega P \operatorname{div} \boldsymbol{\phi} \,\mathrm{d}x\mathrm{d}t + \int_0^T \int_\Omega \lambda \Delta^s \operatorname{div}(\rho \boldsymbol{\phi}) : \Delta^{s+1}(\rho)\mathrm{d}x\mathrm{d}t \\
&+ 2h^2 \int_0^T \int_\Omega \rho \boldsymbol{\nabla}^2 \log \rho \cdot \boldsymbol{\nabla} \boldsymbol{\phi} \mathrm{d}x\mathrm{d}t = 0
\end{aligned} \tag{4.6.25}$$

holds for any test function $\boldsymbol{\phi} \in L^2(0,T;W^{2s+1}(\Omega)) \cap W^{1,2}(0,T;W^{1,2}(\Omega))$ such that $\boldsymbol{\phi}(\cdot,T) = 0$;

(iii) The total energy equation

$$\begin{aligned}
&\int_0^T \int_\Omega \left(\rho \frac{|\boldsymbol{u}|^2}{2} + \rho e + \frac{\lambda}{2} |\boldsymbol{\nabla} \Delta^s \rho|^2 - \delta^2 \rho \Delta \log \rho \right) \partial_t \phi \,\mathrm{d}x\mathrm{d}t \\
&\quad - \int_0^T \int_\Omega \kappa \boldsymbol{\nabla} \theta \cdot \boldsymbol{\nabla} \phi \,\mathrm{d}x\mathrm{d}t + \int_0^T \int_\Omega \left(\rho e + \rho \frac{|\boldsymbol{u}|^2}{2} \boldsymbol{u} \right.
\end{aligned}$$

$$-\delta^2\rho\Delta\log\rho\boldsymbol{u}\Big)\cdot\boldsymbol{\nabla}\phi\,\mathrm{d}x\mathrm{d}t+\int_0^T\int_\Omega P\boldsymbol{u}\cdot\boldsymbol{\nabla}\phi\,\mathrm{d}x\mathrm{d}t$$
$$-\int_0^T\int_\Omega \nu_\rho D(\boldsymbol{u})\boldsymbol{u}\cdot\boldsymbol{\nabla}\phi\mathrm{d}x\mathrm{d}t$$
$$=-\int_0^T\int_\Omega\left(\frac{\varepsilon}{\theta^2}-\varepsilon\theta^5\right)\phi\mathrm{d}x+\int_0^T\int_\Omega R_{\varepsilon,\lambda}\,\mathrm{d}x\mathrm{d}t$$
$$-\int_0^T\int_\Omega\left(\rho\frac{|\boldsymbol{u}|^2}{2}+\rho e+\frac{\lambda}{2}|\boldsymbol{\nabla}\Delta^s\rho|^2\right)(0)\phi\mathrm{d}x\mathrm{d}t \quad (4.6.26)$$

holds for any test function $\phi\in L^2(0,T;W^{2s+1}(\Omega))\cap W^{1,2}(0,T;W^{1,2}(\Omega))$ such that $\phi(\cdot,T)=0$ and

$$R_{\varepsilon,\lambda}(\rho,\theta,\boldsymbol{u},\phi)=\lambda[\Delta(\mathrm{div}(\rho\boldsymbol{u}\phi)\Delta^{s+1}\rho-\Delta\,\mathrm{div}(\rho\boldsymbol{u})\Delta^{s+1}\rho\phi]$$
$$-\lambda\Delta^s\,\mathrm{div}(\rho\boldsymbol{u})\boldsymbol{\nabla}\Delta^{s+1}\rho\cdot\boldsymbol{\nabla}\phi-\lambda[|\Delta^s\boldsymbol{\nabla}(\rho\boldsymbol{u})|^2\phi$$
$$-\Delta^s\boldsymbol{\nabla}(\rho\boldsymbol{u}):\Delta^s\boldsymbol{\nabla}(\rho\boldsymbol{u}\phi)]. \quad (4.6.27)$$

Moreover, using the lower semicontinuity of norm and passing to the limit in (4.4.33),

$$\partial_t(\rho\theta+\beta\theta^4)+\mathrm{div}(\boldsymbol{u}\rho\theta+\beta\boldsymbol{u}\theta^4)+\mathrm{div}(\kappa\boldsymbol{\nabla}\theta)$$
$$\geq-\left(P_m+\frac{\beta}{3}\right)\mathrm{div}\,\boldsymbol{u}+\nu_\rho|D(\boldsymbol{u})|^2+\lambda|\Delta^s\boldsymbol{\nabla}(\rho\boldsymbol{u})|^2$$
$$+2\delta^2\rho|\boldsymbol{\nabla}^2\log\rho|^2 \quad (4.6.28)$$

is satisfied in the sense of distributions on $(0,T)\times\Omega$.

4.6.2 *Limit* $\lambda\to 0$

In this subsection we present the argument for the convergence of a sequence $(\rho_\lambda,\boldsymbol{u}_\lambda,\theta_\lambda)$ to a solution $(\rho,\boldsymbol{u},\theta)$ as specified at the beginning. Some of the arguments here are repetitions from our previous works, so we only recall their formulations.

Strong convergence of the density. The strong convergence of a sequence ρ_λ is guaranteed by the following lemma.

Lemma 4.0.3. *There exists a subsequence ρ_λ such that*

$$\sqrt{\rho_\lambda} \to \sqrt{\rho} \text{ a.e. and strongly in } L^2((0,T)\times\Omega),$$

moreover, $\rho_\lambda \to \rho$ strongly in $C([0,T];L^p(\Omega)), p<3$.

For the proof see [5].

Strong convergence of the convective term. To prove this, we will again take advantage of the special form of the cold component of the pressure close to vacuum (4.1.9). First, we show the necessary condition on γ^- that the convective term $\rho_\lambda|\boldsymbol{u}_\lambda|^3$ is uniformly bounded in $L^p((0,T)\times\Omega)$ for some $p>1$. Following the proof from [5], we may write

$$\rho_\lambda^{1/3}|\boldsymbol{u}_\lambda| = \rho_\lambda^{1/3-\alpha}\rho_\lambda^{\alpha}|\boldsymbol{u}_\lambda|^{2\alpha}|\boldsymbol{u}_\lambda|^{1-2\alpha},$$

and we will use the interpolation inequality for $\rho_\lambda \in L^\infty(0,T;L^{\gamma^+}(\Omega))$, $\rho_\lambda|\boldsymbol{u}_\lambda|^2 \in L^\infty(0,T;L^1(\Omega))$ and $\boldsymbol{u}_\lambda \in L^{\frac{2\gamma^-}{\gamma^-+1}}(0,T;L^{\frac{6\gamma^-}{\gamma^-+1}})$. So, $\rho_\lambda^{1/3}|\boldsymbol{u}_\lambda| \in L^p(0,T;L^q(\Omega))$ with $p,q>3$ if

$$(1-2\alpha)\frac{\gamma^-+1}{2\gamma^-} < \frac{1}{3},$$
$$\left(\frac{1}{3}-\alpha\right)\frac{1}{\gamma^+} + (1-2\alpha)\frac{\gamma^-+1}{6\gamma^-} < \frac{1}{3},$$

meaning that γ^- and γ^+ must satisfy the following relation,

$$\gamma^- > \frac{5\gamma^+-3}{\gamma^+-3}.$$

The above estimate implies that, provided γ^-, γ^+ fulfill the conditions specified above. The convective term $\rho_\lambda\boldsymbol{u}_\lambda^3$ converges weakly to $\rho \boldsymbol{u}^3$ in $L^r((0,T)\times\Omega)$ for some $r>1$. To identify the limit, we prove the following lemma.

Lemma 4.6.4. *We have, up to a subsequence,*

$$\begin{aligned} &\rho_\lambda \boldsymbol{u}_\lambda \rightharpoonup \rho \boldsymbol{u} \ \textit{in}\ C([0,T];L^{\frac{3}{2}}(\Omega)),\\ &\rho_\lambda^{1/3}\boldsymbol{u}_\lambda \rightharpoonup \rho^{1/3} \ \textit{strongly in}\ L^p((0,T)\times\Omega),\\ &\qquad \textit{for some}\ p>3,\\ &\rho_\lambda D(\boldsymbol{u}_\lambda)\boldsymbol{u}_\lambda \rightharpoonup \rho D(\boldsymbol{u}) \ \textit{weakly in}\ L^1((0,T)\times\Omega). \end{aligned} \tag{4.6.29}$$

For the proof see [5].

Strong convergence of the temperature. The difference with respect to previous chapter is that we cannot use the higher order, either for the velocity or for the density, in order to deduce the boundedness of the time derivative of temperature in a appropriate space. However, the idea of proving compactness of the temperature is, as previously, to apply Lemma 4.6.2 with $\varepsilon=\lambda$, $\nu_\lambda=\rho_\lambda\theta_\lambda+\beta\theta_\lambda^4$. Therefore, our next aim is to check that its assumptions are satisfied uniformly with respect to λ.

First, let us note that ν_λ is bounded in $L^2(0,T;L^q(\Omega))$ and in $L^\infty(0,T;L^1(\Omega))$, where $q>\frac{6}{5}$, uniformly with respect to λ. Indeed, it follows directly from (4.4.2) and (4.4.6). Furthermore, from (4.6.27) one deduces that $\partial_t\nu_\lambda\ge g_\lambda$, where g_λ has the form

$$\begin{aligned} g_\lambda &= -\mathrm{div}(\boldsymbol{u}_\lambda\rho_\lambda\theta_\lambda+\beta\boldsymbol{u}_\lambda\theta_\varepsilon^4)+\mathrm{div}(\kappa\boldsymbol{\nabla}\theta_\lambda)\\ &\quad -\left(P_m+\frac{\beta}{3}\theta_\lambda^4\right)\mathrm{div}\,\boldsymbol{u}_\lambda+\nu_{\rho\lambda}|D(\boldsymbol{u}_\lambda)|^2+\lambda|\Delta^s\boldsymbol{\nabla}(\rho_\lambda\boldsymbol{u}_\lambda)|^2\\ &\quad +2h^2\rho_\lambda|\boldsymbol{\nabla}^2\log\rho_\lambda|^2, \end{aligned} \tag{4.6.30}$$

and is bounded in $L^1(0,T;W^{-m,r}(\Omega))$ for some $m\ge 0$, $r>1$ independently of λ. Indeed, this can be estimated, similarly to (4.4.18)–(4.4.22), except for the terms that contains velocity. For them we may write

$$\|\boldsymbol{u}_\lambda \rho_\lambda \theta_\lambda\|_{L^{12/11}((0,T)\times\Omega)} \\ \leq C\|\sqrt{\rho_\lambda}\boldsymbol{u}_\lambda\|_{L^2((0,T)\times\Omega)}\|\sqrt{\rho_\lambda}\|_{L^\infty(0,T;L^6(\Omega))}\|\theta_\lambda\|_{L^\infty(0,T;L^4(\Omega))} \leq C, \tag{4.6.31}$$

on account of (4.4.20) and (4.6.10), and

$$\|\boldsymbol{u}_\lambda \theta_\varepsilon^4\|_{L^{40/31}((0,T)\times\Omega)} \leq C\|\boldsymbol{u}_\lambda\|_{L^{5/2}((0,T)\times\Omega)}\|\theta_\varepsilon^4\|_{L^{8/3}((0,T)\times\Omega)} \leq C. \tag{4.6.32}$$

For the internal pressure we have

$$\|\rho_\lambda \theta_\lambda \operatorname{div} \boldsymbol{u}_\lambda\|_{L^{12/11}((0,T)\times\Omega)} \\ \leq C\|\sqrt{\rho_\lambda}\operatorname{div} \boldsymbol{u}_\lambda\|_{L^2((0,T)\times\Omega)}\|\sqrt{\rho_\lambda}\|_{L^\infty(0,T;L^6(\Omega))} \\ \times \|\theta_\lambda\|_{L^\infty(0,T;L^4(\Omega))} \leq C, \tag{4.6.33}$$

and the term $\theta_\lambda^4 \operatorname{div} \boldsymbol{u}_\lambda$ is bounded in $L^1((0,T)\times\Omega)$ as shown above in (4.5.24). Since the last two terms in (4.6.29) are also uniformly bounded in $L^1((0,T)\times\Omega)$, the assumption of Lemma 4.6.2 is satisfied with $m = 1$, $r > 1$. Therefore, there exists a subsequence ν_λ converging to ν strongly in $L^2(0,T;W^{-1,2}(\Omega))$, which can be used to show the strong convergence of θ_λ exactly as in the previous section.

Passage to the limit in the non-linear terms. The last step in the limit passage $\lambda \to 0$ is verification of convergence in the non-linear terms of the system. The most demanding terms are in the energy equations (4.6.25) and (4.6.26), and we will justify the limit passage only there. The correction of energy $\lambda \boldsymbol{\nabla}^{2s+1}\rho_\lambda \to 0$ strongly in $L^2((0,T)\times\Omega)$, therefore the energy $E_\lambda = \rho_\lambda e_c(\rho_\lambda) + \rho_\lambda\theta_\lambda + \beta\theta_\lambda^4 + \frac{1}{2}\rho_\lambda|\boldsymbol{u}_\lambda|^2 + \frac{\lambda}{2}|\boldsymbol{\nabla}^{2s+1}\rho_\lambda|^2 - \delta^2\rho_\lambda\Delta\log\rho_\lambda$ converges to E due to the strong convergence of ρ_λ, θ_λ and Lemma 4.6.4. Similarly, $\boldsymbol{u}_\lambda\rho_\lambda\theta_\lambda$, $\rho_\lambda\boldsymbol{u}_\lambda^3$ and $P\boldsymbol{u}_\lambda$ converge weakly to $\boldsymbol{u}\rho\theta$, $\rho\boldsymbol{u}^3$ and $P\boldsymbol{u}$ respectively, due to uniform bounds in $L^p((0,T)\times\Omega)$ for $p > 1$ from above, the strong convergence of ρ_λ, θ_λ and Lemma 4.6.4.

Limit passage in the heat flux term $\kappa\boldsymbol{\nabla}\theta_\lambda$ can be performed directly, since it involves only the sequences ρ_λ, θ_λ which are strongly convergent, and a sequence $\nabla\theta_\lambda$ which converges to $\nabla\theta$ in $L^2((0,T)\times\Omega)$. Passage to the last term on the right hand side of (4.6.25) was proven in Lemma 4.6.4.

We are now ready to prove that the corrector term R_λ converges to 0 strongly in $L^1((0,T)\times\Omega)$ as $\lambda\to 0$, or rather that the most demanding terms listed in (4.6.33) vanish when $\lambda\to 0$.

First of all, observe that, due to (4.6.16)–(4.6.18) we have

$$\begin{aligned}
&\int_0^T\int_\Omega |R_\lambda(\rho_\lambda,\theta_\lambda)|\mathrm{d}x\mathrm{d}t\\
&\quad=\lambda\int_0^T\int_\Omega |(\Delta^s\nabla(\rho_\lambda\boldsymbol{u}_\lambda)\Delta^s(\rho_\lambda\boldsymbol{u}_\lambda)+\Delta^s(\rho_\lambda\boldsymbol{u}_\lambda)\Delta^{s+1}\rho_\lambda\\
&\qquad+\Delta^s\operatorname{div}(\rho_\lambda\boldsymbol{u}_\lambda).\\
&\nabla\Delta^s\rho_\lambda)\nabla\phi|\mathrm{d}x\mathrm{d}t\le\|\nabla\phi\|_{L^\infty((0,T)\times\Omega)}\lambda\\
&\qquad\times\|\rho_\lambda\boldsymbol{u}_\lambda\|_{L^2(0,T;W^{2s,2}(\Omega))}\|\rho_\lambda\boldsymbol{u}_\lambda\|_{L^2(0,T;W^{2s+1,2}(\Omega))}.
\end{aligned}\tag{4.6.34}$$

Thus the task is to show that the term after the last inequality symbol converges to 0.

But this is evident, since one can use the Gagliardo–Nirenberg interpolation inequality and the uniform bounds for $\rho_\lambda\boldsymbol{u}_\lambda$ in $L^\infty(0,T;L^{3/2}(\Omega))$ and for ρ_λ in $L^\infty(0,T;L^3(\Omega))$, together with the uniform bounds for $\sqrt{\lambda}\rho_\lambda\boldsymbol{u}_\lambda$ and for $\sqrt{\lambda}\rho_\lambda$ in $L^2(0,T;W^{2s+1,2}(\Omega))$ and $L^2(0,T;W^{2s+2,2}(\Omega))$, respectively. This completes the proof of the main theorem.

Chapter 5

Boundary Problem of Compressible Quantum Euler–Poisson Equations

5.1 Boundary problem for compressible stationary quantum Euler–Poisson equations

In this section, we consider the one-dimensional compressible quantum Euler–Poisson equations:

$$\left(\frac{j^2}{\rho} + TP(\rho)\right)_x - h^2\rho\left(\frac{(\sqrt{\rho})_{xx}}{\sqrt{\rho}}\right)_x$$

$$= \rho\phi_x - \frac{j}{\tau(x)} - \nu\rho(\beta(\rho))_{xx}, \tag{5.1.1}$$

$$\lambda^2\phi_{xx} = \rho - D, \tag{5.1.2}$$

subject to the boundary conditions

$$\rho(0) = \rho_0 > 0, \quad \rho(1) = \rho_1 > 0,$$

$$\phi(0) = \phi_0, \quad \phi_x(0) = -E_0, \tag{5.1.3}$$

$$h^2\frac{(\sqrt{\rho})_{xx}(0)}{\sqrt{\rho_0}} - \nu\beta'(\rho_0)\rho_x(0) = \frac{j^2}{2\rho_0^2} + Th(\rho_0) - V_0 + K. \tag{5.1.4}$$

Here $x \in \Omega := (0, 1)$ the functions ρ, j, ϕ are the particle density, the particle current density and the electric potential respectively; Planck constant h is positive, ν is the viscosity, K is a constant to be fixed; function $h(s)$ defined as $h'(s) = p'(s)/s$, $s > 0$, $h(1) = 0$. For the isothermal fluid, $h(s) = \log(s)(s \geq 0)$; for the isentropic fluid,

$h(s) = \frac{\alpha}{\alpha-1}(s^{\alpha-1} - 1)(s > 0)$, with $\alpha > 1$. We assume that

$$\beta(\rho) = -\frac{1}{\gamma - 1} \cdot \frac{1}{\rho^{(\gamma-1)/2}}, \tag{5.1.5}$$

with $\gamma > 4$. This condition implies the positivity of the particle density. It is well known that the positivity of solutions to higher-order equations is a delicate problem since maximum principle arguments generally do not apply. See [79], [80], [82], [83].

In order to derive the system of the second-order equations, we rewrite (5.1.1) as

$$n\left[\frac{j^2}{2\rho^2} + Th(\rho) - \phi - h^2\frac{(\sqrt{\rho})_{xx}}{\sqrt{\rho}} + j\int_0^x \frac{\mathrm{d}s}{\tau\rho} + \nu\beta(n)_x\right]_x = 0. \tag{5.1.6}$$

Furthermore, if $n > 0$,

$$\frac{j^2}{2\rho^2} + Th(\rho) - V - h^2\frac{(\sqrt{\rho})_{xx}}{\sqrt{\rho}} + j\int_0^x \frac{\mathrm{d}s}{\tau\rho} + \nu\beta(n)_x = -K, \tag{5.1.7}$$

where K is a constant defined in (5.1.4). By the definition of $\beta(\rho)$ and set $w = \sqrt{\rho}$, we obtain

$$\begin{aligned} h^2 w_{xx} &= \frac{j^2}{2w^3} + Twh(w^2) - \phi w + Kw \\ &\quad + jw\int_0^x \frac{\mathrm{d}s}{\tau w^2} + \nu\frac{w_x}{w^\gamma}, \end{aligned} \tag{5.1.8}$$

$$\lambda^2\phi_{xx} = w^2 - D. \tag{5.1.9}$$

These equations have to be solved subject to the boundary conditions

$$w(0) = w_0 > 0, \quad w(1) = w_1 > 0, \quad \phi(0) = \phi_0, \quad \phi_x(0) = -E_0, \tag{5.1.10}$$

where $w_0 = \sqrt{\rho_0}$, $w_1 = \sqrt{\rho_1}$. Choosing

$$K \triangleq \phi_0 + \max(-E_0, 0) + \lambda^{-2}M^2,$$

where

$$M \triangleq \max(w_0, w_1, M_0), \tag{5.1.11}$$

and M_0 satisfies $h(M_0^2) \geq 0$ such that $-\phi(x) + K \geq 0$.

In this section, we impose the following assumptions:

(H1) $h \in C^1(0, \infty)$ and p' are non-decreasing and satisfy $p'(s) = sh'(s)$, $s > 0$, $\lim_{s\to\infty} h(s) > 0$, $\lim_{s\to 0+} h(s) < 0$, $\lim_{s\to 0+} \sqrt{s}h(s) > -\infty$;

(H2) $D \in L^2(\Omega)$, $D \geq 0$, $\tau \in L^\infty(\Omega)$, $\tau(x) \geq \tau_0 > 0$;

(H3) j, w_0, w_1, $h, \lambda, T > 0$, $\nu \geq 0$, ϕ_0, $E_0 \in \mathbf{R}$.

5.1.1 *Existence of solutions if h > 0, ν > 0*

Theorem 5.1.1. *Let the hypotheses* (H1)–(H3) *hold and let* $\nu > 0$. *Then, for any* $j > 0$, *there exists a classical solution* $(w, \phi) \in (C^2(\Omega))^2$ *to* (5.1.8)–(5.1.10) *satisfying*

$$0 < m(\nu) \leq w(x) \leq M, \quad \forall x \in \Omega.$$

Remark 5.1.1. The constant $m(\nu)$ is defined by $m(\nu) = \min(w_0, w_1, m_1, m_2)$, where $h(4m_1^2) \leq 0$, $m_2 \leq \left(\frac{1}{2\gamma+1}\frac{\nu}{j^2/2+j/\tau_0+\max(0,K-k)}\right)^{1/(\gamma-4)}$, $k = \phi_0 - \max(E_0, 0) - \lambda^{-2}\|D\|_{L^1(\Omega)}$. The constant M is defined in (5.1.11).

In order to prove Theorem 5.1.1, define the function

$$r(x) = \delta(2 - x), \quad x \in [0, 1], \quad 0 < \delta < \min(1, M/2), \qquad (5.1.12)$$

and consider the truncated problem

$$\begin{aligned} h^2_{w_{xx}} &= \frac{j^2 w}{2t_\delta(w)^4} + Twh(w^2) - \phi w + Kw \\ &\quad + jw\int_0^x \frac{\mathrm{d}s}{\tau t_\delta(w)^2} + \nu\frac{(t_r(w_M))_x w}{(t_r(w_M))^{\gamma+1}}, \end{aligned} \qquad (5.1.13)$$

$$\lambda^2\phi_{xx} = w^2 - D(x), \qquad (5.1.14)$$

where $t_\delta(w) = \max(\delta, w)$, $t_r(w_M) = \max(r(\cdot), \min(M, w(\cdot)))$. We can show the existence of solutions.

Proposition 5.1.1. *Let the assumptions* (H1)–(H3) *hold and let* $\nu > 0$, $\varepsilon > 0$. *Then, for any* $j > 0$, *there exists a solution* $(w, \phi) \in$

$(H^2(\Omega))^2$ *to* (5.1.13), (5.1.14), (5.1.10) *satisfying*

$$0 \le w(x) \le M, \quad \forall x \in \Omega.$$

For the proof of Proposition 5.1.1, we consider the approximate problem

$$\begin{aligned} h^2_{w_{xx}} &= \frac{j^2 w^+}{2t_\delta(w)^4} + Tw^+ h(w^2) - \phi w^+ + kw^+ \\ &\quad + jw^+ \int_0^x \frac{ds}{\tau t_\delta(w)^2} + \nu \frac{(t_r(w_M))_x w^+}{(t_r(w_M))^{\gamma+1}}, \end{aligned} \tag{5.1.15}$$

$$\lambda^2 \phi_{xx} = w_M^2 - D(x), \tag{5.1.16}$$

where $w^+ = \max(0, w)$, $w_M = \min(M, w)$. Let (w, ϕ) be a weak solution to (5.1.13), (5.1.14), (5.1.10). Then we have the following *a priori* estimates

Lemma 5.1.1. *It holds for all* $x \in \Omega$

$$0 \le w(x) \le M, \quad k \le \phi(x) \le K,$$

where $k = \phi_0 - \max(E_0, 0) - \lambda^{-2}\|C\|_{L^1(\Omega)}$.

Proof. The problem (5.1.16), (5.1.10) is equivalent to

$$\phi(x) = \phi_0 - E_0 x + \lambda^{-2} \int_0^x \int_0^y (w(z)_M^2 - D(z))\mathrm{d}z\mathrm{d}y, \quad x \in [0, 1], \tag{5.1.17}$$

which implies

$$\phi_0 - \max(E_0, 0) - \lambda^{-2}\|C\|_{L^1(\Omega)} \le \phi(x) \le \phi_0 + \max(-E_0, 0) + \lambda^{-2} M^2.$$

This shows the second chain of inequalities of (5.1.1).

Using $w^- = \min(0, w)$ as test function in (5.1.15), immediately we get $w \ge 0$ in Ω. Finally with the test function $(w - M)^+ =$

$\max(0, w - M)$ in (5.1.15), we obtain

$$\begin{aligned}&\int_\Omega ((w-M)_x^+)^2 \mathrm{d}x\\&= \int_\Omega ((w-M)^+)w\left[-\frac{j^2}{2t_\delta(w)^4} - Th(M^2) + (\phi - K)\right.\\&\quad \left. -j\int_0^x \frac{\mathrm{d}s}{\tau t_\delta(w)^2}\right]\mathrm{d}x - \nu\int_\Omega \frac{(t_r(w_M))_x w(w-M)^+}{(t_r(w_M))^{\gamma+1}}\mathrm{d}x,\end{aligned}$$

taking into account the monotonicity of h. The first integral on the right hand side is non-positive since $\phi \le K$ and $h(M^2) \ge h(M_0^2) \ge 0$. Therefore,

$$\int_\Omega ((w-M)_x^+)^2 \mathrm{d}x \le -\nu\int_\Omega \frac{(t_r(w_M))_x w(w-M)^+}{(t_r(w_M))^{\gamma+1}}\mathrm{d}x \le 0.$$

Hence $w \le M$ in Ω. □

Lemma 5.1.2. *There exist constants c_1, c_2 only depending on the given data and on h, δ and M, but not on w and ϕ, such that*

$$\|w\|_{H^1(\Omega)} \le c_1, \quad \|\phi\|_{H^1(\Omega)} \le c_2.$$

Proof. The second assertion follows from Lemma 5.1.1 and

$$\phi_{x(x)} = -E_0 + \lambda^{-2}\int_0^x (w(y)_M^2 - D(y))dy, \quad x \in [0,1].$$

The first assertion follows from Lemma 5.1.1 and (5.1.15), after employing the test function $w - w_D(x)$, where $w_D(x) = (1-x)w_0 + xw_1$. Indeed, we have

$$\begin{aligned}&h^2\int_\Omega (w_x)^2 \mathrm{d}x\\&= h^2\int_\Omega w_x w_D(x)\mathrm{d}x - \nu\int_\Omega \frac{(t_r(w_M))_x w(w-w_D)}{(t_r(w_M))^{\gamma+1}}\mathrm{d}x\\&\quad -\int_\Omega ((w-w_D))w\left[\frac{j^2}{2t_\delta(w)^4} + Th(w^2) - \phi + K\right.\\&\quad \left. +j\int_0^x \frac{\mathrm{d}s}{\tau t_\delta(w)^2}\right]\mathrm{d}x.\end{aligned}$$

Using Young's inequality for the first two integrals on the right hand side and Lemma 5.1.1 for the last integral, we get

$$h^2 \int_\Omega (w_x)^2 \mathrm{d}x \le c.$$

□

Lemma 5.1.3. *There exists a constant c_3, not depending on w, such that*

$$\|w\|_{H^2(\Omega)} \le c_3.$$

Proof. The lemma follows immediately from Lemma 5.1.2, (5.1.15), and the embedding $H^1(\Omega) \hookrightarrow L^\infty(\Omega)$. □

Proof of Proposition 5.1.1. We apply the Leray–Schauder fixed-point theorem. Let $u \in H^1(\Omega)$, $\phi \in H^2(\Omega)$ be the unique solution of

$$\lambda^2 \phi_{xx} = u_M^2 - D(x), \quad \phi(0) = \phi_0, \quad \phi_x(0) = -E_0.$$

Let $w \in H^2(\Omega)$ be the unique solution of

$$\begin{aligned} h^2 w_{xx} &= \sigma g \left[\frac{j^2 u^+}{2t_\delta(u)^4} + Tu^+ h(u^2) - \phi u^+ + Ku^+ \right. \\ &\quad \left. + ju^+ \int_0^x \frac{\mathrm{d}s}{\tau t_\delta(u)^2} g \right] + \sigma\nu \frac{(t_r(u_M))_x u^+}{(t_r(u_M))^{\gamma+1}}, \\ w(0) &= \sigma w_0, \quad w(1) = \sigma w_1, \quad \sigma \in [0,1]. \end{aligned}$$

This defines the fixed-point operator $S : H^1(\Omega) \times [0,1] \to H^1(\Omega)$, $(u,\sigma) \to w$. It holds $S(u,0) = 0$ for all $u \in H^1(\Omega)$. Similarly, as in the proofs of Lemmas 5.1.1–5.1.3, we can show that there exists a constant $c > 0$ independent of w and σ such that

$$\|w\|_{H^2(\Omega)} \le c,$$

for all $w \in H^1(\Omega)$ satisfying $S(w,\sigma) = w$. Noting that the embedding $H^2(\Omega) \hookrightarrow H^1(\Omega)$ is compact, standard arguments show that S is continuous and compact. Thus the fixed- point theorem applies.

Proof of Theorem 5.1.1. We only have to show that w is strictly positive in Ω. Taking $(w-r)^- \in H_0^1(\Omega)$ as test function in (5.1.15), where r is defined in (5.1.12), we have

$$\begin{aligned}
&h^2 \int_\Omega ((w-r)_x^-)^2 \mathrm{d}x \\
&= \int_\Omega (-(w-r)^-) w \left[\frac{j^2}{2t_\delta(w)^4} + Th(w^2) - (\phi - K) \right.\\
&\quad \left. + j \int_0^x \frac{\mathrm{d}s}{\tau t_\delta(w)^2} \right] \mathrm{d}x + \nu \int_\Omega \frac{(t_r(w))_x w}{t_r(w)^{\gamma+1}} \mathrm{d}x \\
&\le \int_\Omega (-(w-r)^-) w \left[\frac{j^2}{2\delta^4} + Th(r^2) - k + K + \frac{j}{\tau_0 \delta^2} + \frac{\nu r_x}{r^{\gamma+1}} \right] \mathrm{d}x \\
&\le \int_\Omega (-(w-r)^-) w \left[\frac{j^2}{2\delta^4} + Th(r^2) - k + K + \frac{j}{\tau_0 \delta^2} \right.\\
&\quad \left. - \frac{\nu \delta}{(2\delta)^{\gamma+1}} \right] \mathrm{d}x.
\end{aligned}$$

Using the monotonicity of h and Lemma 5.1.1, we get

$$\begin{aligned}
&h^2 \int_\Omega ((w-r)_x^-)^2 \mathrm{d}x \\
&\quad \le \int_\Omega (-(w-r)^-) w \left[\frac{j^2}{2\delta^4} + Th(4\delta^4) - k + K + \frac{j}{\tau_0 \delta^2} \right.\\
&\quad\quad \left. - \frac{\nu}{2^{\gamma+1}\delta\gamma} \right] \mathrm{d}x. \qquad (5.1.18)
\end{aligned}$$

Now choose $\delta \in (0,1)$ (see H1) such that

$$j(4\delta^2) \le 0, \quad \delta \le \left(\frac{1}{2^{\gamma+1}} \frac{\nu}{j^2/2 + j/\tau_0 + \max(0, K-k)} \right)^{1/(\gamma-4)}.$$

Notice that $\gamma > 4$, $\delta \le 1$, we obtain

$$\begin{aligned}
&\frac{j^2}{2\delta^4} + Th(4\delta^4) - k + K + \frac{j}{\tau_0 \delta^2} - \frac{\nu}{2^{\gamma+1}\delta\gamma} \\
&\quad \le \frac{1}{\delta^4} \left(\frac{j^2}{2} + \max(0, K-k) + \frac{j}{\tau_0} - \frac{\nu}{2^{\gamma+1}\delta^{\gamma-4}} \right) \le 0.
\end{aligned}$$

From (5.1.18), we have $w(x) \geq r(x) \geq \delta > 0$ for all $x \in \Omega$. Let $m(\nu) = \delta$, we then obtain the result. □

5.1.2 *Existence of small solutions for the isothermal equations if $h > 0$, $\nu = 0$*

In the case of vanishing viscosity we can only expect to get the existence of solutions for sufficiently small $J > 0$. In this section, we assume that there exists $m_0 > 0$ such that

$$\frac{1}{2}T_{p'}(m_0^2) + Th(m_0^2) + \frac{1}{\tau_0}\sqrt{T_{p'}(m_0^2)} + K - k \leq 0. \tag{5.1.19}$$

Since $K-k = -E_0+\lambda^{-2}(M^2+\|D\|_{L^1(\Omega)})$, this assumption is satisfied if for instance, $\lim_{s\to 0^+} h(s) = -\infty$, or $E_0 > 0$ is sufficiently large.

Now, define

$$m = \min(w_0, w_1, m_0). \tag{5.1.20}$$

Theorem 5.1.2. *Let the assumptions* (H1)–(H3) *and* (5.1.19) *hold and let* $\nu = 0$. *Furthermore, let* $j > 0$ *be such that*

$$j \leq j_0 = m^2\sqrt{T_{p'}(m_2)}. \tag{5.1.21}$$

Then there exists a classical solution $(w, \phi) \in (C^2(\overline{\Omega}))^2$ *to* (5.1.8)–(5.1.10) *satisfying*

$$0 < m \leq w(x) \leq M, \quad \forall x \in \Omega,$$

where m *and* M *are defined in* (5.1.20) *and* (5.1.11) *respectively.*

Remark 5.1.2. The condition (5.1.21) can be interpreted as a "subsonic condition" since it implies that the velocity $\frac{j}{n}$ satisfies

$$\frac{j}{n} = \frac{j}{w^2} \leq \frac{j}{m^2} \leq \sqrt{T_{p'}(m^2)} \leq \sqrt{T_{p'}(n)}.$$

Proof. By Proposition 5.1.1, there exists a solution (w, ϕ) to the truncated problem (5.1.15), (5.1.16), (5.1.10) with $\delta = m > 0$, $\nu = 0$.

It remains to show that $w \geq m$ in Ω. Using $(w-m)^-$ as test function gives

$$
\begin{aligned}
h^2 & \int_\Omega ((w-m)_x^-)^2 \mathrm{d}x \\
&= \int_\Omega (-(w-m)^-)w \left[\frac{j^2}{2t_m(w)^4} + Th(w^2) - \phi + K \right. \\
&\quad \left. + j \int_0^x \frac{\mathrm{d}s}{\tau t_m(w)^2} \right] \mathrm{d}x \\
&\leq \int_\Omega (-(w-m)^-)w \left[\frac{j_0^2}{2m^4} + Th(m^2) - k + K + \frac{j}{\tau_0 m^2} \right] \mathrm{d}x \\
&\leq \int_\Omega (-(w-r)^-)w \left[\frac{1}{2} T_{p'}(m^2) + Th(m^2) - k + K \right. \\
&\quad \left. + \frac{1}{\tau_0} \sqrt{T_{p'}(m^2)} \right] \mathrm{d}x \\
&\leq \int_\Omega (-(w-r)^-)w \left[\frac{1}{2} T_{p'}(m_0^2) + Th(m_0^2) - k + K \right. \\
&\quad \left. + \frac{1}{\tau_0} \sqrt{T_{p'}(m_0^2)} \right] \mathrm{d}x \leq 0.
\end{aligned}
$$

This implies that $w \geq m$, $\forall x \in \Omega$. □

Finally, we prove that every weak solution is necessarily strictly positive:

Proposition 5.1.2. *Let (w, ϕ) be a weak solution to (5.1.8)–(5.1.10) with $\nu \geq 0$ and $\frac{1}{w^3} \in L^1(\Omega)$. Then there exists $m > 0$ such that*

$$
w(x) \geq m > 0, \quad \forall x \in \Omega.
$$

Proof. By the definition of weak solutions, we know that each term of the right hand side of (5.1.8) belongs to $L^1(\Omega)$ when $\nu = 0$. In particular, $\frac{1}{w^3} \in L^1(\Omega)$, which implies $w_{xx} \in L^1(\Omega)$, and hence, $w \in W^{2,1}(\Omega) \hookrightarrow W^{1,\infty}(\Omega)$. Suppose that there exists $x_0 \in \Omega$ such that

$w(x_0) = 0$, then we have

$$w(x) = (x - x_0) \int_0^1 w_x(\theta_x + (1-\theta)x_0)\mathrm{d}\theta, \quad x \in \Omega,$$

$$\int_0^1 \frac{\mathrm{d}x}{|w(x)|^3} \geqslant \int_0^1 \left(\int_0^1 |w_x(s)|\mathrm{d}s \right)^{-3} |x - x_0|^{-3}\mathrm{d}x = \infty,$$

which contradicts the integrability of $\frac{1}{w^3}$. Hence, $w > 0$ for all $x \in \Omega$. Since w is continuous in $\bar{\Omega}$, there exists $m > 0$ such that $w(x) \geq m$, $\forall x \in \Omega$. □

5.1.3 *Non-existence of large solutions for the isentropic equations if $\epsilon > 0$, $\nu = 0$*

In this section, we show that, under some condition on h, a weak solution to (5.1.8)–(5.1.10) cannot exist if $j > 0$ is large enough. For this, we have to prove that w is bounded from above independently of J.

Lemma 5.1.4. *Let (w, ϕ) be any weak solution to* (5.1.8)–(5.1.10) *with $\nu \geq 0$, $K \in \mathbf{R}$, $j > 0$. Furthermore, let*

$$h(s) \geq c_0(s^{\alpha-1} - 1), \quad s \geq 0, \ \alpha > 2. \tag{5.1.22}$$

Then there exists $M_1 > 0$, independent of J, such that

$$w(x) \leq M_1, \quad \forall x \in \Omega.$$

Remark 5.1.3. The bound M_1 does not depend on K if $K > 0$.

Proof. Take $(w - \mathbf{\Lambda})^+$ with $\mathbf{\Lambda} \geq \max(w_0, w_1)$ as test function in (5.1.8) to get

$$\begin{aligned} h^2 &\int_\Omega ((w - \mathbf{\Lambda})_x^+)^2 \mathrm{d}x \\ &= \int_\Omega (w - \mathbf{\Lambda})^+ w\phi \mathrm{d}x + \int_\Omega (w - \mathbf{\Lambda})^+ w \left[-\frac{j^2}{2w^4} - h(w^2) \right. \\ &\quad \left. -K - j \int_0^x \frac{\mathrm{d}s}{\tau w^2} \right] \mathrm{d}x. \end{aligned} \tag{5.1.23}$$

The main difficulty is to estimate the last integral. From (5.1.17) it follows that

$$\phi(x) \leq \phi_0 + \max(-E_0, 0) + \lambda^{-2} \int_\Omega w^2 \mathrm{d}x.$$

Since

$$\begin{aligned} w^2 &= (w + \mathbf{\Lambda})(w - \mathbf{\Lambda}) + \mathbf{\Lambda}^2 \leq (w + \mathbf{\Lambda})(w - \mathbf{\Lambda})^+ + \mathbf{\Lambda}^2 \\ &\leq 2w(w - \mathbf{\Lambda})^+ + \mathbf{\Lambda}^2, \end{aligned}$$

we have

$$\phi(x) \leq \phi_0 + \max(-E_0, 0) + 2\lambda^{-2} \int_\Omega w(w - \mathbf{\Lambda})^+ \mathrm{d}x + \lambda^{-2}\mathbf{\Lambda}^2.$$

Setting $\phi_1 = \phi_0 + max(-E_0, 0)$, we get

$$\begin{aligned} &\int_\Omega (w - \mathbf{\Lambda})^+ w\phi \mathrm{d}x \\ &\leq \int_\Omega (w - \mathbf{\Lambda})^+ w(\phi_1 + \lambda^{-2}\mathbf{\Lambda}^2)\mathrm{d}x + 2\lambda^{-2}\left(\int_\Omega (w - \mathbf{\Lambda})^+ w \mathrm{d}x\right)^2 \\ &\leq \phi_1 \int_\Omega (w - \mathbf{\Lambda})^+ w \mathrm{d}x + \lambda^{-2} \int_\Omega (w - \mathbf{\Lambda})^+ w^3 \mathrm{d}x \\ &\quad + 2\lambda^{-2} \int_\Omega (w - \mathbf{\Lambda})^{+2} w^2 \mathrm{d}x \\ &\leq \phi_1 \int_\Omega (w - \mathbf{\Lambda})^+ w \mathrm{d}x + 3\lambda^{-2} \int_\Omega (w - \mathbf{\Lambda})^+ w^3 \mathrm{d}x. \end{aligned}$$

Since $\mathbf{\Lambda} \leq w$ on $\{x : w(x) \geq \mathbf{\Lambda}\}$ and $(w - \mathbf{\Lambda})^+ \leq w$. Therefore we get from (5.1.23) that

$$\begin{aligned} &h^2 \int_\Omega ((w - \mathbf{\Lambda})_x^+)^2 \mathrm{d}x \\ &\quad \leq \int_\Omega (w - \mathbf{\Lambda})^+ w[\phi_1 + 3\lambda^{-2}w^2 - K - c_0(w^{2\alpha-2} - 1)]\mathrm{d}x. \end{aligned}$$

Since $(2\alpha - 2) > 2$, there exists $M_1 > 0$ such that

$$-\phi_1 - 3\lambda^{-2}M_1^2 + K + c_0 M_1^{2\alpha-2} - c_0 \geq 0. \tag{5.1.24}$$

Taking $\mathbf{\Lambda} = M_1$, we obtain that $w \leq M_1$, $\forall x \in \Omega$. □

Theorem 5.1.3. *Let* (H1)–(H3) *and* (5.1.22) *hold and let* $K \in \boldsymbol{R}$ *and* $\nu = 0$. *Then there exists* $j_1 > 0$ *such that if* $j \geq j_1$, *then the problem* (5.1.8)–(5.1.10) *cannot have a weak solution.*

Remark 5.1.4. The constant j_1 is defined by

$$\frac{j_1^2}{2M_1^3} = 8h^2(\max(w_0, w_1) + 1) - Th_0 + M_1(|\phi_0| + |E_0| + \lambda^{-2}M_1^2 - \min(0, K)), \tag{5.1.25}$$

where $h_0 = \inf\{sh(s^2) : 0 < s \leq M_1\} > -\infty$ and $M_1 > 0$ is defined by (5.1.24).

Proof. Suppose that there exists a weak solution (w, ϕ) to (5.1.8)–(5.1.10) for all $j > 0$ and for some $K \in \mathbf{R}$. It follows from Lemma 5.1.4 that

$$w(x) \leq M_1, \quad \forall x \in \Omega.$$

On the other hand, it follows from Proposition 5.1.2 that $w \in H^1(\Omega) \hookrightarrow C^0(\overline{\Omega})$, $w > 0$, $\forall x \in \Omega$. Thus $w_{xx} \in C^0(\overline{\Omega})$. In other words, w is a classical solution. Let $j \geq j_1$, where j_1 is defined in (5.1.25). Then

$$\begin{aligned} h^2 w_{xx} &\geq \frac{j_1^2}{2M_1^3} + Th_0 - w\left(\sup_\Omega \phi - \min(0, K)\right) \\ &\geq \frac{j_1^2}{2M_1^3} + Th_0 - M_1(|\phi_0| + |E_0| + \lambda^{-2}M_1^2 - \min(0, K)) \\ &= 8h^2(\max(w_0, w_1) + 1) \end{aligned}$$

Introduce $q(x) = 4(\max(w_0, w_1)+1)\left(x - \frac{1}{2}\right)^2 - 1$, $x \in [0,1]$, then for $x \in \Omega$,

$$q(0) = q(1) = \max(w_0, w_1),$$
$$q_{xx}(x) = 8(\max(w_0, w_1) + 1),$$

which implies

$$w - q \leq 0, \quad x \in \partial\Omega,$$
$$(w - q)_{xx} \geq 0, \quad x \in \Omega.$$

The maximum principle gives $(w - q) \leq 0$, $x \in \bar{\Omega}$. In particular, $w\left(\frac{1}{2}\right) \leq q\left(\frac{1}{2}\right) = -1 < 0$, which contradicts the positivity of w. □

Corollary 5.1.1. *Let* (H1)–(H3) *hold and let* $\nu = 0$.

(i) *If* (5.1.19) *holds, then there exists* $j_0 > 0$ *such that for all* $j \leq j_0$, *there exists a solution* $(w, \phi) \in (C^3(\bar{\Omega}))^2$ *to* (5.1.1)–(5.1.4) *with strictly positive* n.

(ii) *Let in addition* (5.1.22) *hold. Then there exists* $j_1 > 0$ *such that for all* $j \geq j_1$, *the problem* (5.1.1)–(5.1.4) *cannot have a weak solution.*

Proof. The first part follows from Theorem 5.1.2 by letting $n = w^2$. The second part follows from Theorem 5.1.3 since the constant $K \in \mathbf{R}$ is fixed in (5.1.4). Thus the problems (5.1.1)–(5.1.4) and (5.1.8)–(5.1.10) are equivalent and the non-existence of solutions to (5.1.8)–(5.1.10) implies the non-existence of solutions to (5.1.1)–(5.1.4). □

5.1.4 *Uniqueness of solutions for the isentropic equations if* $h > 0$, $\nu = 0$

This section deals with the uniqueness of 'subsonic' solutions to (5.1.1)–(5.1.4).

Theorem 5.1.4. *Let the assumptions* (H1)–(H3) *hold and let* $\nu = 0$, $\frac{1}{\tau} = 0$. *For any fixed* $K \in \mathbf{R}$, *suppose the function* h' *is non-decreasing and* $\delta > 0$. *Then there exists* $E_0 > 0$ *and* $T_0 > 0$ *such* *that for any* $T \geq T_0$, *there is uniqueness of weak solutions to* (5.1.8)–(5.1.10) *in the class of positive densities satisfying the 'subsonic' condition*

$$\frac{j}{w(x)^2} \leq \sqrt{(1-\delta)T_{p'}(w(x)^2)}, \quad \forall x \in \Omega. \tag{5.1.26}$$

Proof. Let (w_1, ϕ_1) and (w_2, ϕ_2) be two weak solutions to (5.1.8)–(5.1.10) satisfying the condition (5.1.26). Then w_1, w_2 are positive in Ω, by Proposition 5.1.2, and classical solutions. Introduce $\chi(w) = \frac{j^2}{2w^4} + (1-\delta)Th(w^2)$. Then, proceeding similarly as in [81], we write

$$\begin{aligned} -h^2\frac{w_{1,xx}}{w_1} + h^2\frac{w_{2,xx}}{w_2} = &-\frac{\chi(w_1)}{w_1} + \frac{\chi(w_2)}{w_2} \\ &- \delta T(h(w_1^2) - h(w_2^2)) + \phi_1 - \phi_2. \end{aligned}$$

Multiplying this equation by $(w_1^2 - w_2^2)$, integrating over Ω and integrating by parts, we obtain, after elementary computations,

$$\begin{aligned} &\int_\Omega (w_1^2 + w_2^2)(\ln w_1 - \ln w_2)_x^2 \mathrm{d}x \\ &= \int_\Omega \left[\left(w_{1,x} - \frac{w_1}{w_2}w_{2,x}\right)^2 + \left(w_{2,x} - \frac{w_2}{w_1}w_{1,x}\right)^2\right] \mathrm{d}x \\ &= -\int_\Omega (w_1^2 - w_2^2)\left(\frac{\chi(w_1)}{w_1} - \frac{\chi(w_2)}{w_2}\right) \mathrm{d}x \\ &\quad - \delta T \int_\Omega (w_1^2 - w_2^2)(h(w_1^2) - h(w_2^2))\mathrm{d}x \\ &\quad + \int_\Omega (w_1^2 - w_2^2)(\phi_1 - \phi_2)\mathrm{d}x. \end{aligned} \tag{5.1.27}$$

Since the function h' is non-decreasing, we have

$$-\delta T \int_\Omega (w_1^2 - w_2^2)(h(w_1^2) - h(w_2^2))\mathrm{d}x$$

$$= -\delta T \int_{\Omega} (w_1^2 - w_2^2)^2 h'(\theta w_1^2 + (1-\theta) w_2^2) \mathrm{d}x$$
$$\leq -c\delta T \int_{\Omega} (w_1^2 - w_2^2)^2 \mathrm{d}x,$$

where

$$c = \int_0^1 h'(\theta m^2) \mathrm{d}\theta > 0$$

is independent of $T > 0$.

Now multiply the difference of (5.1.9) for ϕ_1 and ϕ_2 by $(\phi_1 - \phi_2)$, integrate over Ω and integrate by parts to get

$$\begin{aligned}
&\int_{\Omega} (w_1^2 - w_2^2)(\phi_1 - \phi_2) \mathrm{d}x \\
&\quad = -\lambda^2 \int_{\Omega} (\phi_1 - \phi_2)_x^2 \mathrm{d}x + \lambda^2 (\phi_1 - \phi_2)(1)(\phi_1 - \phi_2)_x(1) \\
&\quad \leq -\lambda^2 \int_{\Omega} (\phi_1 - \phi_2)_x^2 \mathrm{d}x + \lambda^{-2} \left(\int_0^1 (w_1^2 - w_2^2) \mathrm{d}x \right)^2 \\
&\quad \leq -\lambda^2 \int_{\Omega} (\phi_1 - \phi_2)_x^2 \mathrm{d}x + \lambda^{-2} \int_0^1 (w_1^2 - w_2^2)^2 \mathrm{d}x,
\end{aligned}$$

where we used Jensen's inequality. Finally, we will estimate the first integral of the right hand side of (5.1.27). The function $s \rightarrowtail \chi(s)/s$ is non-decreasing if and only if

$$\begin{aligned}
\frac{\mathrm{d}}{\mathrm{d}s} \frac{\chi(s)}{s} &= -\frac{2j^2}{s^5} + 2(1-\delta) T s h'(S^2) \\
&= \frac{2}{s} \left(-\frac{j^2}{s^4} + T(1-\delta) p'(S^2) \right) \geq 0.
\end{aligned}$$

It follows from (5.1.26) that

$$-\int_{\Omega} (w_1^2 - w_2^2) \left(\frac{\chi(w_1)}{w_1} - \frac{\chi(w_2)}{w_2} \right) \mathrm{d}x \leq 0.$$

Hence, we obtain from (5.1.27) that

$$\int_\Omega (w_1^2 + w_2^2)(\ln w_1 - \ln w_2)_x^2 \mathrm{d}x$$
$$\leq -\lambda^2 \int_\Omega (\phi_1 - \phi_2)_x^2 \mathrm{d}x + (\lambda^{-2} - c\delta T) \int_0^1 (w_1^2 - w_2^2)^2 \mathrm{d}x \leq 0$$

by taking $T \geq T_0$, where $T_0 = 1/\lambda^2 \delta c$. Hence $w_1 = w_2$ and $\phi_1 = \phi_2$ in Ω, $\forall x \in \Omega$. □

5.1.5 *High-dimensional third-order equations*

In this section we show how the methods of the previous sections can be extended to the third-order equations in any space dimension. This is the first important step for performing the existence analysis of the multi-dimensional quantum hydrodynamic equations. Since the proofs are similar to those of the previous sections, we only sketch the proofs. Let $\Omega \subset \mathbf{R}^d (d \geq 1)$ be a bounded domain and consider

$$\nabla(A(\boldsymbol{u})\Delta \boldsymbol{u}) = \mu \nabla(F(\boldsymbol{u})) + \nabla(G(\boldsymbol{u})) + \nu(B(\boldsymbol{u})(1 \cdot \nabla)\boldsymbol{u}),$$
$$\forall x \in \partial\Omega, \tag{5.1.28}$$

$$B\boldsymbol{u} = \boldsymbol{u}_D, \ \forall x \in \partial\Omega, \tag{5.1.29}$$

where $\nu \geq 0$ and $\mu > 0$. Recall that $(1 \cdot \nabla)\boldsymbol{u} = \sum_i \partial i \boldsymbol{u}$. When the boundary conditions are such that the integration of (5.1.28) along streamlines is well defined, we obtain

$$\Delta \boldsymbol{u} = \mu f(\boldsymbol{u}) + g(\boldsymbol{u}) + K a(\boldsymbol{u}) + \nu b(\boldsymbol{u})(1 \cdot \nabla)\boldsymbol{u}, \tag{5.1.30}$$

where $K \in \mathbf{R}$ is a constant and

$$f(\boldsymbol{u}) = \frac{F(\boldsymbol{u})}{A(\boldsymbol{u})}, \quad g(\boldsymbol{u}) = \frac{G(\boldsymbol{u})}{A(\boldsymbol{u})}, \quad a(\boldsymbol{u}) = \frac{1}{A(\boldsymbol{u})}, \quad b(\boldsymbol{u}) = \frac{B(\boldsymbol{u})}{A(\boldsymbol{u})}.$$

Any solution $\boldsymbol{u} \in H^2(\Omega)$ to (5.1.30) and (5.1.29) solves the problem (5.1.28) and (5.1.29) and vice versa. We assume that:

(i) $\partial\Omega \in C^{1,1}$, $\boldsymbol{u}_D \in H^2(\Omega) \cap L^\infty(\Omega)$; $\boldsymbol{u}_D \geq \boldsymbol{u}_0 > 0$, $\forall x \in \partial\Omega$;
(ii) $a, b, f, g \in C(0, \infty)$, a, g are non-decreasing, a, b are non-negative;

(iii) $g(0_+) \le 0$; $g(+\infty) > 0$; $\lim_{s\to 0_+} a(s)/sb(s) = 0$;
(iv) $\lim_{0<s<M} f(s) > 0$, $\forall M > 0$; $\lim_{s\to 0_+} f(s)/sb(s) = 0$.

Admissible functions are, for instance,

$$f(\boldsymbol{u}) = \boldsymbol{u}^{-\alpha}, \quad g(\boldsymbol{u}) = \boldsymbol{u}^{\beta} - 1, \quad a(\boldsymbol{u}) = \boldsymbol{u}^{\gamma}, \quad b(\boldsymbol{u}) = \boldsymbol{u}^{-\varepsilon}$$

with $\alpha, \beta, \gamma > 0$, $\gamma > \varepsilon - 1$, $\varepsilon > 1 + \alpha$.

Theorem 5.1.5. *Let $\nu > 0$ and $K > 0$. Then there exists a solution $\boldsymbol{u} \in H^2(\Omega)$ to (5.1.28) and (5.1.29) for all $\mu > 0$.*

Remark 5.1.5. The assumption $K > 0$ can be weakened by choosing appropriate assumptions on the functions a, f and g.

Proof. Since Ω is bounded, there exists $R > 0$, such that Ω is contained in the ball $B_R(0)$ of radius R and center 0. Introduce the comparison function $\psi(x) = \psi(x_1, \dots, x_d) = \frac{\delta}{R}(2R - x_1)$, $0 < \delta \le \min(\boldsymbol{u}_0, M/3)$. Set $t_\psi(\boldsymbol{u}_M) = \max(\psi(\cdot), \min(M, u(\cdot)))$ and consider the truncated problem

$$\Delta \boldsymbol{u} = \mu \frac{f(t_\psi(\boldsymbol{u}_M))}{t_\psi(\boldsymbol{u}_M)} \boldsymbol{u}^+ + \frac{g(t_\psi(\boldsymbol{u}_M))}{t_\psi(\boldsymbol{u}_M)} \boldsymbol{u}^+ + K \frac{a(t_\psi(\boldsymbol{u}_M))}{t_\psi(\boldsymbol{u}_M)} \boldsymbol{u}^+$$
$$+ \nu \frac{b(t_\psi(\boldsymbol{u}_M))}{t_\psi(\boldsymbol{u}_M)} \boldsymbol{u}^+ (1 \cdot \boldsymbol{\nabla})(t_\psi(\boldsymbol{u}_M)), \quad \forall x \in \Omega, \tag{5.1.31}$$

$$\boldsymbol{u} = \boldsymbol{u}_D, \ \forall x \in \partial\Omega, \tag{5.1.32}$$

where $\boldsymbol{u}^+ = \max(0, \boldsymbol{u})$. Using the fact $Ka(s) \ge 0$ and the methods of the proof of Proposition 5.1.1, we easily get the existence of a solution $\boldsymbol{u} \in H^2(\Omega)$ to (5.1.31)–(5.1.32) for any $\nu \ge 0$. It remains to show that $\delta \leqslant \boldsymbol{u}(x) \leqslant M$, $\forall x \in \Omega$.

First we observe that by using $\boldsymbol{u}^-$ as test function in (5.1.31), we immediately conclude that $\boldsymbol{u}(x) \le 0$ in Ω. In order to prove that $\boldsymbol{u}(x) \le M$ in Ω for some $M > 0$, we use $(\boldsymbol{u} - M)^+$ as test function in (5.1.31) with $M \ge \|\boldsymbol{u}_D\|_{L^\infty(\Omega)}$. Since there exists a constant $M > 0$ such that $g(M) \ge 0$, we can show as in the proof of Lemma 5.1.1 that $(\boldsymbol{u} - M)^+ = 0$ in Ω.

For the lower bound, we use $(\boldsymbol{u}-\psi)^-$ as test function in (5.1.31), observing that $\delta \le \psi(x) \le 3\delta$ in Ω. Choosing δ small enough such that for all $x \in \Omega$, $g(\psi) \le 0$, inequality $f(\psi)/\psi b(\psi) \le \nu/(6\mu R)$, $a(\psi)/\psi b(\psi) \le \nu/(6KR)$ holds, then we have

$$\begin{aligned}
&\int_\Omega |\boldsymbol{\nabla}(\boldsymbol{u}-\psi)^{-2}|\mathrm{d}x \\
&\quad = \int_\Omega (-(\boldsymbol{u}-\psi)^-)\boldsymbol{u}\left[\mu\frac{f(t_\psi(\boldsymbol{u}))}{t_\psi(\boldsymbol{u})} + \frac{g(t_\psi(\boldsymbol{u}))}{t_\psi(\boldsymbol{u})} + K\frac{a(t_\psi(\boldsymbol{u}))}{t_\psi(\boldsymbol{u})}\right. \\
&\qquad \left. +\nu\frac{b(t_\psi(\boldsymbol{u}))}{t_\psi(\boldsymbol{u})}(1\cdot\boldsymbol{\nabla})(t_\psi(\boldsymbol{u}))\right]\mathrm{d}x \\
&\quad = \int_\Omega (-(\boldsymbol{u}-\psi)^-)\boldsymbol{u}\left[\mu\frac{f(\psi)}{\psi} + \frac{g(\psi)}{\psi} + K\frac{a(\psi)}{\psi}\right. \\
&\qquad \left. +\nu\frac{b(\psi)}{\psi}(1\cdot\boldsymbol{\nabla})\psi\right]\mathrm{d}x \\
&\quad \le \int_\Omega (-(\boldsymbol{u}-\psi)^-)\boldsymbol{u}b(\psi)\left(\mu\frac{f(\psi)}{\psi b(\psi)} + \frac{g(\psi)}{\psi b\psi}\right. \\
&\qquad \left. +K\frac{a(\psi)}{\psi b(\psi)} - \frac{\nu}{3R}\right)\mathrm{d}x \le 0.
\end{aligned}$$

Thus, we conclude that $\boldsymbol{u} \ge \psi \ge \delta$, $\forall x \in \Omega$. □

Theorem 5.1.6. *Let $\nu = 0$.*

1. *Let $g(0_+) < 0$, $K > 0$, $\lim_{s\to 0_+} a(s)/s = 0$. Then there exists $\mu_0 > 0$, such that for all $0 < \mu \le \mu_0$, there exists a solution $\boldsymbol{u} \in H^2(\Omega)$ to* (5.1.28), (5.1.29) *satisfying*

$$\Delta\boldsymbol{u} = \mu f(\boldsymbol{u}_D) + g(\boldsymbol{u}_D) + Ka(\boldsymbol{u}_D), \quad \forall x \in \partial\Omega. \tag{5.1.33}$$

2. *Let $g(0_+) > -\infty$. Then there exists $\mu_1 > 0$ such that for all $\mu_1 \ge \mu_1$, there is no weak solution to* (5.1.28), (5.1.29) *and* (5.1.34).

Proof. For the first part of the theorem we only have to show that the solution $\boldsymbol{u}$ to (5.1.31) and (5.1.32) is strictly positive for sufficiently small $\mu > 0$. Take $\delta = m > 0$ and $\varepsilon > 0$ such that $m \le u_0$, $g(\phi)/\phi \le h < 0$ and $a(\phi)/\phi \le h/2K$ (if $K > 0$) in Ω. The existence of m and ε is ensured by condition (iii) and the assumption on a. Let $f_m = \sup_{m<s<3m} f(s)/s$, then $f_m > 0$ in view of assumption (iv). Choose $0 < \mu_0 \le \varepsilon/2f_m$ and let $0 < \mu \le \mu_0$. Using $(\boldsymbol{u} - m)^-$ as test function yields

$$\begin{aligned}
&\int_\Omega |\boldsymbol{\nabla}(\boldsymbol{u} - m)^-|^2 \mathrm{d}x \\
&= \int_\Omega (-(\boldsymbol{u} - m)^-)\boldsymbol{u}\left[\mu\frac{f(t_\psi(\boldsymbol{u}))}{t_\psi(\boldsymbol{u})} + \frac{g(t_\psi(\boldsymbol{u}))}{t_\psi(\boldsymbol{u})} + K\frac{a(t_\psi(\boldsymbol{u}))}{t_\psi(\boldsymbol{u})}\right]\mathrm{d}x \\
&\le \int_\Omega (-(\boldsymbol{u} - m)^-)\boldsymbol{u}\left[\mu_0\frac{f(\psi)}{\psi} + \frac{g(\psi)}{\psi} + K\frac{a(\psi)}{\psi}\right]\mathrm{d}x \\
&\le \int_\Omega (-(\boldsymbol{u} - m)^-)\boldsymbol{u}[\mu_0 f_m - \varepsilon/2]\mathrm{d}x \le 0.
\end{aligned}$$

Thus $\boldsymbol{u} \ge m > 0$, $\forall x \in \Omega$.

To prove the second part of the theorem, let $\boldsymbol{u} \in H^1(\Omega)$ be a solution to (5.1.29) and (5.1.30) with $\boldsymbol{u} \le M$ in Ω and $K \in \mathbf{R}$. Using the positivity of f, we can easily see that the constant M does not depend on μ. Let $X_0 \in \Omega$. Since Ω is open, there exists a ball $B_r(x_0)$ of radius r with center x_0 contained in Ω. Set $f_0 = \inf\{f(s) : 0 < s < M\} > 0$ and $a_0 = \inf\{a(s) : 0 < s < M\} \ge 0$, and choose $\mu_1 > 0$ such that $L\Delta_{q\mu_1} f_0 + g(0_+) + \max(0, Ka_0) \ge 2d(M+1)/r^2$ and $L \ge M$. Notice that we assumed $g(0_+) > -\infty$. We show that $\boldsymbol{u}(x_0) < 0$ which is a contradiction to the non-negativity of $\boldsymbol{u}$. It holds in the sense of distributions

$$\begin{aligned}
\Delta \boldsymbol{u} &\ge \mu f_0 + g(0_+) + \max(0, Ka_0) \ge \mu_1 f_0 + g(0_+) \\
&\quad + \max(0, Ka_0) = L, \quad \forall x \in \partial\Omega.
\end{aligned}$$

Now define

$$q(x) = \frac{L}{2d}|x - x_0|^2 - 1, \quad \forall x \in B_r(x_0).$$

Then

$$\Delta q = L,\ \forall x \in B_r(x_0), \quad q = r^2 L/2d - 1 > 0,\ \forall x \in \partial B_r(x_0).$$

This implies

$$\Delta(u-q) \geq 0,\ \forall x \in B_r(x_0), \quad (u-q) \leq M - r^2L/2d + 1 \leq 0,$$
$$\forall x \in \partial B_r(x_0).$$

By the maximum principle, we conclude that $(\boldsymbol{u}-q) \geq 0$, $\forall x \in B_r(x_0)$. In particular, $\boldsymbol{u}(x_0) \leq q(x_0) = -1 < 0$, we conclude a contradiction. □

5.2 Initial boundary value problem for compressible quantum Euler–Poisson equations

In this section, we consider the existence and the asymptotic stability of a stationary solution to the initial boundary value problem for a one-dimensional quantum hydrodynamic model of semiconductors

$$\rho_t + j_x = 0, \tag{5.2.1}$$

$$j_t + \left(\frac{j^2}{\rho} + P(\rho)\right)_x - h^2\rho\left(\frac{(\sqrt{\rho})_{xx}}{\sqrt{\rho}}\right)_x = \rho\phi_x - j, \tag{5.2.2}$$

$$\phi_{xx} = \rho - D, \tag{5.2.3}$$

where $x \in \Omega := (0,1)$, ρ, j, ϕ stand for the electron density, the electric current and the electrostatic potential, respectively. The positive constant h is the scaled Planck constant. One can refer to [84]–[97] for the background. We consider the isothermal flow,

$$P = P(\rho) = K\rho, \tag{5.2.4}$$

where K is the positive constant, a doping profile $D \in \mathcal{B}(\bar{\Omega})$, $x \in \bar{\Omega} := [0,1]$, satisfies

$$\inf_{x\in\bar{\Omega}} D(x) > 0. \tag{5.2.5}$$

The initial and the boundary conditions are prescribed as

$$(\rho, j)(0,x) = (\rho_0, j_0)(x), \tag{5.2.6}$$

$$\rho(t,0) = \rho_l > 0, \quad \rho(t,1) = \rho_r > 0, \tag{5.2.7}$$

$$(\sqrt{\rho})_{xx}(t,0) = (\sqrt{\rho})_{xx}(t,1) = 0, \tag{5.2.8}$$

$$\phi(t,0) = 0, \quad \phi(t,1) = \phi_r = 0. \tag{5.2.9}$$

The initial condition (5.2.6) and the boundary conditions (5.2.7)–(5.2.9) satisfy the compatibility conditions

$$\begin{aligned} &\rho_0(0) = \rho_l, \quad \rho_0(1) = \rho_r, \quad j_{0x}(0) = j_{0x}(1) = 0, \\ &(\sqrt{\rho_0})_{xx}(0) = (\sqrt{\rho_0})_{xx}(1) = 0, \end{aligned} \tag{5.2.10}$$

using the boundary condition (5.2.9), and integrating (5.2.3) lead to an explicit formula of the electrostatic potential

$$\begin{aligned} \phi(t,x) &= \Phi[\rho](t,x) \\ &:= \int_0^x \int_0^y (\rho - D)(t,z)\mathrm{d}z\mathrm{d}y \\ &\quad + \left(\phi_r - \int_0^1 \int_0^y (\rho - D)(t,z)\mathrm{d}z\mathrm{d}y\right) x. \end{aligned} \tag{5.2.11}$$

The following properties play essential roles in establishing the existence of the solution to (5.2.1)–(5.2.3)

$$\inf_{x\in\Omega} S[\rho,j] > 0, \quad S[\rho,j] := P'(\rho) - \frac{j^2}{\rho^2}, \tag{5.2.12}$$

$$\inf_{x\in\Omega} \rho > 0, \tag{5.2.13}$$

where (5.2.12) is called a subsonic condition. In the next subsection, (5.2.12)–(5.2.13) hold, if the initial condition satisfies

$$\inf_{x\in\Omega} S[\rho_0, j_0] > 0, \inf_{x\in\Omega} \rho_0 > 0. \tag{5.2.14}$$

It is convenient to rewrite (5.2.1)–(5.2.3) and (5.2.6)–(5.2.9) where $\omega := \sqrt{\rho}$, as

$$2\omega\omega_t + j_x = 0, \tag{5.2.15}$$

$$j_t + 2S[\omega^2, j]\omega\omega_x + 2\frac{j}{\omega^2} j_x - h^2\omega^2 \left(\frac{\omega_{xx}}{\omega}\right)_x = \omega^2\phi_x - j, \tag{5.2.16}$$

$$\phi_{xx} = \omega^2 - D \tag{5.2.17}$$

with the initial and the boundary data

$$(\omega, j)(0,x) = (\omega_0, j_0)(x) := (\sqrt{\rho_0}, j_0)(x), \tag{5.2.18}$$

$$\omega(t,0) = \omega_l := \sqrt{\rho_l} > 0, \quad \omega(t,1) = \omega_r := \sqrt{\rho_r} > 0, \tag{5.2.19}$$

$$\omega_{xx}(t,0) = \omega_{xx}(t,1) = 0, \tag{5.2.20}$$

$$\phi(t,0) = 0, \quad \phi(t,1) = \phi_r. \tag{5.2.21}$$

If the density ρ is positive, (5.2.1)–(5.2.3) is equivalent to (5.2.15)–(5.2.17). Thus, if there exists a solution $(\omega, j, \phi), \omega > 0$ to the initial boundary value problem (5.2.15)–(5.2.21), then there exists a solution to the initial boundary value problem (5.2.1)–(5.2.3) and (5.2.6)–(5.2.9).

Next, we consider the asymptotic behavior of (5.2.1)–(5.2.3), i.e. the solution to (5.2.1)– (5.2.3) converges to the stationary solution $(\tilde{\omega}, \tilde{j}, \tilde{\phi})$, which satisfies

$$\tilde{j}_x = 0, \tag{5.2.22}$$

$$S[\tilde{\rho}, \tilde{j}]\tilde{\rho}_x - h^2\tilde{\rho}\left(\frac{(\sqrt{\tilde{\rho}})_{xx}}{\sqrt{\tilde{\rho}}}\right)_x = \tilde{\rho}\tilde{\phi}_x - \tilde{j}, \tag{5.2.23}$$

$$\tilde{\phi}_{xx} = \tilde{\rho} - D, \tag{5.2.24}$$

and the boundary conditions

$$\tilde{\rho}(0) = \rho_l > 0, \quad \tilde{\rho}(1) = \rho_r > 0, \tag{5.2.25}$$

$$(\sqrt{\tilde{\rho}})_{xx}(0) = (\sqrt{\tilde{\rho}})_{xx}(1) = 0, \tag{5.2.26}$$

$$\tilde{\phi}(0) = 0, \quad \tilde{\phi}(1) = \phi_r > 0. \tag{5.2.27}$$

Let $\tilde{\omega} := \sqrt{\tilde{\rho}}$, by (5.2.22)–(5.2.24) and (5.2.25)–(5.2.27), $(\tilde{\omega}, \tilde{j}, \tilde{\phi})$ satisfies

$$2S[\tilde{\omega}^2, \tilde{j}]\tilde{\omega}\tilde{\omega}_x - h^2\tilde{\omega}^2\left(\frac{\tilde{\omega}_{xx}}{\tilde{\omega}}\right)_x = \tilde{\omega}^2\tilde{\phi}_x - \tilde{j}, \tag{5.2.28}$$

$$\tilde{\phi}_{xx} = \tilde{\omega}^2 - D \tag{5.2.29}$$

and (5.2.22) with boundary conditions

$$\tilde{\omega}(0) = \omega_l > 0, \quad \tilde{\omega}(1) = \omega_r > 0, \tag{5.2.30}$$

$$\tilde{\omega}_{xx}(0) = \tilde{\omega}_{xx}(1) = 0 \tag{5.2.31}$$

and (5.2.27). Divide (5.2.28) by $\tilde{\omega}^2$, integrate over $(0, x)$ and use the boundary conditions (5.2.27), (5.2.30) and (5.2.31). Moreover, apply Green formula to (5.2.29) together with the boundary condition (5.2.27), we have

$$h^2 \frac{\tilde{\omega}_{xx}}{\omega} = F(\tilde{\omega}^2, \tilde{j}) - F(\rho_l, \tilde{j}) - \tilde{\phi} + \int_0^x \frac{\tilde{j}}{\tilde{\omega}^2}(y)\mathrm{d}y, \tag{5.2.32}$$

$$\tilde{\phi} = \mathcal{G}[\tilde{\omega}^2] := \int_0^1 G(x, \xi)(\tilde{\omega}^2 - D)(\xi)\mathrm{d}\xi + \phi_r x, \tag{5.2.33}$$

$$F(\xi, \zeta) := \frac{\zeta^2}{2\xi^2} + K \log \xi, \quad G(x, \xi) := \begin{cases} x(\xi - 1), & x < \xi, \\ \xi(x - 1), & x > \xi. \end{cases} \tag{5.2.34}$$

Substituting $x = 1$ in (5.2.32), we have from (5.2.30) and (5.2.31) the current-voltage relationship

$$\phi_r = F(\rho_r, \tilde{j}) - F(\rho_l, \tilde{j}) + \tilde{j} \int_0^1 \frac{1}{\tilde{\rho}}\mathrm{d}x. \tag{5.2.35}$$

Next, we introduce the strength of the boundary data

$$\delta := |\rho_l - \rho_r| + |\phi_r|. \tag{5.2.36}$$

The existence of the stationary solution $(\tilde{\rho}, \tilde{j}, \tilde{\phi})$ is stated in the following lemma.

Lemma 5.2.1. *Let the doping profile and the boundary data satisfy* (5.2.5), (5.2.7) *and* (5.2.9). *For an arbitrary* ρ_l, *there exist positive constants* δ_1 *and* h_1, *such that if* $\delta \le \delta_1$, *and* $h \le h_1$, *then there exists a unique solution* $(\tilde{\rho}, \tilde{j}, \tilde{\phi}) \in \mathcal{B}^4(\overline{\Omega}) \times \mathcal{B}^4(\overline{\Omega}) \times B^2(\overline{\Omega})$ *to the stationary problem* (5.2.22)–(5.2.27) *satisfying the conditions* (5.2.12)–(5.2.13).

This lemma follows from (5.2.5) and (5.2.7).

The electric current $\tilde{j}$ in Lemma (5.2.1) is given by

$$\tilde{j} = \mathcal{J}[\tilde{\rho}] := 2B_b \left[\int_0^1 \tilde{\rho}^{-1} \mathrm{d}x + \sqrt{\left(\int_0^1 \tilde{\rho}^{-1} \mathrm{d}x \right)^2 + 2B_b(\rho_r^{-2} - \rho_l^{-2})} \right]^{-1}, \tag{5.2.37}$$

$$B_b := \phi_r - K\{log\rho_r - \log \rho_l\}.$$

We introduce the function space

$$\bar{\mathfrak{X}}_i^l([0,T]) := \bigcap_{k=0}^{[i/2]} C^k([0,T]; H^{l+i-2k}(\Omega)), \quad i, l = 0, 1, 2, \ldots, \tag{5.2.38}$$

$$\bar{\mathfrak{X}}_i([0,T]) := \bar{\mathfrak{X}}_i^0([0,T]), \quad i = 0, 1, 2, \ldots, \tag{5.2.39}$$

$$\mathfrak{Y} := C^2([0,T]; H^2(\Omega)) \tag{5.2.40}$$

where $[\tilde{\rho}]$ denotes the largest integer which is less than or equal to $\tilde{\rho}$. The stability of the stationary solution is stated in the next theorem.

Theorem 5.2.1. *Let $(\tilde{\rho}, \tilde{j}, \tilde{\phi})$ be the stationary solution of (5.2.22)–(5.2.27). Assume that the initial data $(\rho_0, j_0) \in H^4(\Omega) \times H^3(\Omega)$, and the boundary data ρ_l, ρ_r, ϕ_r satisfy (5.2.7), (5.2.9), (5.2.10) and (5.2.14). Then there exists a positive constant δ_2 such that, if $(\delta+h)+\|(\rho_0-\tilde{\rho}, j_0-\tilde{j})\|_2+\|(h\partial_x^3\{\rho_0-\tilde{\rho}\}, h\partial_x^3\{j_0-\tilde{j}\}, h^2\partial_x^4\{\rho_0-\tilde{\rho}\})\| \leq \delta_2$, the initial boundary value problem (5.2.1)–(5.2.3) and (5.2.6)–(5.2.9) has a unique solution $(\rho, j, \phi) \in \bar{\mathfrak{X}}_4([0,\infty)) \times \mathfrak{X}^3([0,\infty)) \times \mathfrak{Y}([0,\infty))$. Moreover, (ρ, j, ϕ) satisfies $\phi - \tilde{\phi} \in \bar{\mathfrak{X}}_4^2([0,\infty))$ and the decay estimate*

$$\begin{aligned} &\|(\rho-\tilde{\rho}, j-\tilde{j})(t)\|_2 + \|(h\partial_x^3\{\rho-\tilde{\rho}\}, h\partial_x^3\{j-\tilde{j}\}, \\ &\quad h^2\partial_x^4\{\rho-\tilde{\rho}\})(t)\| + \|(\phi-\tilde{\phi})(t)\|_4 \\ &\leq C(\|(\rho_0-\tilde{\rho}, j_0-\tilde{j})\|_2 + \|(h\partial_x^3\{\rho_0-\tilde{\rho}\},\ h\partial_x^3\{j_0-\tilde{j}\}, \\ &\quad h^2\partial_x^4\{\rho_0-\tilde{\rho}\})\|)\mathrm{e}^{-\alpha_1 t}, \end{aligned} \tag{5.2.41}$$

where C and α_1 are positive constants, independent of t and h.

Now, we start to study the singular limit of the solution ($h \to 0$) to the initial boundary value problem (5.2.1)–(5.2.3) and (5.2.6)–(5.2.9). Let (ρ^0, j^0, ϕ^0) be a solution to the hydrodynamic model (5.2.1)–(5.2.3) ($h = 0$). In order to avoid confusion, we write solutions to (5.2.1)–(5.2.3) and (5.2.6)–(5.2.9) with the suffix h as (ρ^h, j^h, ϕ^h). Then we have the hydrodynamic model satisfied by (ρ^0, j^0, ϕ^0)

$$\rho^0_t + j^0_x = 0, \tag{5.2.42}$$

$$j^0_t + \left(\frac{(j^0)^2}{\rho^0} + P(\rho^0)\right)_x = \rho^0\phi^0_x - j^0, \tag{5.2.43}$$

$$\phi^0_{xx} = \rho^0 - D, \tag{5.2.44}$$

the initial and the boundary data are prescribed by (5.2.6), (5.2.7) and (5.2.9). The stationary solution $(\tilde{\rho}^0, \tilde{j}^0, \tilde{\phi}^0)$ to (5.2.42)–(5.2.44) is independent of t, and solves

$$\tilde{j}^0_x = 0, \tag{5.2.45}$$

$$S[\tilde{\rho}^0, \tilde{j}^0]\tilde{\rho}^0_x = \tilde{\rho}^0\tilde{\phi}^0_x - \tilde{j}^0, \tag{5.2.46}$$

$$\tilde{\phi}^0_{xx} = \tilde{\rho}^0 - D, \tag{5.2.47}$$

with the boundary conditions (5.2.25) and (5.2.27).

Next, in the following lemma we give the existence of the stationary solution to the hydrodynamic model.

Lemma 5.2.2. *Let the doping profile and the boundary data satisfy conditions,* (5.2.5), (5.2.7) *and* (5.2.9). *For an arbitrary ρ_l, there exists a positive constant δ_3 such that, if $\delta \le \delta_3$, then the stationary problem* (5.2.25), (5.2.27) *and* (5.2.45)–(5.2.47) *has a unique solution $(\tilde{\rho}^0, \tilde{j}^0, \tilde{\phi}^0)$ satisfying the conditions* (5.2.12)–(5.2.13) *in the space $\mathcal{B}^2(\bar{\Omega})$. Moreover, the stationary solution satisfies the following estimates*

$$0 < c \le \tilde{\rho}^0 \le C, \quad |\tilde{j}^0|_0 \le C\delta, \quad |\tilde{\rho}^0|_2 + |\tilde{\phi}^0|_2 \le C, \tag{5.2.48}$$

where c and C are positive constants independent of ρ_r and ϕ_r.

The stability of the stationary solution $(\tilde{\rho}^0, \tilde{j}^0, \tilde{\phi}^0)$ is stated in the following lemma.

Lemma 5.2.3. *Let* $(\tilde{\rho}^0, \tilde{j}^0, \tilde{\phi}^0)$ *be the stationary solution to* (5.2.25), (5.2.27) *and* (5.2.45)–(5.2.47). *Suppose that the boundary data* ρ_l, ρ_r, ϕ_r *satisfy the conditions* (5.2.7) *and* (5.2.9). *Moreover, assume that the initial data* $(\rho_0, j_0) \in H^2(\Omega)$ *satisfies* (5.2.12)–(5.2.13) *and the compatibility conditions* $\rho_0(0) = \rho_l$, $\rho_0(1) = \rho_r$, $j_{0x}(0) = j_{0x}(1) = 0$. *Then there exists a positive constant* δ_4 *such that, if* $\delta + \|(\rho_0 - \tilde{\rho}^0, j_0 - \tilde{j}^0)\|_2 \leq \delta_4$, *the initial boundary value problem* (5.2.6), (5.2.7), (5.2.9) *and* (5.2.42)–(5.2.44) *has a unique solution* $(\rho^0, j^0, \phi^0) \in \bar{\mathfrak{X}}_2([0,\infty))$. *The solution* (ρ^0, j^0, ϕ^0) *satisfies the regularity* $(\phi - \tilde{\phi}) \in \bar{\mathfrak{X}}_2([0,\infty))$ *and the decay estimate*

$$\begin{aligned} &\|(\rho^0 - \tilde{\rho}^0, j^0 - \tilde{j}^0)(t)\|_2 + \|(\phi^0 - \tilde{\phi}^0)(t)\|_4 \\ &\quad \leq C\|(\rho_0 - \tilde{\rho}^0, j_0 - \tilde{j}^0)\|_2 e^{-\alpha_2 t}, \end{aligned} \tag{5.2.49}$$

where α_2 *and* C *are positive constants independent of* t.

In the above lemma, $\bar{\mathfrak{X}}_2$ and $\bar{\mathfrak{X}}_2^2$ are defined by

$$\bar{\mathfrak{X}}_2([0,T]) := \bigcap_{k=0}^{2} C^k([0,T]; H^{2-k}(\Omega)),$$

$$\bar{\mathfrak{X}}_2^2([0,T]) := \bigcap_{k=0}^{2} C^k([0,T]; H^{4-k}(\Omega)).$$

Naturally, as $h \to 0$, we expect that the solution to (5.2.1)–(5.2.3) converges to the solution to (5.2.42)–(5.2.44). Firstly, we consider the convergence of the stationary solutions, as $h \to 0$. We prove that the solution $(\tilde{\rho}^h, \tilde{j}^h, \tilde{\phi}^h)$ to (5.2.22)–(5.2.27) converges to the stationary solution $(\tilde{\rho}^0, \tilde{j}^0, \tilde{\phi}^0)$ to (5.2.25), (5.2.27) and (5.2.45)–(5.2.47).

Lemma 5.2.4. *Under the assumptions in Lemmas* 5.2.1 *and* 5.2.2, *let* $(\tilde{\rho}^0, \tilde{j}^0, \tilde{\phi}^0)$ *be the stationary solution to* (5.2.25), (5.2.27) *and* (5.2.45)–(5.2.47), *and* $(\tilde{\rho}^h, \tilde{j}^h, \tilde{\phi}^h)$ *be the stationary solution to* (5.2.22)–(5.2.27). *For an arbitrary* ρ_l, *there exists a positive constant* δ_5 *such that, if* $\delta + h \leq \delta_5$, *the solution* $(\tilde{\rho}^h, \tilde{j}^h, \tilde{\phi}^h)$ *to* (5.2.22)–(5.2.27) *converges to the stationary solution* $(\tilde{\rho}^0, \tilde{j}^0, \tilde{\phi}^0)$ *to* (5.2.25), (5.2.27) *and* (5.2.45)–(5.2.47). *More precisely,*

$$\|\tilde{\rho}^h - \tilde{\rho}^0\|_1 + |\tilde{j}^h - \tilde{j}^0| + \|\tilde{\phi}^h - \tilde{\phi}^0\|_3 \le Ch, \tag{5.2.50}$$

$$\|(\partial_x^2\{\tilde{\rho}^h - \tilde{\rho}^0\}, \partial_x^4\{\tilde{\phi}^h - \tilde{\phi}^0\}, h\partial_x^3\tilde{\rho}^h, h^2\partial_x^4\tilde{\rho}^h)\| \to 0, h \to 0, \tag{5.2.51}$$

where C is a positive constant independent of h.

The classical limit of the non-stationary problem is stated in the following theorem.

Theorem 5.2.2. *Under the assumptions of Theorem* 5.2.1 *and Lemma* 5.2.3, *there exists a positive constant δ_6 such that, if*

$$\begin{aligned} &\delta + h + \|(\rho_0 - \tilde{\rho}^0, j_0 - \tilde{j}^0)\|_2 + \|(\rho_0 - \tilde{\rho}^h, j_0 - \tilde{j}^h)\|_2 \\ &\quad + \|(h\partial_x^3\{\rho_0 - \tilde{\rho}^h\}, h\partial_x^3\{j_0 - \tilde{j}^h\}, h^2\partial_x^4\{\rho_0 - \tilde{\rho}^h\})\| \le \delta_6, \end{aligned} \tag{5.2.52}$$

then as $h \to 0$, the global solution (ρ^h, j^h, ϕ^h) to (5.2.1)–(5.2.3), (5.2.6)–(5.2.9) *converges to the solution (ρ^0, j^0, ϕ^0) to* (5.2.6), (5.2.7), (5.2.9) *and* (5.2.42)–(5.2.44), *precisely*

$$\|(\rho^h - \rho^0, j^h - j^0)(t)\|_1 + \|(\phi^h - \phi^0)(t)\|_3 \le \sqrt{h}Ce^{\beta t}, \quad t \in [0,\infty), \tag{5.2.53}$$

$$\sup_{t\in[0,\infty)} \{\|(\rho^h - \rho^0, j^h - j^0)(t)\|_1 + \|(\phi^h - \phi^0)(t)\|_3\} \to 0, \quad h \to 0, \tag{5.2.54}$$

where β and C is a positive constant independent of h and t.

For a non-negative integer $k \ge 0$, $\mathcal{B}^k(\bar{\Omega})$ denotes the space of the functions whose derivatives up to kth-order are continuous and bounded over $\bar{\Omega}$, equipped with the norm

$$|f|_k := \sum_{i=0}^{k} \sup_{x\in\bar{\Omega}} |\partial_x^i f(x)|.$$

5.2.1 *Existence and uniqueness of the stationary solution*

In this section, we consider the unique existence of stationary solution. Firstly, the existence of the stationary solution is obtained by the Leray–Schauder fixed-point theorem. Secondly, we obtain the estimates of the stationary solution. Finally, using the obtained estimates and the energy method, the uniqueness of the stationary solution is proved.

Existence

Obviously, if the density $\tilde{\omega}$ is positive, (5.2.22)–(5.2.24) is equivalent to (5.2.22), (5.2.32)–(5.2.34). Therefore, if the stationary problem (5.2.30), (5.2.32)–(5.2.34) with the current-voltage relationship (5.2.35) has a solution $(\tilde{\omega}, \tilde{j}, \tilde{\phi})$ ($\tilde{\omega} > 0$), then there exists a solution to (5.2.22)–(5.2.24), (5.2.25)–(5.2.27). In fact, substituting $x = 0$, $x = 1$ into (5.2.32), we see that $(\tilde{\omega}, \tilde{j}, \tilde{\phi})$ satisfies the boundary condition (5.2.26). (5.2.23) is obtained by differentiating (5.2.32) and multiplying the resultant equation by $\tilde{\omega}^2$. Moreover, (5.2.29) and the boundary condition (5.2.27) follow from (5.2.33).

The following constants will be frequently used in the sequel

$$B_0 := |D|_0 + \phi_r + \sqrt{K} + \frac{K}{2} + K|\log \frac{\rho_r}{\rho_l}|,$$

$$B_M := \max\{\omega_l, \omega_r\} \exp\left(\frac{B_0}{2K}\right),$$

$$B_m := \min\{\omega_l, \omega_r\} \exp\left(\frac{-1}{2K}\left\{B_M^2 + B_0 + \frac{K}{2}\right\}\right),$$

$$A(x) := \omega_l(1-x) + \omega_r x.$$

Lemma 5.2.5. *Let the doping profile and the boundary data satisfy* (5.2.5), (5.2.7) *and* (5.2.9). *Moreover, suppose that the following inequalities hold:*

$$B_M^{-4} + 2B_b(\rho_r^{-2} - \rho_l^{-2}) > 0, \tag{5.2.55}$$

$$S[B_m^2, \mathcal{J}[B_M^2]] > 0. \tag{5.2.56}$$

Then, the stationary problem (5.2.30), (5.2.32)–(5.2.34) *with the current-voltage relationship* (5.2.35) *has a solution* $(\tilde{\rho}, \tilde{j}, \tilde{\phi}) \in \mathcal{B}^4(\bar{\Omega}) \times \mathcal{B}^4(\bar{\Omega}) \times \mathcal{B}^2(\bar{\Omega})$ *satisfying* (5.2.12)–(5.2.13). *Furthermore, it holds that* $\tilde{j} \lesseqgtr 0$ *if and only if* $B_b \lesseqgtr 0$.

Proof. By (5.2.55)–(5.2.56), there exists a positive constant μ such that

$$(B_M + \mu)^{-4} + 2B_b(\rho_r^{-2} - \rho_l^{-2}) \geq 0, \tag{5.2.57}$$

$$S[(B_m - \mu)^2, \mathcal{J}[(B_M + \mu)^2]] > 0. \tag{5.2.58}$$

Now we define a mapping: $T : v \longmapsto V$ by solving the linear problem

$$\begin{aligned} &\varepsilon^2 V_{xx} = g(v_{\alpha,\beta}), \\ &g(v_{\alpha,\beta}) := v_{\alpha,\beta}\left(F(v_{\alpha,\beta}^2, \mathcal{J}[v_{\alpha,\beta}^2]) - F(\rho_l, \mathcal{J}[v_{\alpha,\beta}^2])\right. \\ &\qquad \left. -\mathcal{G}[v_{\alpha,\beta}^2] + \mathcal{J}[v_{\alpha,\beta}^2]\int_0^x \frac{1}{v_{\alpha,\beta}^2}(y)\mathrm{d}y\right) \\ &v_{\alpha,\beta} := \max\{\beta, \min\{\alpha, v\}\}, \quad \alpha := B_M + \mu, \quad \beta := B_m - \mu \end{aligned} \tag{5.2.59}$$

with the boundary condition (5.2.30), where F, $\mathcal{J}$ and $\mathcal{G}$ are given by (5.2.34), (5.2.37) and (5.2.33).

By (5.2.57), $\mathcal{J}[v_{\alpha,\beta}^2]$ satisfies the current-voltage relationship (5.2.35), where $(\tilde{\omega}^2, j)$ is replaced by $(v_{\alpha,\beta}^2, \mathcal{J}[v_{\alpha,\beta}^2])$. Apparently, the mapping T is well defined by the standard theory of the elliptic equations. In fact, owing to $v_{\alpha,\beta} \in H^1$, $g(v_{\alpha,\beta}) \in H^1$. Therefore, we have the solution $T(v) = V \in H^3$ to the problem (5.2.59) and (5.2.30). Moreover, the mapping T is a continuous and compact mapping from H^1 onto itself. Next, in order to employ Leray–Schauder fixed-point theorem, we show that for any $u \in \{f \in H^1;\ f = \lambda T(f),\ \lambda \in [0,1]\}$, there exists a positive constant M, such that $\|u\|_1 \leq M$. Since $\lambda = 0$ is trivial, we may assume the case $\lambda > 0$. Here it is sufficient to verify that $\|\tilde{\omega}\|_1 \leq M$ for $\tilde{\omega}$ satisfying an equation and a boundary condition

$$h^2\tilde{\omega}_{xx} = \lambda g(\tilde{\omega}_{\alpha,\beta}), \tag{5.2.60}$$

$$\tilde{\omega}(0) = \lambda\omega_l, \quad \tilde{\omega}(1) = \lambda\omega_r. \tag{5.2.61}$$

Multiplying (5.2.60) by $(\tilde{\omega} - \lambda A)$, integrating the resultant equality over the domain Ω and using the estimate $|g(\nu_{\alpha,\beta})| \leq C$, where C is a positive constant depending on $\alpha, \beta, \rho_l, \rho_r, \phi_r, |D|_0$, we have $\|\tilde{\omega}\|_1 \leq M$. Therefore, using Leray–Schauder fixed-point theorem, the mapping T has a fixed-point $\tilde{\omega} = T(\tilde{\omega}) \in H^3$, namely,

$$h^2\tilde{\omega}_{xx} = g(\tilde{\omega}_{\alpha,\beta}). \tag{5.2.62}$$

It is sufficient to show $\tilde{\omega} = \tilde{\omega}_{\alpha,\beta}$. By (5.2.58), $(\tilde{\omega}^2_{\alpha,\beta}, \mathcal{J}[\tilde{\omega}^2_{\alpha,\beta}])$ satisfies the subsonic condition $S[(\tilde{\omega}^2_{\alpha,\beta}, \mathcal{J}[\tilde{\omega}^2_{\alpha,\beta}])] > 0$. For the case $\rho_l \geq \rho_r$, add $-K\tilde{\omega}_{\alpha,\beta}\log\rho_l$ to (5.2.62), and multiply (5.2.62) by $(\log\tilde{\omega}^2_{\alpha,\beta} - \log\rho_l)^n_+$, $n = 1, 2, 3, \ldots$. For the case $\rho_l < \rho_r$, add $-K\tilde{\omega}_{\alpha,\beta}\log\rho_r$ to (5.2.62) and multiply (5.2.62) by $(\log\tilde{\omega}^2_{\alpha,\beta} - \log\rho_r)^n_+$, $n = 1, 2, 3, \ldots$. Here $(\cdot)_+ := \max\{0, \cdot\}$. Then, we consider $\rho_l < \rho_r$ since the case $\rho_l \geq \rho_r$ is easier to handle, the above computations imply that

$$\begin{aligned}
&-h^2\tilde{\omega}_{xx}\left(\log\frac{\tilde{\omega}^2_{\alpha,\beta}}{\rho_r}\right)^n_+ + K\tilde{\omega}_{\alpha,\beta}\left(\log\frac{\tilde{\omega}^2_{\alpha,\beta}}{\rho_r}\right)^{n+1}_+ \\
&= \left(\phi_r x - \int_0^1 GD\mathrm{d}\xi - \int_0^x \frac{\mathcal{J}[\tilde{\omega}^2_{\alpha,\beta}]}{\tilde{\omega}^2_{\alpha,\beta}}\mathrm{d}y + \frac{\mathcal{J}^2[\tilde{\omega}^2_{\alpha,\beta}]}{2\rho_l^2}\right. \\
&\quad \left. + K\log\frac{\rho_l}{\rho_r}\right)\tilde{\omega}_{\alpha,\beta}\left(\log\frac{\tilde{\omega}^2_{\alpha,\beta}}{\rho_r}\right)^n_+ \\
&\quad - \left(-\int_0^1 G\tilde{\omega}^2_{\alpha,\beta}\mathrm{d}\xi + \frac{\mathcal{J}^2[\tilde{\omega}^2_{\alpha,\beta}]}{2\tilde{\omega}^4_{\alpha,\beta}}\right)\tilde{\omega}_{\alpha,\beta}\left(\log\frac{\tilde{\omega}^2_{\alpha,\beta}}{\rho_r}\right)^n_+ \\
&\leq B_0\tilde{\omega}_{\alpha,\beta}\left(\log\frac{\tilde{\omega}^2_{\alpha,\beta}}{\rho_r}\right)^n_+,
\end{aligned} \tag{5.2.63}$$

where we have used the subsonic condition (5.2.12) and $|G| \leq 1$ for the first term of the right hand side of the equality, and the fact that G is non-positive for the second term. The term $K\log\frac{\rho_l}{\rho_r}$ for the case

$\rho_l \geq \rho_r$ vanishes, the first term of (5.2.63) is rewritten as

$$(1st\ term) = 2h^2 n \frac{(\tilde{\omega}_{\alpha,\beta x})^2}{\tilde{\omega}^2_{\alpha,\beta}} [2(2n-1)] \left(\log \frac{\tilde{\omega}^2_{\alpha,\beta}}{\rho_r} \right)^{n-1}_+ - \left[h^2 \tilde{\omega}_x \left(\log \frac{\tilde{\omega}^2_{\alpha,\beta}}{\rho_r} \right)^n_+ \right]_x . \tag{5.2.64}$$

By Young's inequality, the last term of (5.2.63) is estimated as

$$(last\ term) \leq \frac{n}{n+1} K \tilde{\omega}_{\alpha,\beta} \left(\log \frac{\tilde{\omega}^2_{\alpha,\beta}}{\rho_r} \right)^{n+1}_+ + \frac{|K\tilde{\omega}_{\alpha,\beta}|_0}{n+1} \left(\frac{B_0}{K} \right)^{n+1} . \tag{5.2.65}$$

Noting that the first term of (5.2.63) is non-negative, since $(\log \tilde{\omega}^2_{\alpha,\beta} - \log \rho_r) + (0) = (\log \tilde{\omega}^2_{\alpha,\beta} - \log \rho_r) + (1)$, the second term vanishes. Substituting (5.2.64) and (5.2.65) into (5.2.63) and integrating the result, we have

$$\int_0^1 K\sqrt{\rho_r} \left(\log \frac{\tilde{\omega}^2_{\alpha,\beta}}{\rho_r} \right)^{n+1}_+ dx \leq |K\tilde{\omega}_{\alpha,\beta}|_0 \left(\frac{B_0}{K} \right)^{n+1}, \tag{5.2.66}$$

where we have used $\tilde{\omega}_{\alpha,\beta} \geq \sqrt{\rho_r}$. Taking the $(n+1)$th root of (5.2.66) yields

$$\left(\int_0^1 \left(\log \frac{\tilde{\omega}^2_{\alpha,\beta}}{\rho_r} \right)^{n+1}_+ dx \right)^{1/(n+1)} \leq \left(\frac{|\tilde{\omega}_{\alpha,\beta}|_0}{\sqrt{\rho_r}} \right)^{1/(n+1)} \frac{B_0}{K}. \tag{5.2.67}$$

Letting $n \to \infty$ in (5.2.67), we have $\tilde{\omega}^2_{\alpha,\beta} \leq B_M^2$.

We show the lower bound of $\tilde{\omega}^2_{\alpha,\beta}$. For the case $\rho_l \geq \rho_r$, add $-K\tilde{\omega}_{\alpha,\beta} \log \rho_r$ to (5.2.62) and multiply (5.2.62) by $(\log \tilde{\omega}^2_{\alpha,\beta} - \log \rho_r)^{2n-1}_-$, $n = 1, 2, 3, \ldots$. For the case $\rho_l < \rho_r$, add $-K\tilde{\omega}_{\alpha,\beta} \log \rho_l$ to (5.2.62) and multiply (5.2.62) by $(\log \tilde{\omega}^2_{\alpha,\beta} - \log \rho_l)^{2n-1}_-$, $n = 1, 2, 3, \ldots$. Here $(\cdot)_- := \min\{0, \cdot\}$, we treat the former case since the latter case is easier. The above computations yield that

$$
\begin{aligned}
&-h^2\frac{\tilde{\omega}_{xx}}{\tilde{\omega}_{\alpha,\beta}}\left(\log\frac{\tilde{\omega}^2_{\alpha,\beta}}{\rho_r}\right)_-^{2n-1}+K\left(\log\frac{\tilde{\omega}^2_{\alpha,\beta}}{\rho_r}\right)_-^{2n}\\
&=\left(\mathcal{G}[\tilde{\omega}^2_{\alpha,\beta}]-\int_0^x\frac{\mathcal{J}[\tilde{\omega}^2_{\alpha,\beta}]}{2\tilde{\omega}^2_{\alpha,\beta}}\mathrm{d}y+\frac{\mathcal{J}[\tilde{\omega}^2_{\alpha,\beta}]}{2\rho_l^2}-\frac{\mathcal{J}^2[\tilde{\omega}^2_{\alpha,\beta}]}{2\tilde{\omega}^4_{\alpha,\beta}}\right.\\
&\quad\left.+K\log\frac{\rho_l}{\rho_r}\right)\left(\log\frac{\tilde{\omega}^2_{\alpha,\beta}}{\rho_r}\right)_-^{2n-1}\\
&\le\frac{2n-1}{2n}K\left(\log\frac{\tilde{\omega}^2_{\alpha,\beta}}{\rho_r}\right)_-^{2n}+\frac{K}{2n}\left(\frac{B_M^2}{K}+\frac{B_0}{K}+\frac{1}{2}\right)^{2n}.
\end{aligned}
\tag{5.2.68}
$$

We have used the subsonic condition (5.2.12), $|G|\le 1$ and Young's inequality, the first term in (5.2.68) is rewritten as

$$
\begin{aligned}
(1st\ term)=h^2\frac{(\tilde{\omega}_{\alpha,\beta_x})^2}{\tilde{\omega}^2_{\alpha,\beta}}&\left[2(2n-1)\left(\log-\frac{\tilde{\omega}^2_{\alpha,\beta}}{\rho_r}\right)_-^{2n-2}\right.\\
&\left.-\left(\log\frac{\tilde{\omega}^2_{\alpha,\beta}}{\rho_r}\right)_-^{2n-1}\right]-\left[h^2\frac{\tilde{\omega}_x}{\tilde{\omega}_{\alpha,\beta}}\left(\log\frac{\tilde{\omega}^2_{\alpha,\beta}}{\rho_r}\right)_-^{2n-1}\right]_x.
\end{aligned}
\tag{5.2.69}
$$

The first term in (5.2.69) is non-negative, since $(\log\tilde{\omega}^2_{\alpha,\beta}-\log\rho_r)_+(0)=(\log\tilde{\omega}^2_{\alpha,\beta}-\log\rho_r)_+(1)$, the last term in (5.2.69) disappears after integration. Substituting (5.2.69) into (5.2.68), integrating over Ω and taking $(2n)$th root yield

$$
\left[\int_0^1\left(\log\frac{\tilde{\omega}^2_{\alpha,\beta}}{\rho_r}\right)_-^{2n}\mathrm{d}x\right]^{1/2n}\le\frac{1}{K}\left(B_M^2+B_0+\frac{K}{2}\right).
\tag{5.2.70}
$$

Letting $n\to\infty$ in (5.2.70), we have $B_m^2\le\tilde{\omega}^2_{\alpha,\beta}$.

Consequently, we have $B_m\le\tilde{\omega}_{\alpha,\beta}\le B_M$, which implies $\tilde{\omega}=\tilde{\omega}_{\alpha,\beta}$. Thus, $(\tilde{\omega},\mathcal{J}[\tilde{\omega}^2],\mathcal{G}[\tilde{\omega}^2])$ is a solution to the problem (5.2.30), (5.2.32)–(5.2.34) and (5.2.80). Differentiating (5.2.32) and using

$\tilde{\omega} \in H^3$ allow us to obtain the regularity of the stationary solution. Moreover, we deduce from (5.2.37) that $\mathcal{J}[\tilde{\omega}^2] \lesseqqgtr 0$ holds if and only if $B_b \lesseqqgtr 0$. □

Uniqueness

Lemma 5.2.5 shows the existence of the stationary solution. In order to show the uniqueness, an additional assumption is needed (see Lemma 5.2.7). We firstly prove estimates for the stationary solution (see Lemma 5.2.6). The following inequalities are frequently used in the proof of Lemma 5.2.6:

$$|f|_0^2 \leq \|f\|^2 + 2\|f\|\,\|f_x\| \quad f \in H^1(\Omega), \tag{5.2.71}$$

$$\|f\|^2 \leq \frac{1}{4}\|f_x\|^2 \quad f \in H_0^1\Omega, \tag{5.2.72}$$

$$\|f_x\|^2 \leq \frac{1}{2}\|f_{xx}\|^2 \quad f \in \{f \in H^2(\Omega);\ f(0) = f(1)\}. \tag{5.2.73}$$

Lemma 5.2.6. *Let* $(\tilde{\omega}, \tilde{j}, \tilde{\phi}) \in \mathcal{B}^4(\bar{\Omega}) \times \mathcal{B}^4(\bar{\Omega}) \times \mathcal{B}^2(\bar{\Omega})$ *be a stationary solution to* (5.2.22), (5.2.27), (5.2.28)–(5.2.31) *satisfying the condition* (5.2.12)–(5.2.13). *Assume that the conditions in* (5.2.55)–(5.2.56) *and the inequality*

$$\sqrt{K} < |2B_b\mathcal{J}[B_M^2]^{-1}(\rho_r^{-2} - \rho_l^{-2})^{-1}B_M^{-2}| \tag{5.2.74}$$

hold. Then the solution $(\tilde{\omega}, \tilde{j}, \tilde{\phi})$ *solves* (5.2.37) *and verifies*

$$B_m \leq \tilde{\omega} \leq B_M, \tag{5.2.75}$$

$$|\tilde{\phi}|_2 \leq C, \tag{5.2.76}$$

$$\|\tilde{\omega}\|_2 \leq C, \quad \|\partial_x^3\tilde{\omega}\| \leq C\varepsilon^{-1} + C, \quad \|\partial_x^4\tilde{\omega}\| \leq C\varepsilon^{-2} + C, \tag{5.2.77}$$

where C *is a positive constant depending only on* ρ_l, ρ_r, ϕ_r, $|D|_0$ *but independent of* ε.

Proof. The estimate (5.2.75) is established similarly as the derivation of $\tilde{\omega}_{\alpha,\beta} = \tilde{\omega}$ in Lemma 5.2.5. Using (5.2.75), the estimate (5.2.76) follows from (5.2.33). By solving the current-voltage relationship (5.2.35) with respect to $\tilde{j}$, we see that $\tilde{j}$ is given by (5.2.37), owing to (5.2.74). The other solution of the quadratic equation (5.2.35) violates the subsonic condition (5.2.12).

It is sufficient to prove (5.2.77). Multiply (5.2.28) by $\tilde{\omega}_x/\tilde{\omega}^2$, and integrate the resultant equality over Ω by parts with the boundary conditions (5.2.30), (5.2.31) and (5.2.29), we have

$$\int_0^1 \left(2S[\tilde{\omega}^2,\tilde{j}]\frac{\tilde{\omega}_x^2}{\tilde{\omega}} + h^2\frac{\tilde{\omega}_{xx}^2}{\tilde{\omega}}\right)\mathrm{d}x$$
$$= \int_0^1 [-(\tilde{\omega}^2 - D)(\tilde{\omega} - A) + \tilde{\phi}_x A_x]\mathrm{d}x + \tilde{j}\left(\frac{1}{\tilde{\omega}_r} - \frac{1}{\tilde{\omega}_l}\right) \leq C, \tag{5.2.78}$$

where C is a positive constant depending only on ρ_l, ρ_r, ϕ_r, $|D|_0$ but independent of ε. Here, we have used the estimates (5.2.75) and (5.2.76). Since the left hand side of (5.2.78) is estimated by $2S[B_m^2, \mathcal{J}[B_M^2]]\|\tilde{\omega}_x\|^2/B_M$ below, we have $\|\tilde{\omega}_x\| \leqslant C$.

Multiply (5.2.28) by $(\tilde{\omega}_{xx}/\tilde{\omega})_x/\tilde{\omega}^2$, integrate the resultant equality over Ω, apply the integration by parts and then use (5.2.31) and (5.2.29), we have

$$\int_0^1 \left[h^2\left(\frac{\tilde{\omega}_{xx}}{\tilde{\omega}}\right)_x^2 + S[\tilde{\omega}^2,\tilde{j}]\left(\frac{\tilde{\omega}_{xx}}{\tilde{\omega}}\right)^2\right]\mathrm{d}x$$
$$= \int_0^1 -2\left(S[\tilde{\omega}^2,\tilde{j}]\frac{1}{\tilde{\omega}}\right)_x\frac{\tilde{\omega}_x\tilde{\omega}_{xx}}{\tilde{\omega}}\mathrm{d}x$$
$$+ \int_0^1 \left[(\tilde{\omega}^2 - D) - \left(\frac{\tilde{j}}{\tilde{\omega}^2}\right)_x\right]\frac{\tilde{\omega}_{xx}}{\tilde{\omega}}\mathrm{d}x. \tag{5.2.79}$$

By Hölder, Schwarz inequalities and (5.2.71), the first term of the right hand side of (5.2.79) is estimated as

$$\begin{aligned}|(1st\ term)| &\leq C|\tilde{\omega}_x|_0\|\tilde{\omega}_x\|\,\|\tilde{\omega}_{xx}\|\\ &\leq C\sqrt{\|\tilde{\omega}_x\|^2 + 2\|\tilde{\omega}_x\|\,\|\tilde{\omega}_{xx}\|}\|\tilde{\omega}_x\|\,\|\tilde{\omega}_{xx}\|\\ &\leq C(1 + \|\tilde{\omega}_{xx}\|) + \frac{S[B_m^2, \mathcal{J}[B_M^2]]}{B_M^2}\|\tilde{\omega}_{xx}\|^2,\end{aligned}$$

by using (5.2.72), (5.2.75) and Hölder inequality, and we have $|(2nd\ term)| \leq C(1 + \|\tilde{\omega}_{xx}\|)$. Noting that the left hand side of

(5.2.79) is estimated by $2S[B_m^2, \mathcal{J}[B_M^2]]\|\tilde{\omega}_{xx}\|^2/B_M^2$ below. Substituting these estimates in (5.2.79), we have $\|\tilde{\omega}_{xx}\| \leq C$. Hence, the first inequality in (5.2.77) is proved.

Then we prove the second inequality in (5.2.77). Substituting these three inequalities in (5.2.79) yields the estimate $h^2\|(\tilde{\omega}_{xx}/\tilde{\omega})_x\|^2 \leq C$. Using estimates (5.2.71) and (5.2.75) yields $\|(\tilde{\omega}_{xx}/\tilde{\omega})_x\| \geq \|\tilde{\omega}_{xxx}\|\backslash B_M - C$. Owing to these two inequalities, we have $\|\tilde{\omega}_{xxx}\| \leq C + C/h$. Furthermore, differentiating (5.2.28) and multiplying the resultant equality by $1/\tilde{\omega}$, we have

$$h^2\tilde{\omega}_{xxxx} = h^2\frac{\tilde{\omega}_{xx}^2}{\tilde{\omega}} + \frac{2}{\tilde{\omega}}(S[\tilde{\omega}^2, \tilde{j}]\tilde{\omega}\tilde{\omega}_x)_x - 2\tilde{\omega}_x\tilde{\phi}_x - \tilde{\omega}\tilde{\phi}_{xx}. \tag{5.2.80}$$

Estimating the right hand side of (5.2.80) with using (5.2.71) and (5.2.75), we have $\|\tilde{\omega}_{xxxx}\| \leq C + C/h^2$, which completes the proof of the third estimate in (5.2.77). □

Obviously, the next corollary follows from the proof of Lemma 5.2.6.

Corollary 5.2.1. *Under the assumptions of Lemma* 5.2.6, *for an arbitrary* ρ_l, *there exists a positive constant* δ_1 *such that, if* $\delta + h \leqslant \delta_1$, *then the stationary solution satisfies the estimates* (5.2.76) *and* (5.2.77), *where* C *depends only on* ρ_l, $|D|_0$ *but independent of* δ *and* ε.

Next, we start to prove the uniqueness of the stationary solution. Let $\bar{\omega} := \log\tilde{\rho} = \log\tilde{\omega}^2$, and rewrite (5.2.22)–(5.2.27) as

$$\tilde{j}_x = 0, \tag{5.2.81}$$

$$S[e^{\bar{\omega}}, \tilde{j}]\bar{\omega}_x - \tilde{\phi}_x - \frac{h^2}{2}\left(\bar{\omega}_{xx} + \frac{\bar{\omega}_x^2}{2}\right)_x = -\frac{\tilde{j}}{e^{\bar{\omega}}}, \tag{5.2.82}$$

$$\tilde{\phi}_{xx} = e^{\bar{\omega}} - D, \tag{5.2.83}$$

$$\bar{\omega}(0) = \log\, \rho l,\ \bar{\omega}(1) = \log\, \rho_r, \tag{5.2.84}$$

$$\left(\bar{\omega}_{xx} + \frac{\bar{\omega}_x^2}{2}\right)(0) = \left(\bar{\omega}_{xx} + \frac{\bar{\omega}_x^2}{2}\right)(1) = 0, \tag{5.2.85}$$

$$\tilde{\phi}(0) = 0,\ \tilde{\phi}(1) = \phi_r > 0. \tag{5.2.86}$$

Noting that if the uniqueness of the stationary solution to (5.2.81)–(5.2.86) with $\tilde{\rho} > 0$ is proved, the uniqueness of that to (5.2.22)–(5.2.27) immediately follows.

Lemma 5.2.7. *Under the assumptions of Lemma 5.2.6, for an arbitrary* ρ_l, *there exists a positive constant* δ_1, *such that if* $\delta+h \leqslant \delta_1$. *There exists a unique solution* $(\tilde{\omega}, \tilde{j}, \tilde{\phi}) \in \mathcal{B}^4(\overline{\Omega}) \times \mathcal{B}^4(\overline{\Omega}) \times \mathcal{B}^2(\overline{\Omega})$, *satisfying* (5.2.12)–(5.2.13).

Proof. *Owing to Lemma* 5.2.6, $\tilde{j}$ is written by the explicit formula (5.2.37), i.e. $\tilde{j} = \mathcal{J}[e^{\tilde{\omega}}]$. Let $(\tilde{\omega}_1, \tilde{j}_1, \tilde{\phi}_1)$, $(\tilde{\omega}_2, \tilde{j}_2, \tilde{\phi}_2)$ be solutions to the stationary problem (5.2.81)–(5.2.86). Denoting $\tilde{j}_1 = \mathcal{J}[e^{\tilde{\omega}_1}]$, $\tilde{j}_2 = \mathcal{J}[e^{\tilde{\omega}_2}]$, and then using the mean value theorem and (5.2.75), we have

$$|\tilde{j}_1 - \tilde{j}_2| \leqslant C\delta||\tilde{\omega}_1 - \tilde{\omega}_2||, \tag{5.2.87}$$

where C is a positive constant independent of δ and h. Owing to (5.2.82), the difference $\tilde{\omega} := \tilde{\omega}_1 - \tilde{\omega}_2$ satisfies

$$\begin{aligned} &-\frac{h^2}{2}\left(\tilde{\omega}_{xx} + \frac{\tilde{\omega}_{1x}^2}{2} - \frac{\tilde{\omega}_{2x}^2}{2}\right)_x + S[e^{\tilde{\omega}_1}, \tilde{j}_1]\tilde{\omega}_x - (\phi_1 - \phi_2)_x \\ &= \frac{\tilde{j}_1^2}{e^{2\tilde{\omega}_1}} - \frac{\tilde{j}_2^2}{e^{2\tilde{\omega}_2}}\tilde{\omega}_{2x} - \left(\frac{\tilde{j}_1}{e^{\tilde{\omega}_1}} - \frac{\tilde{j}_2}{e^{\tilde{\omega}_2}}\right). \end{aligned} \tag{5.2.88}$$

Multiplying (5.2.88) by $\tilde{\omega}_x$, integrating the resultant equality and using the boundary conditions (5.2.85), (5.2.86) and (5.2.83), we obtain

$$\begin{aligned} &\int_0^1 \left[\frac{h^2}{2}\tilde{\omega}_{xx}^2 + S[e^{\tilde{\omega}_1}, \tilde{j}_1]\right]\tilde{\omega}_x^2 + (e^{\tilde{\omega}_1} - e^{\tilde{\omega}_2})\tilde{\omega}]dx \\ &= \int_0^1 \left[\left(\frac{\tilde{j}_1}{e^{\tilde{\omega}_1}} + \frac{\tilde{j}_2}{e^{\tilde{\omega}_2}}\right)\tilde{\omega}_{2x} - 1\right]\left(\frac{\tilde{j}_1}{e^{\tilde{\omega}_1}} - \frac{\tilde{j}_2}{e^{\tilde{\omega}_2}}\right)\tilde{\omega}_x \mathrm{d}x \\ &\quad - \int_0^1 \frac{h^2}{4}(\tilde{\omega}_1 + \tilde{\omega}_2)_x\tilde{\omega}_x\tilde{\omega}_{xx}\mathrm{d}x. \end{aligned} \tag{5.2.89}$$

Using estimates (5.2.71), (5.2.72), (5.2.75), and (5.2.87), we handle the first term of the right hand side of (5.2.89)

$$\begin{aligned}|(1st\ term)| &\leqslant C(\|\tilde{j}_1(e^{-\tilde{\omega}_1} - e^{-\tilde{\omega}_2})\| + \|e^{-\tilde{\omega}_2}(\tilde{j}_1 - \tilde{j}_2)\|)\|\tilde{\omega}_x\| \\ &\leqslant C\delta\|\tilde{\omega}_x\|^2. \end{aligned} \tag{5.2.90}$$

The second term of the right hand side of (5.2.89) is estimated by using Hölder and the Schwarz inequalities as

$$|(2nd)| \leqslant h^2 C\|\tilde{\omega}_x\|\,\|\tilde{\omega}_{xx}\| \leqslant \frac{h^2}{2}\|\tilde{\omega}_{xx}\|^2 + Ch^2\|\tilde{\omega}_x\|^2, \tag{5.2.91}$$

where we have used (5.2.71) and Corollary 5.2.1. Substituting the equalities (5.2.90) and (5.2.91) in (5.2.89) by virtue of $(e^{\tilde{\omega}_1} - e^{\tilde{\omega}_2})\tilde{\omega} \geqslant 0$ and $S[e^{\tilde{\omega}_1}, \tilde{j}_1] \geqslant S[B_m^2, \mathcal{J}[B_M^2]] > 0$, let δ and h be small enough that $\|\tilde{\omega}_x\|^2 \leqslant 0$, thus $\tilde{\omega}_1 = \tilde{\omega}_2$. The equalities $\tilde{j}_1 = \tilde{j}_2$ and $\tilde{\phi}_1 = \tilde{\phi}_2$ follow from (5.2.83), (5.2.86) and (5.2.87). The proof is completed.

Consequently, since the smallness of $(\delta + h)$ implies that all the assumptions in Lemmas 5.2.5 and 5.2.7 hold, Lemmas 5.2.5 and 5.2.7 imply Lemma 5.2.1.

5.2.2 *Asymptotic stability of the stationary solution*

Unique existence of a solution locally in time

Now, we show the unique existence of the solution locally in time to the initial boundary value problem (5.2.15)–(5.2.21). Then we can prove the unique local existence of the solution with $(\rho > 0)$ to its equivalent problem (5.2.1)–(5.2.3) and (5.2.6)–(5.2.9).

Lemma 5.2.8. *Suppose that the initial data* $(\omega_0, j_0) \in H^4(\Omega) \times H^3(\Omega)$, *and the boundary data* ρ_l, ρ_r, ϕ_r *satisfy* (5.2.7), (5.2.9) *and* $\omega_0 > 0$. *Then there exists a constant* $T_1 > 0$, *such that the initial boundary value problem* (5.2.15)–(5.2.21) *has a weak solution* $(\omega, j, \phi) \in \bar{\mathfrak{X}}_4([0, T_1]) \times \bar{\mathfrak{X}}_3([0, T_1]) \times \mathfrak{Y}([0, T_1])$ *satisfying* $\omega > 0$.

The next corollary follows from Lemma 5.2.8.

Corollary 5.2.2. *Suppose the initial data* $(\rho_0, j_0) \in H^4(\Omega) \times H^3(\Omega)$, *and the boundary data* ρ_l, ρ_r, ϕ_r *satisfy* (5.2.7), (5.2.9), (5.2.10) *and* (5.2.14). *Then there exists a constant* $T_2 > 0$, *such that the initial boundary value problem* (5.2.1)–(5.2.3) *and* (5.2.6)–(5.2.9) *has a unique solution* $(\rho, j, \phi) \in \bar{\mathfrak{X}}_4([0,T_2]) \times \bar{\mathfrak{X}}_3([0,T_2]) \times \mathfrak{Y}([0,T_2])$ *satisfying* (5.2.12)–(5.2.13).

In order to solve the problem (5.2.15)–(5.2.21), we define the successive approximate sequence. For the unknown $(\hat{\omega}, \hat{j})$, we consider the following linearized system

$$2\omega\hat{\omega}_t + \hat{j}_x = 0, \tag{5.2.92}$$

$$\hat{j}_t + 2S[\omega^2, j]_{\omega\hat{\omega}_x} + 2\frac{j}{\omega^2}\hat{j}_x - h^2\omega^2\left(\frac{\hat{\omega}_{xx}}{\omega}\right)_x$$
$$= \omega^2\phi_x - j \tag{5.2.93}$$

with the initial data (5.2.18) and the boundary data (5.2.19) and (5.2.20), where the function ϕ is defined by (5.2.11), i.e. $\phi = \Phi[\omega^2]$. Let the functions (ω, j) in the coefficients in (5.2.92)–(5.2.93) satisfy

$$(\omega, j) \in \bar{\mathfrak{X}}_4([0,T]) \times \bar{\mathfrak{X}}_3([0,T]), (\omega, j)(0,x) = (\omega_0, j_0), \tag{5.2.94}$$

$$\omega(t,x) \geqslant m, (t,x) \in [0,T] \times \Omega, \tag{5.2.95}$$

$$||\omega(t)||_4 + ||\omega_t(t)||_2 + ||\omega_{tt}(t)|| + ||j(t)||_3 + ||j_t(t)||_1$$
$$\leqslant M, t \in [0,T], \tag{5.2.96}$$

where T, m, M are positive constants. We denote by $X(T; m, M)$ the set of functions (w, j) satisfying (5.2.94)–(5.2.96) and abbreviate $X(T; m, M)$ by $X(\cdot).\phi$ satisfying

$$\phi \in \mathfrak{Y}([0,T]), ||\partial_t^i\phi(t)||_2 \leqslant M, \quad i = 0, 1, 2, \quad t \in [0,T].$$

Next, we give the proof of the existence of the solution to the linearized problem (5.2.11), (5.2.18)–(5.2.20) and (5.2.92)–(5.2.93) and we consider the general scalar equation

$$u_{tt} + L_1 u_t + b_1 u_t + b_2 u_x + b_3 u_{xx} + L_2 u = f, \tag{5.2.97}$$

$$L_1 := b\partial_3, \ b \in \mathcal{B}^1([0,1] \times [0,T]), \ L_2 := a\partial_x^4, \ a > 0, \tag{5.2.98}$$

$$b_1, b_2, b_3 \in \mathcal{B}^0([0,1] \times [0,T]) \cap C^1([0,T]; L^2),$$
$$f \in C^1([0,T]; L^2), \qquad (5.2.99)$$

with the initial and the boundary data

$$u(0,x) = u_1(x) \in \mathcal{H}, \quad u_t(0,x) = u_2(x) \in H_0^1 \cap H^2, \quad (5.2.100)$$
$$u(t,0) = u(t,1) = 0, \quad u_{xx}(t,0) = u_{xx}(t,1) = 0, \qquad (5.2.101)$$

where $\mathcal{H} := \{g \in H_0^1 \cap H^4;\ g_{xx}(0) = g_{xx}(1) = 0\}$.

Lemma 5.2.9. *The initial boundary value problem* (5.2.97)–(5.2.101) *has a unique solution* $u \in \bar{\mathfrak{X}}_4([0,T])$.

Proof of Lemma 5.2.9. We use Galerkin method to prove Lemma 5.2.9. First, we consider the problem (5.2.97)–(5.2.101) with the initial data $u(0,x) = u_t(0,x) = 0$. Define the sequence $\{\nu_l(x) := \sqrt{2}\sin l\pi x\}_{l=1}^{\infty}$, which is a complete orthonormal system in L^2, to construct approximate sequence $\{\sum_{l=1}^{n} a_l^n(t)\,\nu_l(x)\}_{n=1}^{\infty}$, where $a_l^n(t)$ solves the ordinary differential equation

$$(u_{tt}^n, \nu_l) + (L_1 u_t^n, \nu_l) + (b_1 u_t^n, \nu_l)$$
$$+ (b_2 u_x^n + b_3 u_{xx}^n, \nu_l) + (L_2 u^n, \nu_l) = (f, \nu_l), \qquad (5.2.102)$$
$$a_l^n(0) = a_{lt}^n(0) = 0, \qquad (5.2.103)$$

for $l = 1, 2, \ldots, n$, where $(\cdot,\cdot)$ denotes a standard L^2-inner product. Owing to the standard theory of the ordinary differential equation, (5.2.102)–(5.2.103) has a unique solution $a_l^n \in \mathcal{B}([0,T])$. Therefore, multiply $u^n \in C^3([0,T]; \mathcal{H})$, (5.2.102) by $a_l^n(t)$, for $l = 1, 2, \ldots, n$, and sum up the resultant equalities to obtain that

$$(u_{tt}^n, u_t^n) + (b u_{xt}^n, u_t^n) + (b_1 u_t^n, u_t^n)$$
$$+ (b_2 u_x^n + b_3 u_{xx}^n, u_t^n + a(u_{xxxx}^n, u_t^n) = (f, u_t^n). \quad (5.2.104)$$

Integrating (5.2.104) by parts, using the boundary condition $\nu_l(0) = \nu_l(1) = \nu_{lxx}(0) = \nu_{lxx}(1) = 0$, (5.2.72), (5.2.73) and Schwarz inequality yield that

$$\frac{d}{dt}(||u_t^n(t)||^2 + ||u_{xx}^n(t)||^2) \leqslant C||(u_t^n, u_{xx}^n, f)(t)||^2. \quad (5.2.105)$$

Differentiating with respect to t, multiplying (5.2.102) by a_{ltt}^n and summing up the resultant equalities for $l = 1, 2, \ldots, n$, we have

$$\begin{aligned}&(u_{ttt}^n, u_{tt}^n) + (bu_{xtt}^n, u_{tt}^n) + (b_1 u_{tt}^n, u_{tt}^n) + (b_2 u_{xt}^n + b_3 u_{xxt}^n, u_{tt}^n \\ &+ a(u_{xxxxt}^n, u_{tt}^n) + (b_t u_{xt}^n, u_{tt}^n) + (b_{1t} u_t^n, u_{tt}^n) \\ &+(b_{2t} u_x^n + b_{3t} u_{xx}^n, u_{tt}^n) = (f_t, u_{tt}^n). \end{aligned} \quad (5.2.106)$$

By virtue of (5.2.71), then the same argument as the derivation of (5.2.105) yields

$$\frac{d}{dt}(||u_{tt}^n(t)||^2 + ||u_{xxt}^n(t)||^2) \leqslant C||(u_{tt}^n, u_{xx}^n, u_{xxt}^n, u_{xxxx}^n, f_t)(t)||^2. \quad (5.2.107)$$

Multiplying (5.2.102) by $(l\pi)^4 a_l^n, l = 1, 2, \ldots, n$, corresponding to $a_l^n \partial_x^4$ summing up the resultants for $l = 1, 2, \ldots, n$, and applying the Schwarz inequality, we have

$$||u_{xxxx}^n(t)||^2 \leqslant C||(u_{tt}^n, u_{xxt}^n, u_{xx}^n)(t)||^2, \quad (5.2.108)$$

where we have used (5.2.72) and (5.2.73), on the other hand, since $\{\nu_l\}_{l=1}^{\infty}$ is a complete orthonormal system in L^2, substituting $t = 0$ in (5.2.97) and using Bessel inequality yield that

$$||u_{tt}^n(0)|| \leqslant ||f(0)||. \quad (5.2.109)$$

Add (5.2.105) to (5.2.107) and use (5.2.108). Moreover, use Gronwall inequality with $u^n(t) = 0$, $u_t^n(t) = 0$, and substitute (5.2.109) to obtain

$$||(u_t^n, u_{xx}^n, u_{tt}^n, u_{xxt}^n, u_{xxxx}^n)(t)|| \leqslant C, \quad (5.2.110)$$

where C is a constant depending on T but independent of t. Consequently, the sequence $\{(u_t^n, u_{xx}^n, u_{tt}^n, u_{xxt}^n)\}_{n=1}^{\infty}$ is bounded in L^2. The inequalities (5.2.72), (5.2.73) and (5.2.110) show that $\{u^n\}_{n=1}^{\infty}$ is bounded in $C([0,T];\mathcal{H}) \cap C^1([0,T];\ H_0^1 \cap H^2) \cap C^2([0,T];\ L^2)$. Thus, since $\mathcal{H}$ and $H_0^1 \cap H^2$ are the Hilbert spaces, then there exists a function u and a subsequence $\{u^n\}_{n=1}^{\infty}$ (not relabeled), as $n \to \infty$, such that

$$
\begin{aligned}
& u^n \to u \text{ strongly in } C([0,T], H_0^1 \cap H^2) \cap C^1([0,T]; L^2), \\
& u^n \rightharpoonup u \text{ weakly}^* \text{ in } L^\infty(0,T;\mathcal{H}), \\
& u_t^n \rightharpoonup u_t \text{ weakly}^* \text{ in } L^\infty(0,T;H_0^1 \cap H^2), \\
& u_{lt}^n \rightharpoonup u_{lt} \text{ weakly}^* \text{ in } L^\infty(0,T;L^2).
\end{aligned}
$$

Passing to the limit in (5.2.102), u solves the problem (5.2.97) and (5.2.101) with the initial data $u(0,x) = u(1,x) = 0$ in the sense of distribution. In fact, since $u \in C([0,T];\, H_0^1 \cap H^2) \cap L^\infty(0,T;\mathcal{H})$, the boundary condition (5.2.101) holds. By the standard theory, we see that $(u_{tt}, u_{xxt}, u_{xxxx})(t)$ is continuous in L^2 at $t = 0$. Mollifying with respect to t shows that $u \in \bar{\mathfrak{X}}_4([0,T])$. Consequently, the proof of Lemma 5.2.9 is completed for the initial data $u(0,x) = u(1,x) = 0$.

Finally, we consider the initial boundary value problem (5.2.97)–(5.2.101) for the general initial data (5.2.100). Since $\{\nu_l(x)/\sqrt{1+(l\pi)^2+(l\pi)^4}\}_{l=1}^{\infty} \subset \mathcal{H}$ is a complete orthonormal system in $H_0^1 \cap H^2$, choosing an approximate sequence $\{u_2^k\}_{k=0}^{\infty} \subset \mathcal{H}$, such that as $k \to \infty$, $u_2^k \to u_2$ converges strongly in $H_0^1 \cap H^2$. Define the function u^k, satisfying the initial boundary value problem (5.2.97)–(5.2.101) with the initial data $u(0,x) = u_1(x)$ and $u_t(0,x) = u_2^k(x)$. For this purposes, let $\bar{u}^k := u^k - u_1 - u_2^k t$, and rewrite this problem as

$$
\begin{aligned}
& \bar{u}_{tt}^k + L_1 \bar{u}_t^k + b_1 \bar{u}_t^k + b_2 \bar{u}_x^k + b_3 \bar{u}_{xx}^k + L_2 \bar{u}^k \\
& \quad = f - b u_{2x}^k - b_1 u_2^k - b_2 (u_2^k t + u_1)_x - b_3 \partial_x^2 (u_2^k t + u_1) \\
& \qquad - a \partial_x^4 (u_2^k t + u_1), \qquad (5.2.111) \\
& \bar{u}^k(0,x) = \bar{u}_t^k(0,x) = 0, \qquad (5.2.112) \\
& \bar{u}^k(t,0) = \bar{u}^k(t,1) = 0, \bar{u}_{xx}^k(t,0) = \bar{u}_{xx}^k(t,1) = 0. \qquad (5.2.113)
\end{aligned}
$$

Noting that, since $u_1, u_2^k \in \mathcal{H}$, the right hand side of (5.2.111) belongs to $C([0,T];L^2)$. Consequently, the initial boundary value problem (5.2.111)–(5.2.113) has a solution $\bar{u}^k \in \bar{\mathfrak{X}}_4([0,T])$. Hence, $u^k = \bar{u}^k +$

$u_1 + u_2^k t$ is a solution to the initial boundary value problem (5.2.97)–(5.2.101) with the initial data $u(0,x) = u_1(x)$ and $u_t(0,x) = u_2^k(x)$. For $u_k - u_l$, $k,l = 0,1,2,\ldots$, by using energy method, the sequence $\{u^k\}_{k=0}^{\infty}$ is the Cauchy sequence in $\bar{\mathfrak{X}}_4([0,T]$. Then, there exists some function $u \in \bar{\mathfrak{X}}_4([0,T])$, such that as $k \to \infty$, $u^k \to u$ converges strongly in $\bar{\mathfrak{X}}_4([0,T])$. Obviously, the function u is a solution to the problem (5.2.97)–(5.2.101), the uniqueness of the solution is derived by the standard energy method. Thus the proof of Lemma 5.2.9 is completed. □

Using Lemma 5.2.9, we construct a solution to the linearized problem (5.2.11), (5.2.18)–(5.2.20) and (5.2.92)–(5.2.93). Differentiating (5.2.93) with respect to x, dividing the resultant equality by 2ω, and then using the equation (5.2.92), let $U := \hat{\omega}$, we deduce the following scalar equation

$$U_{tt} + \bar{b}\partial_x U_t + \bar{b}_1 U_t + \bar{b}_2 U_x + \bar{b}_3 U_{xx} + \bar{a}\partial_x^4 U = \bar{f},$$

$$\bar{a} := \frac{\varepsilon^2}{2}, \bar{b} := \frac{2j}{\omega^2}, \bar{b}_1 := \frac{1}{\omega}\left[\left(\frac{2j}{\omega}\right)_x + \omega_t\right], \bar{b}_2 := -\frac{1}{\omega}(S[\omega^2, j]\omega)_x,$$

$$\bar{b}_3 := -S[\omega^2, j] - \frac{h^2}{2}\frac{\omega_{xx}}{\omega}, \ \bar{f} := -(\omega^2\phi_x - j)_x \frac{1}{2\omega}, \tag{5.2.114}$$

with the initial and the boundary conditions

$$U(0,x) = \omega_0, \ U_t(0,x) = -\frac{j_{0x}}{2\omega_0}, \tag{5.2.115}$$

$$U(t,0) = \omega_l, \ U(t,1) = \omega_r, U_{xx}(t,0) = U_{xx}(t,1) = 0, \tag{5.2.116}$$

where the initial data (5.2.115) follows from (5.2.92), noting that $(\hat{\omega}, \hat{j}) \in \bar{\mathfrak{X}}_4([0,T]) \times \bar{\mathfrak{X}}_3([0,T])$ is a solution to the problem (5.2.11), (5.2.18)–(5.2.20), (5.2.92)–(5.2.93). Then $U = \hat{\omega} \in \bar{\mathfrak{X}}_4([0,T])$ satisfies (5.2.114)–(5.2.116).

Owing to Lemma 5.2.9, the existence of the solution to the problem (5.2.114)–(5.2.116) is ensured. Let $\overline{U} := U - A$, where $A(x) = \omega_l(1-x) + \omega_r x$, we obtain from (5.2.114)–(5.2.116) that

$$\begin{aligned}\overline{U}_{tt} + \bar{b}\partial_x\overline{U}_t + \bar{b}_1\overline{U}_t + \bar{b}_2\overline{U}_x + \bar{b}_3\overline{U}_{xx}\\ + \bar{a}\partial_x^4\overline{U} = \bar{f} + \frac{1}{\omega}(S[\omega^2, j]\omega)_x A_x,\end{aligned} \tag{5.2.117}$$

$$\overline{U}(0,x) = \omega_0 - A,\ \overline{U}_t(0,x) = -\frac{j_{0x}}{2\omega_0}, \tag{5.2.118}$$

$$\overline{U}(t,0) = \overline{U}(t,1) = 0,\ \overline{U}_{xx}(t,0) = \overline{U}_{xx}(t,1) = 0, \tag{5.2.119}$$

Noting that the left hand side of (5.2.117) and the coefficients satisfy the conditions (5.2.98) and (5.2.99), since $(\omega, j) \in \bar{\mathfrak{X}}_4([0,T]) \times \bar{\mathfrak{X}}_3([0,T])$. Moreover, owing to the compatibility conditions (5.2.10), the initial data (5.2.118) verifies (5.2.100). Thus, by using Lemma 5.2.9, there exists a unique solution to the problem (5.2.114)–(5.2.116).

Next, for given constructed U, we construct the solution $(\hat{\omega}, \hat{j})$ to the initial boundary value problem (5.2.11), (5.2.18)–(5.2.20), (5.2.92)–(5.2.93). Define $(\hat{\omega}, \hat{j})$ by

$$\hat{\omega}(t,x) := U(t,x), \tag{5.2.120}$$

$$\hat{j}(t,x) := \int_0^x -2\omega U_t(t,x)\mathrm{d}x + \hat{j}(t,0), \tag{5.2.121}$$

$$\hat{j}(t,0) := \int_0^t \left[\frac{4j}{\omega}U_t - 2S[\omega^2, j]\omega U_x + h^2\omega^2\left(\frac{U_{xx}}{\omega}\right)_x + \phi_x\omega^2 - j\right] \times (t,0)\mathrm{d}t + j_0(0).$$

It is sufficient to prove that $(\omega, \hat{j}) \in \bar{\mathfrak{X}}_4([0,T]) \times \bar{\mathfrak{X}}_3([0,T])$ is a solution to the linearized problem (5.2.11), (5.2.18)–(5.2.20) and (5.2.92)–(5.2.93). Obviously, (5.2.121) yields the equality $\hat{j}_x = -2\omega\hat{\omega}_t$. Moreover, differentiating (5.2.121) with respect to t, and using (5.2.114) lead to the following equality

$$\begin{aligned}
\hat{j}_t(t,x) &= \int_0^x \left[\frac{4j}{\omega}U_t - 2S[\omega^2, j]_\omega U_x + h^2\omega^2\left(\frac{U_{xx}}{\omega}\right)_x + \phi_x\omega^2 - j\right]_x \times (t,x)dx \\
&\quad + \left[\frac{4j}{\omega}U_t - 2S[\omega^2, j]_\omega U_x + h^2\omega^2\left(\frac{U_{xx}}{\omega}\right)_x + \phi_x\omega^2 - j\right](t,0) \\
&= \left[-\frac{2j}{\omega^2}\hat{j}_x - 2S[\omega^2, j]\omega\hat{\omega}_x + h^2\omega^2\left(\frac{\hat{\omega}_{xx}}{\omega}\right)_x + \phi_x\omega^2 - j\right](t,x),
\end{aligned}$$

where we have used $2\omega U_t = -\hat{j}_x$ and $U = \hat{\omega}$. Thus, $(\hat{\omega}, \hat{j})$ satisfies (5.2.92)–(5.2.93). Then, we confirm that (5.2.92)–(5.2.93) satisfies the initial condition (5.2.6). In fact, by (5.2.121) and (5.2.115), the equality $\hat{\omega}(0,x) = U(0,x) = \omega_0(x)$ and $\hat{j}(0,x) = \int_0^x j_{0x}\mathrm{d}x + j_0(0) = j_0(x)$ hold. Moreover, (5.2.116) yields the boundary conditions (5.2.19) and (5.2.20). Consequently, $(\hat{\omega}, \hat{j})$ is a solution to the linearized problem (5.2.11), (5.2.18)–(5.2.20) and (5.2.92)–(5.2.93).

Then the next lemma means that for suitable constants T, m, M under the mapping $(\omega, j) \to (\hat{\omega}, \hat{j})$ defined by solving the problem (5.2.92)–(5.2.93) and (5.2.18)–(5.2.20), the set $X(\cdot)$ is invariant.

Lemma 5.2.10. *Suppose that the initial data $(\omega_0, j_0) \in H^4(\Omega) \times H^3(\Omega)$, and the boundary data ρ_l, ρ_r satisfy* (5.2.7) *and $\omega_0 > 0$. Assume that the compatibility condition* (5.2.10) *holds. Then there exist constants T, m, M satisfying the following properties: If $(\omega, j) \in X(\cdot)$, then the problem* (5.2.11), (5.2.18)–(5.2.20) *and* (5.2.92)–(5.2.93) *has a weak solution $(\hat{\omega}, \hat{j}) \in X(\cdot)$.*

Lemma 5.2.10 follows from the standard energy method, now we can use Lemma 5.2.10 to prove Lemma 5.2.8.

Proof of Lemma 5.2.8. Define the successive approximate sequence $\{(\omega^n, j^n)\}_{n=0}^{\infty}$ by solving $(\omega^0, j^0) = (\omega_0, j_0)$ and

$$2\omega^n \omega_t^{n+1} + j_x^{n+1} = 0, \tag{5.2.122}$$

$$j_t^{n+1} + 2S[(\omega^n)^2, j^n]\omega^n \omega_x^{n+1} + 2\frac{j^n}{(\omega^n)^2} j_x^{n+1} - h^2(\omega^n)^2\left(\frac{\omega_{xx}^{n+1}}{\omega^n}\right)_x = (\omega^n)^2\phi_x^n - j^n, \tag{5.2.123}$$

$$\phi^n = \Phi[(\omega^n)^2] \tag{5.2.124}$$

with the initial and the boundary conditions

$$(\omega^{n+1}, j^{n+1})(0,x) = (\omega_0, j_0)(x), \tag{5.2.125}$$

$$\omega^{n+1}(t,0) = \omega_l, \omega^{n+1}(t,1) = \omega_r, \tag{5.2.126}$$

$$\omega_{xx}^{n+1}(t,0) = \omega_{xx}^{n+1}(t,1) = 0, \tag{5.2.127}$$

for $n = 0, 1, \ldots$, where Φ is defined by (5.2.11). Lemma 5.2.10 implies that the sequence $\{(\omega^n, j^n)\}$ is well defined, and satisfies $\{(\omega^n, j^n)\} \in X(\cdot)$. Moreover, the estimate

$$||\omega^n(t)||_4 + ||\omega_t^n(t)||_2 + ||\omega_{tt}^n(t)|| + ||j_t^n(t)||_3 + ||j_t^n(t)||_1 \leqslant M$$

holds. Thus, applying the standard energy method to the linear system of the equations for the difference $\{(\omega^{n+1} - \omega^n, j^{n+1} - j^n)\}$, we see that $\{(\omega^n, j^n)\}$ is the Cauchy sequence in $\bar{\mathfrak{X}}_2([0, T_2]) \times \bar{\mathfrak{X}}_1([0, T_2])$. Furthermore, the estimates for the higher-order derivatives can be obtained. We estimate the derivatives in t and then rewrite them into those in x by using the linear equations. Consequently, there exists a function $(\omega, j) \in \bar{\mathfrak{X}}_2([0, T_2]) \times \bar{\mathfrak{X}}_1([0, T_2])$, such that as $n \to \infty$, (ω^n, j^n) converges strongly in $\bar{\mathfrak{X}}_2([0, T_2]) \times \bar{\mathfrak{X}}_1([0, T_2])$ to (ω, j). Furthermore, $(\omega, j) \in \bar{\mathfrak{X}}_4([0, T_2]) \times \bar{\mathfrak{X}}_3([0, T_2])$, by using the definition $\phi = \Phi[\omega^2]$, it is easy to see that (ω, j, ϕ) is a solution to the problem (5.2.15)–(5.2.21). This completes the proof of Lemma 5.2.8. □

5.2.3 *A priori estimate*

To show the asymptotic stability of the stationary solution $(\tilde{\omega}, \tilde{j}, \tilde{\phi})$, we introduce a perturbation from the stationary solution $(\tilde{\omega}, \tilde{j}, \tilde{\phi})$

$$\begin{aligned} \psi(t,x) &:= \omega(t,x) - \tilde{\omega}(x), \eta(t,x) := j(t,x) - \tilde{j}(x), \sigma(t,x) \\ &:= \phi(t,x) - \tilde{\phi}(x). \end{aligned}$$

Dividing (5.2.2) by ω^2 yields

$$\left(\frac{j}{\omega^2}\right) t + \frac{j}{\omega^2}\left(\frac{j}{\omega^2}\right)_x + K(\log \omega^2)_x - h^2\left(\frac{\omega_{xx}}{\omega}\right)_x = \phi_x - \frac{j}{\omega^2}. \tag{5.2.128}$$

Similarly, we deduce from (5.2.23) that

$$\frac{\tilde{j}}{\tilde{\omega}^2}\left(\frac{\tilde{j}}{\tilde{\omega}^2}\right)_x + K(\log \tilde{\omega}^2)_x - h^2\left(\frac{\tilde{\omega}_{xx}}{\tilde{\omega}}\right)_x = \tilde{\phi}_x - \frac{\tilde{j}}{\tilde{\omega}^2}. \tag{5.2.129}$$

Subtracting (5.2.15) from (5.2.22), (5.2.128) from (5.2.129), (5.2.17) from (5.2.24), respectively, we derive the equations for the perturbation (ψ, η, σ) as

$$2(\psi + \tilde{\omega})\psi_t + \eta_x = 0, \tag{5.2.130}$$

$$\begin{aligned}&\left(\frac{\eta+\tilde{j}}{(\psi+\tilde{\omega}^2)}\right)t + \frac{1}{2}\left[\left(\frac{\eta+\tilde{j}}{(\psi+\tilde{\omega})^2}\right)^2 - \left(\frac{\tilde{j}}{\tilde{\omega}^2}\right)^2\right]_x \\ &+ K[\log(\psi+\tilde{\omega})^2 - \log\tilde{\omega}^2]_x - h^2\left(\frac{(\psi+\tilde{\omega})_{xx}}{\psi+\tilde{\omega}} - \frac{\tilde{\omega}_{xx}}{\tilde{\omega}}\right)_x \\ &-\sigma_x + \frac{\eta+\tilde{j}}{(\psi+\tilde{\omega})^2} - \frac{\tilde{j}}{\tilde{\omega}^2} = 0,\end{aligned} \tag{5.2.131}$$

$$\sigma_{xx} = (\psi + 2\tilde{\omega})\psi. \tag{5.2.132}$$

The initial and the boundary conditions to the system (5.2.130)–(5.2.132) are derived from (5.2.18)–(5.2.21) and (5.2.25)–(5.2.27) as

$$\begin{aligned}\psi(x,0) &= \psi_0(x) := \omega_0(x) - \tilde{\omega}(x), \eta(x,0) = \eta_0(x) \\ &:= j_0(x) - \tilde{j}(x),\end{aligned} \tag{5.2.133}$$

$$\psi(t,0) = \psi(t,1) = 0, \tag{5.2.134}$$

$$\psi_{xx}(t,0) = \psi_{xx}(t,1) = 0, \tag{5.2.135}$$

$$\sigma(t,0) = \sigma(t,1) = 0. \tag{5.2.136}$$

Since $(\tilde{\omega}, \tilde{j}, \tilde{\phi}) \in \bar{\mathfrak{X}}_4([0,T]) \times \bar{\mathfrak{X}}_3([0,T]) \times \mathfrak{Y}([0,T])$, and σ satisfies (5.2.132), the local existence of the solution (ψ, η, σ) to the initial boundary value problem (5.2.130)–(5.2.136) follows from Lemma 5.2.1 and Corollary 5.2.2.

Corollary 5.2.3. *Suppose that the initial data $(\psi_0, \eta_0) \in H^4(\Omega) \times H^3(\Omega)$, $((\tilde{\omega} + \psi_0)^2,\ \tilde{j} + \eta_0)$ satisfies (5.2.12)–(5.2.13). Then there exists a constant $T_3 > 0$, such that the initial boundary value problem (5.2.130)–(5.2.136) has a unique local solution $(\psi, \eta, \sigma) \in \bar{\mathfrak{X}}_4([0,T_3]) \times \bar{\mathfrak{X}}_3([0,T_3]) \times \bar{\mathfrak{X}}_4^2([0,T_3])$, with the property that $((\tilde{\omega} + \psi)^2, \tilde{j} + \eta)$ satisfies (5.2.12)–(5.2.13).*

By virtue of Corollary 5.2.3, it is sufficient to deduce *a priori* estimate (5.2.137) in order to prove the global existence of the solution. We introduce the following notations

$$N_\varepsilon(t) := \sup_{0\leq\tau\leq t} n_h(\tau), n_h^2(\tau) := ||(\psi,\eta)(\tau)||_2^2$$
$$+ ||(h\partial_x^3\psi, h\partial_x^3\eta, h^2\partial_x^4\psi)(\tau)||^2,$$
$$M^2(t) := \int_0^t (||(\psi,\eta)(\tau)||_1^2 + ||\sigma_x(\tau)||^2)\mathrm{d}\tau.$$

Proposition 5.2.1. *Let* $(\psi,\eta,\sigma)(t,x) \in \bar{\mathfrak{X}}_4([0,T]) \times \bar{\mathfrak{X}}_3([0,T]) \times \bar{\mathfrak{X}}_4^2([0,T])$ *be a solution to* (5.2.130)–(5.2.136)*. Then there exists a positive constant* δ_0 *such that, if* $(N_h(T) + \delta + h) \leqslant \delta_0$*, then for* $t \in [0,T]$*, the following estimate holds*

$$n_h^2(t) + ||\sigma(t)||_4^2 + \int_0^t (n_h^2(\tau) + ||\sigma(\tau)||_4^2)\mathrm{d}\tau \leqslant Cn_h^2(0), \tag{5.2.137}$$

where C *is a positive constant independent of* T *and* h.

Basic estimate

In order to prove the basic estimate, we define an energy form $\mathcal{E}$

$$\mathcal{E} := \frac{1}{2\omega^2}(j-\tilde{j})^2 + \psi(\omega^2,\tilde{\omega}^2) + \frac{1}{2}[(\phi-\tilde{\phi})_x]^2 + h^2(\omega-\tilde{\omega})_x^2,$$
$$\psi(\omega^2,\tilde{\omega}^2) := K\int_{\tilde{\omega}^2}^{\omega^2} (\log\xi - \log\tilde{\omega}^2)\mathrm{d}\xi. \tag{5.2.138}$$

Noting that if $|(\psi,\eta,\sigma,\varepsilon\psi_x)| < c, \mathcal{E}$ is equivalent to $|(\psi,\eta,\sigma,\varepsilon\psi_x)|^2$, i.e. there exist positive constants c_1 and C_1 such that

$$c_1|(\psi,\eta,\sigma,h\psi_x)|^2 \leqslant \mathcal{E} \leqslant C_1|(\psi,\eta,\sigma,h\psi_x)|^2. \tag{5.2.139}$$

Multiplying (5.2.131) by η and integrating by parts yield

$$\mathcal{E}_t + \frac{1}{\tilde{\omega}^2}\eta^2 = R_{1x} + R_2,$$

$$R_1 := \sigma\sigma_{xt} + \sigma\eta - K(\log\eta - \log\tilde{\omega}^2)\eta + h^2\left(\frac{\omega_{xx}}{\omega} - \frac{\tilde{\omega}_{xx}}{\tilde{\omega}}\right)\eta + h^2\psi_x\psi_t,$$

$$R_2 := \left(\frac{\eta}{2\omega^4} - \frac{j}{\omega^4}\right)\eta\eta_x - \frac{1}{2}\left[\left(\frac{j}{\omega^2}\right)^2 - \left(\frac{\tilde{j}}{\tilde{\omega}^2}\right)^2\right]_x \eta + \frac{j(\omega+\tilde{\omega})}{\omega^2\tilde{\omega}^2}\psi\eta$$
$$+ \frac{h^2\tilde{\omega}_{xx}}{\tilde{\omega}\omega}\psi\eta_x, \tag{5.2.140}$$

where we have used the estimates (5.2.130) and (5.2.132). By applying the inequality (5.2.71) on R_2, with (5.2.37), (5.2.35) and Corollary 5.2.1, we have

$$|R_2| \leqslant C(N_h(T) + \delta + h^{3/2})|(\psi, \eta, \psi_x, \eta_x, \sigma_x)|^2. \tag{5.2.141}$$

Similar to Lemma 5.2.10, we have the following lemma.

Lemma 5.2.11. *Under the assumptions of Proposition* 5.2.1, *for* $t \in [0, T]$, *the following estimates hold:*

$$||\partial_t^i\sigma(t)||_2^2 \leqslant C\|\partial_t^i\psi(t)\|^2, \quad i = 0, 1, 2, \tag{5.2.142}$$
$$||\sigma_{xt}(t)||^2 \leqslant C(N_h(T) + \delta)||\psi(t)||^2 + C||\eta(t)||^2, \tag{5.2.143}$$

where C *is a positive constant independent of* T *and* h.

Lemma 5.2.12. *Under the assumptions of Proposition* 5.2.1, *then there exists a positive constant* δ_0, *such that if* $N_h(T) + \delta + h \leqslant \delta_0$, *for* $t \in [0, T]$, *the following estimate holds*:

$$||(\psi, \eta, \sigma_x, h\psi_x)(t)||^2 + \int_0^t ||(\psi, \eta, \sigma_x, h\psi_x)(\tau)||^2 d\tau$$
$$\leqslant C||(\psi, \eta, \sigma_x, h\psi_x)(0)||^2 + C(N_h(T) + \delta + h)M^2(t), \tag{5.2.144}$$

where C *is a positive constant independent of* T *and* h.

Proof. Firstly, integrating (5.2.140) over $[0, t] \times \Omega$, and applying estimate (5.2.141) to handle the integration of R_2, owing to the boundary conditions (5.2.134)–(5.2.136), we have $\int_0^1 R_{1x}\mathrm{d}x = 0$, then

$$\begin{aligned}
&\int_0^1 \mathcal{E}(t,x)\mathrm{d}x + \int_0^t\int_0^1 \frac{1}{\tilde{\omega}^2}\eta^2 \mathrm{d}x\mathrm{d}\tau \\
&= \int_0^1 \mathcal{E}(0,x)\mathrm{d}x + \int_0^t\int_0^1 R_2 \mathrm{d}x\mathrm{d}\tau \qquad (5.2.145)\\
&\leqslant \int_0^1 \mathcal{E}(0,x)\mathrm{d}x + C(N_\varepsilon(T) + \delta + h^{3/2})M^2(t). \qquad (5.2.146)
\end{aligned}$$

Multiplying (5.2.131) by $-\sigma_x$, integrating the resultant equality over $[0,t]\times\Omega$, integrating by parts and using equation (5.2.132) and the boundary conditions (5.2.134), (5.2.135), we have

$$\begin{aligned}
&\int_0^t\int_0^1 [K(\log(\psi+\tilde{\omega})^2 - \log\tilde{\omega}^2)(\psi+2\tilde{\omega})\psi + \sigma_x^2 \\
&\quad -h^2\left(\frac{(\psi+\tilde{\omega})_{xx}}{\psi+\tilde{\omega}} - \frac{\tilde{\omega}_{xx}}{\tilde{\omega}}\right)(\psi+2\tilde{\omega})\psi]\mathrm{d}x\mathrm{d}\tau \\
&= \int_0^1 \left[\left(\frac{\eta+\tilde{j}}{(\psi+\tilde{\omega})^2} - \frac{\tilde{j}}{\tilde{\omega}^2}\right)\sigma_x(t,x) - \left(\frac{\eta+\tilde{j}}{(\psi+\tilde{\omega})^2} - \frac{\tilde{j}}{\tilde{\omega}^2}\right)\sigma_x(0,x)\right]\mathrm{d}x \\
&\quad + \int_0^t\int_0^1 \left\{\frac{1}{2}\left[\left(\frac{\eta+\tilde{j}}{(\psi+\tilde{\omega})^2}\right)^2 - \left(\frac{\tilde{j}}{\tilde{\omega}^2}\right)^2\right]_x \sigma_x \right.\\
&\quad \left. + \left(\frac{\eta+\tilde{j}}{(\psi+\tilde{\omega})^2} - \frac{\tilde{j}}{\tilde{\omega}^2}\right)(\sigma_x - \sigma_{xt})\right\}\mathrm{d}x\mathrm{d}\tau \qquad (5.2.147)\\
&\leqslant C(||(\psi,\eta,\sigma_x)(0)||^2 + ||(\psi,\eta,\sigma_x)(t)||^2) \\
&\quad + \int_0^t (C||\eta(\tau)||^2 + \frac{1}{2}||\sigma_x(\tau)||^2)\mathrm{d}\tau + C(N_\varepsilon(T)+\delta)M^2(t), \qquad (5.2.148)
\end{aligned}$$

where we have used Schwarz inequality and Sobolev inequality as well as (5.2.37), (5.2.75), (5.2.142) and (5.2.143). Now we estimate each term in the left hand side of (5.2.147). The first term is estimated $c||\psi(t)||^2$ below, integrating by parts with using (5.2.134). We rewrite the third term of the left hand side of (5.2.147) as

$$(3rd\ term) = -h^2 \int_0^t\int_0^1 \left(\frac{\psi+2\tilde{\omega}}{\psi+\tilde{\omega}}\psi_{xx}\psi - \frac{\psi+2\tilde{\omega}}{\tilde{\omega}(\psi+\tilde{\omega})}\tilde{\omega}_{xx}\psi^2\right)\mathrm{d}x\mathrm{d}\tau$$

$$
\begin{aligned}
&= h^2 \int_0^t \int_0^1 \left[\left(1 + \frac{\tilde{\omega}}{\psi + \tilde{\omega}}\right) \psi_x^2 + \left(\frac{\tilde{\omega}}{\psi + \tilde{\omega}}\right)_x \psi_x \psi \right. \\
&\quad \left. + \frac{\psi + 2\tilde{\omega}}{\tilde{\omega}(\psi + \tilde{\omega})} \tilde{\omega}_{xx} \psi^2 \right] \mathrm{d}x\mathrm{d}\tau \qquad (5.2.149) \\
&\geqslant \int_0^t ch^2 ||\psi_x(\tau)||^2 \mathrm{d}\tau - Ch^{3/2} M^2(t), \qquad (5.2.150)
\end{aligned}
$$

where we have used Schwarz inequality and Corollary 5.2.1. Here c is a positive constant. Substituting these estimates in (5.2.147)–(5.2.148), multiplying the resultant inequality by μ, which is a positive constant to be determined, adding the resultant inequality to (5.2.145)–(5.2.146), choosing $(N_h(T) + \delta + h)$ to be small enough, we deduce the estimate (5.2.144). □

Higher order estimates

Now we start to derive the higher order estimates, since the regularity of the solution (ψ, η) constructed in Corollary 5.2.3 is not sufficient. We need to use mollifier with respect to t. Hereafter, we use notations

$$
\begin{aligned}
A_i^2(t) &:= ||(\psi, \eta)(t)||^2 + \sum_{k=0}^{i} ||(\partial_t^k \psi_t, \partial_t^k \psi_x, \varepsilon \partial_t^k \psi_{xx})(t)||^2, \quad i \geqslant 0, \\
A_{-1}^2(t) &:= ||(\psi, \eta)(t)||^2.
\end{aligned}
$$

Differentiate (5.2.16) with respect to x, and divide the resultant equality by ω, then we use equation (5.2.15) to rewrite the resultant equality. Differentiate (5.2.28) with respect to x, and divide the resultant equality by $\tilde{\omega}$. Then taking a difference of the resultant equalities, and differentiating the result with respect to t imply that

$$
\begin{aligned}
&2\partial_t^i \psi_{tt} - 2K \partial_t^i \psi_{xx} + h^2 \partial_t^i \psi_{xxxx} + 2\partial_t^i \psi_t \\
&= 2\frac{\eta + \tilde{j}}{(\psi + \tilde{\omega})^3} \partial_t^i \eta_{xx} - 2\frac{(\eta + \tilde{j})^2}{(\psi + \tilde{\omega})^4} \partial_t^i \psi_{xx} \\
&\quad + h^2 \frac{(i+1)\psi_{xx} + 2\tilde{\omega}_{xx}}{\psi + \tilde{\omega}} \partial_t^i \psi_{xx} + \sum_{i=1}^{4} \partial_t^i F_i + G_i, \quad i = 0, 1,
\end{aligned}
$$

$$F_1 = -2\frac{\eta + 2\tilde{j}}{(\psi+\tilde{\omega})^4}\tilde{\omega}_{xx}\eta + 2\tilde{j}\tilde{\omega}_{xx}\frac{(\psi+\tilde{\omega})^4 - \tilde{\omega}^4}{(\psi+\tilde{\omega})^4\tilde{\omega}^4},$$

$$F_2 = \left(\frac{4K}{\tilde{\omega}}\tilde{\omega}_x - 2\tilde{\phi}_x\right)\psi_x,$$

$$F_3 = \frac{2\eta_x^2}{(\psi+\tilde{\omega})^3} - \frac{8(\eta+\tilde{j})(\psi+\tilde{\omega})_x}{\psi+\tilde{\omega}}\eta_x + 2K\frac{\psi_x^2}{\tilde{\omega}} + \frac{6(\eta+\tilde{j})^2(\psi+2\tilde{\omega})_x}{(\psi+\tilde{\omega})^5}\psi_x$$
$$- \frac{2\psi_t^2}{\psi+\tilde{\omega}},$$

$$F_4 = 6\frac{\eta+2\tilde{j}}{(\psi+\tilde{\omega})^5}\tilde{\omega}_x^2\eta - 2K\frac{(\psi+\tilde{\omega})_x^2}{(\psi+\tilde{\omega})\tilde{\omega}}\psi - [(\psi+\tilde{\omega})(\psi+2\tilde{\omega}) + (\tilde{\omega}^2 - D)]\psi$$
$$-6\tilde{j}^2\tilde{\omega}_x^2\frac{(\psi+\tilde{\omega})^5 - \tilde{\omega}^5}{(\psi+\tilde{\omega})^5\tilde{\omega}^5} - 2(\psi+\tilde{\omega})_x\sigma_x - \frac{h^2\tilde{\omega}_{xx}^2}{2\tilde{\omega}(\psi+\tilde{\omega})}\psi,$$

$$G_0 = 0,\ G_1 = 2\left(\frac{\eta+\tilde{j}}{(\psi+\tilde{\omega})^3}\right)_t\eta_{xx} - 2\left(\frac{\eta+\tilde{j}}{(\psi+\tilde{\omega})^2}\right)_t\psi_{xx}$$
$$-h^2\frac{\psi_{xx}+2\tilde{\omega}_{xx}}{(\psi+\tilde{\omega})^2}\psi_t\psi_{xx}. \tag{5.2.151}$$

The L^2-norms of $F_1 - F_4$ are estimated as

$$\begin{aligned}
||F_1|| &\leqslant C(N_\varepsilon(T)+\delta)||\tilde{\omega}_{xx}||(|\eta|_0 + |\psi|_0)\\
&\leqslant C(N_\varepsilon(T)+\delta)(||\eta||_1 + ||\psi_x||),\\
||F_2|| &\leqslant C(h^{1/2}+\delta)||\psi_x||,\ ||F_3|| \leqslant C(N_h(T)+\delta)||(\eta_x,\psi_x,\psi_t)||,\\
||F_4|| &\leqslant C||(\eta,\psi)||,
\end{aligned} \tag{5.2.152}$$

where C is a positive constant independent of T and h. In order to derive the estimate of $||F_2||$, we use equation (5.2.28) and the inequality

$$\left|\frac{4K}{\tilde{\omega}}\tilde{\omega}_x - 2\tilde{\phi}_x\right| = \left| -\frac{2\tilde{j}^2\tilde{\omega}_x}{\tilde{\omega}^5} + 2h^2\left(\frac{\tilde{\omega}_{xx}}{\tilde{\omega}}\right)_x - \frac{2\tilde{j}}{\tilde{\omega}^2}\right| \leqslant C(h^{1/2}+\delta),$$

which follows from (5.2.37), (5.2.71), (5.2.75) and Corollary 5.2.1. The other estimates in (5.2.152) are proved by (5.2.37), (5.2.71),

(5.2.75), (5.2.142) and Corollary 5.2.1. Similarly, we have

$$
\begin{aligned}
&||F_{1t}|| \leqslant C(N_h(T)+\delta)(||\eta_t||_1 + ||\psi_{xt}||), \\
&||F_{2t}|| \leqslant C(h^{1/2}+\delta)||\psi_{xt}||, \quad ||F_{3t}|| \leqslant C(N_h(T)+\delta) \\
&||(\eta_x,\psi_x,\psi_t,\eta_{xt},\psi_{xt},\psi_{tt})||, ||F_{4t}|| \leqslant C||\psi_t|| \\
&\quad + C(N_h(T)+\delta)||(\eta_t,\psi_{xt})||, \\
&||G_1|| \leqslant C(N_h(T)+\delta)||(\eta_{xx},\psi_{xx})|| \qquad (5.2.153)
\end{aligned}
$$

owing to the estimate

$$|(\psi_t,\eta_t)(t)|_0 \leqslant CN_h(T), \qquad (5.2.154)$$

where C is a positive constant independent of T and h. Applying the inequality (5.2.71) on equations (5.2.130) and (5.2.131), the estimate (5.2.154) holds.

Differentiating (5.2.130) with respect to x yields

$$
\begin{aligned}
&\partial_t^i \eta_{xx} = -2(\psi+\tilde{\omega})\partial_t^i \psi_{xt} + H_i, \\
&\quad H_0 = -2(\psi+\tilde{\omega})_x \psi_t, H_1 = -4(\psi+\tilde{\omega})_x \psi_{xt} - 2\psi_t\psi_{xt},
\end{aligned}
\qquad (5.2.155)
$$

for $i = 0, 1$. Owing to equations (5.2.130) and (5.2.155), the following estimates hold

$$||\partial_t^i \eta_x(t)|| \leqslant CA_i(t), ||(\eta_{xx}, h\partial_x^3\eta)(t)|| \leqslant CA_1(t). \qquad (5.2.156)$$

Moreover, owing to (5.2.132) and (5.2.156), it holds that

$$M^2(t) \leqslant C\int_0^t A_0^2(\tau)\mathrm{d}\tau, \ ||\omega(t)||_4 \leqslant ||\psi(t)||_2. \qquad (5.2.157)$$

Lemma 5.2.13. *Under the assumptions of Proposition* 5.2.1, *then the estimate*

$$cA_1(t) \leqslant n_h(t) \leqslant CA_1(t) \qquad (5.2.158)$$

holds, where C and c are positive constants independent of T and h.

Proof. Let $i = 0$ in (5.2.151), multiplying (5.2.151) by ψ_{xx}, and applying integration by parts with using the boundary condition (5.2.135) yield

$$
\begin{aligned}
2K\|\psi_{xx}\|^2 + h^2\|\psi_{xxx}\|^2 = & \int_0^1 \left(2\psi_{tt} + 2\psi_t - 2\frac{\eta+\tilde{j}}{(\psi+\tilde{\omega})^3}\eta_{xx} \right. \\
& \left. + 2\frac{\eta+\tilde{j}}{(\psi+\tilde{\omega})^4}\psi_{xx} \right) \psi_{xx}\mathrm{d}x \\
& - \int_0^1 \left(h^2\frac{\psi_{xx}+2\tilde{\omega}_{xx}}{\psi+\tilde{\omega}}\psi_{xx} + \sum_{l=1}^4 F_l \right) \\
& \times \psi_{xx}\mathrm{d}x. \qquad (5.2.159)
\end{aligned}
$$

Applying the Schwarz inequality to the right hand side of (5.2.159), together with (5.2.37), (5.2.71), (5.2.75), (5.2.152), (5.2.156) and Corollary 5.2.1, we have

$$\|(\psi_{xx}, h\partial_x^3\psi)(t)\| \leqslant CA_1(t). \qquad (5.2.160)$$

Solving (5.2.151) with respect to $h^2\psi_{xxxx}$, taking L^2-norm, and then using the estimates (5.2.152), (5.2.156) and (5.2.160), yield $h^2\|\partial_x^4\psi(t)\| \leqslant CA_1(t)$. Similarly, the estimate $\|\psi_{tt}(t)\| \leqslant C(\|(\eta,\psi)(t)\|_2 + h^2\|\partial_x^4\psi(t)\|)$ holds. Owing to (5.2.131) and (5.2.155), $\|\psi_t(t)\|_l \leqslant C(\|\eta(t)\|_{l+1} + \|\psi(t)\|_l)$, $l = 0, 1, 2$ holds. Subtracting (5.2.16) from (5.2.28), and estimating the L^2-norm of the resultant equality by using (5.2.152), (5.2.156) and (5.2.160) lead to

$$\|\eta_t(t)\| \leqslant CA_1(t). \qquad (5.2.161)$$

Consequently, the estimate (5.2.159) holds. □

In order to derive priori estimates, we need the higher-order estimates.

Lemma 5.2.14. *Under the assumptions of Proposition* 5.2.1, *there exists a positive constant* h_0 *such that, if* $(N_h(T) + \delta + h) \leqslant h_0$, *for*

$t \in [0, T]$, $i = 0, 1$, *the following estimate holds*

$$||(\partial_t^i \psi_t, \partial_t^i \psi_x, h\partial_t^i \psi_{xx})(t)||^2 + \int_0^1 ||(\partial_t^i \psi_t, \partial_t^i \psi_x, h\partial_t^i \psi_{xx})(\tau)||^2 d\tau$$

$$\leqslant C(A_i^2(0) + \int_0^1 A_{i-1}^2(\tau)\mathrm{d}\tau). \tag{5.2.162}$$

Proof. Multiplying (5.2.151) by $\partial_t^i \psi$, integrating the resultant equality by parts over Ω and using the boundary conditions $(\psi, \partial_t^i \psi_t, \partial_t^i \psi_{xx})(t, 0) = (\psi, \partial_t^i \psi_t, \partial_t^i \psi_{xx})(t, 1) = 0$ yield that

$$I_1^{(i)}(t) + \int_0^t \int_0^1 (2K(\partial_t^i \psi_x)^2 + h^2(\partial_t^i \psi_{xx})^2)\mathrm{d}x\mathrm{d}\tau$$

$$= I_1^{(i)}(0) + \int_0^t J_1^{(i)}(\tau)\mathrm{d}\tau + \int_0^t \int_0^1 2(\partial_t^i \psi_t)^2 \mathrm{d}x\mathrm{d}\tau,$$

$$I_1^{(i)}(t) = \int_0^1 2\partial_t^i \psi_t \partial_t^i \psi + (\partial_t^i \psi)^2 \mathrm{d}x,$$

$$J_1^{(i)}(t) = \int_0^1 \left[-2\left(\frac{\eta + \tilde{j}}{(\psi + \tilde{\omega})^3} \partial_t^i \psi \right)_x \partial_t^i \eta_x \right.$$

$$\left. + 2\left(\frac{(\eta + \tilde{j})^2}{(\psi + \tilde{\omega})^4} \partial_t^i \psi \right)_x \partial_t^i \psi_x \right] \mathrm{d}x + \int_0^1 \left[h^2 \frac{(i+1)\psi_{xx} + 2\tilde{\omega}_{xx}}{\psi + \tilde{\omega}} \right.$$

$$\left. \times \partial_t^i \psi_{xx} \partial_t^i \psi + \left(\sum_{l=1}^4 \partial_t^i F_l \right) \partial_t^i \psi + G_i \partial_t^i \psi \right] \mathrm{d}x. \tag{5.2.163}$$

By the Schwarz inequality, we have

$$|I_1^{(i)}(t)| \leqslant C A_i^2(t). \tag{5.2.164}$$

Owing to (5.2.71), Corollary 5.2.1 and the Schwarz inequality, the third term in $J_1^{(i)}(t)$ is estimated as

$$|3rd\ term| \leqslant \int_0^1 \frac{h^2}{4} (\partial_t^i \psi_{xx})^2 \mathrm{d}x + C(N_h(T) + h) A_i^2(t).$$

Since (5.2.37) and (5.2.71), we have $|\eta + \tilde{j}|_1 \leqslant C(N_h(T) + \delta)$, then the other terms in $J_1^{(i)}(t)$ are estimated by using (5.2.152), (5.2.153),

(5.2.156), (5.2.160) and (5.2.161),

$$|J_1^{(i)}(t)| \leqslant CA_{i-1}^2(t) + \int_0^1 \frac{h^2}{4}(\partial_t^i \psi_{xx})^2 \mathrm{d}x + C(N_h(T) + \delta + h^{1/2})$$
$$A_i^2(t). \tag{5.2.165}$$

Substituting (5.2.164) and (5.2.165) in (5.2.163), we have

$$I_1^{(i)}(t) + \int_0^t \int_0^1 [2K(\partial_t^i \psi_x)^2 + \frac{3h^2}{4}(\partial_t^i \psi_{xx})^2]\mathrm{d}x\mathrm{d}\tau - \int_0^t \int_0^1 2(\partial_t^i \eta_t)^2 \mathrm{d}x\mathrm{d}\tau$$
$$\leqslant C[A_i^2(0) + \int_0^t A_{i-1}^2(\tau)\mathrm{d}\tau + (N_h(T) + \delta + h^{1/2}) \int_0^t A_i^2(\tau)d\tau]. \tag{5.2.166}$$

Next, multiplying (5.2.151) by $\partial_t^i \psi_t$, and integrating over Ω yield

$$\int_0^t (2\partial_t^i \psi_{tt} - 2K\partial_t^i \psi_{xx} + h^2 \partial_t^i \psi_{xxxx} + 2\partial_t^i \psi_t)\partial_t^i \psi_t \mathrm{d}x$$
$$= \int_0^1 2\frac{\eta + \tilde{j}}{(\psi + \tilde{\omega})^3} \partial_t^i \eta_{xx} \partial_t^i \psi_t \mathrm{d}x - \int_0^1 2\frac{(\eta + \tilde{j})^2}{(\psi + \tilde{\omega})^4} \partial_t^i \psi_{xx} \partial_t^i \psi_t \mathrm{d}x$$
$$+ \int_0^1 h^2 \frac{(i+1)\psi_{xx} + 2\tilde{\omega}_{xx}}{\psi + \tilde{\omega}} \partial_t^i \psi_{xx} \partial_t^i \psi_t \mathrm{d}x + \int_0^1 \left[\left(\sum_{l=1}^4 \partial_t^i F_l\right)\right.$$
$$\left. \times\, \partial_t^i \psi_t + G_i \partial_t^i \psi_t\right] \mathrm{d}x. \tag{5.2.167}$$

By integrating by parts and the boundary condition:

$$(\partial_t^i \psi_t, \partial_t^i \psi_{xx})(t, 0) = (\partial_t^i \psi_t, \partial_t^i \psi_t)(t, 1) = 0,$$

the left hand side of the equality (5.2.167) is rewritten as

$$\frac{d}{\mathrm{d}t} \int_0^1 (\partial_t^i \psi_t)^2 + K(\partial_t^i \psi_x)^2 + \frac{h^2}{2}(\partial_t^i \psi_{xx})^2 \mathrm{d}x + \int_0^1 2(\partial_t^i \psi_t)^2 \mathrm{d}x. \tag{5.2.168}$$

Substituting (5.2.155) in (5.2.167), integrating by parts and using $\partial_t^i \psi_t(t, 0) = \partial_t^i \psi_t(t, 1) = 0$, the first term of the left hand side of the equality (5.2.167) is rewritten as

$$(1st\ term) = \int_0^1 -2\frac{\eta+\tilde{j}}{(\psi+\tilde{\omega})^3}(-2(\psi+\tilde{\omega})\partial_t^i\psi_{xt} + H_i)\partial_t^i\psi_t \mathrm{d}x$$
$$= -\int_0^1 2\left(\frac{\eta+\tilde{j}}{(\psi+\tilde{\omega})^2}\right)_x (\partial_t^i\psi_t)^2 + 2\frac{\eta+\tilde{j}}{(\psi+\tilde{\omega})^3}H_i\partial_t^i\psi_t \mathrm{d}_x. \tag{5.2.169}$$

Similarly, we have

$$(2nd\, term) = \int_0^1 2\frac{(\eta+\tilde{j})^2}{(\psi+\tilde{\omega})^4}\partial_t^i\psi_x\partial_t^i\psi_{xt} + 2\left(\frac{(\eta+\tilde{j})^2}{(\psi+\tilde{\omega})^4}\right)_x \times \partial_t^i\psi_x\partial_t^i\psi_t \mathrm{d}x$$
$$= \frac{\mathrm{d}}{\mathrm{d}t}\int_0^1 \frac{(\eta+\tilde{j})^2}{(\psi+\tilde{\omega})^4}(\partial_t^i\psi_x)^2 \mathrm{d}x - \int_0^1 \left(\frac{(\eta+\tilde{j})^2}{(\psi+\tilde{\omega})^4}\right)_t \times(\partial_t^i\psi_x)^2\mathrm{d}x + \int_0^1 2\left(\frac{(\eta+\tilde{j})^2}{(\psi+\tilde{\omega})^4}\right)_x \partial_t^i\psi_x\partial_t^i\psi_t \mathrm{d}x. \tag{5.2.170}$$

Substituting (5.2.168)–(5.2.170) in (5.2.167), and integrating over $(0,t)$ imply that

$$I_2^{(i)}(t) + \int_0^t\int_0^1 2(\partial_t^i\psi_t)^2 \mathrm{d}x\mathrm{d}\tau = I_2^{(i)}(0) + \int_0^t J_2^{(i)}(\tau)\mathrm{d}\tau,$$
$$I_2^{(i)}(t) = \int_0^1 ((\partial_t^i\psi_t)^2 + K(\partial_t^i\psi_x)^2 + \frac{h^2}{2}(\partial_t^i\psi_{xx})^2 - \frac{(\eta+\tilde{j})^2}{(\psi+\tilde{\omega})^4} \times (\partial_t^i\psi_x)^2)\mathrm{d}x,$$
$$J_2^{(i)}(t) = -\int_0^1 \left[2\left(\frac{\eta+\tilde{j}}{(\psi+\tilde{\omega})^2}\right)_x (\partial_t^i\psi_t)^2 + 2\frac{\eta+\tilde{j}}{(\psi+\tilde{\omega})^3}H_i\partial_t^i\psi_t + \left(\frac{(\eta+\tilde{j})^2}{(\psi+\tilde{\omega})^4}\right)_t (\partial_t^i\psi_x)^2\right]\mathrm{d}x + \int_0^1 \left[2\left(\frac{(\eta+\tilde{j})^2}{(\psi+\tilde{\omega})^4}\right)_x \partial_t^i\psi_x\partial_t^i\psi_t + h^2\frac{(i+1)\psi_{xx}+2\tilde{\omega}_{xx}}{\psi+\tilde{\omega}}\partial_t^i\psi_{xx}\partial_t^i\psi_t\right]\mathrm{d}x + \int_0^1 \left[\left(\sum_{l=1}^4 \partial_t^i F_l\right)\partial_t^i\psi_t + G_i\partial_t^i\psi_t\right]\mathrm{d}x. \tag{5.2.171}$$

Using (5.2.37), (5.2.75) and Sobolev inequality, the fourth term in $I_2^{(i)}(t)$ is estimated as

$$\left| \int_0^1 -\frac{(\eta+j)^2}{(\psi+\tilde{\omega})^4}(\partial_t^i \psi_x)^2 \mathrm{d}x \right| \leqslant C(N_h(T)+\delta)||\partial_t^i \psi_x(t)||^2. \tag{5.2.172}$$

Moreover, we have

$$\left| \int_0^1 \partial_t^i F_4 \partial_t^i \psi_t \mathrm{d}x \right| \leqslant C_\nu A_{i-1}^2(t) + C(\nu + N_h(T)+\delta)A_i^2(t), \tag{5.2.173}$$

where ν is a positive constant, and C_ν is a positive constant depending only on ν. The other terms in $I_2^{(i)}(t)$ and $J_2^{(i)}(t)$ are estimated similarly as the estimate of $I_1^{1(i)}(t)$ and $J_1^{(i)}(t)$,

$$|I_2^{(i)}(t)| \leqslant CA_i^2(t), \tag{5.2.174}$$

$$\begin{aligned} |J_2^{(i)}(t)| \leqslant{}& C_\nu A_{i-1}^2(t) + \int_0^1 \frac{h^2}{4}(\partial_t^i \psi_{xx})^2 \mathrm{d}x \\ &+ C(N_h(T)+\delta+h^{1/2}+\nu)A_i^2(t), \end{aligned} \tag{5.2.175}$$

where we have used the estimate (5.2.154). Finally, substituting (5.2.172)–(5.2.175) in (5.2.171) yields

$$\begin{aligned} &\int_0^1 ((\partial_t^i \psi_t)^2 + K(\partial_t^i \psi_x)^2 + \frac{h^2}{2}(\partial_t^i \psi_{xx})^2)\mathrm{d}x + \int_0^t \int_0^1 2(\partial_t^i \psi_t)^2 \mathrm{d}x\mathrm{d}\tau \\ &\leqslant CA_i^2(0) + C_\nu \int_0^1 [A_{i-1}^2(\tau)\mathrm{d}\tau + C(N_h(T)+\delta)||\partial_t^i \psi_x(t)||^2]\mathrm{d}x \\ &\quad + \int_0^t \int_0^1 \frac{h^2}{4}(\partial_t^i \psi_{xx})^2 \mathrm{d}x\mathrm{d}\tau + C(N_h(T)+\delta+h^{1/2}+\nu) \\ &\quad \times \int_0^t A_i^2(\tau)\mathrm{d}\tau. \end{aligned} \tag{5.2.176}$$

Multiplying (5.2.176) by 2, adding the resultant equality to (5.2.166), and let both $N_h(T)+\delta+h^{1/2}$ and ν be sufficiently small, we can deduce (5.2.162). □

Proof of Proposition 5.2.1. Combining (5.2.144) and (5.2.162), letting $(N_h(T)+\delta+h)$ be sufficiently small, we have the estimate (5.2.137), in the computation, we have also used the estimates (5.2.157) and (5.2.158). □

Decay estimate

The existence of the global solution to the problem (5.2.1)–(5.2.3) and (5.2.6)–(5.2.9) follows from Corollary 5.2.3 and Proposition 5.2.1. In order to prove Theorem 5.2.1, it is sufficient to verify the decay estimate (5.2.41).

Proof of Theorem 5.2.1. Substituting (5.2.149) in (5.2.147) and multiplying the resultant equality by β, β is a constant to be determined later. Moreover, multiplying (5.2.163) by β^2 with $i=0$, (5.2.171) by $2\beta^2$ with $i=0$, (5.2.153) by β^3 with $i=1$, (5.2.171) by $2\beta^3$ with $i=1$, summing up these results and (5.2.145), we have

$$\hat{E}(t)+\int_0^t \hat{F}(\tau)\mathrm{d}\tau=\hat{E}(0), t\in[0,\infty),$$

$$\hat{E}(t)=\int_0^1 \mathcal{E}-\beta\left(\frac{\eta+\tilde{j}}{(\psi+\tilde{\omega})^2}-\frac{\tilde{j}}{\tilde{\omega}^2}\right)\sigma_x\mathrm{d}_x+\sum_{i=0}^{1}\beta^{i+2}(I_1^{(i)}+2I_2^{(i)})(t),$$

$$\begin{aligned}\hat{F}(t)=&\int_0^1 \frac{\eta^2}{\tilde{\omega}^2}+\beta[K(\log(\psi+\tilde{\omega})^2-\log\tilde{\omega}^2)(\psi+2\tilde{\omega})\psi+\sigma_x^2\\&+h^2\left(1+\frac{\tilde{\omega}}{\psi+\tilde{\omega}}\right)\psi_x^2]\mathrm{d}x+\sum_{i=0}^{1}\beta^{i+2}\int_0^1(2\partial_t^i\psi_t)^2+2K(\partial_t^i\psi_x)^2\\&+h^2(\partial_t^i\psi_{xx})^2\mathrm{d}x-\int_0^1 R_2\mathrm{d}x-\beta\int_0^1\frac{1}{2}\left[\left(\frac{\eta+\tilde{j}}{(\psi+\tilde{\omega})^2}\right)^2-\left(\frac{\tilde{j}}{\tilde{\omega}^2}\right)^2\right]_x\sigma_x\\&+\left(\frac{\eta+\tilde{j}}{(\psi+\tilde{\omega})^2}-\frac{\tilde{j}}{\tilde{\omega}^2}\right)(\sigma_x-\sigma_{xt})\mathrm{d}x+\beta\int_0^1 h^2\left[\left(\frac{\tilde{\omega}}{\psi+\tilde{\omega}}\right)_x\psi_x\psi\right.\\&\left.+\frac{\psi+2\tilde{\omega}}{\tilde{\omega}(\psi+\tilde{\omega})}\tilde{\omega}_{xx}\psi^2\right]\mathrm{d}x-\sum_{i=0}^{1}\beta^{i+2}(j_1^{(i)}+2J_2^{(i)})(t).\end{aligned}\tag{5.2.177}$$

Let β and $(N_h(T)+\delta+h)$ be sufficiently small, such that $0 < N_h(T)+\delta+h \ll \beta^3 \ll \beta^2 \ll \beta \ll 1$. Then $\hat{E}(t)$ and $\hat{F}(t)$ are equivalent to $A_1^2(t)$. Hence, by using (5.2.158), $\hat{E}(t)$ and $\hat{F}(t)$ are also equivalent to $n_h^2(t)$. In fact, we can confirm this assertion by employing the Schwarz inequality, (5.2.71) as well as the estimates (5.2.141), (5.2.152)–(5.2.154) and (5.2.156)–(5.2.161).

Since $\hat{E}(t)$ is equivalent to $\hat{F}(t)$, then there exists a positive constant α, such that $\alpha\hat{E}(t) \leqslant \hat{F}(t)$. Differentiating (5.2.177) and substituting this inequality in the resultant equality yield the following ordinary differential inequality

$$\frac{d}{dt}\hat{E}(t) + \alpha\hat{E}(t) \leqslant 0, t \in [0,\infty). \tag{5.2.178}$$

Since $\hat{E}(t)$ is also equivalent to $n_\varepsilon^2(t)$, we deduce from (5.2.178)

$$n_\varepsilon^2(t) \leqslant Cn_\varepsilon^2(0)e^{-\alpha t}, \tag{5.2.179}$$

where C is positive constant independent of t and ε. The inequality (5.2.179) and the elliptic estimate (5.2.157) lead to the decay estimate (5.2.41). □

5.2.4 *Semiclassical limit*

In this subsection, we consider the semiclassical limit from the quantum hydrodynamic model to the hydrodynamic model. Let (ρ^h, j^h, ϕ^h) be a solution to the problem (5.2.1)–(5.2.3) and (5.2.6)–(5.2.9), $(\tilde{\rho}^h, \tilde{j}^h, \tilde{\phi}^h)$ be a solution to the problem (5.2.22)–(5.2.27), (ρ^0, j^0, ϕ^0) be a solution to the problem (5.2.6), (5.2.7) and (5.2.9), $(\rho^0, \tilde{j}^0, \tilde{\phi}^0)$ be a solution to the problem (5.2.27), (5.2.25) and (5.2.45)–(5.2.47), $(\tilde{\omega}^h, \tilde{j}^h, \tilde{\phi}^h)$ be a solution to the problem (5.2.81)–(5.2.86).

Firstly, we study the semiclassical limit for the stationary solution. Let $\tilde{\omega}^0 := \log \rho^0$, then the solution $(\tilde{\omega}^0, \tilde{j}^0, \tilde{\phi}^0)$ satisfies

$$S[e^{\tilde{\omega}^0}, \tilde{j}^0]\tilde{\omega}_x^0 - \phi_x^0 = -\tilde{j}^0 e^{-\tilde{\omega}^0}, \tag{5.2.180}$$

$$\tilde{\phi}_{xx}^0 = e^{\tilde{\omega}^0} - D. \tag{5.2.181}$$

We introduce the following functions

$$\tilde{W}^h := \tilde{\omega}^h - \tilde{\omega}^0, \tilde{J}^h := \tilde{j}^h - \tilde{j}^0.$$

Proof of Lemma 5.2.4. Firstly, we employ (5.2.82) and (5.2.180) to prove (5.2.50). Noting that if δ is small enough, as the derivation of the formula (5.2.37), $\tilde{j}^0$ is rewritten as $\tilde{j}^0 := \mathcal{J}[e^{\tilde{\omega}^0}]$. The following estimates follow from (5.2.37), (5.2.48) and (5.2.75), then

$$|\tilde{J}^h| \leqslant C|B_b|||\tilde{W}^h|| \leqslant C\delta||\tilde{W}^h||. \tag{5.2.182}$$

Subtracting (5.2.82) from (5.2.180), multiplying the result by $\tilde{W}_x^h$, integrating over Ω, then integrating by parts and using $\tilde{W}^h(0) = \tilde{W}^h(1) = 0$, $(\tilde{\omega}_{xx}^h + (\tilde{\omega}_x^h)^2/2(0) = (\tilde{\omega}_{xx}^h + (\tilde{\omega}_x^h)^2/2)(1) = 0$, (5.2.83) and (5.2.181), we have

$$\begin{aligned}
&\int_0^1 [S[e^{\tilde{\omega}}, \tilde{j}](\tilde{W}_x^h)^2 + (e^{\tilde{\omega}^h} - e^{\tilde{\omega}^0})\tilde{W}^h]\mathrm{d}x \\
&= \int_0^1 \left\{ \left[\left(\frac{\tilde{j}^h}{e^{\tilde{\omega}^h}} + \frac{\tilde{j}^0}{e^{\tilde{\omega}^0}} \right) \tilde{\omega}_x^h - 1 \right] \left(\frac{\tilde{j}^h}{e^{\tilde{\omega}^h}} - \frac{\tilde{j}^0}{e^{\tilde{\omega}^0}} \right) \tilde{W}_x^h \right. \\
&\quad \left. + \frac{h^2}{2} \left(\tilde{\omega}_{xx}^h + \frac{(\tilde{\omega}_x^h)^2}{2} \right) \tilde{W}_{xx}^h \right\} \mathrm{d}x.
\end{aligned} \tag{5.2.183}$$

The right hand side of (5.2.183) is estimated by $(C\delta||\tilde{W}_x^h||^2 + Ch^2)$, by the Hölder inequality, the Poincaré inequality, (5.2.48), (5.2.182) and Corollary 5.2.1. Noting that the second term on the left hand side of the equality is positive. If δ is small enough, $K - (\tilde{j}^0 e^{-\omega^0})^2 - C\delta > 0$, we have $||\tilde{W}_x^h|| \leqslant Ch$. By the Poincaré inequality, $||\tilde{W}^h||_1 \leqslant Ch$. Then $||(\tilde{\rho}^h - \tilde{\rho}^0)||_1 \leqslant Ch$. The other estimates of (5.2.50) are obtained by (5.2.182), (5.2.83) and (5.2.181).

Next we prove (5.2.51). For this purpose, we show that if $h \to 0$, $||\tilde{W}_{xx}^h||$ converges to 0, which gives the convergence of $||(\partial_x^2\{\tilde{\rho}^h - \tilde{\rho}^0\}, \partial_x^4\{\tilde{\phi}^h - \tilde{\phi}^0\})||$. Using the boundedness of $||\tilde{\omega}^h||_2$ and the convergence (5.2.181), we have

$$\tilde{\omega}_{xx}^h \rightharpoonup \tilde{\omega}_{xx}^0 \text{ weakly in } L^2, \text{ as } h \to 0. \tag{5.2.184}$$

Differentiating (5.2.82), and multiplying the resultant equalıty by $(\tilde{\omega}_{xx}^h + (\tilde{\omega}_{xx}^h)^2/2)$, and then integrating over Ω by parts yield

$$\int_0^1 \left\{ S[e^{\tilde{\omega}^h}, \tilde{j}^h](\tilde{\omega}_{xx}^h)^2 + \frac{h^2}{2}\left[\left(\tilde{\omega}_{xx}^h + \frac{(\tilde{\omega}_x^h)^2}{2}\right)_x\right]^2 \right\} dx = Q[\tilde{\omega}^h, \tilde{j}^h, \tilde{\phi}^h],$$

$$Q[\tilde{\omega}^h, \tilde{j}^h, \tilde{\phi}^h] := -\int_0^1 \left\{ S[e^{\tilde{\omega}^h}, \tilde{j}^h]\frac{(\tilde{\omega}_x^h)^2}{2}\tilde{\omega}_{xx}^h \right.$$

$$\left. + \left[S[e^{\tilde{\omega}^h}, \tilde{j}^h]_x \tilde{\omega}_x^h - \tilde{\phi}_{xx}^h + \left(\frac{\tilde{j}^h}{e^{\tilde{\omega}^h}}\right)_x \right] \left(\tilde{\omega}_{xx}^h + \frac{(\tilde{\omega}_x^h)^2}{2}\right) \right\} dx. \tag{5.2.185}$$

Owing to (5.2.50), (5.2.184), Corollary 5.2.1 and the estimate $||\tilde{W}^h||_1 \leqslant Ch$, as $h \to 0$, the quantity $Q[\tilde{\omega}^h, \tilde{j}^h, \tilde{\phi}^h]$ converges to

$$Q[\tilde{\omega}^0, \tilde{j}^0, \tilde{\phi}^0] = \int_0^1 S[e^{\tilde{\omega}^0}, \tilde{j}^0](\tilde{\omega}_{xx}^0)^2 dx. \tag{5.2.186}$$

Differentiating (5.2.180), multiplying the resultant equality by $(\tilde{\omega}_{xx}^0 + (\tilde{\omega}_x^0)^2/2)$, and then integrating by parts lead to the equality (5.2.186). On the other hand, by (5.2.186), Corollary (5.2.186) and the estimate $||\tilde{W}^h||_1 \leqslant\leqslant Ch$,

$$\limsup_{h\to 0} \int_0^1 S[e^{\tilde{\omega}^h}, \tilde{j}^h](\tilde{\omega}_{xx}^h)^2 dx = \limsup_{h\to 0} \int_0^1 S[e^{\tilde{\omega}^0}, \tilde{j}^0](\tilde{\omega}_{xx}^0)^2 dx \tag{5.2.187}$$

holds, then (5.2.185)–(5.2.187) yield the following inequality

$$\limsup_{h\to 0} \int_0^1 S[e^{\tilde{\omega}^0}, \tilde{j}^0](\tilde{\omega}_{xx}^h)^2 dx \leqslant \int_0^1 S[e^{\tilde{\omega}^0}, \tilde{j}^0](\tilde{\omega}_{xx}^0)^2 dx, \tag{5.2.188}$$

since $S[e^{\tilde{\omega}^0}, \tilde{j}^0] > c > 0$, (5.2.184)–(5.2.188), we see that $||\tilde{W}_{xx}^h||$ converges to 0.

We now prove the convergence $||(h\partial_x^3\tilde{\omega}^h, h^2\partial_x^4\tilde{\rho}^h)|| \to 0$, owing to (5.2.75) and Corollary 5.2.1. We have $||(h\partial_x^3\tilde{\rho}^h, h^2\partial_x^4\tilde{\rho}^h)|| \to 0$. In

(5.2.185), let $h \to 0$, $h\|(\tilde{\omega}^h_{xx} + (\tilde{\omega}^h_x)^2/2)_x\| \to 0$ hold. Together with Corollary 5.2.1, this implies that as $h \to 0$, $h\|\partial_x^3 \tilde{\omega}^h\| \to 0$ holds. Differentiating (5.2.82), and taking L^2-norm, we have

$$\frac{h^2}{2}\left\|\left(\tilde{\omega}^h_{xx} + \frac{(\tilde{\omega}^h_x)^2}{2}\right)_{xx}\right\| = \hat{n}_1[\tilde{\omega}^h, \tilde{j}^h, \tilde{\phi}^h]$$

$$:= \|(S[e^{\tilde{\omega}^h}, \tilde{j}^h]\tilde{\omega}^h_x)_x - \tilde{\phi}^h_{xx} + (\tilde{j}^h e^{-\tilde{\omega}^h})_x\|. \quad (5.2.189)$$

Noting that as $h \to 0$, by Corollary 5.2.1, we see that $h^2\|\{(\tilde{\omega}_x h)^2\}_{xx}\|$ converges to 0. Moreover, as $h \to 0$, $\hat{n}_1[\tilde{\omega}^h, \tilde{j}^h, \tilde{\phi}^h]$ converges to $\hat{n}_1[\tilde{\omega}^0, \tilde{j}^0, \tilde{\phi}^0]$. On the other hand, differentiating (5.2.180) with respect to x, we have the equality $\hat{n}_1[\tilde{\omega}^0, \tilde{j}^0, \tilde{\phi}^0] = 0$. Consequently, as $h \to 0$, $h^2\|\partial_x^4 \tilde{\omega}^h\| \to 0$ holds.

In order to study the semiclassical limit for the non-stationary problem, we introduce new functions

$$R^h := \rho^h - \rho^0, J^h := j^h - j^0, \quad \Phi^h := \phi^h - \phi^0.$$

Subtracting (5.2.1)–(5.2.3) from (5.2.42)–(5.2.44), we have

$$R^h_t + J^h_x = 0, \quad (5.2.190)$$

$$J^h_t + KR^h_x - \left[\left(\frac{j^h}{\rho^h}\right)^2 \rho^h_x - \left(\frac{j^0}{\rho^0}\right)^2 \rho^0_x\right] + \left(\frac{2j^h}{\rho^h} j^h_x - \frac{2j^0}{\rho^0} j^0_x\right)$$

$$-(R^h \phi^h_x + \rho^0 \Phi^h_x) + J^h = h^2 \rho^h \left(\frac{(\sqrt{\rho^h})_{xx}}{\sqrt{\rho^h}}\right)_x, \quad (5.2.191)$$

$$\Phi^h_{xx} = R^h. \quad (5.2.192)$$

The boundary condition is obtained from (5.2.7) and (5.2.9),

$$R^h(t,0) = R^h(t,1) = R^h_t(t,0) = R^h_t(t,1) = \Phi^h(t,0)$$
$$= \Phi^h(t,1) = 0. \quad (5.2.193)$$

Differentiating (5.2.191) with respect to x, and using (5.2.190) yield

$$R^h_{tt} - KR^h_{xx} + \left[\left(\frac{j^h}{\rho^h}\right)^2 \rho^h_x - \left(\frac{j^0}{\rho^0}\right)^2 \rho^0_x\right]_x + \left(\frac{2j^h}{\rho^h} j^h_x - \frac{2j^0}{\rho^0} j^0_x\right)_x$$

$$-(R^h\phi^h_x + \rho^0\Phi^h_x)_x + R^h_t = -\left[h^2\rho^h\left(\frac{(\sqrt{\rho^h})_{xx}}{\sqrt{\rho^h}}\right)_x\right]_x. \quad (5.2.194)$$

The following estimates have been obtained

$$||(\rho^0, j^0, \phi^0)(t)||_2 + ||(\rho^0_t, j^0_t)(t)||_1 \leqslant C, \quad (5.2.195)$$

$$\rho^0, S[\rho^0, j^0] > c > 0 \quad (5.2.196)$$

where C and c are positive constants independent of t. □

Proof of Theorem 5.2.2. Multiplying (5.2.191) by J^h, integrating over Ω by parts and using the boundary condition (5.2.8), we have

$$\frac{d}{dt}\int_0^1 \frac{1}{2}(J^h)^2 dx + \int_0^1 (J^h)^2 dx$$

$$\int_0^1 -KR^h_x J^h + \left[\left(\frac{j^h}{\rho^h}\right)^2 \rho^h_x - \left(\frac{j^0}{\rho^0}\right)^2 \rho^0_x\right] J^h dx$$

$$+ \int_0^1 -\left(\frac{2j^h}{\rho^h} j^h_x - \frac{2j^0}{\rho^0} j^0_x\right) J^h + (R^h\phi^h_x + \rho^0\Phi^h_x)J^h - h^2\frac{(\sqrt{\rho^h})_{xx}}{\sqrt{\rho^h}}$$

$$\times (\rho^h J^h)_x dx$$

$$\leqslant C||(R^h, R^h_x, J^h, J^h_x)(t)||^2 + Ch^2. \quad (5.2.197)$$

The last term on the right hand side of (5.2.197) is estimated as

$$|h^2 \int_0^1 \frac{(\sqrt{\rho^h})_{xx}}{\sqrt{\rho^h}}(\rho^h J^h)_x dx|$$

$$\leqslant h^2 \left|\frac{1}{\sqrt{\rho^h}}\right|_0 \left|\left|(\sqrt{\rho^h})_{xx}\right|\right| \left\{\left|\rho^h_x\right|_0 \left|\left|J^h\right|\right| + \left|\rho^h\right|_0 \left|\left|J^h_x\right|\right|\right\} \leqslant Ch^2.$$

(5.2.198)

The other terms can be estimated by (5.2.37), (5.2.75), (5.2.137), (5.2.195)–(5.2.196), Corollary 5.2.1 and $||\Phi^h(t)||_1 \leqslant C||R^h(t)||$.

Multiplying (5.2.194) by R_t^h, integrating over Ω by parts and using the boundary condition (5.2.8) yield

$$\frac{\mathrm{d}}{\mathrm{dt}}\int_0^1 \frac{1}{2}(R_t^h)^2 + \frac{1}{2}S[\rho^0, j^0](R_x^h)^2\mathrm{d}x + \int_0^1 (R_t^h)^2\mathrm{d}x = Q_3(t),$$

$$\begin{aligned} Q_3(t) := &- \int_0^1 \frac{1}{2}\left[\left(\frac{j^0}{\rho^0}\right)^2\right]_x (R_x^h)^2 - \left\{\left[\left(\frac{j^0}{\rho^0}\right)^2 - \left(\frac{j^h}{\rho^h}\right)^2\right]\rho_x^h\right\}_x R_t^h \\ &- \left(\frac{j^0}{\rho^0}\right)_x (R_t^h)^2\mathrm{d}x + \int_0^1 \left[\left(\frac{2j^h}{\rho^h} - \frac{2j^0}{\rho^0}\right) j_x^h\right]_x R_t^h - (R^h\phi_x^h + \rho^0\Phi_x^h)R_t^h \\ &+ h^2\rho^h\left(\frac{(\sqrt{\rho^h})_{xx}}{\sqrt{\rho^h}}\right)_x R_{xx}^h\mathrm{d}x. \end{aligned} \tag{5.2.199}$$

Owing to (5.2.137), (5.2.158), (5.2.195)–(5.2.196) and Corollary 5.2.1, the last term of $Q_3(t)$ is estimated as

$$\begin{aligned} &h^2\int_0^1 \{\sqrt{\rho^h}(\sqrt{\rho^h})_{xxx} - (\sqrt{\rho^h})_{xx}(\sqrt{\rho^h})_x\}R_{xt}^h\mathrm{d}x \\ &\quad \leqslant h^2C(||(\sqrt{\rho^h})_{xxx}(t)|| + ||(\sqrt{\rho^h})_{xx}(t)||)||R_{xt}^h(t)|| \leqslant Ch. \end{aligned} \tag{5.2.200}$$

Substituting (5.2.200) in (5.2.199), the other terms of $Q_3(t)$ can be estimated by the Schwarz inequality, the Sobolev inequality, (5.2.137), (5.2.195)–(5.2.196), Corollary 5.2.1 and $||\Phi^h(t)||_3 \leqslant C||R^h(t)||_1$, we have

$$\begin{aligned} &\frac{\mathrm{d}}{\mathrm{dt}}\int_0^1 \frac{1}{2}(R_t^h)^2 + \frac{1}{2}S[\rho^0, j^0](R_x^h)^2\mathrm{d}x + \int_0^1 (R_t^h)^2\mathrm{d}x \\ &\quad \leqslant C||(R_t^h, R_x^h, R^h, J^h, J_x^h)(t)||^2 + Ch. \end{aligned} \tag{5.2.201}$$

Noting that, due to (5.2.190), $R_t^h(0,x) = J_x^h(0,x) = 0$ and $||R_t^h(t)|| = ||J_x^h(t)||$ hold. By (5.2.72), $||R^h(t)|| \leqslant C||R_x^h(t)||$ holds. Therefore, by adding (5.2.201) to (5.2.197), and applying the Gronwall inequality, we have the estimate (5.2.53). Finally, we prove the estimate (5.2.54), letting $\gamma \in (0, 1/2)$ be fixed and $T_1 := (\log 1/h^\gamma)/\beta$. For $t \leqslant T_1$, the estimate (5.2.53) yields

$$\|(R^h, J^h)(t)\|_1 \leqslant \sqrt{h} C e^{\beta T_1} \leqslant C h^{(1/2)-\gamma}, \tag{5.2.202}$$

For $T_1 \leqslant t$, by using estimates (5.2.41), (5.2.49), (5.2.50), we have

$$\begin{aligned} \|(R^h, J^h)(t)\|_1 &\leqslant C\|(\rho^h - \tilde{\rho}^h, j^h - \tilde{j}^h, \rho^0 - \tilde{\rho}^0, j^0 - \tilde{j}^0, \tilde{\rho}^h \\ &\quad - \tilde{\rho}^0, \tilde{j}^h - \tilde{j}^0)(t)\|_1 \\ &\leqslant C(e^{-\alpha_1 T_1} + e^{-\alpha_2 T_1} + h) \leqslant C(h^{\alpha_1\gamma/\beta} + h^{\alpha_2\gamma/\beta} + h). \end{aligned} \tag{5.2.203}$$

Owing to (5.2.202) and (5.2.203), as $h \to 0$, $\|(R^h, J^h)(t)\|_1$ converges to 0. The other assertion in Theorem 5.2.2 follows from $\|\Phi^h(t)\|_3 \leqslant C\|R^h(t)\|_1$. □

Chapter 6

Asymptotic Limit to the Bipolar Quantum Hydrodynamic Equations

6.1 Semiclassical limit

In this section we will focus on the semiclassical limit to the bipolar quantum hydrodynamic model [99], [101], [107]

$$\partial_t \rho_i + \nabla \cdot (\rho_i \boldsymbol{u}_i) = 0, \tag{6.1.1}$$

$$\begin{aligned} &\partial_t(\rho_i \boldsymbol{u}_i) + \nabla \cdot (\rho_i \boldsymbol{u}_i \otimes \boldsymbol{u}_i) + \nabla P_i(\rho_i) = q_i \rho_i \boldsymbol{E} \\ &\quad + \frac{h^2}{2} \rho_i \nabla \left(\frac{\Delta \sqrt{\rho i}}{\sqrt{\rho i}} \right) - \frac{\rho_i \boldsymbol{u}_i}{\tau_i}, \end{aligned} \tag{6.1.2}$$

$$\begin{aligned} &\lambda^2 \nabla \cdot \boldsymbol{E} = \rho_a - \rho_b - \mathcal{C}, \nabla \times \boldsymbol{E} \\ &\quad = 0, E(x) \to 0, |x| \to +\infty, \end{aligned} \tag{6.1.3}$$

where $(x, t) \in \mathbf{R}^3 \times \mathbf{R}^+$, the index $i = a, b$, and constants $q_a = 1$, $q_b = -1$. The unknown functions $\rho_a > 0$, $\rho_b > 0$, $\boldsymbol{u}_a$, $\boldsymbol{u}_b$ and $\boldsymbol{E}$ are the density, velocities and electric field, respectively. $P_a(\cdot)$ and $P_b(\cdot)$ are the pressure-density functions. The parameters $h > 0$, $\tau_a = \tau_b = \tau > 0$, and $\lambda > 0$ are the scaled Planck constant, momentum relaxation time, and Debye length respectively. $\mathcal{C} = \mathcal{C}(x)$ is doping profile.

Let $h \to 0$ formally in (6.1.1)–(6.1.3), we can obtain the well-known bipolar hydrodynamic(HD) model [102], [105]

$$\partial_t \rho_i + \nabla \cdot (\rho_i \boldsymbol{u}_i) = 0, \tag{6.1.4}$$

$$\partial_t(\rho_i \boldsymbol{u}_i) + \boldsymbol{\nabla} \cdot (\rho_i \boldsymbol{u}_i \otimes \boldsymbol{u}_i) + \boldsymbol{\nabla} P_i(\rho_i) = q_i \rho_i \boldsymbol{E} - \frac{\rho_i \boldsymbol{u}_i}{\tau_i}, \quad (6.1.5)$$

$$\lambda^2 \boldsymbol{\nabla} \cdot \boldsymbol{E} = \rho_a - \rho_b - \mathcal{C}, \quad \nabla \times \boldsymbol{E} = 0, \; E(x) \to 0, |x| \to +\infty. \quad (6.1.6)$$

This limiting process shows the semiclassical approximation of bipolar quantum hydrodynamic model in terms of bipolar HD model for small Planck constant, and describes the relation from quantum mechanics to the classical Newtonian mechanics.

Next we will give some notations: C and c are the generic positive constants. $L^2(\mathbf{R}^3)$ is the space of square integral functions on $\mathbf{R}^3$ with the norm $\|\cdot\|$ or $\|\cdot\|_{L^2(\mathbf{R}^3)}$. $H^k(\mathbf{R}^3)(k \geq 1)$ denoting the usual Sobolev space of function f satisfying $\partial_x^i f \in L^2(\mathbf{R}^3)(0 \leq i \leq k)$ with norm $\|f\|_k = \sqrt{\sum_{0 \leq \|\alpha\| \leq k} \|D^\alpha f\|^2}$. Here and after $\alpha \in \mathbf{N}^3$, $D^\alpha = \partial_{x_1}^{s_1} \partial_{x_2}^{s_2} \partial_{x_3}^{s_3}$ for $|\alpha| = s_1 + s_2 + s_3$. Especially, $\|\cdot\|_0 = \|\cdot\|$. Let $\mathcal{B}$ be a Banach space. $C^k([0,T];\mathcal{B})$ denotes the space of $\mathcal{B}$-valued k-times continuously differentiable functions on $[0,T]$. We can extend the above norm to the vector-valued function $\boldsymbol{u} = (\boldsymbol{u}_1, \boldsymbol{u}_2, \boldsymbol{u}_3)$ with $|D^\alpha \boldsymbol{u}|^2 = \sum_{r=1}^3 |D^\alpha \boldsymbol{u}_r|^2$ and $\|D^k \boldsymbol{u}\|^2 = \int_{\mathbf{R}^3} (\sum_{r=1}^3 \sum_{|\alpha|=k} (D^\alpha \boldsymbol{u}_r)^2) \mathrm{d}x$, and $\|\boldsymbol{u}\|_k = \|\boldsymbol{u}\|_{H^k(\mathbf{R}^3)} = \sum_{i=0}^k \|D^i \boldsymbol{u}\|$, $\|f\|_{L^\infty([0,T];\mathcal{B})} = \sup_{0 \leq t \leq T} \|f(t)\|_{\mathcal{B}}$. We introduce a special space $\mathcal{H}^k(\mathbf{R}^3) = \{f \in L^6(\mathbf{R}^3), \, Df \in H^{k-1}(\mathbf{R}^3)\}(k \geq 1)$. Sometimes we use $\|(\cdot,\cdot,\cdot\cdot,\cdot)\|_{H^k(\mathbf{R}^3)}$ or $\|(\cdot,\cdot,\cdots)\|_k$ to denote the norm of the space $H^k(\mathbf{R}^3) \times H^k(\mathbf{R}^3) \times \cdots \times H^k(\mathbf{R}^3)$ and $\mathcal{H}^k(\mathbf{R}^3)$ as well. At the same time without confusion, we will omit $\mathbf{R}^3$ in this section.

6.1.1 *Main results*

In this section we only consider the initial value problem of the equations (6.1.1)–(6.1.3). The initial data are

$$(\rho_i, \boldsymbol{u}_i)(x,0) = (\rho_{i0}, \boldsymbol{u}_{i0})(x), \rho_{i0}(x) \to \rho_i^*, \quad \boldsymbol{u}_{i0}(x) \to 0,$$
$$|x| \to +\infty, i = a, b. \quad (6.1.7)$$

Here we set the scaled Debye length to be one $\lambda = 1$ for simplicity.

Next we give the theorems about the global existence and uniqueness of the equations (6.1.1)–(6.1.3).

Theorem 6.1.1 (Global existence). *Let the parameters* $h > 0$, $\tau > 0$ *be fixed. Assume* $P_a, P_b, \in C^5(0, +\infty)$ *and* $C = c^*$ *is a constant satisfying for two positive constants* ρ_a^*, ρ_b^* *that*

$$\rho_a^* - \rho_b^* - c^* = 0, \ P_a'(\rho_a^*), P_b'(\rho_b^*) > 0. \tag{6.1.8}$$

Suppose $\rho_{a0} > 0$, $\rho_{b0} > 0$ *and* $(\sqrt{\rho a0} - \sqrt{\rho_a^*}, \sqrt{\rho b0} - \sqrt{\rho_b^*}, \boldsymbol{u}_{a0}, \boldsymbol{u}_{b0}) \in (H^6)^2 \times (\mathcal{H}^5)^2$. *Then, there is* $\boldsymbol{\Lambda}_1 > 0$, *so that if* $\boldsymbol{\Lambda}_0 := \|(\sqrt{\rho a0} - \sqrt{\rho_a^*}, \sqrt{\rho b0} - \sqrt{\rho_b^*}, \boldsymbol{u}_{a0}, \boldsymbol{u}_{b0})\|_{H^6 \times \mathcal{H}^5} \leq \boldsymbol{\lambda}_1$, *the unique solution* $(\rho_a^h, \rho_b^h, \boldsymbol{u}_a^h, \boldsymbol{u}_b^h, \boldsymbol{E}^h)$ *of the IVP problem* (6.1.1)–(6.1.3) *and* (6.1.7) *exists globally in time with* $\rho_a^h > 0$, $\rho_b^h > 0$, *and satisfies*

$$(\rho_i^h - \rho_i^*, \boldsymbol{E}^h) \in C^k(0, T; H^{6-2k}), \boldsymbol{u}_i^h \in C^k(0, T; \mathcal{H}^{5-2k}), k = 0, 1, 2,$$

and

$$\|(\rho_a^h - \rho_a^*, \rho_b^h - \rho_b^*)\|_{L^\infty} + \|\boldsymbol{E}^h\|_{L^\infty} + \|(\boldsymbol{u}_a^h, \ \boldsymbol{u}_b^h)\|_{L^\infty \to 0}, \ t \to \infty.$$

Then we will state the semiclassical limit of the global solution given in Theorem 6.1.1 for any fixed momentum relaxation time $\tau > 0$.

Theorem 6.1.2 (Semiclassical limit). *Let* $\tau = 1$. $(\rho_a^h, \rho_b^h, \boldsymbol{u}_a^h, \boldsymbol{u}_b^h, \boldsymbol{E}^h)$ *is the global solution to IVP* (6.1.1)–(6.1.3) *and* (6.1.7) *given in Theorem* 6.1.1. *Then, there is* $(\rho_a, \rho_b, \boldsymbol{u}_a, \boldsymbol{u}_b, \boldsymbol{E})$ *with* $\rho a > 0$, $\rho b > 0$ *so that as Planck constant* $h \to 0$, *it holds, for any* $T > 0$, $i = a, b$,

$$\rho_i^h \to \rho_i, \ \textit{strongly in } C(0, T; C_b^3 \cap H_{loc}^{5-s}),$$
$$\boldsymbol{u}_i^h \to \boldsymbol{u}_i, \ \textit{strongly in } C(0, T; C_b^3 \cap \mathcal{H}_{loc}^{5-s}),$$
$$\boldsymbol{E}^h \to \boldsymbol{E}, \ \textit{strongly in } C(0, T; C_b^4 \cap H_{loc}^{6-s}), s \in \left(0, \frac{1}{2}\right).$$

Note that $(\rho_i, \boldsymbol{u}_i, \boldsymbol{E})(i = a, b)$ *is the global solution to the bipolar hydrodynamic model* (6.1.4)–(6.1.7).

6.1.2 *Preliminaries*

In order to prove Theorem 6.1.1 and Theorem 6.1.2, we need to introduce some lemmas.

Lemma 6.1.1. *Let $f \in H^s(\boldsymbol{R}^3)$, $s \geq \frac{3}{2}$. There is a unique solution of the divergence equation*

$$\nabla \cdot \boldsymbol{u} = f, \ \nabla \times \boldsymbol{u} = 0, \ \boldsymbol{u}(x) \to 0, |x| \to +\infty,$$

satisfying

$$||\boldsymbol{u}||_{L^6} \leq C||f||_{L^2}, \ ||D\boldsymbol{u}||_{H^s} \leq C||f||_{H^s}.$$

Lemma 6.1.2. *Set $f \in H^s(\boldsymbol{R}^3)$, $s \geq \frac{3}{2}$, and $\nabla \cdot \boldsymbol{f} = 0$. Then there is a unique solution of the vorticity equation*

$$\nabla \times \boldsymbol{u} = \boldsymbol{f}, \ \nabla \cdot \boldsymbol{u} = 0, \boldsymbol{u}(x) \to 0, |x| \to +\infty,$$

satisfying

$$||\boldsymbol{u}||_L^6 \leq C||\boldsymbol{f}||_{L^2}, \quad ||D\boldsymbol{u}||_H^s \leq C||\boldsymbol{f}||_{H^s}.$$

Lemma 6.1.3 (Moser Lemma). *Let $f, g \in H^s(\boldsymbol{R}^3) \cap L^\infty(\boldsymbol{R}^3)$. Then for any $\alpha, \in \mathbb{N}^3$, $1 \leq |\alpha| s$ (s is an integer), there is*

$$||D^\alpha(\boldsymbol{f}g)|| \leq C||g||_{L^\infty} \cdot ||D^\alpha \boldsymbol{f}|| + C||\boldsymbol{f}||_{L^\infty} \cdot ||D^\alpha g||,$$

$$||D^\alpha(\boldsymbol{f}g) - fD^\alpha g|| \leq C||g||_{L^\infty} \cdot ||D^\alpha \boldsymbol{f}|| + C||\boldsymbol{f}||_{L^\infty} \cdot ||D^{|\alpha|-1} g||.$$

Lemma 6.1.4. *Let $\boldsymbol{f} \in H^s(\boldsymbol{R}^3)$ (s is an integer)w, and function $F(\rho)$ is smooth enough, and $F(0) = 0$. Then $F(\boldsymbol{f})(x) \in H^s(\boldsymbol{R}^3)$, and*

$$||F(\boldsymbol{f})||_{H^s} \leq C||\boldsymbol{f}||_{H^s}.$$

6.1.3 *Proof of main results*

In [100], [103], the local existence of the unipolar QHD model has been obtained. The main method is to research an extended problem derived based on a deposition of the original problem. Then construct an approximate solution sequence, and prove the convergence of this approximate solution to the extended problem. This method can be applied to the bipolar model directly. Here we only give the local existence, and omit the details.

Theorem 6.1.3 (Local existence). *Let the parameters $\varepsilon > 0$, $\tau > 0$, $\lambda > 0$ be fixed. Assume that there are constants $\rho_a^* > 0$, $\rho_b^* > 0$, and c^* satisfying $\rho_a^* - \rho_b^* - c^* = 0$ and $\mathcal{C}(x) - c^* \in H^5(\boldsymbol{R}^3)$, $P_a, P_b \in C^5(0,\infty)$. If $(\sqrt{\rho_{i0}} - \sqrt{\rho_i^*}, \boldsymbol{u}_{i0}) \in H^6(\boldsymbol{R}^3) \times \mathcal{H}^5(\boldsymbol{R}^3)(\rho_{i0} > 0)$, then there exists a finite time $T^* > 0$, such that the unique solution $(\rho_a, \rho_b, \boldsymbol{u}_a, \boldsymbol{u}_b, \boldsymbol{E})$ with $\rho_a > 0$, $\rho_b > 0$ to the problem (6.1.1)–(6.1.3) and (6.1.7) exists in $[0, T^*]$, and it satisfies: $i = a, b$,*

$$\rho_i - \rho_i^* \in C^k([0,T^*]; H^{6-2k}(\boldsymbol{R}^3)),\ \ \boldsymbol{u}_i \in C^k([0,T^*];\ \mathcal{H}^{5-k}(\boldsymbol{R}^3)),$$
$$k = 0,1,2,\ \ \boldsymbol{E} \in C^k([0,T^*];\ \mathcal{H}^{6-k}(\boldsymbol{R}^3)),\ \ k = 0,1.$$

According to the statement of Section 6.2, we make a scale transformation to the equations (6.1.1)–(6.1.3) [107]

$$x \to x,\ t \to \frac{t}{\tau}, (\rho_i^\tau, \boldsymbol{u}_i^\tau, \boldsymbol{E}^\tau)(x,t) = \left(\rho_i, \frac{\boldsymbol{u}_i}{\tau}, \boldsymbol{E}\right)\left(x, \frac{t}{\tau}\right). \tag{6.1.9}$$

Then we can rewrite the equations (6.1.1)–(6.1.3) as

$$\partial_t \rho_i^\tau + \operatorname{div}(\rho_i^\tau \boldsymbol{u}_i^\tau) = 0, \tag{6.1.10}$$

$$\tau^2 \partial_t(\rho_i^\tau \boldsymbol{u}_i^\tau) + \tau^2 \operatorname{div}(\rho_i^\tau \boldsymbol{u}_i^\tau \otimes \boldsymbol{u}_i^\tau \otimes \boldsymbol{u}_i^\tau) + \boldsymbol{\nabla} P_i(\rho_i^\tau)$$
$$= q_i \rho_i^\tau \boldsymbol{E}^\tau + \frac{h^2}{2} \rho_i^\tau \nabla \left(\frac{\Delta \sqrt{\rho_i^T}}{\sqrt{\rho_i^T}} \right) - \rho_i^\tau \boldsymbol{u}_i^\tau, \tag{6.1.11}$$

$$\lambda^2 \boldsymbol{\nabla} \cdot \boldsymbol{E}^\tau = \rho_a^\tau - \rho_b^\tau - \mathcal{C}(x),\ \ \boldsymbol{\nabla} \times \boldsymbol{E}^\tau = 0,$$
$$\boldsymbol{E}^\tau(x) \to 0, |x| \to +\infty. \tag{6.1.12}$$

For simplicity, we set $\lambda = 1$, and we omit the superscribe τ in this section. Denoting $\psi_i = \sqrt{\rho_i^\tau}$ and $\boldsymbol{u}_i = \boldsymbol{u}_i^\tau (i = a, b)$, from the equations (6.1.10)–(6.1.12), we can get

$$\tau^2 \psi_{itt} + \psi_{it} + \frac{h^2 \Delta^2 \psi_i}{4} + \frac{q_i}{2\psi_i} \operatorname{div}(\psi_i^2 \boldsymbol{E}) - \frac{1}{2\psi_i} \boldsymbol{\nabla}^2 (\psi_i^2 \boldsymbol{u}_i \otimes u_i)$$
$$- \frac{1}{2\psi_i} \Delta P_i(\psi_i^2) + \frac{\psi_{it}^2}{\psi_i} - \frac{h^2 |\Delta \psi_i|^2}{4\psi_i} = 0. \tag{6.1.13}$$

The initial data are:

$$\psi_i(x,0) = \psi_{i0}(x) = \psi_{i0}^{\tau}(x) = \sqrt{\rho_{i0}(x)},$$
$$\psi_{it}(x,0) = \psi_{i1}(x) = -\frac{1}{2}\psi_{i0}^{\tau}\nabla\cdot\boldsymbol{u}_{i0}^{\tau} - \boldsymbol{u}_{i0}^{\tau}\cdot\nabla\psi_{i0}^{\tau}.$$

Since $(\boldsymbol{u}_i\cdot\nabla)\boldsymbol{u}_i = \frac{1}{2}\nabla(|\boldsymbol{u}_i|^2) - \boldsymbol{u}_i\times(\nabla\times\boldsymbol{u}_i)$, from (6.1.10)–(6.1.12), we can obtain the equations about $\boldsymbol{u}_i = \boldsymbol{u}_i^{\tau}(i=a,b)$:

$$\begin{aligned}\tau^2\boldsymbol{u}_{it} + \boldsymbol{u}_i + \frac{\tau^2}{2}\nabla(|\boldsymbol{u}_i|^2) - \tau^2\boldsymbol{u}_i\times\boldsymbol{\phi}_i + \frac{\nabla(\psi_i^2)}{\psi_i^2} \\ = q_i\boldsymbol{E} + \frac{h^2}{2}\nabla\left(\frac{\Delta\psi_i}{\psi_i}\right),\end{aligned} \tag{6.1.14}$$

where $\boldsymbol{\phi}_i = \nabla\times\boldsymbol{u}_i$ is the vorticity of $\boldsymbol{u}_i$. Taking curl of (6.1.14), we have

$$\tau^2\boldsymbol{\phi}_{it} + \boldsymbol{\phi}_i + \tau^2(\boldsymbol{u}_i\cdot\nabla)\boldsymbol{\phi}_i + \tau^2\boldsymbol{\phi}_i\nabla\cdot\boldsymbol{u}_i - \tau^2(\boldsymbol{\phi}_i\cdot\nabla)\boldsymbol{u}_i = 0. \tag{6.1.15}$$

Introduce new variables $\omega_i = \psi_i - \sqrt{\rho_i^*}(i=a,b)$, the equations about $(\omega_a,\,\omega_b,\,\boldsymbol{\phi}_a,\,\boldsymbol{\phi}_b,\,\boldsymbol{E})$ are

$$\begin{aligned}\tau^2\omega_{att} + \omega_{at} + \frac{h^2\Delta^2\omega_a}{4} + \frac{1}{2}(\omega_a + \sqrt{\rho_a^*}) \\ \times\nabla\cdot\boldsymbol{E} - P_a'(\rho_a^*)\Delta\omega_a = f_{a1},\end{aligned} \tag{6.1.16}$$

$$\begin{aligned}\tau^2\omega_{btt} + \omega_{bt} + \frac{h^2\Delta^2\omega_b}{4} - \frac{1}{2}(\omega_b + \sqrt{\rho_b^*}) \\ \times\nabla\cdot\boldsymbol{E} - P_b'(\rho_b^*)\Delta\omega_b = f_{b1},\end{aligned} \tag{6.1.17}$$

$$\tau^2\boldsymbol{\phi}_{at} + \boldsymbol{\phi}_a = f_{a2}, \tag{6.1.18}$$

$$\tau^2\boldsymbol{\phi}_{bt} + \boldsymbol{\phi}_b = f_{b2}, \tag{6.1.19}$$

$$\nabla\cdot\boldsymbol{E} = \omega_a^2 - \omega_b^2 + 2\sqrt{\rho_a^*}\omega_a - 2\sqrt{\rho_b^*}\omega_b,\ \nabla\times\boldsymbol{E} = 0, \tag{6.1.20}$$

where

$$
\begin{aligned}
f_{i1} := f_{i1}(x,t) = & \frac{-\tau^2\omega_{it}^2}{\omega_i + \sqrt{\rho_i^*}} - q_i \nabla\omega_i \boldsymbol{E} + (P_i'((\omega + \sqrt{\rho_i^*})^2) \\
& - P_i'(\rho_i^*))\Delta\omega_i + 2(\omega_i + \sqrt{\rho_i^*})P_i''((\omega_i + \sqrt{\rho_i^*})^2)|\nabla\omega_i|^2 \\
& + P_i'((\omega_i + \sqrt{\rho_i^*})^2)\frac{|\nabla\omega_i|^2}{\omega_i + \sqrt{\rho_i^*}} + \frac{h^2(\Delta\omega_i)^2}{4(\omega_i + \sqrt{\rho_i^*})} \\
& + \frac{\tau^2\nabla^2((\omega_i + \sqrt{\rho_i^*})^2\boldsymbol{u}_i \otimes \boldsymbol{u}_i)}{2(\omega_i + \sqrt{\rho_i^*})},
\end{aligned} \tag{6.1.21}
$$

$$
\begin{aligned}
f_{i2} := f_{i2}(x,t) = & \tau^2((\boldsymbol{\phi}_i \cdot \nabla)\boldsymbol{u}_i - (\boldsymbol{u}_i \cdot \nabla)\boldsymbol{\phi}_i - \boldsymbol{\phi}_i \nabla \cdot \boldsymbol{u}_i), \\
& i = a, b.
\end{aligned} \tag{6.1.22}
$$

The last term in (6.1.21) can be decomposed by using equation (6.1.10) as

$$
\begin{aligned}
& \frac{\tau^2\nabla^2((\omega_i + \sqrt{\rho_i^*})^2\boldsymbol{u}_i \otimes \boldsymbol{u}_i)}{2(\omega_i + \sqrt{\rho_i^*})} = \tau^2\Bigg[- \omega_{it}\nabla \cdot \boldsymbol{u}_i \\
& - 2\boldsymbol{u}_i \cdot \nabla\omega_{it} - \frac{\omega_{it}\boldsymbol{u}_i \cdot \nabla\omega_i}{2(\omega_i + \sqrt{\rho_i^*})} + \nabla_{\omega_i} \cdot ((\boldsymbol{u}_i \cdot \nabla)\boldsymbol{u}_i), \\
& \frac{\omega_i + \sqrt{\rho_i^*}}{2}\sum_{k,l=1}^{3}|\partial_k \boldsymbol{u}_i^l|^2 - \frac{\omega_i + \sqrt{\rho_i^*}}{2}|\boldsymbol{\phi}_i|^2 - \boldsymbol{u}_i \cdot \nabla(\boldsymbol{u}_i \cdot \nabla\omega_i) \\
& + \frac{1}{2(\omega_i + \sqrt{\rho_i^*}}(\omega_{it} + \boldsymbol{u}_i \cdot \nabla\omega_i)(\boldsymbol{u}_i \cdot \nabla\omega_i)\Bigg], \ i = a, b.
\end{aligned} \tag{6.1.23}
$$

The initial conditions are

$$
\begin{aligned}
& \omega_i(x,0) := \omega_{i0} = \psi_{i0} - \sqrt{\rho_i^*}, \ \boldsymbol{\phi}_i(x,0) := \boldsymbol{\phi}_{i0}(x) = \frac{1}{\tau}\nabla \times \boldsymbol{u}_{i0}(x), \\
& \omega_{it}(x,0) := \omega_{i1}(x) = \frac{1}{\tau}\left(-\boldsymbol{u}_{i0} \cdot \nabla\omega_{i0} - \frac{1}{2}(\omega_{i0} + \sqrt{\rho_i^*})\nabla \cdot \boldsymbol{u}_{i0}(x)\right), \\
& \qquad i = a, b.
\end{aligned}
$$

We will also use the relation between $\nabla \cdot \boldsymbol{u}_i$ and $\nabla\omega$, ω_{it}:

$$
2\omega_{it} + 2\boldsymbol{u}_i \cdot \nabla\omega_i + (\omega_i + \sqrt{\rho_i^*})\nabla \cdot \boldsymbol{u}_i = 0, \ i = a, b. \tag{6.1.24}
$$

In this section, we will mainly study the reformulated equations (6.1.16)–(6.1.20) in order to obtain *a priori* estimate of ω_a, ω_b, $\boldsymbol{\phi}_a$, $\boldsymbol{\phi}_b$, $\boldsymbol{E}$.

Set the workspace as

$$X(T) = \{(\omega_a, \omega_b, \boldsymbol{u}_a, \boldsymbol{u}_b) \in L^\infty([0,T]; (H^6(\mathbf{R}^3))^2 \times (\mathcal{H}^5(\mathbf{R}^3))^2)\}$$

and assume the quantity

$$\begin{aligned}\delta_T = \max_{0\le t\le T}\{&\|(\omega_a,\omega_b)(\cdot,t)\|_4^2 + \|\tau(\partial_t\omega_a,\ \partial_t\omega_b)(\cdot,t)\|_3^2 \\ &+ \|\tau(\boldsymbol{u}_a,\boldsymbol{u}_b)(\cdot,t)\|_{\mathcal{H}^4}^2\} + \int_0^{\mathrm{T}} \|(\boldsymbol{u}_a,\boldsymbol{u}_b)(.,t)\|_{\mathcal{H}^3}^2 \\ &+\|(\omega_a,\omega_b)(\cdot,t)\|_5^2 + \|E(\cdot,t)\|_{\mathcal{H}}^2 \mathrm{d}t \end{aligned} \tag{6.1.25}$$

is small. Then by Sobolev embedding theorem we know that the sufficiently small δ_T can assure the positivity of ψ_a, ψ_b as

$$\frac{\sqrt{\rho_a^*}}{2} \le \omega_a + \sqrt{\rho_a^*} \le \frac{3}{2}\sqrt{\rho_a^*},\ \frac{\sqrt{\rho_b^*}}{2} \le \omega_b + \sqrt{\rho_b^*} \le \frac{3}{2}\sqrt{\rho_b^*}.$$

By Sobolev embedding theorem, from the assumption for δ_T, we also have

$$\begin{aligned}&\|(D^\alpha\omega_a, D^\alpha\omega_b, \tau D^\beta\omega_{at}, \tau D^\beta\omega_{bt})\|_{L^\infty(\mathbf{R}^3\times[0,T])} \le c\delta_T, \\ &\quad |\alpha| \le 2, |\beta| \le 1, \end{aligned}\tag{6.1.26}$$

$$\|(\tau D^\alpha\boldsymbol{u}_a, \tau D^\alpha\boldsymbol{u}_b)\|_{L^\infty(\mathbf{R}^3\times[0,T])} \le c\delta_T, |\alpha| \le 2, \tag{6.1.27}$$

$$\begin{aligned}&\int_0^{\mathrm{T}} \|(D^\alpha\boldsymbol{u}_a, D^\alpha\boldsymbol{u}_b, \tau^2\boldsymbol{u}at, \tau^2\boldsymbol{u}_{bt})(\cdot,t)\mathrm{d}t\|_{L^\infty(\mathbf{R}^3\times[0,T])}^2 dt \\ &\quad \le c\delta_T, \|\alpha\| \le 1. \end{aligned}\tag{6.1.28}$$

Using Lemma 6.1.1, from the Poisson equation (6.1.20), we have

$$\|\boldsymbol{E}\|_{L^\infty([0,T];\mathcal{H}^5(\mathbf{R}^3))} \le c\delta_T, \|D^\alpha\boldsymbol{E}\|_{L^\infty(\mathbf{R}^3\times[0,T])} \le c\delta_T. \tag{6.1.29}$$

We have the main priori estimate lemma.

Lemma 6.1.5. *Suppose* $(\omega_a, \omega_b, \boldsymbol{u}_a, \boldsymbol{u}_b, \boldsymbol{E})$ *is local solution with* $\delta_T \ll 1$, *then it holds*

$$\boldsymbol{E}_1(t) + \int_0^t \boldsymbol{E}_2(s)\mathrm{d}s \leq c\boldsymbol{\Lambda}_0, \ t \in (0,T), \tag{6.1.30}$$

and c *is a constant independent of* ε *and* τ, *The* Λ_0 *is defined in Theorem* 6.1.1, *and here*

$$\begin{aligned}
\boldsymbol{E}_1(t) := \{&\|(\omega_a, \omega_b)(\cdot,t)\|_4^2 + (\tau + h^2)\|(D^5\omega_a, D^5\omega_b)(\cdot,t)\|^2 \\
&+ \tau h^2\|(D^6\omega_a, D^6\omega_b)(\cdot,t)\|^2 + \tau^2\|(\omega_{at}, \omega_{bt})(\cdot,t)\|_3^2 \\
&+ \tau^3\|(D^4\omega_{at}, D^4\omega_{bt})\|^2 + \tau^2\|(\boldsymbol{u}_a, \boldsymbol{u}_b)(\cdot,t)\|_{\mathcal{H}^4}^2 \\
&+ \|(D^5\boldsymbol{u}_a, D^5\boldsymbol{u}_b)(\cdot,t)\|^2 + \|\boldsymbol{E}(\cdot,t)\|_{\mathcal{H}^5}^2\}, \\
\boldsymbol{E}_2(t) := \{&\|(\nabla\omega_a, \nabla\omega_b)(\cdot,t)\|_4^2 + h^2\|(D^6\omega_a, D^6\omega_b)(\cdot,t)\|^2 \\
&+ \|(\omega_{at}, \omega_{bt})(\cdot,t)\|_3^2 + \tau(D^4\omega_{at}, D^4\omega_{bt})(\cdot,t)\|^2 \\
&+ \|(\boldsymbol{u}_a, \boldsymbol{u}_b)(\cdot,t)\|_{\mathcal{H}^4}^2 + \tau\|(D^5\boldsymbol{u}_a, D^5\boldsymbol{u}_b)(\cdot,t)\|^2 + \boldsymbol{E}(\cdot,t)\|_{\mathcal{H}^5}^2\}.
\end{aligned}$$

Proof.

1. The estimate of ω_a, ω_b. Assume $\tau < 1$ for simplicity. Multiplying (6.1.16) by $(\omega_a + 2\omega_{at})$, and (6.1.17) by $(\omega_a + 2\omega_{at})$, integrating by parts the resulted equations over $\mathbf{R}^3$, summing the resulted two equalities, we have

$$\begin{aligned}
&\frac{\mathrm{d}}{\mathrm{d}t}\int_{\mathbf{R}^3}(\tau^2\omega_{at}^2 + \tau^2\omega_a\omega_{at} + \frac{\omega_a^2}{2} + \tau^2\omega_{bt}^2 + \tau^2\omega_b\omega_{bt} + \frac{\omega_b^2}{2} \\
&\quad + P_a'(\rho_a^*)|\nabla\omega_a|^2 + P_b'(\rho_b^*)|\nabla\omega_b|^2 + \frac{h^2}{4}(|\Delta\omega_a|^2 \\
&\quad + |\Delta\omega_b|^2) + \frac{1}{4}|\nabla\cdot\boldsymbol{E}|^2)\mathrm{d}x + \int_{\mathbf{R}^3}[(2-\tau^2)(\omega_{at}^2 + \omega_{bt}^2)
\end{aligned}$$

$$+P_a'(\rho_a^*)|\boldsymbol{\nabla}\omega_a|^2+P_b'(\rho_b^*)|\boldsymbol{\nabla}\omega_b|^2\frac{h^2}{4}(|\Delta\omega_a|^2+|\Delta\omega_b|^2)$$
$$+\frac{1}{4}|\boldsymbol{\nabla}\cdot\boldsymbol{E}|^2]\mathrm{d}x=\frac{1}{2}\int_{\mathbf{R}^3}(\omega_a\boldsymbol{\nabla}\omega_a-\omega_b\boldsymbol{\nabla}\omega_b)\cdot\boldsymbol{E}\mathrm{d}x$$
$$+\int_{\mathbf{R}^3}f_{a1}(x,t)(\omega_a+2\omega_{at})\mathrm{d}x+\int_{\mathbf{R}^3}f_{b1}(x,t)(\omega_b+2\omega_{bt})\mathrm{d}x. \tag{6.1.31}$$

Here we have used the following facts

$$\int_{\mathbf{R}^3}\left[\left(\frac{1}{2}(\omega_a+\sqrt{\rho_a^*})\boldsymbol{\nabla}\cdot\boldsymbol{E}\right)\omega_a-\left(\frac{1}{2}(\omega_b+\sqrt{\rho_b^*})\boldsymbol{\nabla}\cdot\boldsymbol{E}\right)\omega_b\right]\mathrm{d}x$$
$$=\frac{1}{4}\int_{\mathbf{R}^3}|\boldsymbol{\nabla}\cdot\boldsymbol{E}|^2\mathrm{d}x-\frac{1}{4}\int_{\mathbf{R}^3}\boldsymbol{\nabla}(\omega_a^2-\omega_b^2)\cdot E\mathrm{d}x,$$

and

$$\int_{\mathbf{R}^3}\left[\left(\frac{1}{2}(\omega_a+\sqrt{\rho_a^*})\boldsymbol{\nabla}\cdot\boldsymbol{E}\right)2\omega_{at}-\left(\frac{1}{2}(\omega_b+\sqrt{\rho_b^*})\boldsymbol{\nabla}\cdot\boldsymbol{E}\right)2\omega_{bt}\right]\mathrm{d}x$$
$$=\frac{1}{4}\frac{\mathrm{d}}{\mathrm{d}t}\int_{\mathbf{R}^3}|\boldsymbol{\nabla}\cdot\boldsymbol{E}|^2\mathrm{d}x.$$

Next we will analyze the right hand of (6.1.31). By Sobolev embedding theorem and Hölder inequality, Young's inequality

$$\int_{\mathbf{R}^3}\omega_i\boldsymbol{\nabla}\omega_i\cdot\boldsymbol{E}\mathrm{d}x\le\|\omega_i\|_{L^3}\|\nabla\omega_i\|_{L^2}\cdot\|\boldsymbol{E}\|_{L^6}$$
$$\le c(\|\omega_i\|_{L^2}+\|\boldsymbol{\nabla}\omega_i\|_{L^2})(\|\boldsymbol{\nabla}\omega_i\|_{L^2})\cdot\|\boldsymbol{E}\|_{L^6})$$
$$\le c(\delta_T)^{\frac{1}{2}}(\|\boldsymbol{\nabla}\omega_i\|^2+\|\boldsymbol{\nabla}\cdot\boldsymbol{E}\|^2),\ i=a,b. \tag{6.1.32}$$

Here we have used Lemma 6.1.1 to estimate $\|D\boldsymbol{E}\|^2$ by $\boldsymbol{\nabla}\cdot\boldsymbol{E}\|^2$. Other terms of right hand side of (6.1.31) are analyzed as:

$$\int_{\mathbf{R}^3}\{P_i'[(\omega_i+\sqrt{\rho_i^*})^2]-P_i'(\rho_i^*)\}\Delta\omega_i\cdot(2\omega_{it})\mathrm{d}x$$
$$\le(\delta_T)^{\frac{1}{2}}(\|\Delta\omega_i\|^2+\|\omega_{it}\|^2), \tag{6.1.33}$$

$$\int_{\mathbf{R}^3} \tau^2 \boldsymbol{u}_i \cdot \boldsymbol{\nabla}_{\omega_{it}} (2\omega_{it}) \mathrm{d}x = -\int_{\mathbf{R}^3} \tau^2 \boldsymbol{\nabla} \cdot \boldsymbol{u}_i (\omega_{it})^2 \mathrm{d}x$$

$$\leq c(\delta_T)^{\frac{1}{2}} \|\omega_{it}\|^2, \tag{6.1.34}$$

$$\int_{\mathbf{R}^3} \tau^2 \boldsymbol{u}_i \nabla(\boldsymbol{u}_i \cdot \boldsymbol{\nabla}\omega_i)(2\omega_{it}) \mathrm{d}x \leq c(\delta_T)^{\frac{1}{2}} \|(\nabla\omega_i, \Delta\omega_i, \omega_{it}, \boldsymbol{\phi}_i)^2\|. \tag{6.1.35}$$

In (6.1.35) we have used $\|D\boldsymbol{u}_i\|^2 \leq c(\|\boldsymbol{\nabla} \cdot \boldsymbol{u}_i\|^2 + \|\boldsymbol{\nabla} \times \boldsymbol{u}_i\|^2)$, and $\|\boldsymbol{\nabla}\omega_i\|^2$, $\|\omega_{it}\|^2$ to estimate $\boldsymbol{\nabla} \cdot \boldsymbol{u}$ through (6.1.24). Then, by using (6.1.31)–(6.1.35), integration by parts, Hölder inequality, Young's inequality, and Moser type Lemmas 6.1.3 and 6.1.4 to estimate the other terms of the right hand side of (6.1.31), we can arrive at

$$\begin{aligned}
&\frac{\mathrm{d}}{\mathrm{d}t} \int_{\mathbf{R}^3} (\tau^2 \omega_{at}^2 + \tau^2 \omega_a \omega_{at} + \frac{\omega_a^2}{2} + \tau^2 \omega_{bt}^2 + \tau^2 \omega_b \omega_{bt} + \frac{\omega_b^2}{2} \\
&\quad + P_a'(\rho_a^*)|\boldsymbol{\nabla}\omega_a|^2 \\
&\quad + P_b'(\rho_b^*)|\boldsymbol{\nabla}\omega_b|^2 + \frac{h^2}{4}(|\Delta\omega_a|^2 + |\Delta\omega_b|^2) + \frac{1}{4}|\boldsymbol{\nabla} \cdot \boldsymbol{E}|^2) dx \\
&\int_{\mathbf{R}^3} [(2-\tau^2)(\omega_{at}^2 + \omega_{bt}^2) + P_a'(\rho_a^*)|\boldsymbol{\nabla}\omega_a|^2 + P_b'(\rho_b^*)|\boldsymbol{\nabla}\omega_b|^2 \\
&\quad + \frac{h^2}{4}(|\Delta\omega_a|^2 + |\Delta\omega_b|^2) + \frac{1}{4}|\boldsymbol{\nabla} \cdot \boldsymbol{E}|^2] \mathrm{d}x \\
&\leq c(\delta_T)^{\frac{1}{2}} \|(\boldsymbol{\nabla}\omega_a, \boldsymbol{\nabla}\omega_b, \omega_{at}, \omega_{bt}, \boldsymbol{\nabla} \cdot \boldsymbol{E}, \boldsymbol{\phi}_a, \boldsymbol{\phi}_b)\|^2 \\
&\quad + c(\delta_T)^{\frac{1}{2}} \|\Delta\omega_a, \Delta\omega_b\|^2.
\end{aligned} \tag{6.1.36}$$

The right hand side of estimate (6.1.36) will be used later in the closure of a priori estimate.

Next we will give the higher-order estimates for ω_a, ω_b. Differentiate (6.1.16) and (6.1.17) with respect to x, then the functions $\tilde{\omega}_a = D^\alpha \omega_a, \tilde{\omega}_b = D^\alpha \omega_b$ and $\tilde{\boldsymbol{E}} = D^\alpha \boldsymbol{E}$ $(1 \leq \|\alpha\| \leq 3)$ satisfy

$$\begin{aligned}
&\tau^2 \tilde{\omega}_{itt} + \tilde{\omega}_{it} + \frac{h^2}{4}\Delta^2 \tilde{\omega}_i + \frac{q_i}{2}(\omega_i + \sqrt{\rho_i^*})\boldsymbol{\nabla} \cdot \tilde{\boldsymbol{E}} - P_i'(\rho_i^*)\Delta\tilde{\omega}_i \\
&\quad = D^\alpha f_{i1}(x,t) - D^\alpha \left(\frac{q_i}{2}(\omega_i + \sqrt{\rho_i^*}\right) \boldsymbol{\nabla} \cdot \boldsymbol{E})
\end{aligned}$$

$$
\begin{aligned}
&+\frac{q_i}{2}(\omega_i+\sqrt{\rho_i^*})\boldsymbol{\nabla}\cdot\boldsymbol{E}\\
&\underset{=}{\text{def}}\, Fi(x,t),(i=a,b,q_a=1,q_b=-1)
\end{aligned}
\tag{6.1.37}
$$

Multiplying (6.1.37) for $i = a$ by $(\tilde{\omega}_a + 2\tilde{\omega}_{at})$, and for $i = b$ by $(\tilde{\omega}_b + 2\tilde{\omega}_{bt})$, integrating by parts over $\mathbf{R}^3$, also noting the facts

$$
\begin{aligned}
&\int_{\mathbf{R}^3}\left[\left(\frac{1}{2}(\omega_a+\sqrt{\rho_a^*})\boldsymbol{\nabla}\cdot\tilde{\boldsymbol{E}}\right)(\tilde{\omega}_a+2\tilde{\omega}_{at})\right.\\
&\qquad\left.-\left(\frac{1}{2}(\tilde{\omega}_b+\sqrt{\rho_b^*})\boldsymbol{\nabla}\cdot\tilde{\boldsymbol{E}}\right)(\tilde{\omega}_b+2\tilde{\omega}_{bt})\right]\mathrm{d}x\\
&\quad=\frac{1}{4}\int_{\mathbf{R}^3}|\boldsymbol{\nabla}\cdot\tilde{\boldsymbol{E}}|^2\mathrm{d}x+\frac{1}{4}\frac{\mathrm{d}}{\mathrm{d}t}\int_{\mathbf{R}^3}|\boldsymbol{\nabla}\cdot\tilde{\boldsymbol{E}}|^2\mathrm{d}x-\frac{1}{4}\int_{\mathbf{R}^3}\boldsymbol{\nabla}\cdot\tilde{\boldsymbol{E}}D^\alpha\\
&\qquad\times(\omega_a^2-\omega_b^2)\mathrm{d}x-\frac{1}{2}\int_{\mathbf{R}^3}\boldsymbol{\nabla}\cdot\tilde{\boldsymbol{E}}D^\alpha(\omega_a^2-\omega_b^2)_t\mathrm{d}x\\
&\qquad+\int_{\mathbf{R}^3}\frac{1}{2}\omega_a\boldsymbol{\nabla}\cdot\tilde{\boldsymbol{E}}(\tilde{\omega}_a+2\tilde{\omega}_{at})\mathrm{d}x\\
&\qquad-\frac{1}{2}\int_{\mathbf{R}^3}\omega_b\boldsymbol{\nabla}\cdot\tilde{\boldsymbol{E}}(\tilde{\omega}_b+2\tilde{\omega}_{bt})\mathrm{d}x,
\end{aligned}
\tag{6.1.38}
$$

after a tedious but straightforward computation, one can get

$$
\begin{aligned}
&\frac{\mathrm{d}}{\mathrm{d}t}\int_{\mathbf{R}^3}[\tau^2\tilde{\omega}_{at}^2+\tau^2\tilde{\omega}_a\tilde{\omega}_{at}+\frac{\tilde{\omega}_a^2}{2}+\tau^2\tilde{\omega}_{bt}^2+\tau^2\tilde{\omega}_b\tilde{\omega}_{bt}+\frac{\tilde{\omega}_b^2}{2}\\
&\qquad+P_a'(\rho_a^*)|\boldsymbol{\nabla}\tilde{\omega}_a|^2+P_b'(\rho_b^*)|\boldsymbol{\nabla}\tilde{\omega}_b|^2+\frac{h^2}{4}(|\Delta\tilde{\omega}_a|^2+|\Delta\tilde{\omega}_b|^2)\\
&\qquad+\frac{1}{4}|\boldsymbol{\nabla}\cdot\tilde{\boldsymbol{E}}|^2]\mathrm{d}x+\int_{\mathbf{R}^3}[(2-\tau^2)(\tilde{\omega}_{at}^2+\tilde{\omega}_{bt}^2)+P_a'(\rho_a^*)|\boldsymbol{\nabla}\tilde{\omega}_a|^2\\
&\qquad+P_b'(\rho_b^*)|\boldsymbol{\nabla}\tilde{\omega}_b|^2+\frac{h^2}{4}(\Delta\tilde{\omega}_a|^2+|\Delta\tilde{\omega}_b|^2+\frac{1}{4}|\boldsymbol{\nabla}\cdot\tilde{\boldsymbol{E}}|^2]\mathrm{d}x
\end{aligned}
$$

$$
\begin{aligned}
&= \int_{\mathbf{R}^3} [F_a \cdot (\tilde{\omega}_a + 2\tilde{\omega}_{at}) + F_b \cdot (\tilde{\omega}_b + 2\tilde{\omega}_{bt})] \mathrm{d}x \\
&\quad + \frac{1}{4} \int_{\mathbf{R}^3} \boldsymbol{\nabla} \cdot \tilde{\boldsymbol{E}} D^\alpha (\omega_a^2 - \omega_b^2) \mathrm{d}x \\
&\quad + \frac{1}{2} \int_{\mathbf{R}^3} \boldsymbol{\nabla} \cdot \tilde{\boldsymbol{E}} D^\alpha (\omega_a^2 - \omega_b^2)_t \mathrm{d}x \\
&\quad - \int_{\mathbf{R}^3} \frac{1}{2} \omega_a \boldsymbol{\nabla} \cdot \tilde{\boldsymbol{E}} (\tilde{\omega_a} + 2\tilde{\omega}_{at}^2) \mathrm{d}x + \int_{\mathbf{R}^3} \frac{1}{2} \omega_b \boldsymbol{\nabla} \cdot \tilde{\boldsymbol{E}} \\
&\quad \times (\tilde{\omega}_b + 2\tilde{\omega}_{bt}) \mathrm{d}x. \qquad (6.1.39)
\end{aligned}
$$

Similarly, the analysis of the above estimate for ω_a, ω_b, and using Hölder inequality, Young's inequality, Moser type Lemmas 6.1.3 and 6.1.4, we can arrive at

$$
\begin{aligned}
&\frac{\mathrm{d}}{\mathrm{d}t} \int_{\mathbf{R}^3} [\tau^2 \tilde{\omega}_{at}^2 + \tau^2 \tilde{\omega}_a \tilde{\omega}_{at} + \frac{\tilde{\omega}_a^2}{2} + \tau^2 \tilde{\omega}_{bt}^2 + \tau^2 \tilde{\omega}_b \tilde{\omega}_{bt} + \frac{\tilde{\omega}_b^2}{2} \\
&\quad + P_a'(\rho_a^*) |\boldsymbol{\nabla} \tilde{\omega}_a|^2 + P_b'(\rho_b^*) |\boldsymbol{\nabla} \tilde{\omega}_b|^2 + \frac{h^2}{4} (|\Delta \tilde{\omega}_a|^2 + |\Delta \tilde{\omega}_b|^2) \\
&\quad + \frac{1}{4} |\boldsymbol{\nabla} \cdot \tilde{\boldsymbol{E}}|^2] \mathrm{d}x + \int_{\mathbf{R}^3} [(2 - \tau^2)(\tilde{\omega}_{at}^2 + \tilde{\omega}_{bt}^2) + P_a'(\rho_a^*) |\boldsymbol{\nabla} \tilde{\omega}_a|^2 \\
&\quad + P_b'(\rho_b^*) |\boldsymbol{\nabla} \tilde{\omega}_b|^2 + \frac{h^2}{4} (|\Delta \tilde{\omega}_a|^2 + |\Delta \tilde{\omega}_b|^2) + \frac{1}{4} |\boldsymbol{\nabla} \cdot \tilde{\boldsymbol{E}}|^2] \mathrm{d}x \\
&\leq c(\delta_T)^{\frac{1}{2}} \| (\boldsymbol{\nabla} \omega_a, \boldsymbol{\nabla} \omega_b, \omega_{at}, \omega_{bt}, \boldsymbol{\nabla} \cdot \boldsymbol{E}, \boldsymbol{\phi}_a, \boldsymbol{\phi}_b) \|_3^2 \\
&\quad + c(\delta_T)^{\frac{1}{2}} |(D^5 \omega_a, D^5 \omega_b)|^2. \qquad (6.1.40)
\end{aligned}
$$

Note that we cannot deal with the last term in (6.1.40) by the energy of left-hand side now, so we have to do the highest order estimates in a different way in order to overcome the difficulty.

Then we give the highest-order estimates for ω_a, ω_b. Taking $|\alpha| = 4$, we can get the equations for $\tilde{\omega}_a := D^\alpha \tilde{\omega}_a$, $\tilde{\omega}_b := D^\alpha \omega_b$ and $\tilde{\boldsymbol{E}} := D^\alpha \boldsymbol{E}$. Multiplying (6.1.37) for $i = a$ by $(\tilde{\omega}_a + 2\tau\tilde{\omega}_{at})$, and for $i = b$ by $(\tilde{\omega}_b + 2\tau\tilde{\omega}_{bt})$ but for $|\alpha| = 4$, we can get as previously

$$\begin{aligned}
&\frac{\mathrm{d}}{\mathrm{d}t}\int_{\mathbf{R}^3}[\tau^3\tilde{\omega}_{at}^2+\tau^2\tilde{\omega}_a\tilde{\omega}_{at}+\frac{\tilde{\omega}_a^2}{2}+\tau^3\tilde{\omega}_{bt}^2+\tau^2\tilde{\omega}_b\tilde{\omega}_{bt}+\frac{\tilde{\omega}_b^2}{2}\\
&\quad+\tau P_a'(\rho_a^*)|\boldsymbol{\nabla}\tilde{\omega}_a|^2+\tau P_b'(\rho_b^*)|\boldsymbol{\nabla}\tilde{\omega}_b|^2+\frac{\tau h^2}{4}(|\Delta\tilde{\omega}_a|^2+|\Delta\tilde{\omega}_b|^2)\\
&\quad+\frac{\tau}{4}|\boldsymbol{\nabla}\cdot\tilde{\boldsymbol{E}}|^2]\mathrm{d}x\int_{\mathbf{R}^3}[(2\tau-\tau^2)(\tilde{\omega}_{at}^2+\tilde{\omega}_{bt}^2)+P_a'(\rho_a^*)|\boldsymbol{\nabla}\tilde{\omega}|^2\\
&\quad+P_b'(\rho_b^*)|\boldsymbol{\nabla}\tilde{\omega}_b|^2+\frac{h^2}{4}(|\Delta\tilde{\omega}_a|^2+|\Delta\tilde{\omega}_b|^2+\frac{1}{4}|\boldsymbol{\nabla}\cdot\tilde{\boldsymbol{E}}|^2]\mathrm{d}x\\
&=\int_{\mathbf{R}^3}[F_a\cdot(\tilde{\omega}_a+2\tau\tilde{\omega}_{at})+F_b\cdot(\tilde{\omega}_b+2\tau\tilde{\omega}_{bt})]\mathrm{d}x\\
&\quad+\frac{1}{4}\int_{\mathbf{R}^3}\boldsymbol{\nabla}\cdot\tilde{\boldsymbol{E}}D^\alpha(\omega_a^2-\omega_b^2)\mathrm{d}x+\frac{1}{2}\int_{\mathbf{R}^3}\tau\boldsymbol{\nabla}\cdot\tilde{\boldsymbol{E}}D^\alpha\\
&\quad\times(\omega_a^2-\omega_b^2)_t\mathrm{d}x-\int_{R^3}\frac{1}{2}\omega_a\boldsymbol{\nabla}\cdot\tilde{\boldsymbol{E}}(\tilde{\omega}_a+2\tau\tilde{\omega}_{at})\mathrm{d}x\\
&\quad+\int_{\mathbf{R}^3}\frac{1}{2}\omega_b\boldsymbol{\nabla}\cdot\tilde{\boldsymbol{E}}(\tilde{\omega}_b+2\tau\tilde{\omega}_{bt})\mathrm{d}x. \qquad (6.1.41)
\end{aligned}$$

On the right hand side of (6.1.41), the terms multiplied by $2\tau\tilde{\omega}_{at}$, $2\tau\tilde{\omega}_{bt}$ need a special analysis. Taking $i=a$ for example, the key terms are analyzed as

$$\begin{aligned}
&\int_{\mathbf{R}^3}[P_a'((\omega_a+\sqrt{\rho_a^*})^2)-P_a'(\rho_a^*)]\Delta\tilde{\omega}_a\cdot 2\tau\tilde{\omega}_{at}\mathrm{d}x\\
&\quad\le-\frac{\mathrm{d}}{\mathrm{d}t}\int_{\mathbf{R}^3}\tau[P_a'((\omega_a+\sqrt{\rho_a^*})^2)-P_a'(\rho_a^*)]|\boldsymbol{\nabla}\tilde{\omega}_a|^2\mathrm{d}x \qquad (6.1.42)\\
&\quad+c\delta_T^{\frac{1}{2}}\|\boldsymbol{\nabla}\tilde{\omega}_a\|^2+c\delta_T^{\frac{1}{2}}\tau\|\tilde{\omega}_{at}\|^2,\ \int_{\mathbf{R}^3}\tau^2\boldsymbol{u}_a\boldsymbol{\nabla}\tilde{\omega}_{at}\cdot 2\tau\tilde{\omega}_{at}\mathrm{d}x\\
&=-\int_{\mathbf{R}^3}\boldsymbol{\nabla}\cdot\boldsymbol{u}_a|\tilde{\omega}_{at}|^2\mathrm{d}x\le c\delta_T^{\frac{1}{2}}\tau\|\tilde{\omega}_{at}\|^2, \qquad (6.1.43)
\end{aligned}$$

$$\int_{\mathbf{R}^3}\tau^2\boldsymbol{u}_a\boldsymbol{\nabla}(\boldsymbol{u}_a\cdot\boldsymbol{\nabla}\tilde{\omega}_a)$$

$$\cdot 2\tau\tilde{\omega}_{at}\mathrm{d}x \le -\frac{\mathrm{d}}{\mathrm{d}t}\int_{\mathbf{R}^3}\tau(\tau\boldsymbol{u}_a\cdot\boldsymbol{\nabla}\tilde{\omega}_a)^2\mathrm{d}x + \int_{\mathbf{R}^3}2\tau^3$$
$$\times(\boldsymbol{u}_a\cdot\boldsymbol{\nabla}\tilde{\omega}_a)\boldsymbol{u}_{at}\boldsymbol{\nabla}\tilde{\omega}_a\mathrm{d}x + c\delta_T^{\frac{1}{2}}\|\boldsymbol{\nabla}\tilde{\omega}_a\|^2 + c\delta_T^{\frac{1}{2}}\tau\|\tilde{\omega}_{at}\|^2. \tag{6.1.44}$$

The other terms on the right hand side of (6.1.41) can be analyzed just using Moser type Lemma 6.1.3 and 6.1.4, the Hölder inequality, Sobolev inequality and Young's inequality. These estimates will lead to

$$\frac{\mathrm{d}}{\mathrm{d}t}\int_{\mathbf{R}^3}[\tau^3\tilde{\omega}_{at}^2 + \tau^2\tilde{\omega}_a\tilde{\omega}_{at} + \frac{\tilde{\omega}_a^2}{2} + \tau^3\tilde{\omega}_{bt}^2 + \tau^2\tilde{\omega}_b\tilde{\omega}_{bt} + \frac{\tilde{\omega}_b^2}{2}$$
$$+\tau P_a'(\rho_a^*)|\boldsymbol{\nabla}\tilde{\omega}_a|^2 + \tau P_b'(\rho_b^*)|\boldsymbol{\nabla}\tilde{\omega}_b|^2 + \frac{\tau h^2}{4}(|\Delta\tilde{\omega}_a|^2$$
$$+|\Delta\tilde{\omega}_b|^2) + \frac{\tau}{4}|\boldsymbol{\nabla}\cdot\tilde{\boldsymbol{E}}|^2]dx + \frac{d}{\mathrm{d}t}\int_{\mathbf{R}^3}\{\tau[P_a'((\omega_a+\sqrt{\rho_a^*})^2)$$
$$-P_a'(\rho_a^*)]|\boldsymbol{\nabla}\tilde{\omega}_a|^2\mathrm{d}x\} + \frac{\mathrm{d}}{\mathrm{d}t}\int_{\mathbf{R}^3}\{\tau(\tau\boldsymbol{u}_a\cdot\boldsymbol{\nabla}\tilde{\omega}_a)^2\mathrm{d}x$$
$$+\frac{\mathrm{d}}{\mathrm{d}t}\int_{\mathbf{R}^3}\tau[P_b'((\omega_b+\sqrt{\rho_b^*})^2) - P_b'(\rho_b^*)]|\boldsymbol{\nabla}\tilde{\omega}_b|^2\}\mathrm{d}x$$
$$+\frac{\mathrm{d}}{\mathrm{d}t}\int_{\mathbf{R}^3}\tau(\tau\boldsymbol{u}_b\cdot\boldsymbol{\nabla}\tilde{\omega}_b)^2\mathrm{d}x + \int_{\mathbf{R}^3}[(2\tau-\tau^2)(\tilde{\omega}_{at}^2+\tilde{\omega}_{bt}^2)$$
$$+P_a'(\rho_a^*)|\boldsymbol{\nabla}\tilde{\omega}_a|^2 + P_b'(\rho_b^*)|\boldsymbol{\nabla}\tilde{\omega}_b|^2 + \frac{h^2}{4}(\|\Delta\tilde{\omega}_a^2 + |\Delta\tilde{\omega}_b|^2)$$
$$+\frac{1}{4}|\boldsymbol{\nabla}\cdot\tilde{\boldsymbol{E}}|^2 + 2\tau^3(\boldsymbol{u}_a\cdot\boldsymbol{\nabla}\tilde{\omega}_a)\boldsymbol{u}_{at}\boldsymbol{\nabla}\tilde{\omega}_a]\mathrm{d}x$$
$$\le c\delta_T^{\frac{1}{2}}|(\boldsymbol{\nabla}\omega_a,\boldsymbol{\nabla}\omega_b)\|_4^2 + c\delta_T^{\frac{1}{2}}h^2|(D^6\omega_a,D^6\omega_b)\|^2$$
$$+c\delta_T^{\frac{1}{2}}\|\boldsymbol{\nabla}\cdot\boldsymbol{E}\|_4^2 + c\delta_T^{\frac{1}{2}}\tau\|(D^4\omega_{at},D^4\omega_{bt})\|^2$$
$$+c\delta_T^{\frac{1}{2}}\|(\boldsymbol{\phi}_a,\boldsymbol{\phi}_b)\|_4^2. \tag{6.1.45}$$

2. The estimates for $\boldsymbol{\phi}_a, \boldsymbol{\phi}_b$. Differentiating (6.1.18) and (6.1.19) for ϕ_a, ϕ_b respect to x, then $\boldsymbol{\phi}_a = D^\alpha\boldsymbol{\phi}_a, \tilde{\boldsymbol{\phi}}_b = D^\alpha\boldsymbol{\phi}_b(|\alpha|\le 4)$, will

satisfy (take ϕ_a for example)

$$\tau^2\tilde{\phi}_{at} + \tilde{\phi}_a = D^\alpha f_{a2}. \tag{6.1.46}$$

Taking inner product between (6.1.46) and $2\tilde{\phi}_a$, integrating over $\mathbf{R}^3$, we obtain

$$\tau\frac{\mathrm{d}}{\mathrm{d}t}\int_{\mathbf{R}^3}|\tilde{\phi}_a|^2\mathrm{d}x + 2\int_{\mathbf{R}^3}|\tilde{\phi}_a|^2\mathrm{d}x = \int_{\mathbf{R}^3}D^\alpha f_{a2}\cdot 2\tilde{\phi}_a\mathrm{d}x. \tag{6.1.47}$$

The terms on the right hand side of (6.1.47) can be estimated by using Young's inequality, Moser type Lemma 6.1.3, 6.1.4 and the inequality $\|D\boldsymbol{u}\| \le c(\|\nabla\cdot\boldsymbol{u}\|+\|\nabla\times\boldsymbol{u}\|)$ and also the presentation of $\nabla\cdot\boldsymbol{u}$ by ω_{at}, $\nabla\omega_a$. Then we can deduce

$$\begin{aligned}\tau^2\frac{\mathrm{d}}{\mathrm{d}t}\int_{\mathbf{R}^3}|\tilde{\phi}_a|^2\mathrm{d}x + 2\int_{\mathbf{R}^3}\|\tilde{\phi}_a|^2\mathrm{d}x &\le c\delta_T^{\frac{1}{2}}\|\tilde{\phi}_a\|_4^2 + c\delta_T^{\frac{1}{2}}\tau\|\omega_{at}\|_4^2\\ &+ c\delta_T^{\frac{1}{2}}\|\boldsymbol{\nabla}\omega_a\|_4^2.\end{aligned} \tag{6.1.48}$$

3. The closure of energy estimate. The assumption $\delta_T \ll 1$ and the combination of the estimates (6.1.36) and (6.1.40) for all $|\alpha| \le 3$, and equation (6.1.45) for all $|\alpha| = 4$, (6.1.48) for all $|\alpha| \le 4$ can give us

$$\frac{\mathrm{d}}{\mathrm{d}t}H_1(t) + H_2(t) \le \sum_{i=a,b}\tau\|\boldsymbol{u}_i(\cdot,t)\|_{L^\infty}\cdot\|D^5\omega_i\|^2, \tag{6.1.49}$$

where $H_1(t)$, $H_2(t)$ are two terms satisfying

$$\begin{aligned}&0 < c_1\boldsymbol{E}_1(t) < H_1(t) < c_2\boldsymbol{E}_1(t),\ 0 < c_3\boldsymbol{E}_2(t) < H_2(t)\\ &< c_4\boldsymbol{E}_2(t), t\in[0,T],\end{aligned}$$

and c_1, c_2, c_3, c_4 are positive constants independent of h, τ. $\boldsymbol{E}_1(t)$, $\boldsymbol{E}_2(t)$ are the terms defined in Lemma 6.1.5. From (6.1.49) we can write

$$\frac{\mathrm{d}}{\mathrm{d}t}H_1(t) + H_2(t) \leq cg(t)H_1(t), \ t \in [0,T], \tag{6.1.50}$$

with

$$g(t) = \sum_{i=a,b} \|\boldsymbol{u}_i(.,t)\|_{L^\infty} \cdot 2\|\tau^2 \boldsymbol{u}_{it}(\cdot,t)\|_{L^\infty}.$$

Using Gronwall inequality, we have

$$H_1(t) \leq ce^{\int_0^t g(s)ds} H_1(0) \leq ce^{c\delta_T} H_1(0) \leq cH_1(0), \ t \in [0,T], \tag{6.1.51}$$

provided $\delta_T \leq 1$. Integrating (6.1.50) respect to t on $[0,T]$, using (6.1.51), we derive

$$\int_0^t H_2(s)ds \leq H_1(0) + H_1(t) + c\delta_T H_1(0) \leq C' H_1(0). \tag{6.1.52}$$

The above constants c and C' denote the positive constant independent of the parameters $\varepsilon > 0$, $\tau > 0$.

It follows from (6.1.51), (6.1.52), and the equivalence between $H_1(t)$ and $\boldsymbol{E}_1(t)$, between $H_2(t)$ and $\boldsymbol{E}_2(t)$. The conclusion was stated in Lemma 6.1.5. Thus the proof of Lemma 6.1.5 is completed. □

Proof. The proof of the global existence (Theorem 6.1.1) is a direct conclusion of the combination of the local existence theory Lemma 6.1.4 and global priori estimates Lemma 6.1.5 in terms of the variable transformation presented above and the standard continuity argument. We omit the details. □

Proof. The proof of semiclassical limit (Theorem 6.1.2) starts from Lemma 6.1.5 using a continuity argument. One can easily prove the existence of the global-in-time solutions of the original problem (6.1.10)–(6.1.12) with any small ε and τ which provided the $\Lambda_1 > 0$ then $\Lambda_0 > 0$ small enough.

Let $(\psi_a^h, \psi_b^h, \boldsymbol{u}_a^h, \boldsymbol{u}_b^h, \boldsymbol{E}^h)$ be the solution of (6.1.10)–(6.1.12). Then from Lemma 6.1.5 and the Poisson equation (6.1.20), the uniform estimates to ε hold:

$$\sum_{k=0}^{1} \|(\partial_t^k(\psi_a^\varepsilon - \sqrt{\rho_a^*}), \partial_t^k(\psi_b^\varepsilon - \sqrt{\rho_b^*})(\cdot,t)\|_{5-i}^2$$
$$+\sum_{k=0}^{1} |(\partial_t^k \boldsymbol{u}_a^h, \partial_t^k \boldsymbol{u}_b^h)(\cdot,t)\|_{\mathcal{H}^{5-2i}}^2 + \|\boldsymbol{E}^h(\cdot,t)\|_{\mathcal{H}^6}^2 \le c\Lambda_0, \quad (6.1.53)$$

$$\int_0^t \{\|(\psi_a^h - \sqrt{\rho_a^*}), (\psi_b^h - \sqrt{\rho_b^*})(\cdot,s)\|_5^2 + \|(\partial_t\psi_a^h, \partial_t\psi_b^h)(\cdot,s)\|_4^2\}$$
$$\times \mathrm{d}s \le \Lambda_0 t, \quad (6.1.54)$$

$$\int_0^t \left\{\sum_{k=0}^{1} \|(\partial_t^k \boldsymbol{u}_a^h, \partial_t^k \boldsymbol{u}_b^h)(\cdot,s)\|_{\mathcal{H}^{5-2i}}^2 + \sum_{k=0}^{1} \|(\partial_t^k \boldsymbol{E}^h)(\cdot,s)\|_{\mathcal{H}^{6-i}}^2\right\}$$
$$\times \mathrm{d}s \le c\Lambda_0. \quad (6.1.55)$$

The right hand side of the above inequalities are independent of ε. Thus these uniform estimates and Aubin–Lions Lemma imply the existence of subsequence denoted also by $(\psi_a^h, \psi_b^h, \boldsymbol{u}_a^h, \boldsymbol{u}_b^h, \boldsymbol{E}^h)$ such that

$$\psi_a^h \to \psi_a, \ \psi_b^h \to \psi_b, \ in\ C(0,t; C_b^3 \cap H_{loc}^{5-s}(\mathbf{R}^3)), \quad (6.1.56)$$
$$\boldsymbol{u}_a^h \to \boldsymbol{u}_a, \ \boldsymbol{u}_b^h \to \boldsymbol{u}_b, \ in\ C(0,t; C_b^3 \cap \mathcal{H}_{loc}^{5-s}(\mathbf{R}^3)), \quad (6.1.57)$$
$$\boldsymbol{E}^h \to \boldsymbol{E}, \ in\ C(0,t; C_b^4 \cap \mathcal{H}_{loc}^{6-s}(\mathbf{R}^3)), \quad (6.1.58)$$

with $s \in (0, \frac{1}{2})$, as $h \to 0$. We also have

$$\frac{h^2}{2}\boldsymbol{\nabla}\left(\frac{\Delta\psi_i^h}{\psi_i^h}\right) \to 0, \ in\ L^2(0,t; H_{loc}^3(\mathbf{R}^3)), \varepsilon \to 0.$$

Thus (6.1.52)–(6.1.58) allow $\varepsilon \to 0$, and the limiting solutions satisfy

$$2\psi_a\partial_t\psi_a + \mathrm{div}(\psi_a^2\boldsymbol{u}_a) = 0,$$
$$\tau^2\partial_t(\psi_a^2\boldsymbol{u}_a) + \tau^2\mathrm{div}(\psi_a^2\boldsymbol{u}_a \otimes \boldsymbol{u}_a) + \boldsymbol{\nabla}P_a(\psi_a^2) + \psi_a^2\boldsymbol{u}_a - \psi_a^2\boldsymbol{E} = 0,$$
$$2\psi_b\partial_t\psi_b + \mathrm{div}(\psi_b^2\boldsymbol{u}_b) = 0,$$

$$\tau^2\partial_t(\psi_b^2\boldsymbol{u}_b) + \tau^2\text{div}(\psi_b^2\boldsymbol{u}_b\otimes\boldsymbol{u}_b) + \boldsymbol{\nabla}P_b(\psi_b^2) + \psi_b^2\boldsymbol{u}_b + \psi_b^2\boldsymbol{E} = 0,$$

$$\lambda^2\boldsymbol{\nabla}\cdot\boldsymbol{E} = \psi_a^2 - \psi_b^2 - \mathcal{C},\ \boldsymbol{\nabla}\times E = 0.$$

Let $\rho_a = \psi_a^2$, $\rho_b = \psi_b^2$. It is easy to verify that $(\rho_a, \rho_b, \boldsymbol{u}_a, \boldsymbol{u}_b, \boldsymbol{E})$ solves the bipolar HD model (6.1.4)–(6.1.6). The convergence of the bipolar QHD model to bipolar HD model is established, and the proof of Theorem 6.1.2 is completed. □

6.2 The relaxation time limit

In this section, we will consider the relaxation limit of the bipolar QHD model (6.1.1)–(6.1.3) (the asymptotic limit as $\tau\to 0$). In Section 6.1, for simplicity, we have made a scaled transformation (6.1.9), and obtained the equations (6.1.10)–(6.1.12). Therefore we only need to research the limit of the equations (6.1.10)–(6.1.12) as $\tau\to 0$. Formally, if $\tau\to 0$, we can get that the solution of (6.1.10)–(6.1.12) satisfies the bipolar quantum drift-diffusion (QDD) equations

$$\partial_t\rho_i + \boldsymbol{\nabla}\left[q_i\rho_i\boldsymbol{E} - \boldsymbol{\nabla}P_i(\rho_i) + \frac{\varepsilon^2}{2}\rho_i\boldsymbol{\nabla}\left(\frac{\Delta\sqrt{\rho_i}}{\sqrt{\rho_i}}\right)\right] = 0, \qquad (6.2.1)$$

$$\lambda^2\boldsymbol{\nabla}\cdot\boldsymbol{E} = \rho_a - \rho_b - \mathcal{C}, \quad \boldsymbol{\nabla}\times\boldsymbol{E} = 0,$$

$$E(x)\to 0, \quad |x|\to+\infty. \qquad (6.2.2)$$

Then we will give the strict proof of the relaxation limit.

Theorem 6.2.1 (Relaxation limit). *Let h be a fixed constant. Set $(\rho_a^{(\tau,h)}, \rho_b^{(\tau,h)}, \boldsymbol{u}_a^{(\tau,h)}, \boldsymbol{u}_b^{(\tau,h)}, \boldsymbol{E}^{(\tau,h)})$ to be the solution of the IVP to the equations* (6.1.10)–(6.1.12) *and* (6.1.7) *given in Theorem* 6.1.1. *Then, as $\tau\to 0$, there exists $(\rho_a, \rho_b, \boldsymbol{u}_a, \boldsymbol{u}_b, \boldsymbol{E})$ with $\rho_a > 0$, $\rho_b > 0$, and for any $T > 0$, $i = a, b, s \in \left(0, \frac{1}{2}\right)$,*

$$\rho_i^{(\tau,h)} \to \rho_i,\ \textit{strongly in } C(0,T;C_b^2\cap H_{loc}^{4-s}),$$

$$\boldsymbol{u}_i^{(\tau,h)} \rightharpoonup \boldsymbol{u}_i,\ \textit{weakly in } L^2(0,T;\mathcal{H}^4),$$

$$\boldsymbol{E}^{(\tau,h)} \to \boldsymbol{E},\ \textit{strongly in } C(0,T;C_b^3\cap H_{loc}^{5-s}),$$

where $(\rho_i, \boldsymbol{u}_i, \boldsymbol{E})$ $(i = a, b)$ *is the global solution to the bipolar QDD equations* (6.2.1)–(6.2.2).

Proof. The proof of this theorem is similar to the proof of Theorem 6.1.2. From the uniform priori estimates Lemma 6.1.5 about τ in Section 6.1, we know that for any $t > 0$, the following holds:

$$\begin{aligned}&\|(\psi_a^{(\tau,h)} - \sqrt{\rho_a^*}, \psi_b^{(\tau,h)} - \sqrt{\rho_b^*})(\cdot,t)\|_4^2\\&\quad + \|(\tau \boldsymbol{u}_a^{(\tau,h)}, \tau \boldsymbol{u}_b^{(\tau,h)})(\cdot,t)\|_{\mathcal{H}^4} \le c\boldsymbol{\nabla}_0,\end{aligned} \tag{6.2.3}$$

$$\|(\tau\partial_t\psi_a^{(\tau,h)}, \tau\partial_t\psi_b^{(\tau,h)})(\cdot,t)\|_3^2 + \|\boldsymbol{E}^{(\tau,h)}(\cdot,t)\|_{\mathcal{H}^5}^2 \le c\boldsymbol{\nabla}_0, \tag{6.2.4}$$

$$\begin{aligned}&\int_0^t [\|(\psi_a^{(\tau,h)} - \sqrt{\rho_a^*}, \psi_b^{(\tau,h)} - \sqrt{\rho_b^*})(\cdot,s)\|_5^2\\&\quad \|(\partial_t\psi_a^{(\tau,h)}, \partial_t\psi_b^{(\tau,h)})(\cdot,s)\|_3^2]\mathrm{d}s \le c\boldsymbol{\nabla}_0,\end{aligned} \tag{6.2.5}$$

$$\int_0^t [\|(\boldsymbol{u}_a^{(\tau,h)}, \boldsymbol{u}_b^{(\tau,h)})(\cdot,s)\|_{\mathcal{H}^4}^2 + \|E^{(\tau,h)}(\cdot,s)\|_{\mathcal{H}^5}^2]\mathrm{d}s \le c\boldsymbol{\nabla}_0. \tag{6.2.6}$$

Similarly, the Aubin–Lions Lemma and Sobolev embedding theorem imply the existence of subsequence denoted also by $(\psi_a, \psi_b, \boldsymbol{u}_a, \boldsymbol{u}_b, \boldsymbol{E})$ such that, as $\tau \to 0$,

$$\rho_i^{(\tau,h)} \to \rho_i, \ \textit{strongly in } C(0,T; C_b^2 \cap H_{loc}^{4-s}), \tag{6.2.7}$$

$$\boldsymbol{u}_i^{(\tau,h)} \to \boldsymbol{u}_i, \ \textit{weakly in } L^2(0,T;\mathcal{H}^4), \tag{6.2.8}$$

$$\boldsymbol{E}^{(\tau,h)} \to \boldsymbol{E}, \ \textit{strongly in } C(0,T; C_b^3 \cap H_{loc}^{5-s}), \tag{6.2.9}$$

and $s \in \left(0, \frac{1}{2}\right)$.

From (6.2.3)–(6.2.4), we know $\psi_a > 0$, $\psi_b > 0$ in $(0,t) \times \mathbf{R}^3$, and also

$$\tau^2|\boldsymbol{u}_a^{(\tau,h)}|^2 \to 0, \quad \tau^2|\boldsymbol{u}_b^{(\tau,h)}|^2 \to 0, \ \textit{in } L^1(0,t; W_{loc}^{3,3}(\mathbf{R}^3)), \ \textit{as } \tau \to 0. \tag{6.2.10}$$

Thus let $\tau \to 0$ in equations (6.1.10)–(6.1.12), we can get the convergence of the solutions of the bipolar QHD model to the solutions of the bipolar QDD model (6.2.1)–(6.2.2). The proof of this theorem is completed. □

In fact *a priori* estimate in Section 6.1 implies that we can study the combined limits as both τ and ε tend to 0 freely. Thus we can get the limit equation bipolar QDD model

$$\partial_t \rho_i + \nabla[q_i \rho_i \boldsymbol{E} - \boldsymbol{\nabla} P_i(\rho_i)] = 0, \tag{6.2.11}$$

$$\lambda^2 \boldsymbol{\nabla} \cdot \boldsymbol{E} = \rho_a - \rho_b - \mathcal{C}(x), \quad \boldsymbol{\nabla} \times \boldsymbol{E} = 0,$$

$$E(x) \to 0, \quad |x| \to +\infty. \tag{6.2.12}$$

Theorem 6.2.2 (Relaxation time limit and Semiclassical limit). *Let* $(\rho_i^{(\tau,h)}, \boldsymbol{u}_i^{(\tau,h)}, \boldsymbol{E}^{(\tau,h)})$ $(i = a, b)$ *be the unique global solution to the bipolar QHD model* (6.1.10)–(6.1.12) *given in Theorem* 6.1.1. *Then there exists* $(\rho_a, \rho_b, \boldsymbol{E})$ *such that, as* $h \to 0$ *and* $\tau \to 0$,

$$\rho_i^{(\tau,h)} \to \rho_i, \text{ strongly in } C(0,T; C_b^2 \cap H_{loc}^{4-s}),$$

$$\boldsymbol{u}_i^{(\tau,h)} \to \boldsymbol{u}_i, \text{ weakly in } C(0,T; C_b^3 \cap \mathcal{H}_{loc}^{5-s}),$$

$$\boldsymbol{E}^{(\tau,h)} \to \boldsymbol{E}, \text{ strongly in } L^1(0,T; W_{loc}^{3,3}), \quad S \in \left(0, \frac{1}{2}\right),$$

and $(\rho_a, \rho_b, \boldsymbol{E})$ *is the strong solution of the bipolar QDD model* (6.2.11)–(6.2.12) *and* (6.1.7).

Proof. We can easily get this theorem by referring to Theorem 6.2.1. Here we only state the main step. From Lemma 6.1.5 we can get the uniform priori estimates about h, τ,

$$\|(\psi_a^{(\tau,h)} - \sqrt{\rho_a^*}, \psi_b^{(\tau,h)} - \sqrt{\rho_b^*})(\cdot,t)\|_4^2$$
$$+ \|(\tau \boldsymbol{u}_a^{(\tau,h)}, \tau \boldsymbol{u}_b^{(\tau,h)})(\cdot,t)\|_{\mathcal{H}^4} \le c\boldsymbol{\Lambda}_0, \tag{6.2.13}$$

$$\|(\tau \partial_t \psi_a^{(\tau,h)}, \tau \partial_t \psi_b^{(\tau,h)})(\cdot,t)\|_3^2 + \|\boldsymbol{E}^{(\tau,h)}(\cdot,t)\|_{\mathcal{H}^5}^2 \le c\boldsymbol{\Lambda}_0, \tag{6.2.14}$$

$$\int_0^t [\|(\psi_a^{(\tau,h)} - \sqrt{\rho_a^*} \psi_b^{(\tau,h)} - \sqrt{\rho_b^*})(\cdot,s)\|_5^2$$
$$+ \|(\partial_t \psi_a^{(\tau,h)}, \partial_t \psi_b^{(\tau,h)})(\cdot,s)\|_3^2] \mathrm{d}s \le c\boldsymbol{\Lambda}_0, \tag{6.2.15}$$

$$\int_0^t (\|(\boldsymbol{u}_a^{(\tau,h)}, \boldsymbol{u}_b^{(\tau,h)})(\cdot,s)\|_{\mathcal{H}^4}^2 + \|\boldsymbol{E}^{(\tau,h)}(\cdot,s)\|_{\mathcal{H}^5}^2) \mathrm{d}s \le c\boldsymbol{\Lambda}_0. \tag{6.2.16}$$

Similarly, using Aubin–Lions Lemma and Sobolev embedding theorem, we know that there exists a subsequence denoting itself and functions $(\psi_a, \psi_b, \boldsymbol{u}_a, \boldsymbol{u}_b, \boldsymbol{E})$ satisfying that, as $\tau \to 0$,

$$\rho_i^{(\tau,h)} \to \rho_i, \ \textit{strongly in } C(0,T;C_b^2 \cap H_{loc}^{4-s}), \tag{6.2.17}$$

$$\boldsymbol{u}_i^{(\tau,h)} \to \boldsymbol{u}_i, \ \textit{weakly in } L^2(0,T;\mathcal{H}^4), \tag{6.2.18}$$

$$\boldsymbol{E}^{(\tau,h)} \to \boldsymbol{E}, \ \textit{strongly in } C(0,T;C_b^3 \cap H_{loc}^{5-s}). \tag{6.2.19}$$

Since (6.2.13)–(6.2.14), it holds that $\psi_a > 0$, $\psi_b > 0$ in $(0,t) \times \mathbf{R}^3$, and as $\tau \to 0$,

$$\tau^2 |\boldsymbol{u}_a^{(\tau,h)}|^2 \to 0, \quad \tau^2 |\boldsymbol{u}_b^{(\tau,h)}|^2 \to 0, \quad \textit{in } L^1(0,t;W_{loc}^{3,3}(\mathbf{R}_3^2)). \tag{6.2.20}$$

Thus as $\tau \to 0$ in equations (6.1.10)–(6.1.12), we obtain that the solutions of bipolar QHD model converge to the solutions of bipolar QDD equations (6.2.11)–(6.2.12). The proof of this theorem is completed. □

6.3 Quasineutral limit

In this section, we will study the quasineutral limit of the bipolar QHD model [106]:

$$\partial_t \rho_i^\lambda + \operatorname{div}(\rho_i^\lambda \boldsymbol{u}_i^\lambda), \tag{6.3.1}$$

$$\partial t(\rho_i^\lambda \boldsymbol{u}_i^\lambda) + \operatorname{div}(\rho_i^\lambda \boldsymbol{u}_i^\lambda \otimes \boldsymbol{u}_i^\lambda) + \boldsymbol{\nabla} P_i^\lambda(\rho_i^\lambda)$$

$$= q_i \rho_i^\lambda \boldsymbol{\nabla} \phi^\lambda + \frac{h^2}{2} \rho_i^\lambda \boldsymbol{\nabla} \left(\frac{\Delta \sqrt{\rho_i^\lambda}}{\sqrt{\rho_i^\lambda}} \right) - \frac{\rho_i^\lambda \boldsymbol{u}_i^\lambda}{\tau_i}, \tag{6.3.2}$$

$$\lambda^2 \Delta \phi^\lambda = \rho_a^\lambda - \rho_b^\lambda - \mathcal{C}(x), \tag{6.3.3}$$

where $i = a, b$.

The initial data are

$$\rho_i^\lambda(x,0) = \rho_{i0}^\lambda, \quad \boldsymbol{u}_i^\lambda(x,0) = \boldsymbol{u}_{i0}^\lambda, \quad x \in \mathbb{T}^3, \tag{6.3.4}$$

where $\mathbb{T}^3$ is a torus in $\mathbf{R}_3$, and $q_a = 1$, $q_b = -1$. Unknown functions $\rho_a^\lambda \geq 0$, ρ_b^λ, $\boldsymbol{u}_a^\lambda$, $\boldsymbol{u}_b^\lambda$ and ϕ^λ represent the density, velocity and electrostatic potential. P_a^λ and P_b^λ are pressure-density function. $\mathcal{C}(x) \geq 0$ is the given doping profile. The parameters $h > 0$, τ_a, τ_b and $\lambda > 0$ are the Planck constant, relaxation time and Debye length respectively.

The purpose of this section is to investigate the quasineutral limit of the system (6.3.1)–(6.3.3). Formally, if we set $\mathcal{C} \equiv 0$, $\lambda = 0$, and $\tau_a = \tau_b = \tau > 0$, $P_a^\lambda = A(\rho_a^\lambda)^\gamma$ and $P_b^\lambda = A(\rho_b^=\lambda)^\gamma$ $(A > 0, \gamma > 1$ are fixed constants $\rho_a^\lambda = \rho_b^\lambda$ holds. Thus the bipolar QHD model will become the QHD model [98] (if ρ_i^λ and $\boldsymbol{u}_i^\lambda$, the limits of ρ and $\boldsymbol{u}, i = a, b$ exist, respectively):

$$\partial_t \rho + \operatorname{div}(\rho \boldsymbol{u}) = 0, \tag{6.3.5}$$

$$\partial_t(\rho \boldsymbol{u}) + \operatorname{div}(\rho \boldsymbol{u} \otimes \boldsymbol{u}) + \boldsymbol{\nabla} P(\rho) = \frac{h^2}{2} \rho \boldsymbol{\nabla} \left(\frac{\Delta \sqrt{\rho}}{\sqrt{\rho}} \right) - \frac{\rho \boldsymbol{u}}{\tau}, \tag{6.3.6}$$

where $P(\rho) = A\rho^\gamma$. Moreover, we set $h \to 0$ in equations (6.3.5)–(6.3.6), we obtain the compressible Euler equations with damping,

$$\partial_t \rho + \operatorname{div}(\rho \boldsymbol{u}) = 0, \tag{6.3.7}$$

$$\partial t(\rho \boldsymbol{u}) + \operatorname{div}(\rho \boldsymbol{u} \otimes \boldsymbol{u}) + \boldsymbol{\nabla} P(\rho) = -\frac{\rho \boldsymbol{u}}{\tau}. \tag{6.3.8}$$

Denote $C, \bar{C}, \bar{\bar{C}}, C_i$ $(i = 1, 2, \ldots)$ and C_T are the different constants independent on λ, and C_T may be dependent on T. For simplicity, $\int f = \int_{\mathbb{T}^3} f dx$.

In Section 6.1, we have proved the global existence of solutions to the system (6.3.1)–(6.3.3) in R^3. Here we only state the global existence theorem of the model (6.3.1)–(6.3.3) in $\mathbb{T}^3$, and we left the detail proof to the reader.

Theorem 6.3.1 (Global existence). *Let $\mathcal{C}(x) \equiv 0$, $\tau_a = \tau_b = \tau$ and*

$$P_a^\lambda(\rho_a^\lambda) = A(\rho_a^\lambda)^\gamma, \quad P_b^\lambda(\rho_b^\lambda) = A(\rho_b^\lambda)^\gamma, \tag{6.3.9}$$

where $A > 0$, $\gamma > 1$ are given constants. Assume the initial data satisfy $(\rho_{i0}^\lambda, \boldsymbol{u}_{i0}^\lambda) \in H^6(\mathbb{T}^3) \times \mathcal{H}^\nabla(\mathbb{T})^3$,

$$\int (\rho_{a0}^\lambda(x) - \rho_{b0}^\lambda(x)) = 0, \quad \min_{x \in \mathbb{T}^3} \rho_{i0}^\lambda(x) > 0,$$

with $\mathbf{\Lambda}_{i0} := \|(\rho_{i0}^\lambda, \boldsymbol{u}_{i0}^\lambda)\|_{H^6\times\mathcal{H}^5}(i = a, b)$. *If there is a* $\mathbf{\Lambda} > 0$ *such that* $\max\{\mathbf{\Lambda}_{a0}, \mathbf{\Lambda}_{b0}\} \leq \mathbf{\Lambda}$, *the system* (6.3.1)–(6.3.4) *has a unique solution* $(\rho_i^\lambda, \boldsymbol{u}_i^\lambda, \phi^\lambda)$ $(i = a, b)$, *where* $\rho_i^\lambda > 0$, *and for any* $0 < T < +\infty$,

$$\rho_i^\lambda \in C^k(0, T; H^{6-2k}(\mathbb{T})^3), \quad \boldsymbol{u}_i^\lambda \in C^6(0, T; \mathcal{H}^{5-2k}(\mathbb{T}^3),$$
$$\boldsymbol{\nabla}\phi^\lambda \in C^k(0, T; \mathcal{H}^{6-2k}(\mathbb{T}^3)).$$

Here $\mathcal{H}^k(\mathbb{T}^3) = \{f \in L^6(\mathbb{T}^3), Df \in H^{k-1}(\mathbb{T}^3), k \geq 1\}$, *and* $D^k f$ *is the kth-order space derivate.*

The proof of the local existence of (6.3.5)–(6.3.6) is similar to [103]. We omit the proof here.

Theorem 6.3.2 (Local existence). *Assume that the initial data*

$$(\rho(x, 0), \boldsymbol{u}(x, 0)) = (\rho_0(x), \boldsymbol{u}_0(x)) \in H^6(\mathbb{T}^3) \times \mathcal{H}^5(\mathbb{T}^3),$$

with $\min_{x\in\mathbb{T}^3} \rho_0(x) > 0$, *then there exists a constant* $T^* > 0$ *such that, for any* $0 < T < T^*$, *the unique solution* $(\rho, \boldsymbol{u})$ *of the system* (6.3.5)–(6.3.6) *exists on* $[0, T]$, *satisfying on* $[0, T]$ *for any* $0 < T < T^*$, $\inf_{x\in\mathbb{T}^3} \rho(x, t) > 0$,

$$\rho \in C^k(0, T; H^{6-2k}(\mathbb{T}^3)), \quad \boldsymbol{u} \in C^k(0, T; \mathcal{H}^{5-2k}(\mathbb{T}^3)), \quad k = 0, 1, 2.$$

It is well know that the compressible Euler equations (6.3.7)–(6.3.8) have a unique local smooth solution.

Theorem 6.3.3 (Local existence of the Euler equation). *Assume that the initial density* $\rho(x, 0) = \rho_0(x)$, *and the initial velocity* $\boldsymbol{u}(x, 0) = \boldsymbol{u}_0(x)$ *satisfies* ρ_0, $\boldsymbol{u}_0 \in H^3(\mathbb{T}^3)$ *and* $\min_{x\in\mathbb{T}^3} \rho_0(x) > 0$. *Then there exists a constant* $T^{**} \in (0, +\infty)$ *such that the compressible Euler equations* (6.3.7)–(6.3.8) *have a unique solution* $(\rho, \boldsymbol{u}) \in L^\infty_{loc}([0, T^{**}];\ H^3(\mathbb{T}^3))$ *satisfying on* $[0, T]$ *for any* $0 < T < T^{**}$, $\inf_{x\in\mathbb{T}^3} \rho(x, t) > 0$, *and*

$$\sup_{0\leq t\leq T} (\|\rho\|_{H^3} + \|\boldsymbol{u}\|_{H^3} + \|\partial_t \boldsymbol{u}\|_{H^2} + \|\boldsymbol{\nabla}_\rho\|_{H^3} + \|\partial_t \boldsymbol{\nabla}_\rho\|_{H^2}) \leq C_T, \tag{6.3.10}$$

where C_T *is a constant dependent on* T.

Next we will give out the main results about the quasineutral limit of the bipolar QHD model.

Theorem 6.3.4 (Quasineutral limit). *Suppose the assumption in Theorem* 6.3.1 *still holds. Let* $(\rho_i^\lambda, \boldsymbol{u}_i^\lambda, \phi^\lambda)(i = a, b)$, $\rho_i^\lambda > 0$ *be the unique global smooth solution to the system* (6.3.1)–(6.3.4). *In addition, we assume that* $\gamma \geq 2$ *and initial data* $(\rho_{i0}^\lambda, \boldsymbol{u}_{i0}^\lambda)(i = a, b)$ *satisfy that* $\boldsymbol{u}_{a0}^\lambda$ $\boldsymbol{u}_{b0}^\lambda$ *converge to* $\boldsymbol{u}_0$ *in* $H^s(\mathbb{T}^3)(s \geq 3)$, *and*

$$\rho_{a0}^\lambda = \rho_{b0}^\lambda + \lambda^2 \Delta \phi_0^\lambda, \tag{6.3.11}$$

$$\begin{aligned}&\frac{1}{2}\sum_{i=a,b}\int \rho_{i0}^\lambda|\boldsymbol{u}_{i0}^\lambda - \boldsymbol{u}_0|^2 + \frac{h^2}{2}\sum_{i=a,b}\int |\nabla\sqrt{\rho_{i0}^\lambda} - \nabla\sqrt{\rho_0}|^2\\&\quad + \frac{\lambda^2}{2}\int |\nabla\phi_0^\lambda|^2 + \frac{A}{\gamma-1}\sum_{i=a,b}\int [(\rho_{i0}^\lambda)^\gamma - \rho_0^\gamma - \gamma\rho_0^{\gamma-1}\\&\quad \times (\rho_{i0}^\lambda - \rho_0)] \to 0, \quad (\lambda \to 0).\end{aligned} \tag{6.3.12}$$

Then, for all $0 < T < T^*$, *the densities* $\rho_i^\lambda(i = a, b)$ *converge strongly to* ρ *in* $L^\infty(0, T; L^\gamma(\mathbb{T}^3))$, *and the velocities* $\boldsymbol{u}_i^\lambda(i = a, b)$ *converge strongly to* $\boldsymbol{u}$ *in* $L^\infty(0, T; L^2(\mathbb{T}^3))$, *where* $(\rho, \boldsymbol{u})$ *is the strong solution of the QHD equations* (6.3.5)–(6.3.6) *with initial data* $(\rho(x, 0), \boldsymbol{u}(x, 0)) = (\rho_0(x), \boldsymbol{u}_0(x))$, *and* T^* *denotes the maximal existence time of the strong solution given in Theorem* 6.3.2.

Remark 6.3.1. Condition (6.3.12) can be easily satisfied. Indeed, the strong convergence of $\boldsymbol{u}_{i0}^\lambda$ to $\boldsymbol{u}$ in L^2 implies the convergence of the first term in (6.3.12). For given $\phi_0^\lambda \in H^s(\mathbb{T}^3)$, $s > 3 + \frac{3}{2}$, Choosing ρ_0^λ by (6.3.12), it is very easy to verify that

$$\frac{h^2}{2}\int |\nabla\sqrt{\rho_{a0}^\lambda} - \nabla\sqrt{\rho_{b0}^\lambda}|^2 \leq C\lambda^2 h^2 \|\nabla(\Delta\phi_0^\lambda)\|_{L^2(\mathbb{T}^2)}^3,$$

and the convergence of the third and fourth terms is also satisfied.

If we make more assumptions on the initial data $(\rho_{i0}^\lambda, \boldsymbol{u}_{i0}^\lambda)(i = a, b)$, then we can obtain the convergence rate of the quasineutral limit.

Theorem 6.3.5. *Under the assumption of Theorem 6.3.4, if we further assume that*

$$\begin{aligned}
&\frac{1}{2}\sum_{i=a,b}\int \rho_{i0}^\lambda|\boldsymbol{u}_{i0}^\lambda - \boldsymbol{u}_0|^2 + \frac{h^2}{2}\sum_{i=a,b}\int |\nabla\sqrt{\rho_{i0}^\lambda} - \nabla\sqrt{\rho_0}|^2 \\
&\quad + \frac{\lambda^2}{2}\int |\nabla\phi_0^\lambda|^2 + \frac{A}{\gamma-1}\sum_{i=a,b}\int [(\rho_{i0}^\lambda)^\gamma - \rho_0^\gamma - \gamma\rho_0^{\gamma-1}(\rho_{i0}^\lambda - \rho_0)] \\
&\quad \le \bar{C}\lambda, \qquad (6.3.13)
\end{aligned}$$

then it holds that, for any $0 < T < T^*$,

$$\begin{aligned}
&\frac{1}{2}\sum_{i=a,b}\int \rho_i^\lambda|\boldsymbol{u}_i^\lambda - \boldsymbol{u}|^2 + \frac{h^2}{2}\sum_{i=a,b}\int |\nabla\sqrt{\rho_i^\lambda} - \nabla\sqrt{\rho}|^2 + \frac{\lambda^2}{2}\int |\nabla\phi^\lambda|^2 \\
&\quad + \frac{A}{\gamma-1}\sum_{i=a,b}\int [(\rho_i^\lambda)^\gamma - \rho^\gamma - \gamma\rho^{\gamma-1}(\rho_i^\lambda - \rho)] \le M_T\lambda, \qquad (6.3.14)
\end{aligned}$$

where $M_T > 0$ *is a constant independent of* λ, *and* $(\rho, \boldsymbol{u})$ *is the strong solution to the quantum hydrodynamic model* (6.3.5)–(6.3.6).

If we assume that the Debye length λ and the Planck constant h go to 0 simultaneously, then we obtain the convergence of the bipolar QHD model (6.3.1)–(6.3.3) to the compressible Euler equation (6.3.7)–(6.3.8). Here we take $h = \lambda$ for convenience.

Theorem 6.3.6 (Relaxation limit and quasineutral limit). *Suppose the assumptions on Theorem 6.3.1 and Theorem 6.3.4 hold. Then for any* $0 < T < T^{**}$, *the densities* $\rho_i^\lambda(i = a, b)$ *converge strongly to* ρ *in* $L^\infty(0, T; L^\gamma(\mathbb{T}^3))$, *and the velocities* $\boldsymbol{u}_i^\lambda(i = a, b)$ *converge strongly to* $\boldsymbol{u}$ *in* $L^\infty(0, T; L^2(\mathbb{T}^3))$. *Here* $(\rho, \boldsymbol{u})$ *is the solution to the compressible Euler equation* (6.3.7)–(6.3.8) *with initial data* $(\rho(x, 0), \boldsymbol{u}(x, 0)) = (\rho_0(x), \boldsymbol{u}_0(x))$ *and* T^{**} *is the maximal existence time of the strong solution given in Theorem* 6.3.2.

Similarly, we can get the convergence rate of the quasineutral limit.

Theorem 6.3.7. *Under the assumption of Theorem 6.3.6, if*

$$
\begin{aligned}
&\frac{1}{2}\sum_{i=a,b}\int \rho_{i0}^{\lambda}|\boldsymbol{u}_{i0}^{\lambda}-\boldsymbol{u}_0|^2+\frac{h^2}{2}\sum_{i=a,b}\int|\nabla\sqrt{\rho_{i0}^{\lambda}}-\nabla\sqrt{\rho_0}|^2\\
&\quad+\frac{\lambda^2}{2}\int|\nabla\phi_0^{\lambda}|^2+\frac{A}{\gamma-1}\sum_{i=a,b}\int[(\rho_{i0}^{\lambda})^{\gamma}-\rho_0^{\gamma}-\gamma\rho_0^{\gamma-1}(\rho_{i0}^{\lambda}-\rho_0)]\\
&\quad\le\bar{\bar{C}}\lambda,
\end{aligned}
\tag{6.3.15}
$$

then it holds that, for any $0<T<T^{**}$,

$$
\begin{aligned}
&\frac{1}{2}\sum_{i=a,b}\int \rho_{i}^{\lambda}|\boldsymbol{u}_{i}^{\lambda}-\boldsymbol{u}|^2+\frac{h^2}{2}\sum_{i=a,b}\int|\nabla\sqrt{\rho_{i}^{\lambda}}-\nabla\sqrt{\rho}|^2+\frac{\lambda^2}{2}\int|\nabla\phi^{\lambda}|^2\\
&\quad+\frac{A}{\gamma-1}\sum_{i=a,b}\int[(\rho_{i}^{\lambda})^{\gamma}-\rho^{\gamma}-\gamma\rho^{\gamma-1}(\rho_{i}^{\lambda}-\rho)]\le M_T'\lambda,
\end{aligned}
\tag{6.3.16}
$$

where $M_T'>0$ *is a constant independent on* λ, *and* $(\rho,\boldsymbol{u})$ *is the strong solution of the compressible Euler equation* (6.3.5)–(6.3.6).

Next we will give the proof of Theorem 6.3.4 and Theorem 6.3.5.

The proof of Theorem 6.3.4. We will divide the proof in three steps:

1. Basic energy estimates. for the smooth solutions to problem (6.3.1)–(6.3.4), we define the total energy $\mathcal{E}^{\lambda,h}(t)$ as follows:

$$
\begin{aligned}
\mathcal{E}^{\lambda,h}(t)&=\frac{1}{2}\sum_{i=a,b}\int\rho_i^{\lambda}|\boldsymbol{u}_i^{\lambda}|^2+\frac{h^2}{2}\sum_{i=a,b}\int|\nabla\sqrt{\rho_i^{\lambda}}|^2\\
&\quad+\frac{A}{\gamma-1}\sum_{i=a,b}\int(\rho_i^{\lambda})^{\gamma}+\frac{\lambda^2}{2}\int|\nabla\phi^{\lambda}|^2.
\end{aligned}
\tag{6.3.17}
$$

Through a direct computation, we know that

$$
\frac{\mathrm{d}\mathcal{E}^{\lambda,h}(t)}{\mathrm{d}t}=-\frac{1}{\tau}\sum_{i=a,b}\int\rho_i^{\lambda}|\boldsymbol{u}_i^{\lambda}|^2<0. \tag{6.3.18}
$$

Moreover, the following inequality holds:

$$\mathcal{E}^{\lambda,h}(t) \le \mathcal{E}^{\lambda,h}(0), \tag{6.3.19}$$

where

$$\mathcal{E}^{\lambda,h}(0) = \frac{1}{2}\sum_{i=a,b}\int \rho_{i0}^{\lambda}|\boldsymbol{u}_{i0}^{\lambda}|^2 + \frac{h^2}{2}\sum_{i=a,b}\int |\nabla\sqrt{\rho_{i0}^{\lambda}}|^2 + \frac{A}{\gamma-1}\sum_{i=a,b}\int (\rho_{i0}^{\lambda})^{\gamma} + \frac{\lambda^2}{2}\int |\nabla\phi_0^{\lambda}|^2.$$

For the smooth solution to the QHD equations (6.3.5)–(6.3.6), we can define the total energy $\mathcal{G}(t)$ as follows:

$$\mathcal{G} = \frac{1}{2}\int \rho|\boldsymbol{u}|^2 + \frac{h^2}{2}\int |\nabla\sqrt{\rho}|^2 + \frac{A}{\gamma-1}\int \rho^{\gamma}. \tag{6.3.20}$$

Then, a direct computation implies that

$$\frac{\mathrm{d}\mathcal{G}}{\mathrm{d}t}(t) = -\frac{1}{\tau}\int \rho|\boldsymbol{u}|^2 < 0. \tag{6.3.21}$$

Therefore, the following inequality holds:

$$\mathcal{G}(t) \le \mathcal{G}(0) = \frac{1}{2}\int \rho_0|\boldsymbol{u}_0|^2 + \frac{h^2}{2}\int |\nabla\sqrt{\rho_0}|^2 + \frac{A}{\gamma-1}\int \rho_0^{\gamma}. \tag{6.3.22}$$

2. The modulated energy functional and uniform estimates.
Based on the total energy inequalities (6.3.18) and (6.3.21), we introduce the following modulated energy functional:

$$\mathcal{H}^{\lambda,h}(t) = \frac{1}{2}\sum_{i=a,b}\int \rho_i^{\lambda}|\boldsymbol{u}_i^{\lambda} - \boldsymbol{u}|^2 + \Pi^{\lambda}(t) + \frac{h^2}{2}\sum_{i=a,b}\int |\nabla\sqrt{\rho_i^{\lambda}} - \nabla\sqrt{\rho}|^2 + \frac{\lambda^2}{2}\int |\nabla\phi^{\lambda}|^2, \tag{6.3.23}$$

where

$$\Pi^\lambda(t) = \frac{A}{\gamma - 1}\sum_{i=a,b}\int [(\rho_i^\lambda)^\gamma - \rho^\gamma - \gamma\rho^{\gamma-1}(\rho_i^\lambda - \rho)]. \quad (6.3.24)$$

Using (6.3.17) and (6.3.22), we have

$$\begin{aligned}
\frac{\mathrm{d}\mathcal{H}^{\lambda,h}(t)}{\mathrm{d}t} &= \frac{\mathrm{d}\mathcal{E}^{\lambda,h}(t)}{\mathrm{d}t} - \sum_{i=a,b}\int \partial_t(\rho_i^\lambda \boldsymbol{u}_i^\lambda \cdot \boldsymbol{u}) + \frac{1}{2}\sum_{i=a,b}\int \partial_t(\rho_i^\lambda |\boldsymbol{u}|^2) \\
&\quad - h^2 \sum_{i=a,b}\int \partial_t(\boldsymbol{\nabla}\sqrt{\rho_i^\lambda}\cdot \boldsymbol{\nabla}\sqrt{\rho}) + h^2 \int \partial(\boldsymbol{\nabla}\sqrt{\rho}^2) \\
&\quad - \frac{A\gamma}{\gamma - 1}\sum_{i=a,b}\int \partial_t(\rho^{\gamma-1}\rho_i^\lambda) + A\int \partial_t(\rho^\gamma) \\
&:= -\frac{1}{\tau}\sum_{i=a,b}\int \rho_i^\lambda |\boldsymbol{u}_i^\lambda|^2 + \sum_{j=1}^{6} I_j. \quad (6.3.25)
\end{aligned}$$

Next we will estimate $I_j, j = 1, \ldots, 6$. Using the momentum equation (6.3.2), we obtain

$$\begin{aligned}
I_1 &= -\sum_{i=a,b}\int \rho_i^\lambda \boldsymbol{u}_i^\lambda \cdot \partial_t \boldsymbol{u} - \sum_{i=a,b}\int \partial_t(\rho_i^\lambda \boldsymbol{u}_i^\lambda)\cdot \boldsymbol{u} \\
&= -\sum_{i=a,b}\int \{\rho_i^\lambda \boldsymbol{u}_i^\lambda \cdot \partial_t \boldsymbol{u} - \boldsymbol{u}\cdot [\mathrm{div}(\rho_i^\lambda \boldsymbol{u}_i^\lambda \otimes \boldsymbol{u}_i^\lambda) \\
&\quad + A\boldsymbol{\nabla}(\rho_i^\lambda)^\gamma - q_i \rho_i^\lambda \boldsymbol{\nabla}\phi^\lambda]\} \\
&\quad - \sum_{i=a,b}\int \boldsymbol{u}\cdot \left[\frac{h^2}{2}\rho_i^\lambda \boldsymbol{\nabla}\left(\frac{\Delta\sqrt{\rho_i^\lambda}}{\sqrt{\rho_i^\lambda}}\right) - \frac{\rho_i^\lambda \boldsymbol{u}_i^\lambda}{\tau}\right] \\
&= -\sum_{i=a,b}\left[\int \rho_i^\lambda \boldsymbol{u}_i^\lambda \cdot \partial_t \boldsymbol{u} + \int (\rho_i^\lambda \boldsymbol{u}_i^\lambda \otimes \boldsymbol{u}_i^\lambda) : \boldsymbol{\nabla} u \right. \\
&\quad \left. - \int \frac{\rho_i^\lambda \boldsymbol{u}_i^\lambda \cdot \boldsymbol{u}}{\tau}\right] + II, \quad (6.3.26)
\end{aligned}$$

where: denotes the summation over both matrix indices, and

$$II = A\sum_{i=a,b}\int \nabla(\rho_i^\lambda)^\gamma\cdot\boldsymbol{u} - \sum_{i=a,b}\int q_i\rho_i^\lambda\nabla\phi^\lambda\cdot\boldsymbol{u}$$

$$+\frac{h^2}{2}\int\frac{\Delta\sqrt{\rho_i^\lambda}}{\sqrt{\rho_i^\lambda}}\operatorname{div}(\rho_i^\lambda\boldsymbol{u}). \tag{6.3.27}$$

We rewrite term I_2 as

$$I_2 = \frac{1}{2}\sum_{i=a,b}\int[\partial_t\rho_i^\lambda|\boldsymbol{u}|^2 + \rho_i^\lambda\partial_t|\boldsymbol{u}|^2] := I_{21} + I_{22}. \tag{6.3.28}$$

Noticing

$$-\int(\rho_i^\lambda\boldsymbol{u}_i^\lambda\otimes\boldsymbol{u}_i^\lambda):\nabla\boldsymbol{u}$$

$$= -\int(\rho_i^\lambda(\boldsymbol{u}_i^\lambda-\boldsymbol{u})\otimes(\boldsymbol{u}_i^\lambda-\boldsymbol{u})):\nabla\boldsymbol{u} + \int(\rho_i^\lambda\boldsymbol{u}\otimes\boldsymbol{u}):\nabla\boldsymbol{u}$$

$$-\int(\rho_i^\lambda\boldsymbol{u}_i^\lambda\otimes\boldsymbol{u}):\nabla\boldsymbol{u} - \int(\rho_i^\lambda\boldsymbol{u}\otimes\boldsymbol{u}_i^\lambda):\nabla\boldsymbol{u},$$

it holds that

$$I_{22} - \sum_{i=a,b}\int\{\rho_i^\lambda\boldsymbol{u}_i^\lambda\cdot\partial_t\boldsymbol{u} + (\rho_i^\lambda\boldsymbol{u}\otimes\boldsymbol{u}_i^\lambda):\nabla u - (\rho_i^\lambda\boldsymbol{u}\otimes\boldsymbol{u}):\nabla\boldsymbol{u}\}$$

$$= \sum_{i=a,b}\int[\boldsymbol{u}_t + (u\cdot\nabla)\boldsymbol{u}]\cdot(\rho_i^\lambda\boldsymbol{u} - \rho_i^\lambda\boldsymbol{u}_i^\lambda).$$

Since

$$-\sum_{i=a,b}\int(\rho_i^\lambda\boldsymbol{u}_i^\lambda\otimes\boldsymbol{u}):\nabla\boldsymbol{u} = \frac{1}{2}\sum_{i=a,b}\int\operatorname{div}(\rho_i^\lambda\boldsymbol{u}_i^\lambda)|\boldsymbol{u}|^2$$

$$= -\frac{1}{2}\sum_{i=a,b}\int\partial_t\rho_i^\lambda|\boldsymbol{u}|^2 = -I_{21},$$

we have

$$\begin{aligned}
I_1 + I_2 = &-\sum_{i=a,b}\int(\rho_i^\lambda(\boldsymbol{u}_i^\lambda - \boldsymbol{u})\otimes(\boldsymbol{u}_i^\lambda - \boldsymbol{u})) : \boldsymbol{\nabla}_u \\
&+\sum_{i=a,b}\int\frac{\rho_i^\lambda \boldsymbol{u}_i^\lambda\cdot\boldsymbol{u}}{\tau} + \sum_{i=a,b}\int[\boldsymbol{u}_t + (\boldsymbol{u}\cdot\boldsymbol{\nabla})\boldsymbol{u}] \\
&\cdot(\rho_i^\lambda\boldsymbol{u} - \rho_i^\lambda\boldsymbol{u}_i^\lambda) + II \\
= &-\sum_{i=a,b}\int(\rho_i^\lambda(\boldsymbol{u}_i^\lambda - \boldsymbol{u})\otimes(\boldsymbol{u}_i^\lambda - \boldsymbol{u})) : \boldsymbol{\nabla}_u \\
&+\sum_{i=a,b}\int\frac{\rho_i^\lambda \boldsymbol{u}_i^\lambda\cdot\boldsymbol{u}}{\tau} + \sum_{i=a,b}\int(\rho_i^\lambda\boldsymbol{u} - \rho_i^\lambda\boldsymbol{u}_i^\lambda) \\
&\cdot\left[-\boldsymbol{A}_\gamma\rho^{\gamma-2}\boldsymbol{\nabla}_\rho + \frac{h^2}{2}\boldsymbol{\nabla}\left(\frac{\Delta\sqrt{\rho}}{\sqrt{\rho}}\right) - \frac{\boldsymbol{u}}{\tau}\right] + II. \qquad (6.3.29)
\end{aligned}$$

Using the equation of the conversation of mass (6.3.1) and (6.3.5), we calculate the terms $I_j, j = 3, \ldots, 6$ as follows

$$\begin{aligned}
I_3 = &-h^2\sum_{i=a,b}\int\partial_t(\boldsymbol{\nabla}\sqrt{\rho_i^\lambda})\cdot\boldsymbol{\nabla}\sqrt{\rho} - h^2\sum_{i=a,b}\int\boldsymbol{\nabla}\sqrt{\rho_i^\lambda}\cdot\partial_t(\boldsymbol{\nabla}\sqrt{\rho}) \\
= &-\frac{h^2}{2}\sum_{i=a,b}\int\boldsymbol{\nabla}\left(\frac{\partial_t\rho_i^\lambda}{\sqrt{\rho_i^\lambda}}\right)\cdot\boldsymbol{\nabla}\sqrt{\rho} - \frac{h^2}{2} \\
&\times\sum_{i=a,b}\int\boldsymbol{\nabla}\left(\frac{\partial_t\rho}{\sqrt{\rho}}\right)\cdot\boldsymbol{\nabla}\sqrt{\rho_i^\lambda} \\
= &-\frac{h^2}{2}\sum_{i=a,b}\int\operatorname{div}(\rho_i^\lambda\boldsymbol{u}_i^\lambda)\frac{\Delta\sqrt{\rho}}{\sqrt{\rho_i^\lambda}} - \frac{h^2}{2}\sum_{i=a,b}\int\operatorname{div}(\rho\boldsymbol{u})\frac{\Delta\sqrt{\rho_i^\lambda}}{\sqrt{\rho}}, \\
&\qquad (6.3.30)
\end{aligned}$$

$$I_4 = 2h^2 \int \nabla\sqrt{\rho}\cdot\nabla(\partial_t\sqrt{\rho}) = h^2 \int \nabla\sqrt{\rho}\cdot\nabla\left(\frac{\partial_t\rho}{\sqrt{\rho}}\right)$$
$$= h^2 \int \operatorname{div}(\rho\boldsymbol{u})\frac{\Delta\sqrt{\rho}}{\sqrt{\rho}}, \tag{6.3.31}$$

$$I_5 = -A_\gamma \sum_{i=a,b} \int \rho^{\gamma-2}\partial_t\rho\rho_i^\lambda - \frac{A_\gamma}{\gamma-1}\sum_{i=a,b}\int \rho^{\gamma-1}\partial_t\rho_i^\lambda$$
$$= A_\gamma \sum_{i=a,b}\int \rho^{\gamma-2}\operatorname{div}(\rho\boldsymbol{u})\rho_i^\lambda + \frac{A_\gamma}{\gamma-1}\sum_{i=a,b}\int \rho^{\gamma-1}\operatorname{div}(\rho_i^\lambda\boldsymbol{u}_i^\lambda)$$
$$= -\sum_{i=a,b}\int \Big\{(A_{\gamma\rho}\rho_i^\lambda\boldsymbol{u}\cdot\nabla\rho^{\gamma-2} + \rho^{\gamma-2}\boldsymbol{u}\cdot\nabla\rho_i^\lambda)$$
$$+\frac{A_\gamma}{\gamma-1}\rho_i^\lambda\boldsymbol{u}_i^\lambda\cdot\nabla\rho^{\gamma-1}\Big\}, \tag{6.3.32}$$

$$I_6 = \gamma\boldsymbol{A}\int \rho^{\gamma-1}\partial_{t\rho} = -\gamma\boldsymbol{A}\int \rho^{\gamma-1}\operatorname{div}(\rho\boldsymbol{u}) = \gamma\boldsymbol{A}\int \rho\boldsymbol{u}\cdot\nabla\rho^{\gamma-1}. \tag{6.3.33}$$

Plugging (6.3.27) and (6.3.29)–(6.3.33) into equation (6.3.25), and using the Poisson equation (6.3.3), we have

$$\frac{\mathrm{d}\mathcal{H}^{\lambda,h}}{\mathrm{d}t} = -\frac{1}{\tau}\sum_{i=a,b}\int \rho_i^\lambda|\boldsymbol{u}_i^\lambda - \boldsymbol{u}|^2 - \sum_{i=a,b}$$
$$\times\int(\rho_i^\lambda(\boldsymbol{u}_i^\lambda - \boldsymbol{u})\otimes(\boldsymbol{u}_i^\lambda - \boldsymbol{u})) : \nabla\boldsymbol{u}$$
$$- A\sum_{i=a,b}\int[(\rho_i^\lambda)^\gamma - \rho^\gamma - \gamma\rho^{\gamma-1}(\rho_i^\lambda - \rho)]\operatorname{div}\boldsymbol{u}$$
$$- \lambda^2\int \Delta\phi^\lambda\nabla\phi^\lambda\cdot\boldsymbol{u} + \frac{h^2}{2}\sum_{i=a,b}\int\frac{\Delta\sqrt{\rho_i^\lambda}}{\sqrt{\rho_i^\lambda}}\operatorname{div}(\rho_i^\lambda\boldsymbol{u})$$
$$- \frac{h^2}{2}\sum_{i=a,b}\int\frac{\Delta\sqrt{\rho}}{\sqrt{\rho}}\operatorname{div}(\rho_i^\lambda\boldsymbol{u}) + \frac{h^2}{2}\sum_{i=a,b}$$

$$\times \int \frac{\Delta\sqrt{\rho}}{\sqrt{\rho}} \operatorname{div}(\rho_i^\lambda \boldsymbol{u}_i^\lambda) + h^2 \int \frac{\Delta\sqrt{\rho}}{\sqrt{\rho}} \operatorname{div}(\rho \boldsymbol{u})$$

$$- \frac{h^2}{2} \sum_{i=a,b} \int \frac{\Delta\sqrt{\rho}}{\sqrt{\rho_i^\lambda}} \operatorname{div}(\rho_i^\lambda \boldsymbol{u}_i^\lambda) - \frac{h^2}{2} \sum_{i=a,b}$$

$$\times \int \frac{\Delta\sqrt{\rho_i^\lambda}}{\sqrt{\rho}} \operatorname{div}(\rho^\lambda \boldsymbol{u}). \tag{6.3.34}$$

Next we will estimate the right hand side of (6.3.34). Firstly, combining the regularity of ρ and $\boldsymbol{u}$, $\|\operatorname{div} \boldsymbol{u}\|_{L^\infty}$ and $\|\boldsymbol{\nabla} u\|_{L^\infty}$ are bounded. Thus we have

$$- \sum_{i=a,b} \int (\rho_i^\lambda (\boldsymbol{u}_i^\lambda - \boldsymbol{u}) \otimes (\boldsymbol{u}_i^\lambda - \boldsymbol{u})) : \boldsymbol{\nabla} u$$

$$\leq C_1 \|\boldsymbol{\nabla} u\|_{L^\infty} \sum_{i=a,b} \int \rho_i^\lambda |\boldsymbol{u}_i^\lambda - \boldsymbol{u}|^2, \tag{6.3.35}$$

$$- A \sum_{i=a,b} \int [(\rho_i^\lambda)^\gamma - \rho^\gamma - \gamma \rho^{\gamma-1}(\rho_i^\lambda - \rho)] \operatorname{div} \boldsymbol{u}$$

$$\leq (\gamma - 1) \|\operatorname{div} \boldsymbol{u}\|_{L^\infty} \Pi^\lambda(t). \tag{6.3.36}$$

For the fourth term, we have

$$- \lambda^2 \int \Delta\phi^\lambda \boldsymbol{\nabla}\phi^\lambda \cdot \boldsymbol{u}$$

$$= \lambda^2 \int (\boldsymbol{\nabla}\phi^\lambda \otimes \boldsymbol{\nabla}\phi^\lambda) : \boldsymbol{\nabla} u - \frac{\lambda^2}{2} \int |\boldsymbol{\nabla}\phi^\lambda|^2 \operatorname{div} \boldsymbol{u}$$

$$\leq C_2 (\|\operatorname{div} \boldsymbol{u}\|_{L^\infty} + \|\boldsymbol{\nabla} u\|_{L^\infty}) \lambda^2 \int |\boldsymbol{\nabla}\phi^\lambda|^2. \tag{6.3.37}$$

Finally, we estimate the last six terms (6.3.34). Using the positivity of ρ and $\rho_i^\lambda (i = a, b)$, the regularity of $\rho, \boldsymbol{u}, \rho_i^\lambda, \boldsymbol{u}_i^\lambda (i = a, b)$, and Cauchy inequality, we can get

$$
\begin{aligned}
&\frac{h^2}{2}\sum_{i=a,b}\int \frac{\Delta\sqrt{\rho_i^\lambda}}{\sqrt{\rho_i^\lambda}}\operatorname{div}(\rho_i^\lambda \boldsymbol{u}) - \frac{h^2}{2}\sum_{i=a,b}\int \frac{\Delta\sqrt{\rho}}{\sqrt{\rho}}\operatorname{div}(\rho_i^\lambda \boldsymbol{u}) \\
&\quad + \frac{h^2}{2}\sum_{i=a,b}\int \frac{\Delta\sqrt{\rho}}{\sqrt{\rho}}\operatorname{div}(\rho_i^\lambda \boldsymbol{u}_i^\lambda) + h^2\int \frac{\Delta\sqrt{\rho}}{\sqrt{\rho}}\operatorname{div}(\rho^\lambda \boldsymbol{u}) \\
&\quad - \frac{h^2}{2}\sum_{i=a,b}\int \frac{\Delta\sqrt{\rho}}{\sqrt{\rho_i^\lambda}}\operatorname{div}(\rho_i^\lambda \boldsymbol{u}_i^\lambda) - \frac{h^2}{2}\sum_{i=a,b}\int \frac{\Delta\sqrt{\rho_i^\lambda}}{\sqrt{\rho}}\operatorname{div}(\rho^\lambda \boldsymbol{u}) \\
&= -\frac{h^2}{2}\sum_{i=a,b}\int (\Delta\sqrt{\rho_i^\lambda} - \Delta\sqrt{\rho})\frac{\operatorname{div}(\rho\boldsymbol{u})}{\sqrt{\rho}} \\
&\quad + \frac{h^2}{2}\sum_{i=a,b}\int \frac{\sqrt{\rho}\Delta\sqrt{\rho_i^\lambda} - \Delta\sqrt{\rho}\sqrt{\rho_i^\lambda}}{\sqrt{\rho}\sqrt{\rho_i^\lambda}}\operatorname{div}(\rho_i^\lambda \boldsymbol{u}) \\
&\quad + \frac{h^2}{2}\sum_{i=a,b}\int \frac{\Delta\sqrt{\rho}(\sqrt{\rho_i^\lambda} - \sqrt{\rho})}{\sqrt{\rho_i^\lambda}\sqrt{\rho}}\operatorname{div}(\rho_i^\lambda \boldsymbol{u}_i^\lambda) \\
&= \frac{h^2}{2}\sum_{i=a,b}\int (\boldsymbol{\nabla}\sqrt{\rho_i^\lambda} - \boldsymbol{\nabla}\sqrt{\rho})\boldsymbol{\nabla}\left(\frac{\operatorname{div}(\rho\boldsymbol{u})}{\sqrt{\rho}}\right) \\
&\quad - \frac{h^2}{2}\sum_{i=a,b}\int (\boldsymbol{\nabla}\sqrt{\rho_i^\lambda} - \boldsymbol{\nabla}\sqrt{\rho})\boldsymbol{\nabla}\left(\frac{\operatorname{div}(\rho_i^\lambda\boldsymbol{u})}{\sqrt{\rho_i^\lambda}}\right) \\
&\quad - \frac{h^2}{2}\sum_{i=a,b}\int \rho_i^\lambda(\boldsymbol{u}_i^\lambda - \boldsymbol{u})\cdot\boldsymbol{\nabla}\left((\sqrt{\rho_i^\lambda} - \sqrt{\rho})\frac{\Delta\sqrt{\rho}}{\sqrt{\rho_i^\lambda}\sqrt{\rho}}\right) \\
&= -\frac{h^2}{2}\sum_{i=a,b}\int (\boldsymbol{\nabla}\sqrt{\rho_i^\lambda} - \boldsymbol{\nabla}\sqrt{\rho})\boldsymbol{\nabla}((\sqrt{\rho_i^\lambda} - \sqrt{\rho})\operatorname{div}\boldsymbol{u}) \\
&\quad + \frac{h^2}{2}\sum_{i=a,b}\int (\boldsymbol{\nabla}\sqrt{\rho_i^\lambda} - \boldsymbol{\nabla}\sqrt{\rho})\boldsymbol{\nabla}\left((\sqrt{\rho_i^\lambda} - \sqrt{\rho})\frac{\boldsymbol{\nabla}\sqrt{\rho}\cdot\boldsymbol{u}}{\sqrt{\rho_i^\lambda}\sqrt{\rho}}\right)
\end{aligned}
$$

$$
\begin{aligned}
&+\frac{h^2}{2}\sum_{i=a,b}\int(\nabla\sqrt{\rho_i^\lambda}-\nabla\sqrt{\rho})\nabla\left((\nabla\sqrt{\rho_i^\lambda}-\nabla\sqrt{\rho})\cdot\frac{\boldsymbol{u}}{\sqrt{\rho_i^\lambda}}\right)\\
&+\frac{h^2}{2}\sum_{i=a,b}\int\rho_i^\lambda(\boldsymbol{u}_i^\lambda-\boldsymbol{u})\cdot\nabla\left((\sqrt{\rho_i^\lambda}-\sqrt{\rho})\frac{\Delta\sqrt{\rho}}{\sqrt{\rho_i^\lambda}\sqrt{\rho}}\right)\\
&\le C_2h^2\sum_{i=a,b}\int|\sqrt{\rho_i^\lambda}-\sqrt{\rho}|^2+C_4\Pi^\lambda(t)\\
&+C_5\sum_{i=a,b}\int\rho_i^\lambda|\boldsymbol{u}_i^\lambda-\boldsymbol{u}|^2.
\end{aligned}
\tag{6.3.38}
$$

Here, we have used the condition $\gamma \geq 2$ and the following inequality

$$|\sqrt{\rho_i^\lambda}-\sqrt{\rho}|^2 \le \tilde{C}_1|\rho_i^\lambda-\rho|^2 \le \tilde{C}_2\Pi^\lambda(t),$$

where $\tilde{C}_1$ and $\tilde{C}_2$ are two positive constants.
Therefore, by plugging (6.3.35)–(6.3.38) into equation (6.3.34), we obtain

$$\frac{\mathrm{d}\mathcal{H}^{\lambda,h}(t)}{\mathrm{d}t} \le C_6\mathcal{H}^{\lambda,h}(t). \tag{6.3.39}$$

3. Convergence of modulated energy functional and the end of proof. By applying Gronwall inequality to (6.3.39), we obtain

$$\mathcal{H}^{\lambda,h}(t) \le \mathcal{H}^{\lambda,h}(0)\mathrm{e}^{-C_6t}. \tag{6.3.40}$$

Thus, the initial assumption (6.3.12) implies that

$$\lim_{\lambda\to 0}\mathcal{H}^{\lambda,h}(t) = 0. \tag{6.3.41}$$

From (6.3.41), we obtain $\rho_i^\lambda(i = a,b)$ converges to ρ in $C([0,T];\mathbb{T}^3)$. Therefore, the positivities of $\rho_i^\lambda(i=a,b)$ imply that $\boldsymbol{u}_i^\lambda(i=a,b)$ converges to $\boldsymbol{u}$ in $L^\infty([0,T];L^2(\mathbb{T}^3))$. This completes the proof of Theorem 6.3.4.

The proof of Theorem 6.3.5. From the proof of Theorem 6.3.4, we get

$$\mathcal{H}^{\lambda,h}(t) \le \mathcal{H}^{\lambda,h}(0)e^{-C_6 t}.$$

Using condition (6.3.13), we have $\mathcal{H}^{\lambda,h}(0) \le \bar{C}\lambda$. Thus, we can get $\mathcal{H}^{\lambda,h}(t) \le M_T\lambda$ for some positive constant M_T, which completes the proof.

6.4 Time decay

In this section, we will focus on the long time behavior of the solution to Cauchy problem of the bipolar QHD model Cauchy [104].

$$\partial_t\rho_i + \operatorname{div}(\rho_i \boldsymbol{u}_i) = 0, \tag{6.4.1}$$

$$\begin{aligned} &\partial_t(\rho_i \boldsymbol{u}_i) + \operatorname{div}(\rho_i \boldsymbol{u}_i \otimes \boldsymbol{u}_i) + \boldsymbol{\nabla} P_i(\rho_i) \\ &\quad = q_i\rho_i \boldsymbol{E} + \frac{h^2}{2}\rho_i\left(\frac{\Delta\sqrt{\rho_i}}{\sqrt{\rho_i}}\right) - \frac{\rho_i \boldsymbol{u}_i}{\tau}, \end{aligned} \tag{6.4.2}$$

$$\begin{aligned} &\lambda^2\boldsymbol{\nabla}\cdot\boldsymbol{E} = \rho_a - \rho_b - \mathcal{C}(x), \\ &\quad \boldsymbol{\nabla}\times\boldsymbol{E} = 0, \quad \boldsymbol{E}(x)\to 0, \quad |x|\to+\infty. \end{aligned} \tag{6.4.3}$$

The initial data are

$$(\rho_i \boldsymbol{u}_i)(x,0) = (\rho_{i0}, \boldsymbol{u}_{i0})(x), \tag{6.4.4}$$

where $i = a, b$, $q_a = 1$, $q_b = -1$. The variables $\rho_a > 0$, $\rho_b > 0$ and $\boldsymbol{u}_a$, $\boldsymbol{u}_b$, $\boldsymbol{E}$ are the particle densities, velocities, and electric field, respectively. We can define the usual momentum $J_a = \rho_a \boldsymbol{u}_a$, $J_b = \rho_b \boldsymbol{u}_b$. $P_a(\cdot)$ and $P_b(\cdot)$ are the pressure functions. The parameters $h > 0$, $\tau_a = \tau_b = \tau > 0$ and $\lambda > 0$ represent the Planck constant, relaxation time and Debye length. $\mathcal{C} = \mathcal{C}(x)$ is the doping profile. Formally, $(\rho_b, \rho_b \boldsymbol{u}_b) \equiv (0,0)$, the equations (6.4.1)–(6.4.3) become the unipolar QHD model.

The main results of this section is as follows.

Theorem 6.4.1. *Let $\mathcal{C}(x) = c^* > 0$ (c^* is a constant). Assume that the constant ρ_a^*, $\rho_b^* > 0$ satisfy $\rho_a^* - \rho_b^* - c^* = 0$. Furthermore, $P_a, P_b \in C^6$ and $P_a'(\rho_a^*)$, $P_b'(\rho_b^*) > 0$. Let the initial data satisfy $(\rho_{i0} - \rho_i^*, \boldsymbol{u}_{i0}) \in$*

$H^6(\boldsymbol{R}^3)\times\mathcal{H}^5(\boldsymbol{R}^3)(i=a,b)$, *and* $\boldsymbol{\Lambda}_0 := \|(\rho_{i0}-\rho_i^*, \boldsymbol{u}_{i0})\|_{H^6(\boldsymbol{R}^3)\times\mathcal{H}^5(\boldsymbol{R}^3)}$. *If there exists* $\boldsymbol{\Lambda}_1 > 0$ *such that* $\boldsymbol{\Lambda}_0 \leq \boldsymbol{\Lambda}_1$, *then the unique solution* $(\rho_i, \boldsymbol{u}_i, \boldsymbol{E})(\rho_i > 0)$ *to the IVP* (6.4.1)–(6.4.4) *exists globally in time, and satisfies*

$$\begin{cases} (\rho_i-\rho_i^*) \in C^k(0,T;H^{6-2k}(\mathbf{R}^3)), & \boldsymbol{u}_i \in C^k(0,T;\mathcal{H}^{5-2k}(\mathbf{R}^3)), \\ \boldsymbol{E} \in C^k(0,T;\mathcal{H}^{6-2k}(\mathbf{R}^3)), & k=0,1,2. \end{cases} \tag{6.4.5}$$

Moreover, $(\rho_i, \boldsymbol{u}_i, \boldsymbol{E})$ *tends to the equilibrium state* $(\rho_i^*, 0, 0)$ *at an algebraic time-decay rate, for* $0 \leq k \leq 5$,

$$(1+t)^k\|D^k(\rho_i-\rho_i^*)\|^2 + (1+t)^5\|hD^6(\rho_i-\rho_i^*)\|^2 \leq c\boldsymbol{\Lambda}_0, \tag{6.4.6}$$

for $1 \leq k \leq 5$,

$$(1+t)^k\|D^k(\boldsymbol{u}_i, J_i)\|^2 + (1+t)^k\|D^k\boldsymbol{E}\|^2 + (1+t)^6\|D^6\boldsymbol{E}\|^2 \leq c\boldsymbol{\Lambda}_0, \tag{6.4.7}$$

where c *is a constant independent on* h. *The space* $\mathcal{H}^k(\mathbf{R}^3) = \{f \in L^6(\mathbf{R}^6),\ Df \in H^{k-1}(\mathbf{R}^3)\}(k \geq 1)$, *and* $D^k f$ *denotes the* k*-times spatial derivative of* f.

Remark 6.4.1. By (6.4.6)–(6.4.7) and Nirenberg inequality

$$\|\boldsymbol{u}\|_{L^\infty(\mathbf{R}^3)} \leq c\|D^2\boldsymbol{u}\|^{\frac{1}{2}}_{L^2(\mathbf{R}^3)}\|\boldsymbol{u}\|^{\frac{1}{2}}_{L^6(\mathbf{R}^3)} \leq c\|D^2\boldsymbol{u}\|^{\frac{1}{2}}_{L^2(\mathbf{R}^3)}\|D_{\boldsymbol{u}}\|^{\frac{1}{2}}_{L^2(\mathbf{R}^3)}, \tag{6.4.8}$$

we can get the optimal L^∞ time-decay rate of the solution

$$\|(\rho_i-\rho_i^*, \boldsymbol{u}_i, \boldsymbol{E})(t)\|_{L^\infty(\mathbf{R}^3)} \leq c(1+t)^{-\frac{3}{4}}. \tag{6.4.9}$$

Firstly, we linearize the equations (6.4.1)–(6.4.3) at the equilibrium $(\rho_a, \boldsymbol{u}_a, \rho_b, \boldsymbol{u}_b, \boldsymbol{E}) = (\rho_a^*, 0, \rho_b^*, 0, 0)$,

$$\begin{cases} W_{at} + \boldsymbol{\nabla}\cdot\boldsymbol{J}_a = 0, \\ \boldsymbol{J}_{at} + P_a'(\rho_a^*)\boldsymbol{\nabla}W_a - \dfrac{h^2}{4}\boldsymbol{\nabla}\Delta_{W_a} + \boldsymbol{J}_a - \rho_a^*\boldsymbol{E} = 0, \\ W_{bt} + \boldsymbol{\nabla}\cdot\boldsymbol{J}_b = 0, \\ \boldsymbol{J}_{bt} + P_b'(\rho_b^*)\boldsymbol{\nabla}W_b - \dfrac{h^2}{4}\boldsymbol{\nabla}\Delta W_b + \boldsymbol{J}_b + \rho_b^*\boldsymbol{E} = 0, \\ \boldsymbol{\nabla}\cdot\boldsymbol{E} = W_a - W_b, \boldsymbol{\nabla}\times\boldsymbol{E} = 0, \boldsymbol{E} \to 0, \text{as}|x| \to \infty. \end{cases} \tag{6.4.10}$$

For simplicity, we assume the constant $\tau = 1, \lambda = 1$. The initial data of the equation (6.4.11) are

$$(W_a, \boldsymbol{J}_a, W_b, \boldsymbol{J}_b)(x,0) = (W_{a0}, \boldsymbol{J}_{a0}, W_{b0}, \boldsymbol{J}_{b0})(x). \tag{6.4.11}$$

From the Poisson equation $(6.4.10)_5$ about $\boldsymbol{E}$, we have

$$\boldsymbol{E} = \nabla\Delta^{-1}(W_a - W_b). \tag{6.4.12}$$

Assume the initial data (6.4.11) satisfy

$$W_{a0}, W_{b0} \in H^6(\mathbf{R}^3) \cap L^1(\mathbf{R}^3), J_{a0}, J_{b0} \in H^5(\mathbf{R}^3), \tag{6.4.13}$$

so that the initial electric field $\boldsymbol{E}_0$ obtained from Poisson equation (6.4.12) at initial time has the regularity

$$\boldsymbol{E}_0 = \boldsymbol{\nabla}\Delta^{-1}(W_{a0} - W_{b0}) \in H^5(\mathbf{R}^3). \tag{6.4.14}$$

For simplicity, we just consider the IVP (6.4.10)–(6.4.11) for following case

$$\frac{h^2}{4} = 1, \rho_a^* = 2, \rho_b^* = 1, c^* = 1, P_a'(2) = P_a'(1) = 1. \tag{6.4.15}$$

Next we give the algebraic time-decay of the linearized equations (6.4.10)–(6.4.11).

Theorem 6.4.2. *Suppose that* (6.4.13)–(6.4.15) *hold. Furthermore, assume that the Fourier transformation* $(\hat{W}_{a0}, \hat{W}_{b0})$ *of initial density* *satisfies for some constants $m_0 > 0, r > 0$ that*

$$\inf_{\xi \in B(0,r)} |(\hat{W}_{a0} + 2\hat{W}_{b0})(\xi)| \geq m_0, \tag{6.4.16}$$

and the initial perturbation of momentum satisfies

$$\nabla \cdot (\boldsymbol{J}_{a0} + 2\boldsymbol{J}_{b0}) = 0. \tag{6.4.17}$$

Then, the unique global solution to (6.4.10)–(6.4.11) *exists and satisfies,*

$$\begin{aligned} &W_a, W_b \in C([0, +\infty); H^6(\mathbf{R}^3)), \boldsymbol{J}_a, \boldsymbol{J}_b \in C([0, +\infty); H^5(\mathbf{R}^3)), \\ &\boldsymbol{E} \ \in C([0, +\infty); H^5(\mathbf{R}^3)), \end{aligned} \tag{6.4.18}$$

and

$$\begin{aligned} &c_1(1+t)^{-\frac{k}{2}-\frac{3}{4}} \leq \|(\partial_x^k W_a, \partial_x^k W_b)(t)\|_{L^2(\boldsymbol{R}^3)} \leq c_2(1+t)^{-\frac{k}{2}}, \\ &\quad 0 \leq k \leq 6, \end{aligned} \tag{6.4.19}$$

$$\begin{aligned} &c_1(1+t)^{-\frac{k}{2}-\frac{5}{4}} \leq \|(\partial_x^k \boldsymbol{J}_a, \partial_x^k \boldsymbol{J}_b)(t)\|_{L^2(\boldsymbol{R}^3)} \\ &\quad \leq c^2(1+t)^{-\frac{k}{2}}, 0 \leq k \leq 5, \end{aligned} \tag{6.4.20}$$

$$\|\boldsymbol{E}(t)\|_{H^5(\boldsymbol{R})^3} \leq C e^{-c\frac{t}{3}}, \tag{6.4.21}$$

where $i = a, b, c_1, c_2, c_3$ *are the constants independent on* $m_0, \|U_0\|_{H^6(\mathbf{R}^3)\times H^5(\mathbf{R}^3)}$ *and* $\|(W_{a0}, W_{b0})\|$.

Remark 6.4.2. The regularity of electric field $\boldsymbol{E}$ can be established as follows. In fact, the norm $\|D^k \boldsymbol{E}\|_{L^2} (k > 0)$ of $\boldsymbol{E}$ with integer $k > 0$ can be obtained directly by Lemma 6.1.3 through the Poisson equation $(6.4.10)_5$. To estimate the norm $\|\boldsymbol{E}\|_{L^2}$ and the exponential time-decay rate, we need to apply the energy method to the equation for electric field $\boldsymbol{E}$, which is derived from (6.4.10), (6.4.12), and (6.4.18), as

$$\boldsymbol{E}_{tt} - \Delta \boldsymbol{E} + \Delta^2 \boldsymbol{E} + \boldsymbol{E}_t + 3\boldsymbol{E} = 0.$$

And integrating by parts the above equation in $\mathbf{R}^3$, then using the Gronwall inequality, we can get (6.4.21).

Remark 6.4.3. Theorem 6.4.2 shows that the solution of the linearized bipolar QHD system (6.4.10) decays in the algebraic rate to the equilibrium state. Since the non-linear QHD can be viewed as a small perturbation to the corresponding linearized system, the solution of the bipolar quantum hydrodynamic models (6.4.1)–(6.4.3) is also decaying at the algebraic rate to the equilibrium state.

Next, we will provide *a priori* estimate of the solution to the bipolar QHD models (6.4.1)–(6.4.3).

For simplicity, we assume $\lambda = 1, \tau = 1$. Denoting $\psi_i = \sqrt{\rho_i}(i = a, b)$, and it holds that

$$\begin{aligned}\psi_{itt} + \psi_{it} + \frac{h^2\Delta^2\psi_i}{4} + \frac{qi}{2\psi_i}\boldsymbol{\nabla}\cdot(\psi_i^2\boldsymbol{E}) - \frac{1}{2\psi_i}\boldsymbol{\nabla}^2(\psi_i^2\boldsymbol{u}_i\otimes\boldsymbol{u}_i)\\ -\frac{1}{2\psi_i}\Delta P_i(\psi_i^2) + \frac{\psi_{it}^2}{\psi_i} - \frac{h^2|\Delta\psi_i|^2}{4\psi_i} = 0.\end{aligned} \tag{6.4.22}$$

The initial data are

$$\begin{aligned}\psi_i(x,0) : \psi_{i0}(x) = \sqrt{\rho_{i0}(x)}, \psi_{it}(x,0) := \psi_1(x)\\ = -\frac{1}{2}\psi_{i0}\boldsymbol{\nabla}\cdot\boldsymbol{u}_{i0} - \boldsymbol{u}_{i0}\cdot\boldsymbol{\nabla}\psi_{i0}.\end{aligned} \tag{6.4.23}$$

Taking curl of (6.4.2), using $(\boldsymbol{u}_i\cdot\boldsymbol{\nabla})\boldsymbol{u}_i = \frac{1}{2}\boldsymbol{\nabla}(|\boldsymbol{u}_i|^2) - \boldsymbol{u}_i\times(\boldsymbol{\nabla}\times\boldsymbol{u}_i)$, we can get the equation about $\boldsymbol{\phi}_i := \boldsymbol{\nabla}\times\boldsymbol{u}_i$:

$$\boldsymbol{\phi}_{it} + \boldsymbol{\phi}_i + (\boldsymbol{u}_i\cdot\boldsymbol{\nabla})\boldsymbol{\phi}_i + \boldsymbol{\phi}_i\boldsymbol{\nabla}\cdot\boldsymbol{u}_i - (\boldsymbol{\phi}_i\cdot\boldsymbol{\nabla})\boldsymbol{u}_i - \boldsymbol{u}_i(\boldsymbol{\nabla}\cdot\boldsymbol{\phi}_i) = 0, \tag{6.4.24}$$

here we have used $\boldsymbol{\nabla}\cdot\boldsymbol{\phi} = 0$.

Denote

$$\omega_a = \psi_a - \sqrt{\rho_a^*}, \quad \omega_b = \psi_b - \sqrt{\rho_b^*}.$$

From (6.4.22), we can derive the system about $(\omega_a, \omega_b, \boldsymbol{\phi}_a, \boldsymbol{\phi}_b, \boldsymbol{E})$,

$$\begin{aligned}
&\omega_{att} + \omega_{at} + \frac{h^2\Delta^2\omega_a}{4} + \frac{1}{2}(\omega_a + \sqrt{\rho_a^*})\boldsymbol{\nabla}\cdot\boldsymbol{E} \\
&\quad - P_a'(\rho_a^*)\Delta\omega_a = f_{a1}, && (6.4.25)\\
&\omega_{btt} + \omega_{bt} + \frac{h^2\Delta^2\omega_b}{4} - \frac{1}{2}(\omega_b + \sqrt{\rho_b^*})\boldsymbol{\nabla}\cdot\boldsymbol{E} \\
&\quad -P_b'(\rho_b^*)\Delta\omega_b = f_{b1}, && (6.4.26)\\
&\boldsymbol{\phi}_{at} + \boldsymbol{\phi}_a = f_{a2}, && (6.4.27)\\
&\boldsymbol{\phi}_{bt} + \boldsymbol{\phi}_b = f_{b2}, && (6.4.28)\\
&\boldsymbol{\nabla}\cdot\boldsymbol{E} = \omega_a^2 - \omega_b^2 + 2\sqrt{\rho_a^*}\omega_a - 2\sqrt{\rho_b^*}\omega_b, \\
&\quad \boldsymbol{\nabla}\times\boldsymbol{E} = 0, && (6.4.29)
\end{aligned}$$

with the initial conditions given for $i = a, b$

$$\begin{aligned}
&\omega_i(x,0) := \omega_{i0}(x) = \psi_{i0} - \sqrt{\rho_i^*}, \quad \boldsymbol{\phi}_i(x,0) := \boldsymbol{\phi}_{i0}(x) \\
&\qquad = \boldsymbol{\nabla}\times\boldsymbol{u}_{i0}(x), && (6.4.30)\\
&\omega_i t(x,0) := \omega_{i1}(x) = \left(-\boldsymbol{u}_{i0}\cdot\boldsymbol{\nabla}\omega_{i0} - \frac{1}{2}(\omega_{i0} + \sqrt{\rho_i^*}\right) \\
&\qquad \times\boldsymbol{\nabla}\cdot\boldsymbol{u}_{i0}), && (6.4.31)
\end{aligned}$$

and

$$\begin{aligned}
f_{i1} &:= f_{i1}(x,t) \\
&= \frac{-\omega_{it}^2}{\omega_i + \sqrt{\rho_i^*}} - qi\boldsymbol{\nabla}\omega_i\boldsymbol{E} + (P_i'((\omega_i + \sqrt{\rho_i^*})^2) - P_i'(\rho_i^*))\Delta\omega_i \\
&\quad +2(\omega_i + \sqrt{\rho_i^*})P_i''((\omega_i + \sqrt{\rho_i^*})^2)|\boldsymbol{\nabla}\omega_i|^2 + P_i'((\omega_i + \sqrt{\rho_i^*})^2) \\
&\quad \times\frac{|\boldsymbol{\nabla}\omega_i|^2}{\omega_i + \sqrt{\rho_i^*}} + \frac{h^2(\Delta\omega_i)^2}{4(\omega_i + \sqrt{\rho_i^*})} + \frac{\boldsymbol{\nabla}^2((\omega_i + \sqrt{\rho_i^*})^2\boldsymbol{u}_i\otimes\boldsymbol{u}_i)}{2(\omega_i + \sqrt{\rho^*}_i)},
\end{aligned} \qquad (6.4.32)$$

$$f_{i2} := f_{i2}(x,t) = ((\boldsymbol{\phi}_i\cdot\boldsymbol{\nabla})\boldsymbol{u}_i - (\boldsymbol{u}_i\cdot\boldsymbol{\nabla})\boldsymbol{\phi}_i - \boldsymbol{\phi}_i\boldsymbol{\nabla}\cdot\boldsymbol{u}_i). \qquad (6.4.33)$$

From (6.4.1) we can get the relation between $\nabla\cdot\boldsymbol{u}_i$, $\nabla\omega_i$ and ω_{it}

$$2\omega_{it} + 2\boldsymbol{u}_i\cdot\nabla\omega_i + (\omega_i + \sqrt{\omega_i} + \sqrt{\rho_i^*})\nabla\cdot\boldsymbol{u}_i = 0. \tag{6.4.34}$$

Assume that the classical solution $(\omega_i, \boldsymbol{u}_i, \boldsymbol{E})$ satisfies *a priori* estimate

$$\begin{aligned}\delta_T \triangleq \max_{0\le t\le T}\Bigg\{&\sum_{k=0}^{5}(1+t)^k\|D^k\omega_i\|^2 + \sum_{k=1}^{5}(1+t)^k\|D^k\boldsymbol{u}_i\|^2\\ &+\sum_{k=0}^{3}(1+t)^{k+2}\|D^k\omega_{it}\|^2 + \sum_{k=1}^{3}(1+t)^{2+k}\|D^k\boldsymbol{u}_{it}\|^2\\ &+(1+t)^5\|D^4_{\omega_{it}}\|^2 + \sum_{k=1}^{5}(1+t)^k\|D^kE\|^2\\ &+\sum_{k=0}^{2}(1+t)^{3+k}\|D^k\omega_{itt}\|^2\Bigg\} \ll 1.\end{aligned} \tag{6.4.35}$$

From (6.4.35), we know that for the sufficiently small δ_T the positivity of density $\psi_i (i = a, b)$ can be sure that

$$\frac{\sqrt{\rho_i^*}}{2} \le \omega_i + \sqrt{\rho_i^*} \le \frac{3}{2}\sqrt{\rho_i^*}.$$

Using Nirenberg inequality for 3D case and (6.4.35), we have

$$\begin{aligned}&\sum_{k=0}^{3}(1+t)^{k+1}\|D^k\omega_i\|^2_{L^\infty} + \sum_{k=0}^{2}(1+t)^{k+3}\|D^k\omega_{it}\|^2_{L^\infty}\\ &\quad + (1+t)^4\|\omega_{itt}\|^2_{L^\infty}] \le \delta_T,\end{aligned} \tag{6.4.36}$$

$$\sum_{k=0}^{2}(1+t)^{k+1}\|D^k \boldsymbol{u}_i\|_{L^\infty}^2+\sum_{k=0}^{1}(1+t)^{k+3}\|D^k \boldsymbol{u}_{it}\|_{L^\infty}^2$$

$$+\sum_{k=0}^{3}(1+t)^{k+1}\|D^k E\|_{L^\infty}^2 \le c\delta_T. \tag{6.4.37}$$

From *a priori* estimate (6.4.35), we have the following estimates.

Lemma 6.4.1. *For the local strong solution* $(\omega_i, \boldsymbol{u}_i, \boldsymbol{E})$, *and, for sufficiently small* $t \in [0,T], \delta_T$, *it holds that*

$$\sum_{k=0}^{5}(1+t)^k\|D^k_{\omega_i}\|^2+(1+t)^5\|hD^6\omega_i\|^2$$

$$+\sum_{k=0}^{3}(1+t)^{k+2}\|D^k\omega_{it}\|^2+(1+t)^5\|D^4\omega_{it}\|^2$$

$$+\sum_{k=0}^{2}(1+t)^{k+3}\|D^k\omega_{itt}\|^2 \le c\boldsymbol{\Lambda}_0. \tag{6.4.38}$$

$$\sum_{k=1}^{5}(1+t)^k\|D^k\boldsymbol{u}_i\|^2+\sum_{k=1}^{3}(1+t)^{2+k}\|D^k\boldsymbol{u}_{it}\|^2 \le c\boldsymbol{\Lambda}_0. \tag{6.4.39}$$

$$\sum_{k=1}^{5}(1+t)^k\|D^k\boldsymbol{E}\|^2+\int_0^t\sum_{k=1}^{5}(1+s)^{k-1}\|D^k\boldsymbol{E}\|^2\mathrm{d}s$$
$$\le c\boldsymbol{\Lambda}_0. \tag{6.4.40}$$

$$\int_0^t\left\{\sum_{k=1}^{5}(1+s)^{k-1}\|D^k\omega_i\|^2+\sum_{k=0}^{4}(1+s)^{k+1}\|D^k\omega_{it}\|^2\right\}\mathrm{d}s$$
$$\le c\boldsymbol{\Lambda}_0. \tag{6.4.41}$$

$$\int_0^t\left\{\sum_{k=1}^{5}(1+s)^k\|D^k\boldsymbol{u}_i\|^2+\sum_{k=1}^{3}(1+s)^{k+2}\|D^k\boldsymbol{u}_{it}\|^2\right\}\mathrm{d}s$$
$$\le c\boldsymbol{\Lambda}_0. \tag{6.4.42}$$

Proof. Firstly, we give the basic energy estimates. Multiplying (6.4.25) by $(\omega_a + 2\omega_{at})$, (6.4.26) by $(\omega_b + 2\omega_{bt})$, integrating by parts on R^3 (Here we usually omit R^3), and summing up the above results, noting the fact that

$$\begin{aligned}
&\int \left\{ (\frac{1}{2}(\omega_a + \sqrt{\rho_a^*})\boldsymbol{\nabla}\cdot\boldsymbol{E})(\omega_a + 2\omega_{at}) \right. \\
&\quad \left. - \left[\frac{1}{2}(\omega_b + \sqrt{\rho_b^*})\boldsymbol{\nabla}\cdot\boldsymbol{E}\right](\omega_b + 2\omega_{bt}) \right\} \mathrm{d}x \\
&= \frac{1}{4}\frac{\mathrm{d}}{\mathrm{d}t}\int |\boldsymbol{\nabla}\cdot\boldsymbol{E}|^2 \mathrm{d}x + \frac{1}{4}\int |\boldsymbol{\nabla}\cdot\boldsymbol{E}|^2 \mathrm{d}x - \frac{1}{4}\int \boldsymbol{\nabla}(\omega_a^2 - \omega_b^2)\cdot\boldsymbol{E}\mathrm{d}x,
\end{aligned}$$

we have

$$\begin{aligned}
&\frac{\mathrm{d}}{\mathrm{d}t}\int \left[\omega_{at}^2 + \omega_a\omega_{at} + \frac{\omega_a^2}{2} + \omega_{bt}^2 + \omega_b\omega_{bt} + \frac{\omega_b^2}{2} + P_a'(\rho_a^*)|\boldsymbol{\nabla}\omega_a|^2 \right. \\
&\quad \left. + P_b'(\rho_b^*)|\boldsymbol{\nabla}\omega_b|^2 + \frac{h^2}{4}(|\Delta\omega_a|^2 + |\Delta\omega_b|^2) + \frac{1}{4}|\boldsymbol{\nabla}\cdot\boldsymbol{E}|^2\right] \mathrm{d}x \\
&\quad + \int \left[(\omega_{at}^2 + \omega_{bt}^2) + P_a'(\rho_a^*)|\boldsymbol{\nabla}\omega_a|^2 + P_b'(\rho_b^*)|\boldsymbol{\nabla}\omega_b|^2 \right. \\
&\quad \left. + \frac{h^2}{4}(|\Delta\omega_a|^2 + |\Delta\omega_b|^2) + \frac{1}{4}|\boldsymbol{\nabla}\cdot\boldsymbol{E}|^2\right] \mathrm{d}x \\
&= \int \left\{\frac{1}{4}(\boldsymbol{\nabla}(\omega_a^2 - \omega_b^2)\cdot\boldsymbol{E}) + [f_{a1}(x,t)(\omega_a + 2\omega_{at}) \right. \\
&\quad \left. + f_{b1}(x,t)(\omega_b + 2\omega_{bt})]\right\} \mathrm{d}x \qquad (6.4.43)
\end{aligned}$$

From the assumption (6.4.35), using Sobolev embedding theorem, Hölder inequality, Young's inequality and integrating by parts, we can estimate the right hand side of (6.4.43) as follows

$$\begin{aligned}
\int \omega_i\boldsymbol{\nabla}\omega_i\cdot\boldsymbol{E}\mathrm{d}x &\le \|\omega_i\|_{L^3}\|\boldsymbol{\nabla}_{\omega_i}\|_{L^2}\|\boldsymbol{E}\|_{L^6} \\
&\le c(\|\omega_i\|_{L^2} + \|\boldsymbol{\nabla}_{\omega_i}\|_{L^2})\|\boldsymbol{\nabla}\omega_i\|_{L^2}\cdot\|\boldsymbol{E}\|_{L^6} \\
&\le c\delta_T(\|\boldsymbol{\nabla}\omega_i\|^2 + \|\boldsymbol{\nabla}\cdot\boldsymbol{E}\|^2), \qquad (6.4.44)
\end{aligned}$$

$$\int\left\{P_i'\left[\left(\omega_i+\sqrt{\rho_i^*}\right)^2\right]-P_i'\left(\rho_i^*\right)\right\}\Delta\omega_i\cdot(2\omega_{it})\mathrm{d}x$$
$$\leq-\frac{\mathrm{d}}{\mathrm{d}t}\int\{P_i'[(\omega_i+\sqrt{\rho_i^*})^2]-P_i'(\rho_i^*)\}|\nabla\omega_i|^2\mathrm{d}x$$
$$+c\delta_T\|(\nabla_{\omega_i,\omega_{it}})\|^2\mathrm{d}x, \tag{6.4.45}$$

$$\int \boldsymbol{u}_i\cdot\nabla\omega_{it}(2\omega_{it})\mathrm{d}x=$$
$$=-\int\nabla\cdot\boldsymbol{u}_i(\omega_i t)^2\mathrm{d}x\leq c\delta_T\|\omega_{it}\|^2, \tag{6.4.46}$$

$$\int \boldsymbol{u}_i\nabla(\boldsymbol{u}_i\cdot\nabla\omega_i)\cdot 2\omega_{it}\mathrm{d}x$$
$$\leq-\frac{\mathrm{d}}{\mathrm{d}t}\int(\boldsymbol{u}_i\cdot\nabla_{\omega_i})^2\mathrm{d}x+c\delta_T\|(\nabla\omega_i,\omega_{it})\|^2. \tag{6.4.47}$$

Here we use the inequality $\|D\boldsymbol{u}_i\|^2\leq c(\|\nabla\cdot\boldsymbol{u}_i\|^2+\|\nabla\times\boldsymbol{u}_i\|^2)$. The other terms on the right hand side of (6.4.43) can also be estimated easily by integration by parts, Hölder inequality, Young's inequality and Lemma 6.1.3 and 6.1.4. We can get that

$$\frac{\mathrm{d}}{\mathrm{d}t}\int\left[\omega_{at}^2+\omega_a\omega_{at}+\frac{\omega_a^2}{2}+\omega_{bt}^2+\omega_b\omega_{bt}+\frac{\omega_b^2}{2}+P_a'(\rho_a^*)|\nabla\omega_a|^2\right.$$
$$+P_b'(\rho_b^*)|\nabla\omega_b|^2+\frac{h^2}{4}(|\Delta\omega_a|^2+|\Delta\omega_b|^2)+\frac{1}{4}|\nabla\cdot\boldsymbol{E}|^2$$
$$+(P_a'((\omega_a+\sqrt{\omega_a^*})^2)-P_a'(\rho_a^*))|\nabla\omega_a|^2$$
$$+[P_b'((\omega_b+\sqrt{\omega_b^*})^2)-P_b'(\rho_b^*)]|\nabla\omega_b|^2$$
$$+(\boldsymbol{u}_a\cdot\nabla\omega_a)^2+(\boldsymbol{u}_b\cdot\nabla\omega_b)^2]\mathrm{d}x$$
$$+\int[(\omega_{at}^2+\omega_{bt}^2)+P_a'(\rho_a^*)|\nabla\omega_a|^2+P_b'(\rho_b^*)|\nabla\omega_b|^2$$
$$\left.+\frac{h^2}{4}(|\Delta\omega_a|^2+|\Delta\omega_b|^2)+\frac{1}{4}|\nabla\cdot\boldsymbol{E}|^2\right]\mathrm{d}x$$
$$\leq c\delta_T\|(\nabla\omega_a,\nabla\omega_b,\omega_{at},\omega_{bt},\nabla\cdot\boldsymbol{E},\phi_a,\phi_b)\|^2. \tag{6.4.48}$$

Multiplying (6.4.27) by $2\phi_a$, and (6.4.28) by $2\phi_b$, and integrating by parts on $\mathbf{R}^3$, we obtain

$$\begin{aligned}
&\frac{\mathrm{d}}{\mathrm{d}t}\int(|\phi_a|^2+|\phi_b|^2)\mathrm{d}x+2\int(|\phi_a|^2+|\phi_b|^2)\mathrm{d}x \\
&\quad =\int(f_{a2}\cdot 2\phi_a+f_{b2}\cdot 2\phi_b)\mathrm{d}x. \qquad (6.4.49)
\end{aligned}$$

From (6.4.35), we have

$$\begin{aligned}
&\frac{\mathrm{d}}{\mathrm{d}t}\int(|\phi_a|^2+|\phi_b|^2)\mathrm{d}x+2\int(|\phi_a|^2+|\phi_b|^2)\mathrm{d}x \\
&\quad \le c\delta_T\|(\phi_a,\phi_b,\nabla_{\omega_a},\nabla\omega_b,\omega_{at},\omega_{bt})\|^2. \qquad (6.4.50)
\end{aligned}$$

Integrating (6.4.48) and (6.4.50) on $[0,t]$, and using

$$P_i'(\rho_i^*)>0,\quad \frac{1}{6}(x^2+y^2)\le x^2+xy+\frac{y^2}{2}\le 2(x^2+y^2),$$

we can get

$$\begin{aligned}
&\|(\omega_a,\omega_b)\|_1^2+\|(hD^2\omega_a,hD^2_{\omega_b})\|^2+\|(\omega_{at},\omega_{bt},\phi_a,\phi_b,D\boldsymbol{E})\|^2 \\
&\quad +\int_0^t[\|(\nabla\omega_a,\nabla\omega_b,hD^2\omega_a,hD^2\omega_b,\omega_{at},\omega_{bt},\phi_a,\phi_b)\|^2 \\
&\quad +\|D\boldsymbol{E}\|^2]\mathrm{d}s\le c\boldsymbol{\Lambda}_0. \qquad (6.4.51)
\end{aligned}$$

On the other hand, through calculating $\int[(6.3.25)\times 2(1+t)\omega_{at}+(6.4.26)\times 2(t+1)\omega_{bt}]\mathrm{d}x$ and $(6.4.49)\times(1+t)$, we have

$$\begin{aligned}
&\frac{\mathrm{d}}{\mathrm{d}t}\Big\{(1+t)\int\big[\omega_{at}^2+\omega_{bt}^2+P_a'(\rho_a^*)|\nabla\omega_a|^2 \\
&\quad +P_b'(\rho_b^*)|\nabla\omega_b|^2+\frac{h^2}{4}(|\Delta\omega_a|^2+|\Delta\omega_b|^2) \\
&\quad +\frac{1}{4}|\nabla\cdot\boldsymbol{E}|^2+|\phi_a|^2+|\phi_b|^2+[P_a'((\omega_a+\sqrt{\rho_a^*})^2)
\end{aligned}$$

$$- P_a'(\rho_a^*)]|\boldsymbol{\nabla}\boldsymbol{\omega_a}|^2 + \left[P_b'((\omega_b + \sqrt{\rho_b^*})^2)\right.$$
$$\left. - P_b'(\rho_b^*)\right] |\boldsymbol{\nabla}\omega_b|^2 + (\boldsymbol{u}_a \cdot \omega_a)^2 + (\boldsymbol{u}_b \cdot \omega_b)^2\right] \mathrm{d}x\}$$
$$\times 2(1+t)\|(\omega_{at}, \omega_{bt}, \boldsymbol{\phi}_a, \boldsymbol{\phi}_b)\|^2$$
$$\leq +(1+t)\|(\omega_{at}, \omega_{bt}, \boldsymbol{\phi}_a, \boldsymbol{\phi}_b)\|^2$$
$$+ c\delta_T \|(\boldsymbol{\nabla}\omega_a, \boldsymbol{\nabla}\omega_b, hD^2\omega_a, hD^2\omega_b, \boldsymbol{\phi}_a, \boldsymbol{\phi}_b, D\boldsymbol{E})\|^2, \tag{6.4.52}$$

where we have used the *a priori* assumptions (6.4.35) on time-decay rate, Holder inequality, and Young's inequality to estimate the right hand side terms as follows:

$$\frac{1}{4}\int (1+t)\boldsymbol{\nabla}(\omega_a^2 - \omega_b^2) \cdot \boldsymbol{E}\mathrm{d}x + \int (1+t)[f_{a1}(\omega_a + 2\omega_{at})$$
$$+ f_{b1}(\omega_b + 2\omega_{bt})]\mathrm{d}x + \int (1+t)(f_{a2} \cdot 2\boldsymbol{\phi}_a + f_{b2} \cdot 2\boldsymbol{\phi}_b)\mathrm{d}x$$
$$\leq (1+t)\|(\omega_{at}, \omega_{bt}, \boldsymbol{\phi}_a, \boldsymbol{\phi}_b)\|^2 + c\delta_T \|(\boldsymbol{\nabla}\omega_a, \boldsymbol{\nabla}\omega_b, hD^2\omega_a, hD^2\omega_b,$$
$$\times \boldsymbol{\phi}_a, \boldsymbol{\phi}_b, D\boldsymbol{E})\|^2.$$

Integrating (6.4.52) on $[0, t]$, it holds that

$$(1+t)\|(\boldsymbol{\nabla}\omega_a, \boldsymbol{\nabla}\omega_b, hD^2\omega_a, hD^2\omega_b, \omega_{at}, \omega_{bt}, \boldsymbol{\phi}_a, \boldsymbol{\phi}_b, D\boldsymbol{E})\|^2$$
$$+ \int_0^t (1+s)\|(\omega_{at}, \omega_{bt}, \boldsymbol{\phi}_a, \boldsymbol{\phi}_b)\|^2 \mathrm{d}s \leq c\boldsymbol{\Lambda}_0. \tag{6.4.53}$$

Then combing (6.4.51), we can get the basic estimates in Lemma 6.4.1,

$$\|\omega_i\|^2 + (1+t)\|D\omega_i\|^2 + (1+t)\|D\boldsymbol{E}\|^2 + (1+t)\|D\boldsymbol{u}_i\|^2 \leq c\boldsymbol{\Lambda}_0, \tag{6.4.54}$$

$$\int_0^t \|(\boldsymbol{\nabla}\omega_i, D\boldsymbol{E})\|^2 \mathrm{d}s + \int_0^t (1+s)(\|D_{\boldsymbol{u}_i}\|^2 + \|\omega_{it}\|^2)\mathrm{d}s \leq c\boldsymbol{\Lambda}_0. \tag{6.4.55}$$

Next we give some higher-order estimates.

Denote $\tilde{\omega}_i := D^\alpha \omega_i, \tilde{\boldsymbol{\phi}}_i := D^\alpha \boldsymbol{\phi}_i, \tilde{\boldsymbol{E}} := D^\alpha \boldsymbol{E} (i = a, b, 1 < |\alpha| \leq 4)$. Differentiating (6.4.25)–(6.4.29) respect to x, we can obtain the equations for $(\tilde{\omega}_i, \tilde{\boldsymbol{\phi}}_i, \tilde{\boldsymbol{E}})$:

$$\begin{aligned}
&\tilde{\omega}_{itt} + \tilde{\omega}_{it} + \frac{h^2}{4}\Delta^2 \tilde{\omega}_i - P_i'(\rho_i^*)\Delta\tilde{\omega}_i + \frac{qi}{2}(\omega_i + \sqrt{\rho_i^*})\boldsymbol{\nabla}\cdot\tilde{\boldsymbol{E}} \qquad (6.4.56)\\
&\quad = D^\alpha f_{i1}(x,t) - D^\alpha(\frac{qi}{2}(\omega_i + \sqrt{\rho_i^*})\boldsymbol{\nabla}\cdot\boldsymbol{E})\\
&\quad\quad + \frac{qi}{2}(\omega_i + \sqrt{\rho_i^*})\boldsymbol{\nabla}\cdot\tilde{\boldsymbol{E}},\\
&\tilde{\boldsymbol{\phi}}_{it} + \tilde{\boldsymbol{\phi}}_i = D^\alpha f_{i2}, \qquad (6.4.57)\\
&\boldsymbol{\nabla}\cdot\tilde{\boldsymbol{E}} = D^\alpha(\omega_a^2 - \omega_b^2 + 2\sqrt{\rho_a^*\omega_a} - 2\sqrt{\rho_b^*\omega_b}). \qquad (6.4.58)
\end{aligned}$$

Similar to the basic estimates, for $|\alpha| = k, k = 1, 2, 3, 4$, with the following integrals

$$\begin{aligned}
&\int_0^t \int \sum_{l=0}^{|\alpha|} (6.4.56)_{i=a} \times (1+s)^l (D^\alpha + 2D^\alpha \omega_{at})\\
&\quad + [(6.4.56)_{i=b} \times (1+s)^l (D^\alpha \omega_b + 2D^\alpha \omega_{bt})\Big] \, dxds\\
&\int_0^t \int \sum_{l=0}^{|\alpha|} \Big[(6.4.57)_{i=a} \cdot 2(1+s)^l D^\alpha \boldsymbol{\phi}_a\\
&\quad + (6.4.57)_{i=b} \cdot 2(1+s)^l D^\alpha \boldsymbol{\phi}_b\Big] \, dxds\\
&\int_0^t \int \Big[(6.4.56)_{i=a} \cdot 2(1+s)^l D^{|\alpha|+1} \omega_{at}\\
&\quad + (6.4.56)_{i=b} \cdot 2(1+s)^l D^{|\alpha|+1} \omega_{bt}\Big] \, dxds\\
&\int_0^t \int \Big[(6.4.57)_{i=a} \cdot 2(1+s)^l D^{|\alpha|+1} \boldsymbol{\phi}_a\\
&\quad + (6.4.57)_{i=b} \cdot 2(1+s)^l D^{|\alpha|+1} \boldsymbol{\phi}_b\Big] \, dxds,
\end{aligned}$$

we can get after a straightforward computation

$$
\begin{aligned}
&(1+t)^{k+1}\left\|\left(D^{k+1}\omega_a, D^{k+1}\omega_b, hD^{k+2}\omega_a, hD^{k+2}\omega_b, D^k\omega_{at}, D^k\omega_{bt}\right)\right\|^2\\
&+(1+t)^{k+1}\|(D^k\boldsymbol{\phi}_a, D^k\boldsymbol{\phi}_b, D^{k+1}\boldsymbol{E})\|^2\\
&+\int_0^t (1+s)^k\|(D^{k+1}\omega_a, D^{k+1}\omega_b, hD^{k+2}\omega_a, hD^{k+2}\omega_b, D^{k+1}\boldsymbol{E})\|^2 \mathrm{d}s\\
&+\int_0^t (1+s)^{k+1}\|(D^k\omega_{at}, D^k\omega_{bt}, D^k\boldsymbol{\phi}_a, D^k\boldsymbol{\phi}_b)\|^2 \mathrm{d}s \le c\boldsymbol{\Lambda}0. \quad (6.4.59)
\end{aligned}
$$

By (6.4.59) we can get the expected decay rates in Lemma 6.4.1 as

$$
\begin{aligned}
&(1+t)^k\|D^k\omega_i\|^2 + (1+t)^5\|hD^6\omega_i\|^2 \le c\boldsymbol{\Lambda}_0, 0 \le k \le 5, \quad (6.4.60)\\
&(1+t)^k\|D^k\boldsymbol{u}_i\|^2 + (1+t)^k\|D^k\boldsymbol{E}\|^2 \le c\Lambda_0, 1 \le k \le 5, \quad (6.4.61)\\
&\int_0^t [(1+s)^{k-1}\|(D^k\omega_i, D^k\boldsymbol{E})\|^2 + (1+s)^k\|D^k\boldsymbol{u}_i\|^2]\mathrm{d}s \le c\boldsymbol{\Lambda}_0,\\
&\quad 1 \le k \le 5, \quad (6.4.62)\\
&(1+t)^{k+1}\|D^k\omega_{it}\|^2 + \int_0^t (1+s)^{k+1}\|D^k\omega_{it}\|^2 \mathrm{d}s \le c\boldsymbol{\Lambda}_0,\\
&\quad 1 \le k \le 4. \quad (6.4.63)
\end{aligned}
$$

The higher-order estimate $(1+t)^6\|D^6E\|^2 \le c\boldsymbol{\Lambda}_0$ can be obtained by the Poisson equation (6.4.29) and Lemma 6.1.2.

To complete the proof we still need to estimate the decay rate of the higher-order time derivatives about $(\omega_a, \omega_b, \boldsymbol{u}_a, \boldsymbol{u}_b)$. Set $\overline{\omega}_i := D^\alpha\omega_{it}, \overline{\boldsymbol{\phi}}_i := D^\alpha\boldsymbol{\phi}_{it}, \overline{\boldsymbol{E}} := D^\alpha\boldsymbol{E}_t (0 \le |\alpha| \le 2)$, then we can get the equations for $\overline{\omega}_i, \overline{\boldsymbol{\phi}}_i, \overline{\boldsymbol{E}} (i = a, b)$

$$
\begin{aligned}
&\overline{\omega}_{itt} + \overline{\omega}_{it} + \frac{h^2}{4}\Delta^2\overline{\omega}_i - P_i'(\rho_i^*)\Delta\overline{\omega}_i\\
&+\frac{qi}{2}(\omega_i + \sqrt{\rho_i^*})\boldsymbol{\nabla}\cdot\overline{\boldsymbol{E}}
\end{aligned}
$$

$$= D^\alpha f_{i1}(x,t)_t - D^\alpha \left(\frac{qi}{2}\left(\omega_i + \sqrt{\rho_i^*}\right)\nabla\cdot \boldsymbol{E}\right)_t$$

$$+\frac{qi}{2}(\omega_i + \sqrt{\rho_i^*})\nabla\cdot \overline{\boldsymbol{E}}, \tag{6.4.64}$$

$$\overline{\phi}_{it} + \overline{\phi}_i = D^\alpha (f_{i2})t, \tag{6.4.65}$$

$$\nabla\cdot\overline{\boldsymbol{E}} = D^\alpha(\omega_a^2 - \omega_b^2 + 2\sqrt{\rho_a^*}\omega_a - 2\sqrt{\rho_b^*}\omega_b)t. \tag{6.4.66}$$

Based on the time-decay rate derived in (6.4.60)–(6.4.63), we can get from (6.4.64)– (6.4.66) the faster time-decay rate for $\overline{\omega}_i, \overline{\phi}_i, \overline{\boldsymbol{E}}$. In fact, for $|\alpha| = 0$, 1, 2, estimating the integrals

$$\int_0^t \int \sum_{l=0}^{|\alpha|+2} \Big[(6.4.64)_{i=a} \cdot (1+s)^l (D^\alpha\omega_{at} + 2D^\alpha\omega_{att})$$

$$+ (6.4.64)_{i=b} \cdot (1+s)^l \; (D^\alpha\omega_{bt} + 2D^\alpha\omega_{btt})]\,\mathrm{d}x\mathrm{d}s,$$

$$\int_0^t \int \sum_{l=0}^{|\alpha|+2} [(6.4.65)_{i=a} \cdot 2(1+s)^l D^\alpha\phi_{at} + (6.4.65)_{i=b}$$

$$\cdot 2(1+s)^l D^\alpha\phi_{bt}]\mathrm{d}x\mathrm{d}s, \int_0^t \int \Big[(6.4.64)_{i=a} \cdot 2(1+s)^{|\alpha|+3} D^\alpha\omega_{att}$$

$$+ (6.4.64)_{i=b} \cdot 2(1+s)^{|\alpha|+3} D^\alpha\omega_{btt} + (6.4.65)_{i=a}$$

$$\cdot 2(1+s)^{|\alpha|+3} D^\alpha\phi_{at} + (6.4.65)_{i=b} \cdot 2(1+s)^{|\alpha|+3} D^\alpha\phi_{bt}\Big]\,\mathrm{d}x\mathrm{d}s,$$

then combining (6.4.60)–(6.4.63), we have

$$(1+t)^{k+2}\|D^k\omega_{it}\|^2 + \int_0^t (1+s)^{1+k}\|D^k\omega_{it}\|^2\mathrm{d}s \le c\boldsymbol{\Lambda}_0,$$

$$0 \le k \le 3, \tag{6.4.67}$$

$$(1+t)^{k+3}\|D^k\omega_{itt}\|^2 + \int_0^t (1+s)^{3+k}\|D^k\omega_{itt}\|^2\mathrm{d}s \le c\boldsymbol{\Lambda}_0,$$

$$0 \le k \le 2, \tag{6.4.68}$$

$$(1+t)^{k+3}\|D^k\boldsymbol{\phi}_{it}\|^2 + \int_0^t (1+s)^{3+k}\|D^k\boldsymbol{\phi}_{it}\|^2\mathrm{d}s \le c\boldsymbol{\Lambda}_0,$$
$$0 \le k \le 2, \tag{6.4.69}$$
$$(1+t)^{k+2}\|D^k\boldsymbol{E}_t\|^2 + \int_0^t (1+s)^{1+k}\|D^k\boldsymbol{E}_t\|^2\mathrm{d}s \le c\boldsymbol{\Lambda}_0,$$
$$1 \le k \le 3. \tag{6.4.70}$$

Here we use the inequality $\|D\boldsymbol{u}_t\|^2 \le c(\|\boldsymbol{\nabla}\cdot\boldsymbol{u}_t\|^2 + \|\boldsymbol{\nabla}\times\boldsymbol{u}_t\|^2)$. Then with the help of (6.4.67)–(6.4.70) and the relation of $\boldsymbol{\nabla}\cdot\boldsymbol{u}_i$ and $\boldsymbol{\nabla}\omega_i, \omega_{it}$ through equation (6.4.24), we get

$$(1+t)^{2+k}\|D^k\boldsymbol{u}_{it}\|^2 + \textstyle\int_0^t (1+s)^{2+k}\|D^k\boldsymbol{u}_{it}\|^2\mathrm{d}s \le c\boldsymbol{\Lambda}_0, \quad 1 \le k \le 3. \tag{6.4.71}$$

Then all the estimates in Lemma 6.4.1 follow from (6.4.54)–(6.4.55), (6.4.60)–(6.4.63) and (6.4.67)–(6.4.71). The proof is completed. □

The proof of Theorem 6.4.1. Based on Lemma 6.4.1, for sufficient small $\boldsymbol{\Lambda}_0$, we can extend the local solution to the global solution, and the estimates (6.4.38)–(6.4.42) for $t \in [0, T]$ still hold. Specially, it holds that

$$(1+t)^k\|D^k\omega_i\|^2 + (1+t)^5\|hD^6\omega_i\|^2 c\boldsymbol{\Lambda}_0,$$
$$0 \le k \le 5, \tag{6.4.72}$$
$$(1+t)^{k+2}\|D^{k+2}\omega_{it}\|^2 + (1+t)^5\|D^4\omega_{it}\|^2 \le c\boldsymbol{\Lambda}_0,$$
$$0 \le k \le 3, \tag{6.4.73}$$
$$(1+t)^k\|D^k\boldsymbol{u}_i\|^2 + (1+t)^k\|D^k E\|^2 \le c\boldsymbol{\Lambda}_0,$$
$$1 \le k \le 5, \tag{6.4.74}$$
$$(1+t)^{k+2}\|D^k\boldsymbol{u}_{it}\|^2 \le c\boldsymbol{\Lambda}_0, \quad 1 \le k \le 3. \tag{6.4.75}$$

where the coefficient c is a constant independent on h and $t > 0$. Since $\rho_i = (\omega_i + \sqrt{\rho_i^*})^2$, we can get the conclusion of

Theorem 6.4.1 as

$$(1+t)^k\|D^k(\rho_i-\rho_i^*)\|^2+(1+t)^5\|hD^6(\rho_i-\rho_i^*)\|^2\le c\mathbf{\Lambda}_0,\quad 0\le k\le 5. \tag{6.4.76}$$

$$(1+t)^k\|D^k u_i\|^2+(1+t)^k\|D^k\boldsymbol{E}\ \|^2\le c\mathbf{\Lambda}_0,\quad 1\le k\le 5. \tag{6.4.77}$$

Thus, the proof of Theorem 6.4.1 is completed.

Next we will show the algebraic time-decay of the linearized system (6.4.10).

By (6.4.15), the equation (6.4.10) for $U_a=(W_a,W_b,\boldsymbol{J}_a,\boldsymbol{J}_b)$ can be rewritten as

$$\begin{cases} W_{at}+\boldsymbol{\nabla}\cdot\boldsymbol{J}_a=0,\\ J_{at}+\boldsymbol{\nabla}W_a-\boldsymbol{\nabla}\Delta W_a+\boldsymbol{J}_a-2\boldsymbol{\nabla}\Delta^{-1}(W_a-W_b)=0,\\ W_{bt}+\boldsymbol{\nabla}\cdot\boldsymbol{J}_b=0,\\ J_{bt}+\boldsymbol{\nabla}W_b-\boldsymbol{\nabla}\Delta W_b+\boldsymbol{J}_b+\boldsymbol{\nabla}\Delta^{-1}(W_a-W_b)=0, \end{cases} \tag{6.4.78}$$

with the initial data

$$U(x,0)=U_0(x):=(W_{a0},\boldsymbol{J}_{a0},W_{b0},\boldsymbol{J}_{a0}). \tag{6.4.79}$$

Let us represent the solution of the linear problem (6.4.78)–(6.4.79) formally as

$$U=\mathrm{e}^{At}U_0, \tag{6.4.80}$$

where U is the inverse Fourier transform of $\hat{U}=(\hat{W}_a,\hat{\boldsymbol{J}}_a,\hat{W}_b,\hat{\boldsymbol{J}}_b)$ whose equation can be derived by taking Fourier transform with respect to x on (6.4.78) as

$$\begin{cases} \hat{U}_t=\hat{A}\hat{U},\\ \hat{U}(\xi,0)=(\hat{W}_{a0},\hat{\boldsymbol{J}}_{a0},\hat{W}_{b0},\hat{\boldsymbol{J}}_{b0}), \end{cases} \tag{6.4.81}$$

where $\hat{\boldsymbol{J}}_a = (\hat{\boldsymbol{J}}_a(1), \hat{\boldsymbol{J}}_a^{(2)}, \hat{\boldsymbol{J}}_a^{(3)}), \hat{\boldsymbol{J}}_b = (\hat{\boldsymbol{J}}_b(1), \hat{\boldsymbol{J}}_b^{(2)}, \hat{\boldsymbol{J}}_b^{(3)})$, and the matrix $\hat{\boldsymbol{A}}$ is given by

$$\hat{\boldsymbol{A}} = \begin{pmatrix} 0 & -\mathrm{i}\xi^{\mathrm{T}} & 0 & 0 \\ -\mathrm{i}\xi b_1 & -I_3 & \mathrm{i}\xi d_1 & 0 \\ 0 & 0 & 0 & -\mathrm{i}\xi^{\mathrm{T}} \\ \mathrm{i}\xi d_2 & 0 & -\mathrm{i}\xi b_2 & -I_3 \end{pmatrix},$$

with i being the notation of the imaginary unit and

$$b_1 = 1 + |\xi|^2 + \frac{2}{|\xi|^2}, d_1 = \frac{2}{|\xi|}, b_2 = 1 + |\xi|^2 + \frac{1}{|\xi|^2},$$

$$d_2 = \frac{1}{|\xi|^2}, \boldsymbol{I}_3 = \mathrm{diag}(1, 1, 1).$$

We can solve the IVP problem (6.4.81) straightforward by the standard existence theory for linear ODEs. In fact, one can easily verify that the solution can be written as

$$\hat{U} = e^{\hat{A}t}\hat{U}_0, \tag{6.4.82}$$

where $\hat{U} = (\hat{W}_a, \hat{\boldsymbol{J}}_a, \hat{W}_b, \hat{\boldsymbol{J}}_b)$ with

$$\begin{aligned} \hat{W}_a(\xi, t) = &\frac{1}{6}\hat{W}_{a0}[F_1 + 2F_2 + e_1^- + e_1^+ + 2(e_2^- + e_2^+)] \\ &+\frac{1}{3}\hat{W}_{b0}[F_1 - F_2 + e_1^- + e_1^+ - (e_2^- + e_2^+)] \\ &-\frac{1}{3}(\hat{\boldsymbol{J}}_{a0} \cdot \xi)(F_1 + 2F_2) \\ &-\frac{1}{3}(\hat{\boldsymbol{J}}_{b0} \cdot \xi)(2F_1 - 2F_2), \end{aligned} \tag{6.4.83}$$

$$\begin{aligned} \hat{W}_b(\xi, t) = &\frac{1}{6}\hat{W}_{b0}[2F_1 + F_2 + 2(e_1^- + e_1^+) + (e_2^- + e_2^+)] \\ &+\frac{1}{3}\hat{W}_{a0}[F_1 - F_2 + e_1^- + e_1^+ - (e_2^- + e_2^+)] \\ &-\frac{\mathrm{i}}{3}(\hat{\boldsymbol{J}}_{b0} \cdot \xi)(2F_1 + F_2) - \frac{\mathrm{i}}{3}(\hat{\boldsymbol{J}}_{a0} \cdot \xi)(F_1 - F_2), \end{aligned} \tag{6.4.84}$$

and for $k = 1, 2, 3$,

$$\hat{\boldsymbol{J}}_a^{(k)}(\xi,t) = \frac{\hat{\boldsymbol{J}}_{a0}^{(k)}}{|\xi|^2}(|\xi|^2 - \xi_k^2)e^{-t} - \frac{\xi k}{|\xi|^2}\left(\sum_{\substack{l=1\\ l\neq k}}^{3} \xi_l \hat{\boldsymbol{J}}_{a0}^{(l)}\right)e^{-t}$$

$$\frac{\xi_k}{6|\xi|^2}(\xi\cdot\hat{\boldsymbol{J}}_{a0})[2F_2 + F_1 - 2(e_2^- + e_2^+) - (e_1^+ + e_1^-)]$$

$$+\frac{\xi k}{3|\xi|^2}(\xi\cdot\hat{\boldsymbol{J}}_{b0})(F_2 - F_1 + e_1^- + e_1^+ - e_2^- - e_2^+)$$

$$-2(\hat{W}_{a0} - \hat{W}_{b0})\frac{\mathrm{i}\xi_k}{|\xi|^2}F_2 - \frac{\mathrm{i}}{3}W_{a0}\xi_k(1+|\xi|^2)(2F_2 + F_1)$$

$$-\frac{2\mathrm{i}}{3}\hat{W}_{b0}\xi_k(1+|\xi|^2)(F_1 - F_2), \tag{6.4.85}$$

$$\hat{\boldsymbol{J}}_b^{(k)}(\xi,t) = \frac{\hat{\boldsymbol{J}}_{b0}^{(k)}}{|\xi|^2}(|\xi|^2 - \xi_k^2)e^{-t} - \frac{\xi_k}{|\xi|^2}\left(\sum_{\substack{l=1\\ l\neq k}}^{3} \xi_l \hat{\boldsymbol{J}}_{b0}^{(l)}\right)e^{-t}$$

$$-\frac{\xi_k}{6|\xi|^2}(\xi\cdot\hat{\boldsymbol{J}}_{b0})[F_2 + 2F_1 - (e_2^- + e_2^+)$$

$$-2(e_1^+ + e_1^-)] + \frac{\xi_k}{6|\xi|^2}(\xi\cdot\hat{\boldsymbol{J}}_{a0})$$

$$(F_2 - F_1 + e_1^- + e_1^+ - e_2^- - e_2^+)$$

$$+(\hat{W}_{a0} - \hat{W}_{b0})\frac{\mathrm{i}\xi_k}{|\xi|^2}F_2 - \frac{\mathrm{i}}{3}W_{b0}\xi_k(1+|\xi|^2)(F_2 + 2F_1)$$

$$-\frac{\mathrm{i}}{3}\hat{W}_{a0}\xi_k(1+|\xi|^2)(F_1 - F_2). \tag{6.4.86}$$

Here

$$e_1^- = e^{-\frac{t}{2}(1-I_1)}, e_1^+ = e^{-\frac{t}{2}(1+I_1)}, e_2^- = e^{-\frac{t}{2}(1-I_2)}, e_2^+ = e^{-\frac{t}{2}(1+I_2)}, \tag{6.4.87}$$

with

$$I_1 = \sqrt{1 - 4|\xi|^2(1+|\xi|^2)}, \quad I_2 = \sqrt{1 - 4(3 + |\xi|^2(1+|\xi|^2))} \tag{6.4.88}$$

and

$$F_1 = \frac{e_1^- - e_1^+}{I_1}, \quad F_2 = \frac{e_2^- - e_2^+}{I_2}. \tag{6.4.89}$$

By $\boldsymbol{E}_0 = \boldsymbol{\nabla}\Delta^{-1}(W_{a0} - W_{b0}) \in L^2(\mathbf{R}^3)$, we have $(\hat{W}_{a0} - \hat{W}_{b0})\frac{\mathrm{i}\xi_k}{|\xi|^2} \in L^2(\mathbf{R}^3)$.

This shows the existence of the inverse Fourier transform of $\hat{U}$, and thus the global solvability of U for (6.4.78)–(6.4.79).

The proof of Theorem 6.4.2. Firstly, we focus on the estimates of the lower bound in (6.4.19)–(6.4.20). The idea is to analyze the Fourier transform of U with the help of the Plancherel theorem. In view of (6.4.87)–(6.4.89) we are able to show some properties of the terms contained in $W_a, \boldsymbol{J}_a, W_b, \boldsymbol{J}_b$ given by (6.4.83)–(6.4.86). In fact, we have the following estimates:

$$|e_1^+| + |e_2^-| + |e_2^+| + |F_2| + |\xi|^2|F_2| < ce^{ct}, \quad \xi \in R^3, \tag{6.4.90}$$

$$|e_1^-| \le e^{-\frac{t}{2}}, |F_1| \le \frac{t}{2}e^{-\frac{1}{2}t}, |\xi|^2|F_1| < c\frac{t}{2}e^{-\frac{1}{t}t},$$

$$|\xi|^2 \ge \frac{\sqrt{2}-1}{2}, \tag{6.4.91}$$

$$e_1^- \ge e^{-c|\xi|^2 t}, \quad |\xi|^2 \le \frac{\sqrt{2}-1}{2}. \tag{6.4.92}$$

Hereafter $c > 0$ is a generic positive constant.

Next we show the derivation of (6.4.90)–(6.4.92). The estimate (6.4.90) is gained by a direct computation. The estimates (6.4.91)–(6.4.92) can be obtained as follows. It holds for $|\xi|^2 \ge \frac{\sqrt{2}-1}{2}$, that

$$\begin{aligned} 1 - 4|\xi|^2(1 + |\xi|^2) \le 0, I_1 &= \sqrt{1 - 4|\xi|^2(1 + |\xi|^2)} \\ &= \mathrm{i}\sqrt{4|\xi|^2(1 + |\xi|^2) - 1} = \mathrm{i}|I_1|, \end{aligned}$$

and

$$|e_1^-| = |e^{-\frac{t}{2}(1-I_1)}| \le e^{-\frac{t}{2}}.$$

Since

$$|F_1| = \left| \frac{e^{-\frac{t}{2}(1-I_1)} - e^{-\frac{t}{2}(1+I_1)}}{I_1} \right| = \left| \frac{t}{2} e^{-\frac{t}{2}} \left(\frac{e^{i\frac{t|I_1|}{2}} - e^{-i\frac{t|I_1|}{2}}}{i\frac{t|I_1|}{2}} \right) \right|,$$

and $|(e^{is})'| \leq 1$,

we have

$$|F_1| \leq \frac{t}{2} e^{-\frac{1}{2}t}.$$

For $|\xi|^2|F_1|$, it holds for $\frac{\sqrt{2}-1}{2} \leq |\xi|^2 < \frac{\sqrt{3}-1}{2}$ that

$$|\xi|^2|F_1| \leq c\frac{t}{2} e^{-\frac{1}{2}t}.$$

When $\frac{\sqrt{3}-1}{2} \leq |\xi|^2$, we can directly compute

$$\begin{aligned} |\xi|^2|F_1| &= |\xi|^2 \left| e^{-\frac{t}{2}} \left(\frac{e^{i\frac{t|I_1|}{2}} - e^{-i\frac{t|I_1|}{2}}}{i|I_1|} \right) \right| \\ &= e^{-\frac{t}{2}} \left| \frac{|\xi|^2}{i|I_1|} \right| \left| \left(e^{i\frac{t|I_1|}{2}} - e^{-i\frac{t|I_1|}{2}} \right) \right| \leq ce^{-\frac{t}{2}}. \end{aligned}$$

By the fact that, for $0 \leq s \leq \frac{\sqrt{2}-1}{2}, 1 - \sqrt{1-4s(1+s)} \leq 2(\sqrt{2}+1)s$, we can obtain (6.4.92) easily since

$$e_1^- = e^{-\frac{t}{2}\left(1-\sqrt{1-4|\xi|^2(1+|\xi|^2)}\right)} \geq e^{-c|\xi|^2 t}.$$

With the help of (6.4.90)–(6.4.92), we can get the time-decay rates of $\hat{W}_a, \hat{W}_b, \hat{\boldsymbol{J}}_a, \hat{\boldsymbol{J}}_b$. Take $\hat{W}_a, \hat{\boldsymbol{J}}_a$ for simplicity. Set

$$\hat{W}_a = T_1 + R_1, \quad \hat{\boldsymbol{J}}_a^{(k)} = T_2^{(k)} + R_2^{(k)}, k = 1, 2, 3,$$

with

$$T_1 = \frac{1}{6}(\hat{W}_{a0} + 2\hat{W}_{b0})(F_1 + e_1^-), \tag{6.4.93}$$

$$R_1 = W_a - T_1, (\textit{remainder term}), \tag{6.4.94}$$

$$T_2^{(k)} = -\frac{\mathrm{i}}{3}(\hat{W}_{a0} + 2\hat{W}_{b0})\xi_k(1 + |\xi|^2)F_1, \tag{6.4.95}$$

$$R_2^{(k)} = \hat{\boldsymbol{J}}_a^{(k)} - T_2^{(k)}, (\textit{remainder term}). \tag{6.4.96}$$

By (6.4.90)–(6.4.92) we know

$$\begin{aligned}
\|\hat{W}_a(\cdot,t)\|^2 &\geq \frac{1}{2}\int_{\mathbf{R}^3} |T_1|^2 \mathrm{d}\xi - \int_{\mathbf{R}^3} |R_1|^2 \mathrm{d}\xi \\
&\geq \frac{1}{2}\int_{|\xi|^2<\frac{\sqrt{2}-1}{2}} \left|\frac{1}{6}(\hat{W}_{a0} + 2\hat{W}_{b0})(F_1 + e_1^-)\right|^2 \mathrm{d}\xi \\
&\quad -c(1+t)\mathrm{e}^{-ct} \\
&\geq \frac{1}{2}\int_{|\xi|^2<\frac{\sqrt{2}-1}{2}} \left|\frac{1}{6}(\hat{W}_{a0} + 2\hat{W}_{b0})e_1^-\right|^2 \mathrm{d}\xi - c(1+t)\mathrm{e}^{-ct} \\
&\geq \int_{|\xi|^2<\min\{\frac{\sqrt{2}-1}{2}, r^2\}} c\mathrm{e}^{-2c|\xi|^2 t} \mathrm{d}\xi - c(1+t)\mathrm{e}^{-ct} \\
&\geq c(1+t)^{-\frac{3}{2} - c(t+1)\mathrm{e}^{-ct}},
\end{aligned} \tag{6.4.97}$$

where we have used the assumption in Theorem 6.4.2 that $|\hat{W}_{a0} + 2\hat{W}_{b0}| > m_0 > 0$ in $B(0,r)$, and the fact $F_1 > 0$ for $|\xi|^2 < \frac{\sqrt{2}-1}{2}$. We also used $(|\xi|^n\hat{W}_{a0}, |\xi|^n\hat{W}_{b0}, |\xi|^l\hat{\boldsymbol{J}}_{a0}, |\xi|^l\hat{\boldsymbol{J}}_{b0}) \in L^2(\mathbf{R}^3)$ for $0 \leq n \leq 6, 0 \leq l \leq 5$, and $\int_{\mathbf{R}^3}|R_1|^2\mathrm{d}\xi < c\mathrm{e}^{-ct}$. The above $c > 0$ denotes the generic positive constant depending on the norm of initial data and m_0, not necessarily being the same.

The combination of Plancherel theorem and inequality (6.4.97) implies for $t \gg 1$ that

$$\|W_a(\cdot,t)\| = \|\hat{W}_a(\cdot,t)\| \geq c_1(1+t)^{-\frac{3}{4}}. \tag{6.4.98}$$

Similarly, from (6.4.90)–(6.4.92), we have

$$\begin{aligned}\|\mathrm{i}\xi_k\hat{W}_a(\cdot,t)\|^2 &\geq \frac{1}{2}\int_{\mathbf{R}^3}|\mathrm{i}\xi_k T_1|^2\mathrm{d}\xi - \int_{\mathbf{R}^3}|\mathrm{i}\xi_k R_1|^2\mathrm{d}\xi \\ &\geq \frac{1}{2}\int_{|\xi|<\frac{\sqrt{2}-1}{2}}\left|\frac{\xi_k}{6}(W_{a0}+2W_{b0})(F_1+e_1^-)\right|^2 \\ &\quad\times \mathrm{d}\xi - c(1+t)\mathrm{e}^{-ct} \\ &\geq c\int_{|\xi|^2<\min\{\frac{\sqrt{2}-1}{2},r^2\}}|\xi_k|^2|\mathrm{e}^{-2c|\xi|^2t}|\mathrm{d}\xi - c(1+t)\mathrm{e}^{-ct} \\ &\geq c(1+t)^{-\frac{5}{2}} - c(1+t)e^{-ct}. \end{aligned} \tag{6.4.99}$$

It follows from (6.4.99) that, for $t \gg 1$,

$$\|\partial_{xk}W_a(\cdot,t)\|^2 = \|\mathrm{i}\xi_k\hat{W}_a(\cdot,t)\|^2 \geq c_1(1+t)^{-\frac{5}{4}}. \tag{6.4.100}$$

Repeating the similar procedure as above, we can estimate the higher-order term $\|\mathrm{i}^{|\alpha|}\xi_1^{\alpha 1}\ \ \xi_2^{\alpha 2}\xi_3^{\alpha 3}\ \ \hat{W}_a\|^2(|\alpha| \leq 6)$, which is together with the Plancherel theorem leading to the algebraic time-decay rate for W_a from below,

$$\|\partial_x^l W_a(\cdot,t)\|_{L^2(\mathbf{R}^3)} \geq c(1+t)^{-\frac{l}{2}-\frac{3}{4}}, \quad 0 \leq l \leq 6. \tag{6.4.101}$$

Again, we can repeat the similar argument as above to establish the corresponding algebraic time-decay rate for $\hat{\boldsymbol{J}}_a$. In fact, by (6.4.90)–(6.4.92) we have

$$\begin{aligned}\|\hat{\boldsymbol{J}}_a^{(k)}(\cdot,t)\|^2 &\geq \frac{1}{2}\int_{\mathbf{R}^3}|T_2^{(k)}|^2\mathrm{d}\xi - \int_{\mathbf{R}^3}|R_2^{(k)}|^2\mathrm{d}\xi \\ &\geq \frac{1}{2}\int_{\mathbf{R}^3}|T_2^{(k)}|^2\mathrm{d}\xi - c(1+t)\mathrm{e}^{-ct} \\ &\geq c\int_{|\xi|^2<\min\{\frac{\sqrt{2}-1}{2},r^2\}}|\xi_k F_1|^2\mathrm{d}\xi - c(1+t)e^{-ct}, k=1,2,3. \end{aligned} \tag{6.4.102}$$

Since for $0 \le |\xi|^2 < \frac{1}{2}(\frac{\sqrt{2}-1}{2})$, it holds that $\frac{5-2\sqrt{2}}{4} < I_1 < 1$, we have

$$|F_1| = \left|\frac{e^{-\frac{t}{2}(1-I_1)}-e^{-\frac{t}{2}(1+I_1)}}{I_1}\right| = \frac{1}{|I_1|}|e^{-\frac{t}{2}(1-I_1)}(1-e^{-tI_1})| \ge ce^{-\frac{t}{2}(1-I_1)} \tag{6.4.103}$$

for $t > 1$ and $|\xi|^2 < \frac{1}{2}(\frac{\sqrt{2}-1}{2})$. By (6.4.92) and the fact $e^{-\frac{t}{2}(1-I_1)} \ge ce^{-c|\xi|^2 t}$ for $|\xi|^2 \le \frac{1}{2}(\frac{\sqrt{2}-1}{2})$, we finally obtain that for $|\xi|^2 < \frac{1}{2}(\frac{\sqrt{2}-1}{2})$,

$$|F_1| \ge ce^{-c|\xi|^2 t}. \tag{6.4.104}$$

Set $r_1^2 = \min\{r^2, \frac{1}{2}(\frac{\sqrt{2}-1}{2})\}$, and let $t > 1$. By (6.4.102) and (6.4.104), we get

$$\begin{aligned}\|\hat{\boldsymbol{J}}_a^{(k)}(\cdot,t)\|^2 &\ge c\int_{|\xi|^2<r_1^2} |\xi_k|^2 e^{-2c|\xi|^2 t} d\xi - c(1+t)e^{-ct}\\ &\ge c(1+t)^{-\frac{5}{3}} - c(1+t)e^{-ct}, \quad k = 1,2,3.\end{aligned} \tag{6.4.105}$$

This gives rise to the time-decay rate of $\boldsymbol{J}_a = (\boldsymbol{J}_a^{(1)}, \boldsymbol{J}_a^{(2)}, \boldsymbol{J}_a^{(3)})$ for $t \gg 1$ that

$$\|\boldsymbol{J}_a(\cdot,t)\| = \|\hat{\boldsymbol{J}}_a(\cdot,t)\| \ge c_1(1+t)^{-\frac{5}{4}}. \tag{6.4.106}$$

The higher-order estimates of $\boldsymbol{J}_a$ can be established in a similar argument as obtaining (6.4.99) for W_a, and finally we can have for $t \gg 1$ that

$$\|D_x^l \boldsymbol{J}_a(\cdot,t)\| \ge c(1+t)^{-\frac{l}{2}-\frac{5}{4}}, \quad l = 1,2,3,4,5. \tag{6.4.107}$$

The above estimates are valid for W_b and $\boldsymbol{J}_b^{(k)} (k = 1,2,3)$ due to the symmetry between W_a and W_b, $\boldsymbol{J}_a$ and $\boldsymbol{J}_b$. Thus the proof of the lower bound estimate in Theorem 6.4.2 is finished. On the other hand, the upper bound estimate can be easily estimated by (6.4.83)–(6.4.86) and Plancherel theorem. Thus we complete the proof of Theorem 6.4.2.

References

[1] ANTONELLI P and MARCATI P. On the finite energy weak solutions to a system in quantum fluid dynamics[J]. Communication in mathematical physics, 2009, **287**(2): 657–686.

[2] BRESCH D, DESJARDINS B and LIN C K. On some compressible fluid models: Korteweg, lubrication, and shallow water systems[J]. Communications in partial differential equations, 2003, **28**(3-4): 843–868.

[3] BACCARANI G and WORDEMAN M R. An investigation of steady-state velocity overshoot in silicon[J]. Solid state electronics, 1985, **28**(4): 407–416.

[4] BRESCH D and DESJARDINS B. Quelques modèles diffusifs capillaires de type Korteweg[J]. Comptes rendus mécanique, 2004, **332**(11): 881–886.

[5] CASTELLA F, ERDÖS L, FROMMLET F and MARKOWICH P. Fokker–Planck equations as scaling limits of reversible quantum systems[J]. Journal of statistical physics, 2000, **100**(3–4): 543–601.

[6] CALDEIRA A O and LEGGETT A J. Path integral approach to quantum Brownian motion[J]. Physica a statistical mechanics and its applications, 1983, **121**(3): 587–616.

[7] CHEN L and DREHER M. The viscous model of quantum hydrodynamics in several dimensions[J]. Mathematical models and methods applied sciences, 2007, **17**(7): 1065–1093.

[8] DIÓSI L. On high-temperature Markovian equation for quantum Brownian motion[J]. Europhysics letters, 2007, **22**(1): 1–3.

[9] FEIREISL E. Dynamics of viscous compressible fluids[M]. Oxford: Oxford University Press, 2004.

[10] GUO B L. Viscosity elimination method and viscosity of difference scheme[M]. Beijing: Science Press, 1993.

[11] GUO B L and XI X Y. Global weak solutions to one dimensional compressible viscous hydrodynamic equations[J]. Acta mathematica scientia, 2017, **37**(3): 573–583.
[12] GAMBA I M and JÜNGEL A. Positive solutions to singular second and third order differential equations for quantum fluids[J]. Archive for rational mechanics and analysis, 2001, **156**(3): 183–203.
[13] GAMBA I M and JÜNGEL A. Asymptotic limits in quantum trajectory models[J]. Communications in partial differential equations, 2002, **27**(3–4): 669–691.
[14] GAMBA I M, JÜNGEL A and VASSEUR A. Global existence of solutions to one dimensional viscous quantum hydrodynamic equations[J]. Journal of differential equations, 2009, **247**(11): 3117–3135.
[15] GUALDANI M and JÜNGEL A. Analysis of viscous quantum hydrodynamic equations for semiconductors[J]. European journal of applied mathematics, 2004, **15**(5): 577–595.
[16] LIONS J L, GUO B L and WANG L R. Some methods for solving the nonlinear boundary value problems[M]. Guangzhou: Zhongshan University Press, 1992.
[17] HSIAO L and LI H L. Dissipation and dispersion approximation to hydrodynamic equations and asymptotic limit[J]. Journal of partial differential equations, 2008(2): 59–76.
[18] HAAS F. A magnetohydrodynamic model for quantum plasmas[J]. Physics of plasmas, 2005, **12**(6): 062117.
[19] JÜNGEL A. Transport equations for semiconductors[M]. Berlin: Springer, 2009.
[20] JÜNGEL A. A steady-state quantum Euler-Poisson system for potential flows[J]. Communications in mathematical physics, 1998, **194**(2): 463–479.
[21] JÜNGEL A and MILIŠIĆ J P. Physical and numerical viscosity for quantum hydrodynamics[J]. Communication in mathematical sciences, 2007, **5**(2): 447–471.
[22] LEFLOCH P and SHELUKHIN V. Symmetries and global solvability of the isothermal gas dynamic equations[J]. Archive for rational mechanics and analysis, 2005, **175**(3): 389–430.
[23] NISHIBATA S and SUZUKI M. Initial boundary value problems for a quantum hydrodynamic model of semiconductors: Asymptotic behaviors and classical limits[J]. Journal of differential equations, 2008, **244**(4): 836–874.
[24] BRESCH D and DESJARDINS B. Existence of global weak solutions for a 2D viscous shallow water equations and convergence

to the quasi-geostrophic model[J]. Communications in mathematical physics, 2003, **238**(1–2): 211–223.

[25] BRESCH D and DESJARDINS B. On the construction of approximate solutions for the 2D viscous shallow water model and for compressible Navier–Stokes models[J]. Journal de mathématiques pures et appliquées, 2006, **86**(4): 362–368.

[26] BRENNER H. Navier–Stokes revisited[J]. Physica A: statistical mechanics and its applications, 2005, **349**(1): 60–132.

[27] BENZONI-GAVAGE S, DANCHIN R and DESCOMBES S. On the well-posedness for the Euler–Korteweg model in several space dimensions[J]. Indiana university mathematics journal, 2007, **56**(4): 1499–1580.

[28] BRESCH D and DESJARDINS B. On the existence of global weak solutions to the Navier–Stokes equations for viscous compressible and heat conducting fluids[J]. Journal de mathématiques pures et appliquées, 2007, **87**(1): 57–90.

[29] BRESCH D, DESJARDINS B and GÉRARD-VARET D. On compressible Navier–Stokes equations with density dependent viscosities in bounded domains[J]. Journal de mathématiques pures et appliquées, 2007, **87**(2): 227–235.

[30] BRULL S and MÉHATS F. Derivation of viscous correction terms for the isothermal quantum Euler model[J]. ZAMM-Journal of applied mathematics and mechanics zeitschrift für ange-wandte mathematik und mechanik, 2010, **90**(3): 219–230.

[31] CHEN L and DREHER M. Viscous quantum hydrodynamics and parameter-elliptic systems[J]. Mathematical methods in the applied sciences, 2011, **34**(5): 520–531.

[32] CHEN L and DREHER M. Partial differential equations and spectral theory[M]. Berlin: Springer, 2011.

[33] DONG J. A note on barotropic compressible quantum Navier–Stokes equations[J], Nonlinear analysis: theory, methods and applications, 2010, **73**(4): 854–856.

[34] DANCHIN R and DESJARDINS B. Existence of solutions for compressible fluid models of Korteweg type[C]. Annales de l'IHP analyse non linéaire. 2001, **18**(1): 97–133.

[35] DREHER M. The transient equations of viscous quantum hydrodynamics[J]. Mathematical methods in the applied sciences, 2008, **31**(4): 391–414.

[36] FEIREISL E and VASSEUR A. New perspectives in fluid dynamics: mathematical analysis of a model proposed by Howard Brenner[M]. Switzerland: Birkhäuser Verlag Basel, 2009.

[37] FERRY D K and ZHOU J R. Form of the quantum potential for use in hydrodynamic equations for semiconductor device modeling[J]. Physical review B, 1993, **48**(11): 7944.
[38] GARDNER C L. The quantum hydrodynamic model for semiconductor devices[J]. SIAM journal on applied mathematics, 1994, **54**(2): 409–427.
[39] GRANT J. Pressure and stress tensor expressions in the fluid mechanical formulation of the Bose condensate equations[J]. Journal of physics A: mathematical, nuclear and general, 1973, **6**(11): L151.
[40] HATTORI H and LI D. Global solutions of a high dimensional system for Korteweg materials[J]. Journal of mathematical analysis and applications, 1996, **198**(1): 84–97.
[41] HASPOT B. Existence of weak solutions for compressible fluid models of Korteweg type[J]. Journal of mathematical fluid mechanics, 2011, **13**(2): 223–249.
[42] HOFF D and SMOLLER J. Non-formation of vacuum states for compressible Navier–Stokes equations[J]. Communications in mathematical physics, 2001, **216**(2): 255–276.
[43] HUANG F and WANG Z. Convergence of viscosity solutions for isothermal gas dynamics[J]. SIAM journal on mathematical analysis, 2002, **34**(3): 595–610.
[44] JÜNGEL A. Effective velocity in compressible Navier–Stokes equations with third-order derivatives[J]. Nonlinear analysis: theory, methods and applications, 2011, **74**(8): 2813–2818.
[45] JÜNGEL A and LI H. Quantum Euler–Poisson systems: global existence and exponential decay[J]. Quarterly of applied mathematics, 2004, **62**(3): 569.
[46] JÜNGEL A and MATTHES D. An algorithmic construction of entropies in higher-order non-linear PDEs[J]. Nonlinearity, 2006, **19**(3): 633.
[47] JÜNGEL A and MATTHES D. The Derrida–Lebowitz–Speer–Spohn equation: existence, non-uniqueness, and decay rates of the solutions[J]. SIAM journal on mathematical analysis, 2008, **39**(6): 1996–2015.
[48] JOSEPH D D, HUANG A and HU H. Non-solenoidal velocity effects and Korteweg stresses in simple mixtures of incompressible liquids[J]. Physica D: nonlinear phenomena, 1996, **97**(1): 104–125.
[49] JIANG F. A remark on weak solutions to the barotropic compressible quantum Navier–Stokes equations[J]. Nonlinear analysis: real world applications, 2011, **12**(3): 1733–1735.

[50] JÜNGEL A. Global weak solutions to compressible Navier–Stokes equations for quantum fluids[J]. SIAM journal on mathematical analysis, 2010, **42**(3): 1025–1045.

[51] LI H L, LI J and XIN Z. Vanishing of vacuum states and blow up phenomena of the compressible Navier–Stokes equations[J]. Communications in mathematical physics, 2008, **281**(2): 401–444.

[52] LOFFREDO M I and MORATO L M. On the creation of quantized vortex lines in rotating He II[J]. IL nuovo cimento B, 1993, **108**(2): 205–215.

[53] LUNARDI A. Analytic semigroups and optimal regularity in parabolic problems[M]. Berlin: Springer Science and Business Media, 2012.

[54] YAO L and WANG W J. Compressible Navier–Stokes equations with density-dependent viscosity, vacuum and gravitational force in the case of general pressure[J]. Acta mathematica scientia, 2008, **28**(4): 801–817.

[55] MADELUNG E. Quantentheorie in hydrodynamischer form[J]. Zeitschrift für physik a hadrons and nuclei, 1927, **40**(3): 322–326.

[56] MELLET A and VASSEUR A. On the barotropic compressible Navier–Stokes equations[J]. Communications in partial differential equations, 2007, **32**(3): 431–452.

[57] WYATT R E. Quantum dynamics with trajectories: introduction to quantum hydrodynamics[M]. Berlin: Springer Science and Business Media, 2006.

[58] ZEIDLER E. Nonlinear functional analysis and its applications[M]. Berlin: Springer, 1990.

[59] ZATORSKA E. Fundamental problems to equations of compressible chemically reacting flows[D/OL]. Warszawski: Uniwersytet Warszawski, 2013.

[60] GISCLON M and LACROIX-VIOLET I. About the barotropic compressible quantum Navier–Stokes equations[J]. Nonlinear analysis: theory, methods and applications, 2015, **128**: 106–121.

[61] JÜNGEL A and MILISIK J P. Quantum Navier–Stokes equations: Progress in industrial mathematics at ECMI 2010[C]. Berlin: Springer, 2012.

[62] MUCHA P B, POKORNÝ M and ZATORSKA E. Chemically reacting mixtures in terms of degenerated parabolic setting[J]. Journal of mathematical physics, 2013, **54**(7): 071501.

[63] ORON A, DAVIS S H and BANKOFF S G. Long-scale evolution of thin liquid films[J]. Reviews of modern physics, 1997, **69**(3): 931.

[64] SIMON J. Compact sets in the space $L^p(O,T;\ B)$[J]. Annali di Matematica pura ed applicata, 1986, **146**(1): 65–96.
[65] DEGOND P, GALLEGO S and MEHATS F. On quantum hydrodynamic and quantum energy transport models[J]. Communications in mathematical sciences, 2007, **5**(4): 887–908.
[66] GASSER I and MARKOWICH P A. Quantum hydrodynamics wigner transforms and the classical limit[J]. Asymptotic analysis, 1997, **14**(2): 97–116.
[67] GINIBRE J and VELO G. The global Cauchy problem for the non linear Schrödinger equation revisited[C]. Annales de l'IHP analyse non linéaire. 1985, **2**(4): 309–327.
[68] JÜNGEL A, MATTHES D and MILIŠIĆ J P. Derivation of new quantum hydrodynamic equations using entropy minimization[J]. SIAM journal on applied mathematics, 2006, **67**(1): 46–68.
[69] KHALATNIKOV I M. An introduction to the theory of superfluidity[M]. New Jersey: Addison-Wesley Pub. Co, 1988.
[70] MARKOWICH P and RINGHOFER C. Quantum hydrodynamics for semiconductors in the high-field case[J]. Applied mathematics letters, 1994, **7**(5): 37–41.
[71] NELSON E. Quantum fluctuations[M]. Princeton, NJ:Princeton University Press, 1984.
[72] Wigner E. On the quantum correction for thermodynamic equilibrium[J]. Physical review, 1932, **40**(5): 749–759.
[73] WEIGERT S. How to determine a quantum state by measurements: the Pauli problem for a particle with arbitrary potential[J]. Physical review A, 1996, **53**(4): 2078.
[74] FEIREISL E, NOVOTNÝ A and PETZELTOVÁ H. On the existence of globally defined weak solutions to the Navier–Stokes equations[J]. Journal of mathematical fluid mechanics, 2001, **3**(4): 358–392.
[75] FEIREISL E and NOVOTNÝ A. Singular limits in thermodynamics of viscous fluids[M]. Berlin: Springer Science and Business Media, 2009.
[76] LIONS P L. Mathematical topics in fluid mechanics, Vol. 2, Compressible models[M]. Oxford: Oxford University Press, 2013.
[77] LI J and XIN Z. Global existence of weak solutions to the barotropic compressible Navier–Stokes flows with degenerate viscosities[J]. Mathematics, 2015.
[78] MUCHA P B, POKORNÝ M and ZATORSKA E. Heat-conducting, compressible mixtures with multicomponent diffusion: construction of a weak solution[J]. SIAM journal on mathematical analysis, 2015, **47**(5): 3747–3797.

[79] BERTOZZI A and PUGH M. Long-wave instabilities and saturation in thin film equations[J]. Communications on pure and applied mathematics, 1998, **51**: 625–661.

[80] BLEHER P, LEBOWITZ J and SPEER E. Existence and positivity of solutions of a fourth-order nonlinear PDE describing interface fluctuations[J]. Communications on pure and applied mathematics, 1994, **47**: 923–942.

[81] BREZIS H and OSWALD L. Remarks on sublinear elliptic equations[J]. Nolinear analysis theory methods and applications, 1986, **10**(1): 55–64.

[82] JÜNGEL A and PINNAU R. Global non-negative solutions of a nonlinear fourth-order parabolic equation for quantum systems[J]. SIAM journal on mathematical analysis, 2000, **32**(4): 760–777.

[83] PASSO R, GARCKE H and GRÜN G. On a fourth-order degenerate parabolic equation: global entropy estimates, existence, and qualitative behavior of solutions[J]. SIAM journal on mathematical analysis, 1998, **29**(2): 321–342.

[84] ANCONA M G and IAFRATE G J. Quantum correction to the equation of state of an electron gas in a semiconductor[J]. Physical review B, 1989, **39**(13): 9536–9540.

[85] BLOTEKJAER K. Transport equations for electrons in two-valley semiconductors[J]. IEEE transactions on electron devices, 1970, **17**(1): 38–47.

[86] DEGOND P and MARKOWICH P A. On a one dimensional steady-state hydrodynamic model for semiconductors[J]. Applied mathematics letters, 1990, **3**(3): 25–29.

[87] GUO Y and STRAUSS W. Stability of semiconductor states with insulating and contact boundary conditions[J]. Archive for rational mechanics and analysis, 2006, **179**(1): 1–30.

[88] GILBARG D and TRUDINGER N S. Elliptic partial differential equations of second order[M]. Berlin: Springer, 2015.

[89] KAWASHIMA S, NIKKUNI Y and NISHIBATA S. The initial value problem for hyperbolic-elliptic coupled systems and applications to radiation hydrodynamics[J]. Analysis of systems of conservation laws, 1999, **99**.

[90] KAWASHIMA S, NIKKUNI Y and NISHIBATA S. Large-time behavior of solutions to hyperbolic-elliptic coupled systems[J]. Archive for rational mechanics and analysis, 2003, **170**(4): 297–329.

[91] LI H, MARKOWICH P and MEI M. Asymptotic behavior of solutions of the hydrodynamic model of semiconductors[J]. Proceedings of the royal society of Edinburgh: section a mathematics, 2002, **132**(02): 359–378.

[92] MATSUMURA A and MURAKAMI T. Asymptotic behavior of solutions for a fluid dynamical model of semiconductor equation[J]. Kyoto University, RIMS kokyuroku, 2006, **1495**: 60–70.

[93] NISHIBATA S and SUZUKI M. Asymptotic stability of a stationary solution to a hydrodynamic model of semiconductors[J]. Osaka journal of mathematics, 2007, **44**(3): 639–665.

[94] NISHIBATA S and SUZUKI M. Initial boundary value problems for a quantum hydrodynamic model of semiconductors: asymptotic behaviors and classical limits[J]. Journal of differential equations, 2007, **244**(4): 836–874.

[95] PINNAU R. A note on boundary conditions for quantum hydrodynamic equations[J]. Applied mathematics letters, 1999, **12**(5): 77–82.

[96] RACKE R. Lectures on nonlinear evolution equations: Initial value problems[J]. Aspect of mathematics E, 1992, **19**.

[97] TEMAM R. Infinite-dimensional dynamical systems in mechanics and physics[M]. Berlin: Springer Science and Business Media, 1988.

[98] GAMBA I M, GUALDANI M P and ZHANG P. On the blowing up of solutions to the quantum hydrodynamic models on a bounded domains[J]. Monatshefte für mathematik, 2009, **157**(1): 37–54.

[99] GASSER I, HSIAO L and LI H L. Large time behavior of solutions of the bipolar hydrodynamic model for semiconductors[J]. Journal of differential equations, 2003, **192**(2): 326–359.

[100] HUANG F, LI H, MATSUMURA A, *et al.* Well-posedness and stability of multidimensional quantum hydrodynamics for semiconductors in R^3. Series in contemporary applied mathematics CAM 15[M]. Beijing: High Education Press, 2010.

[101] ALI G and JÜNGEL A. Global smooth solutions to the multi-dimensional hydrodynamic model for two-carrier plasmas[J]. Journal of differential equations, 2003, **190**(2): 663–685.

[102] JÜNGEL A, LI H L and MATSUMURA A. The relaxation-time limit in the quantum hydrodynamic equations for semiconductors[J]. Journal of differential equations, 2005, **225**(2): 440–464.

[103] LI H L and MARCATI P. Existence and asymptotic behavior of multi-dimensional quantum hydrodynamic model for semiconductors[J]. Communication in mathematical physics, 2004, **245**(2): 215–247.

[104] LI H L, ZHANG G J and ZHANG K. Algebraic time decay for the bipolar quantum hydrodynamic model[J]. Mathematical models and methods in applied sciences, 2008, **18**(6): 859–881.

[105] MARCATI P and NATALINI R. Weak solutions to a hydrodynamic model for semiconductors and relaxation to the drift-diffusion

equations[J]. Archive for rational mechanics and analysis, 1995, **129**(2): 129–145.

[106] YANG X H. Quasineutral limit of bipolar quantum hydrodynamic model for semiconductors[J]. Frontiers of mathematics in China, 2011, **6**(2): 349–362.

[107] ZHANG G, LI H L and ZHANG K. Semiclassical and relaxation limits of bipolar quantum hydrodynamic model for semiconductors[J]. Journal of differential equations, 2008, **245**(6): 1433–1453.

CPSIA information can be obtained
at www.ICGtesting.com
Printed in the USA
JSHW061356090723
43825JS00005B/102

9 789811 260834